RESPONSE SURFACES

STATISTICS: Textbooks and Monographs

A Series Edited by

D. B. Owen, Founding Editor, 1972–1991

W. R. Schucany, Coordinating Editor
Department of Statistics
Southern Methodist University
Dallas, Texas

S. C. Chow, Associate Editor
for Biostatistics
Bristol-Myers Squibb Company

W. J. Kennedy, Associate Editor
for Statistical Computing
Iowa State University

A. M. Kshirsagar, Associate Editor
for Multivariate Analysis and
Experimental Design
University of Michigan

E. G. Schilling, Associate Editor
for Statistical Quality Control
Rochester Institute of Technology

1. The Generalized Jackknife Statistic, *H. L. Gray and W. R. Schucany*
2. Multivariate Analysis, *Anant M. Kshirsagar*
3. Statistics and Society, *Walter T. Federer*
4. Multivariate Analysis: A Selected and Abstracted Bibliography, 1957–1972, *Kocherlakota Subrahmaniam and Kathleen Subrahmaniam*
5. Design of Experiments: A Realistic Approach, *Virgil L. Anderson and Robert A. McLean*
6. Statistical and Mathematical Aspects of Pollution Problems, *John W. Pratt*
7. Introduction to Probability and Statistics (in two parts), Part I: Probability; Part II: Statistics, *Narayan C. Giri*
8. Statistical Theory of the Analysis of Experimental Designs, *J. Ogawa*
9. Statistical Techniques in Simulation (in two parts), *Jack P. C. Kleijnen*
10. Data Quality Control and Editing, *Joseph I. Naus*
11. Cost of Living Index Numbers: Practice, Precision, and Theory, *Kali S. Banerjee*
12. Weighing Designs: For Chemistry, Medicine, Economics, Operations Research, Statistics, *Kali S. Banerjee*
13. The Search for Oil: Some Statistical Methods and Techniques, *edited by D. B. Owen*
14. Sample Size Choice: Charts for Experiments with Linear Models, *Robert E. Odeh and Martin Fox*
15. Statistical Methods for Engineers and Scientists, *Robert M. Bethea, Benjamin S. Duran, and Thomas L. Boullion*
16. Statistical Quality Control Methods, *Irving W. Burr*
17. On the History of Statistics and Probability, *edited by D. B. Owen*
18. Econometrics, *Peter Schmidt*
19. Sufficient Statistics: Selected Contributions, *Vasant S. Huzurbazar (edited by Anant M. Kshirsagar)*
20. Handbook of Statistical Distributions, *Jagdish K. Patel, C. H. Kapadia, and D. B. Owen*

RESPONSE SURFACES

Designs and Analyses
Second Edition, Revised and Expanded

ANDRÉ I. KHURI
JOHN A. CORNELL

University of Florida
Gainesville, Florida

CRC Press
Taylor & Francis Group
Boca Raton London New York

CRC Press is an imprint of the
Taylor & Francis Group, an **informa** business

CRC Press
Taylor & Francis Group
6000 Broken Sound Parkway NW, Suite 300
Boca Raton, FL 33487-2742

First issued in paperback 2019

ISBN-13: 978-0-8247-9741-6 (hbk)
ISBN-13: 978-0-367-40125-2 (pbk)

**Visit the Taylor & Francis Web site at
http://www.taylorandfrancis.com**

**and the CRC Press Web site at
http://www.crcpress.com**

To
Ronnie, Marcus, and Roxanne,
Ninette, Paul, and Pierre,
Natalie, Johnny, and Ken,
and in memory of our parents

Preface to the Second Edition

The purpose of the second edition is to update the coverage of response surface methodology (RSM) by including recent advances in this area. In preparing this edition, we have taken into account the reviews of the first edition and incorporated most of the suggestions made.

The second edition includes two entirely new chapters: Chapter 8 entitled "Response Surface Models with Block Effects," and Chapter 11, "Taguchi's Robust Parameter Design." Several new sections have also been added to the remaining chapters.

The book is divided into two parts. Part I contains Chapters 1–5 and covers core response surface material that any user of RSM should be acquainted with. In Chapter 1, we have added a section on the strategy for carrying out an experimental program. We also discuss transformations on the response variable in Chapter 2. The analysis of a screening experiment through the use of a Plackett–Burman design is presented in Chapter 3. Section 5 of Chapter 4 has been expanded by the addition of a numerical example of a fitted second-order model from a Box–Behnken design. Sections 7 and 8 have been removed from this chapter and incorporated into Chapter 8. Chapter 5 now contains a modification to the method of ridge analysis.

Part II consists of Chapter 6–12 and covers more specialized topics in RSM. Chapter 6 is the same as in the first edition. Three new sections have been added to Chapter 7, "Analysis of Multiresponse Experiments." The first of these addresses the effect of scaling the responses on the detection of linear dependencies

v

among multiresponse data. The second new section extends the rotatability property to a multiresponse design, and the third section briefly describes the testing for a general linear hypothesis. Chapter 7 also includes references to recent design criteria for the multiresponse case. Chapter 8 is new. It discusses the analysis of a response surface model in the presence of block effects, which can be either fixed or random. A procedure is presented for determining optimum conditions when the model contains interaction effects involving fixed polynomial effects and random block effects. This chapter provides a significantly more in-depth treatment of the subject of blocking in RSM than was given in Sections 4.7 and 4.8 in the first edition. The analysis of multiresponse mixture experiments through the application of biplots has been added to Chapter 9. A three-component plastics compounding experiment involving five characteristics (responses) considered simultaneously is presented.

Chapter 10 is, for the most part, the same as Chapter 8 in the first edition and covers nonlinear response surface models. Chapter 11, on Taguchi's parameter designs, is new. Although only a short chapter, it discusses the basic elements of the Taguchi approach to parameter design. This chapter has been added because considerable attention was given in the 1980s, and continues today, to the ideas expressed by Genichi Taguchi. Chapter 11 also discusses process performance measures called signal-to-noise ratios. Response surface alternatives, such as robust parameter designs and the modeling of both the mean response and the process variance, are presented. Finally, in Chapter 12 we combine the material of Chapter 10 in the first edition with a study of the various measures of rotatability for a response surface design.

The added material described here amounts to a major revision of the first edition. The book should therefore serve as a useful reference for anyone interested in learning or using RSM. If used as a text, the entire book would be suitable for a two-semester course with Part I covered in the first semester. The second edition contains more examples, and exercises with solutions to some selected exercises are given. A general bibliography is provided at the end of the book.

André I. Khuri
John A. Cornell

Preface to the First Edition

The roots of response surface methodology (RSM) can be traced back to the works of J. Wishart, C.P. Winsor, E.A. Mitscherlich, F. Yates, and others in the early 1930s or even earlier. However, it was not until 1951 that RSM was formally developed by G.E.P. Box and K.B. Wilson and other colleagues at Imperial Chemical Industries in England. Their objective was to explore relationships such as those between the yield of a chemical process and a set of input variables presumed to influence the yield. Since the pioneering work of Box and his co-workers, RSM has been successfully used and applied in many diverse fields such as chemical engineering, industrial development and process improvement, agricultural and biological research, even computer simulation, to name just a few.

In the 16 years that have elapsed since the publication of *Response Surface Methodology* by R.H. Myers, many new developments have taken place in RSM. Myers' book has proved to be quite useful to researchers and students alike, and has been cited extensively over the years. This year a second text appeared, authored by G.E.P. Box and N.R. Draper, entitled *Empirical Model-Building and Response Surfaces*. Beginning with the first principles of RSM, features of this text include a detailed analysis and explanation of maxima and ridge systems, discussions of the considerations that motivate the choice of an experimental design, and the use of transformations. The present text is similar to Myers' book, but includes more up-to-date material like that found in the appendices of the later chapters of Box and Draper's book, in addition to topics not previously addressed in many of the textbooks that briefly discuss response surface methods.

This book is intended for research workers and students, at advanced undergraduate and graduate levels, both in the fields of statistics and other related areas. Persons engaged in applied research in the chemical industry, agriculture, operations research, and biometry will undoubtedly find the contents of the book to be relevant and useful. Graduate students who plan to do research in RSM or those who just need to apply response surface techniques to their research can greatly benefit from this book.

The reader is assumed to be acquainted with basic statistical methods. Some knowledge of matrix algebra and the method of least squares will be helpful, but not absolutely necessary. A short review of matrix algebra and an introduction to the method of least squares are given in Chapter 2.

The book is divided into 10 chapters. Chapter 1 gives an introduction to RSM and acquaints the reader with its basic terminology. Chapter 2 provides an introduction to the method of least squares, matrix algebra, and the analysis of variance associated with fitting the linear model. Introductory comments are provided concerning the design region, the use of coded variables, and of properties, such as orthogonality and rotatability, of a response surface design.

In Chapter 3, we begin our study of RSM by discussing first-order models and designs. Two-level factorials and fractional two-level factorials are presented along with the class of simplex designs. Further comments are made on checking the lack of fit of first-order models. Chapter 4 discusses second-order models and designs. Three-level factorials, central composite designs and various other types of rotatable second-order designs are presented. Chapter 5 presents techniques for the determination of optimum conditions when a first-order or a second-order model is fitted to a data set. These last three chapters have traditionally formed the core topics making up RSM.

The integrated mean squared error of fitted first- and second-order models is presented as a design criterion in Chapter 6. Two rival methods to standard least squares for estimating response surfaces are also introduced: minimum bias estimation and generalized least squares estimation. Procedures for minimizing the integrated mean squared error of the minimum bias estimator and the generalized least squares estimator for the single independent variable case are illustrated with examples.

Chapter 7 is devoted to a discussion of some multivariate aspects of RSM with regard to parameter estimation, design, lack of fit, and simultaneous optimization. This chapter can be useful in experimental situations in which several responses are observed for each setting of a group of input variables. In Chapter 8, the basic ideas concerning designs for nonlinear models are developed. This chapter also includes a discussion concerning confidence intervals and regions for the parameters of a nonlinear model.

Chapter 9 contains an introduction to mixture experiments in addition to a study of the various designs and models for fitting mixture data, including models with process variables. Variance minimizing design criteria contrasting the ideas of discrete versus continuous design measures are discussed briefly in Chapter 10. Different types of robust designs are also included in this chapter.

The book contains numerous examples, some of which are worked out in detail while others serve only to illustrate or clarify certain points. There are also exercises at the end of each chapter, except for Chapters 1 and 10. The book can, therefore, be used as a text or just as a reference. Chapters 1–6 can be covered in a typical one-semester college course. The topics in Chapters 7–10 are more specialized and cover more contemporary work in response surface methodology.

André I. Khuri
John A. Cornell

Contents

RESPONSE SURFACES

1

Introduction to Response Surface Methodology

1.1 INTRODUCTION

Quite often scientific investigations can be characterized as "hit or miss" experiences, unless certain procedures are followed. In most investigations where there is an ever-present element of uncertainty regarding the success or failure of the outcome, some information (however little) may be acquired from the experience if care is taken to properly plan and execute the program of experimentation. This book presents some procedures that help to ensure successful experimentation.

Most exploratory-type investigations are set up with a twofold purpose:

1. To determine and quantify the relationship between the values of one or more measurable response variable(s) and the settings of a group of experimental factors presumed to affect the response(s) and
2. To find the settings of the experimental factors that produce the best value or best set of values of the response(s)

The following example is taken from a product development laboratory where a cleaning solvent is being developed and tested for its maximum level of abrasiveness. Solvent abrasion is measured by rubbing a solvent-soaked cloth against a ceramic tile for 30 minutes and recording the amount of wear of the tile (wear equals thickness of tile prior to rubbing minus thickness of tile after rubbing). The primary ingredient in the solvent is lime, and the abrasiveness of the solvent is known to be an increasing function of the percentage of lime in the solvent up to some unknown level (see Figure 1.1).

Not only is the maximum amount of abrasion caused by rubbing the solvent on the tile unknown to the scientist developing the solvent, but the percentage of lime in the solvent that produces the maximum amount of abrasion is unknown as well. The scientist knows that too much lime in the solvent not only is costly but also can reduce the level of abrasiveness of the solvent. To determine the values of maximum abrasion and percentage of lime for a fixed rubbing time of 30 min., a program of experiments is set up consisting of solvents with different lime percentages. An abrasion value is recorded for each of the solvent formulations, and a mathematical equation is developed that expresses abrasion as a function of the percentage of lime. Estimates of the maximum abrasion value and of the percentage of lime in the solvent resulting in the maximum abrasion value are acquired from the equation.

Another example of seeking an optimal value of the response is in drug manufacturing. Combinations of two drugs, each known to reduce blood pressure in humans, are to be studied. A series of clinical trials involving 100 high blood pressure patients is set up, and each patient is given some predetermined combination of the two drugs. Here the purpose of administering the different

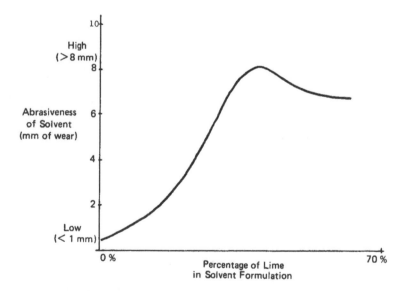

Figure 1.1   The theoretical relationship between the abrasiveness of the solvent and the percentage of lime in the solvent. Abrasion is measured as the amount of wear of ceramic tiles in millimeters.

combinations of the drugs to the individuals is to find the specific combination that results in the greatest reduction in the patient's blood pressure reading within some specified interval of time.

Response surface methodology (RSM) is a set of techniques that encompasses:

1. Setting up a series of experiments (designing a set of experiments) that will yield adequate and reliable measurements of the response of interest
2. Determining a mathematical model that best fits the data collected from the design chosen in (1), by conducting appropriate tests of hypotheses concerning the model's parameters
3. Determining the optimal settings of the experimental factors that produce the maximum (or minimum) value of the response

If discovering the best value, or values, of the response is beyond the available resources of the experiment, then response surface methods are aimed at obtaining at least a better understanding of the overall system. When the behavior of the measured response of interest is governed by certain laws which lead to a deterministic relationship between the response and the set of experimental factors chosen, it should then be possible to determine the best conditions (levels) of the factors to optimize a desired output. Quite often, however, because the relationship is either too complex or unknown, an empirical approach is necessary. The strategy employed, outlined in the preceding list, is the basis of RSM.

1.2 THE SIMILARITY OF REGRESSION ANALYSIS AND RSM

In any system in which variable quantities change, the interest might be in assessing the effects of the factors on the behavior of some measurable quantity (the response). Such an assessment is possible through regression analysis. Using data collected from a set of experimental trials, regression helps to establish empirically (by fitting some form of mathematical model) the type of relationship that is present between the response variable and its influencing factors. The response variable is the dependent variable and is called the *response*, and the influencing factors are called *explanatory*, *regressor*, or *input* variables. Regression analysis is one of the most widely used tools for investigating cause-and-effect relationships having applications in the physical, biological, and social sciences, as well as in engineering and in many other fields.

As mentioned previously, response surface methods are additional techniques employed before, while, and after a regression analysis is performed on the data. Preceding the regression analysis, the experiment must be designed, that is, the input variables must be selected and their values during the actual experimentation designated. After the regression analysis is performed, certain model testing procedures and optimization techniques are applied. Thus, the subject of RSM includes the application of regression as well as other techniques in an attempt to gain a better understanding of the characteristics of the response system under study.

We now review very briefly some of the terminology that will be used throughout this book as an introduction to the subjects of regression and RSM. For simplicity of presentation we shall assume that there is only one response variable to be studied although in practice there can be several response variables that are under investigation simultaneously. Multiple response, or multiresponse, experiments are discussed in Chapter 7.

1.3 TERMINOLOGY

1.3.1 Factors

Factors are processing conditions or input variables whose values or settings can be controlled by the experimenter. Presumably, if one changes the settings of the factors, the value of the response variable varies as well. In the cleaning solvent example mentioned previously, the presence of lime in the formulation was the only factor considered and the amount of abrasion differed for different percentages of lime. If in addition to lime, a second ingredient, calcium chloride, is also present and is known to affect the abrasiveness property of the solvent, then both the percentage of lime and the percentage of calcium chloride are factors to consider.

To use another example, suppose an agricultural experiment is designed to study the yield of soybeans as affected by applying different amounts of nitrogen (N), phosphorus (P), and potassium (K) to the experimental plots. The factors are N, P, and K, and their settings are the amounts of each assigned to the plots. The frequency of application of the different fertilizer blends of N, P, and K to the plots would be another factor if the frequency is varied as well.

Factors in a regression analysis can be qualitative, such as the type of fertilizer, the sex of an applicant seeking a position, catalyst type, or the vendor supplying the material. On the other hand, factors can be quantitative, such as the level of temperature or the amount of fertilizer applied where the levels are defined and arranged on a numerical scale. In most of the response surface investigations discussed in this book, if there are qualitative factors, they are considered blocking variables (see Chapter 8). The specific factors whose levels are to be studied in detail are those that are quantitative in nature, and their levels (or settings) are assumed to be fixed or controlled (without error) by the experimenter. Factors and their levels will be denoted by $X_1, X_2, \ldots, X_k$, respectively.

1.3.2 Response

The *response* variable is the measured quantity whose value is assumed to be affected by changing the levels of the factors. In the cleaning solvent example, the response was the abrasiveness of the solvent in units of millimeters of tile wear. In the agricultural experiment, the response was the yield of soybeans, measured in units of pounds harvested per plot. The true value of the response corresponding to any particular combination of the factor levels and in the absence of experimental error of any kind is denoted by η. (Note that experimental error consists of random measurement error caused by sources such as the production equipment, the testing equipment, and the people who run the equipment, as well as nonrandom error caused by factors that have not been included in the experiment.) However, because experimental error is present in all experiments involving measurements, the response value that is actually observed or measured for any particular combination of the factor levels differs from η. This difference from the true value is written as $Y = \eta + \varepsilon$, where Y represents the observed value of the response and ε denotes experimental error.

1.3.3 The Response Function

When we say that the value of the true response η depends upon the levels X_1, $X_2, \ldots, X_k$ of k quantitative factors, we are saying that there exists some function of $X_1, X_2, \ldots, X_k$, the value of which for any given combination of factor levels supplies the corresponding value of η, that is,

$$\eta = \phi(X_1, X_2, \ldots, X_k)$$

The function ϕ is called the true response function and is assumed to be a continuous function of the X_i.

1.3.4 The Polynomial Representation of a Response Surface

Let us consider the response function $\eta = \phi(X_1)$ for a single factor. If $\phi(X_1)$ is a continuous, smooth function, then it is possible to represent it locally to any required degree of approximation with a Taylor series expansion about some arbitrary point X_{10}, that is,

$$\eta = \phi(X_{10}) + (X_1 - X_{10})\phi'(X_{10}) + \tfrac{1}{2}(X_1 - X_{10})^2 \phi''(X_{10}) + \ldots \tag{1.1}$$

where $\phi'(X_{10})$ and $\phi''(X_{10})$ are, respectively, the first and second derivatives of $\phi(X_1)$ with repsect to X_1 evaluated at X_{10}. The expansion (Eq. 1.1) reduces to a polynomial of the form

$$\eta = \phi(X_1) = \beta_0 + \beta_1 X_1 + \beta_{11} X_1^2 + \ldots \tag{1.2}$$

where the coefficients β_0, β_1, and β_{11} are parameters which depend on X_{10} and the derivates of $\phi(X_1)$ at X_{10}. The successive terms β_0, $\beta_1 X_1$, and $\beta_{11}X_1^2$ of the polynomial are said to be of degree 0, 1, 2, and so on. By taking terms only up to degree 1 we obtain the equation of a straight line, $\eta = \beta_0 + \beta_1 X_1$. This is referred to as a first-order model in X_1. By taking terms up to degree 2, we obtain the equation of a parabola,

$$\eta = \beta_0 + \beta_1 X_1 + \beta_2 X_1^2$$

which is referred to as a second-order model in X_1. A straight line and a parabola are shown in Figure 1.2.

For two factors, X_1 and X_2, a polynomial equation in the factor levels is

$$\eta = \phi(X_1, X_2) = \beta_0 + \beta_1 X_1 + \beta_2 X_2 + \beta_{11}X_1^2 + \beta_{22}X_2^2$$
$$+ \beta_{12}X_1 X_2 + \ldots \tag{1.3}$$

If Eq. 1.3 contains only the first three terms, that is, $\eta = \beta_0 + \beta_1 X_1 + \beta_2 X_2$, then Eq. 1.3 represents a first-order model in X_1 and X_2 and describes a plane positioned directly above the two-dimensional space defined by the values of X_1 and X_2 (see Figure 1.3). When curvature is present in the shape of the surface and the first six terms of Eq. 1.3 are required to describe η, we have a second-order model in X_1 and X_2 which represents what we refer to as a *second-order*

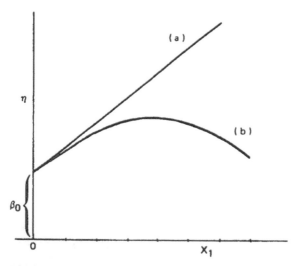

Figure 1.2   Polynomial models, (a) straight line, $\eta = \beta_0 + \beta_1 X_1$, and (b) parabola, $\eta = \beta_0 + \beta_1 X_1 + \beta_{11}X_1^2$, where $\beta_1 > 0$ and $\beta_{11} < 0$.

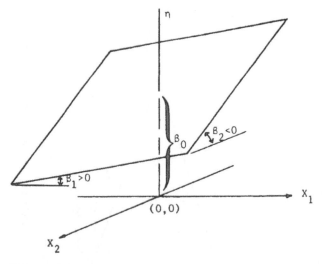

Figure 1.3 The planar surface, $\eta = \beta_0 + \beta_1 X_1 + \beta_2 X_2$, where $\beta_0 > 0$, $\beta_1 > 0$; and $\beta_2 < 0$.

response surface. For example, the surface in Figure 1.4 represents the yield of peanuts, in pounds per experimental plot, defined as a function of the amounts, X_1 and X_2, of two fertilizers which were applied to the plots in the field.

The parameters $\beta_1, \beta_2, \ldots, \beta_{12}, \ldots$ in Eq. 1.3 are called *regression coefficients* or *parameters*. The variables X_1 and X_2 are explanatory or input variables in the regression function. If the polynomial Eq. 1.3 exactly represents the response function $\phi(X_1, X_2)$ in the region of the levels of the two factors under study, then β_0 is the response at $X_1 = 0$ and $X_2 = 0$, and β_0 is only meaningful if the combination $X_1 = 0$ and $X_2 = 0$ is contained within the experimental region. The coefficients β_1 and β_2 are the values of the first-order partial derivatives, $\partial\phi/\partial X_1$ and $\partial\phi/\partial X_2$, of ϕ with respect to X_1 and X_2 evaluated at $X_1 = X_2 = 0$, and are referred to as *first-order effects*. In other words, β_i represents the rate of change of η with respect to X_i ($i = 1, 2$) only, evaluated at $X_1 = X_2 = 0$. In the event of a first-order model in X_1 and X_2, β_i, therefore, represents the slope of a cross-section of the plane $\eta = \beta_0 + \beta_1 X_1 + \beta_2 X_2$ with a plane parallel to the $X_i\eta$ plane (see Figure 1.3). This can be referred to as the tilt of the plane in the direction of the X_i ($i = 1, 2$) axis. The coefficients $\beta_{11}, \beta_{22},$ and β_{12} in Eq. 1.3 are defined as the values of the second-order partial derivatives, $\frac{1}{2}\partial^2\phi/\partial X_1^2$, $\frac{1}{2}\partial^2\phi/\partial X_2^2$, and $\partial^2\phi/\partial X_1\partial X_2$, respectively, at $X_1 = X_2 = 0$, and are called the *second-order effects*. The same can be said of higher-order coefficients, such as β_{111}, β_{112}, and so on.

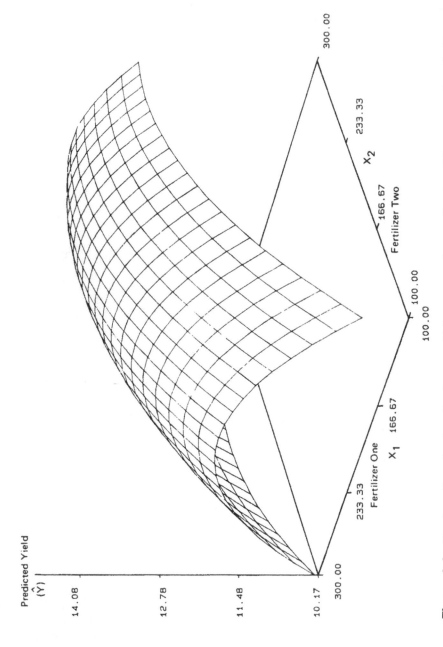

Figure 1.4   The three-dimensional predicted peanut yield response surface as a function of the amounts of two fertilizers (X_1 and X_2) applied to the plots.

1.3.5 The Predicted Response Function

The structural form of ϕ is usually unknown and therefore an approximating form is sought using a polynomial (Eq. 1.2 or 1.3) or some other type of empirical model equation. The steps taken in obtaining the approximating model are as follows: First, an assumed form of model equation in the k input variables is proposed. Then, associated with the proposed model, some number of combinations of the levels $X_1, X_2, \ldots, X_k$ of the k factors are selected for use as the design. At each factor level combination chosen, one or more observations are collected. The observations are used to obtain estimates of the parameters in the proposed model as well as to obtain an estimate of the experimental error variance. Tests are then performed on the magnitudes of the coefficient estimates as well as on the model form itself, and if the fitted model is considered to be satisfactory, it can be used as a prediction equation.

To illustrate the steps taken in obtaining the form of the approximating equation, let us assume the true response function, written in terms of the levels of two factors, can be expressed locally using the first-order model

$$\eta = \beta_0 + \beta_1 X_1 + \beta_2 X_2 \tag{1.4}$$

After collecting some number, $N \geq 3$, of observed response values, Y, from at least three distinct combinations of the levels, X_1 and X_2, estimates of the parameters β_0, β_1, and β_2 are obtained using the method of least squares (see Chapter 2). If these estimates, denoted by b_0, b_1, and b_2, respectively, are used instead of the unknown parameters β_0, β_1, and β_2, we obtain the prediction equation

$$\hat{Y} = b_0 + b_1 X_1 + b_2 X_2 \tag{1.5}$$

where $\hat{Y}$, called "Y hat," denotes the predicted response value for given values of X_1 and X_2. Before any predictions are made with Eq. 1.5, however, one must determine that the values obtained through Eq. 1.5 at the data collection points are "reasonably close" to the observed data values. In Section 2.6 in Chapter 2 we shall discuss ways of testing the adequacy of empirical models fitted to data.

1.3.6 The Response Surface

The relationship $\eta = \phi(X_1, X_2, \ldots, X_k)$ between η and the levels of k factors may be represented by a *hypersurface*. For example, with $k = 2$, an estimated second-order yield surface is plotted in Figure 1.4, where $\hat{Y}$ represents the predicted yield of peanuts measured in pounds per acre and X_1 and X_2 are the amounts of the two fertilizers. The second-order prediction equation that produced the surface in Figure 1.4 is

$$\hat{Y} = 13.85 - 0.90x_1 + 0.56x_2 - 0.57x_1x_2 - 1.94x_1^2 - 0.78x_2^2 \qquad (1.6)$$

The variables x_1 and x_2 in Eq. 1.6 are coded variables defined as

$$x_1 = \frac{X_1 - 200}{100} \qquad x_2 = \frac{X_2 - 200}{100}$$

The use of these coded variables greatly simplifies the numerical calculations used in obtaining the parameter estimates in Eq. 1.6. The predicted yield response, expressed in terms of the levels X_1 and X_2 of the two fertilizers, is obtained by replacing x_1 and x_2 in Eq. 1.6 in the following manner:

$$
\begin{aligned}
\hat{Y} = 13.85 &- 0.90 \left(\frac{X_1 - 200}{100} \right) + 0.56 \left(\frac{X_2 - 200}{100} \right) \\
&- 0.57 \left(\frac{X_1 - 200}{100} \right) \left(\frac{X_2 - 200}{100} \right) \\
&- 1.94 \left(\frac{X_1 - 200}{100} \right)^2 - 0.78 \left(\frac{X_2 - 200}{100} \right)^2 \\
= 1.385 &+ 0.08X_1 + 0.048X_2 - 0.57 \times 10^{-4}X_1X_2 \\
&- 1.94 \times 10^{-4}X_1^2 - 0.78 \times 10^{-4}X_2^2
\end{aligned}
\qquad (1.7)
$$

The measured peanut yield values and the levels of the two fertilizers are listed in Table 1.1.

With k factors, the response surface is a subset of a $(k + 1)$-dimensional Euclidean space, that is, the curve in Figure 1.2 is depicted in a two-dimensional space, whereas the solid surface in Figure 1.4 is visualized in a three-dimensional space, the third dimension representing the height of the surface above the two-dimensional plane of the values of X_1 and X_2. Figure 1.5 plots a three-dimensional cotton yield surface represented as a function of the amounts of nigrogen ($X_1 = N$) and of phosphorus ($X_2 = P$) applied to the experimental plots. The levels X_1 and X_2 used to generate the predicted yield (kilograms/hectare)

Table 1.1 Peanut Yields in Pounds per Acre

		Fertilizer 1 (X_1)		
		100 lb/acre	200 lb/acre	300 lb/acre
Fertilizer	100	11.2, 10.9, 11.0	12.0, 12.4, 12.8	10.9, 9.8, 10.0
2 (X_2)	200	12.6, 13.2, 12.7	14.0, 12.9, 13.7	11.3, 11.7, 10.9
	300	12.8, 12.3, 13.9	13.4, 14.7, 14.1	10.2, 9.7, 9.9

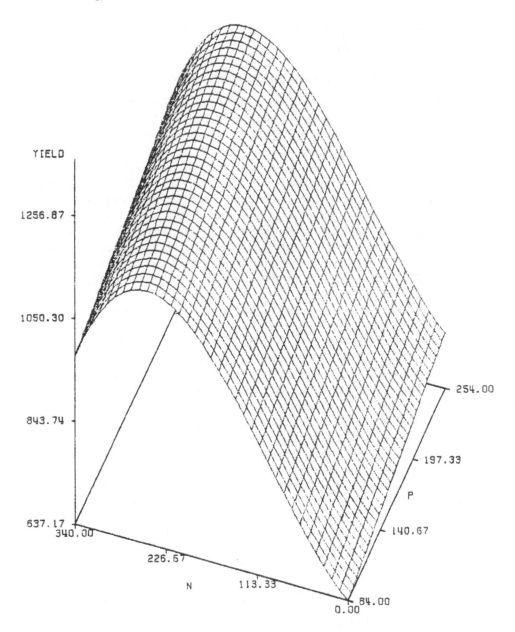

Figure 1.5 A predicted cotton yield (kilograms/hectare) surface generated by the cubic by linear polynomial $\widehat{\text{YIELD}} = 581.73 + 1.53N + 0.012N^2 - 4.0 \times 10^{-5}N^3 + 0.66P$, where N is the level of nitrogen and P is the level of phosphorus applied to the experimental plots.

equation were X_1 = 0, 112, 224 and 340 kg/hectare and X_2 = 84, 168 and 254 kg/hectare, respectively.

1.3.7 Contour Representation of a Response Surface

A technique used to help visualize the shape of a three-dimensional response surface is to plot the contours of the response surface. In a contour plot, lines or curves of constant response values are drawn on a graph or plane whose coordinate axes represent the levels, X_1 and X_2, of the factors. The lines (or curves) are known as contours of the surface. Each contour represents a specific value for the height of the surface (i.e., a specific value of $\hat{Y}$) above the plane defined for combinations of the levels of the factors. Geometrically, each contour is a projection onto the X_1X_2 plane of a cross-section of the response surface made by a plane, parallel to the X_1X_2 plane, cutting through the surface, as in Figure 1.6a. The plotting of different surface height values enables one to focus attention on the levels of the factors at which the changes occur in the surface shape. A contour plot of the predicted peanut yield surface in Figure 1.4 is illustrated in Figure 1.6b.

The contour plot of the predicted peanut yield surface shows the surface to be mound shaped. Each contour curve represents an infinite number of combinations of the amounts of the two fertilizers assumed to produce predicted yields of 12.1, 12.7, 13.3, and 13.9 lb/acre. The maximum predicted yield of 14.12 lb/acre is located at the center of the smallest ellipse which corresponds to the levels, X_1 = 170.06 and X_2 = 246.65 (lb/acre), respectively. The location of the maximum yield value is known as the *point of maximum response*. In Section 5.5 in Chapter 5, a method is presented for locating such a point as well as for obtaining the predicted value of the response at the point.

Contour plotting is not limited to three-dimensional surfaces. The geometrical representation for two and three factors enables the general situation for $k > 3$ factors to be more readily understood, although they cannot be visualized geometrically. For example, if in addition to the levels of the two fertilizers a third factor, the frequency (X_3) of application of the fertilizers had been included, then contour curves of constant predicted yield values could be drawn in the plane of X_1 and X_2 for several values of X_3, assuming the prediction equation is not used for extrapolation purposes when performing these projections.

1.3.8 The Operability Region and the Experimental Region

Let us call the region in the factor space in which the experiments can actually be performed the *operability region* (O). For some applications the experimenter may wish to explore the whole region O, but this is usually rare. Instead, a particular group of experiments is set up to explore only a *limited region of interest*, R, which is entirely contained within the operability region O (see

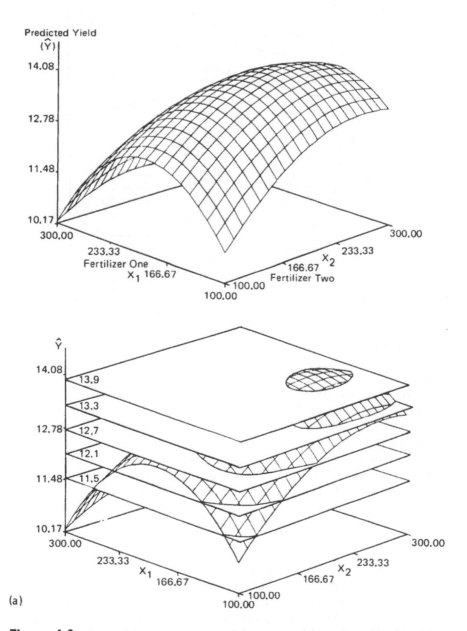

Figure 1.6 Generating a contour plot of the peanut yield surface: (a) slicing the surface at predicted yields of 11.5, 12.1, 12.7, 13.3, and 13.9 lb/acre; (b) the contour plot of the yield surface.

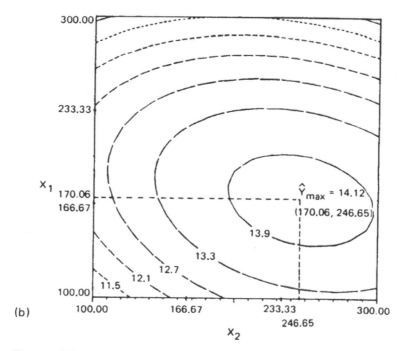

(b)

Figure 1.6 (continued)

Figure 1.7). Although experimental points are not necessarily restricted to the region R and in fact may lie outside it, we shall assume that in most experimental programs, the design points are positioned inside or on the boundary of the region R.

Typically R is defined as, (1) a cuboidal region, or (2) a spherical region, as drawn in Figure 1.7. For example, in studying the effects of the two fertilizers on peanut yield, the range of interest of each of the fertilizers was 100 to 300 lb/acre. The region of interest is therefore the square region containing all combinations of the values of X_1 and X_2 satisfying $100 \leq X_1 \leq 300$ and $100 \leq X_2 \leq 300$. Since yield values were collected at the extreme combinations of X_1 and X_2 defined as $(X_1, X_2) = (100, 100), (100, 300), (300, 100)$, and $(300, 300)$, and these four combinations are the vertices of the square, then the experimental region is the square. With the cotton yield experiment (Figure 1.5), the experimental region is the rectangle defined by the intervals $0 \leq X_1 \leq 340$ and $84 \leq X_2 \leq 254$ kg/ha. During the experimentation only those values of X_1 and X_2 that fall within these ranges are used unless it is discovered during the initial set of experiments that one probably needs to explore levels of X_1 and X_2 that extend beyond the boundaries of the original region.

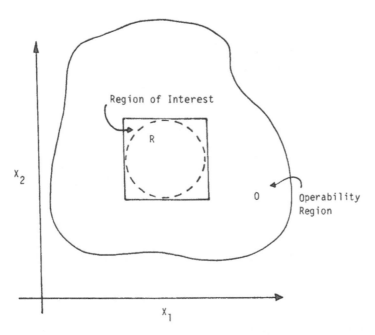

Figure 1.7 The operability region and the square (solid line) and circular (dashed curve) regions of interest in two dimensions.

In Chapters 3 through 9 of this book, we focus our attention primarily on the fitting of polynomial models. In doing so, we assume one's interest is focused on only a *local approximation*, that is, on approximating the shape of the unknown surface directly above a reasonably well-defined neighborhood we have called the region of interest R. The class of lower-degree polynomials (specifically, degrees 1, 2, and at most 3) have stood the test of time by proving their versatility when it comes to approximating successfully a wide variety of surface shapes over such localized regions (see Figure 1.8).

However, when we are faced with modeling a surface over a region whose boundaries extend beyond the localized neighborhood R, such as the region defined by the dashed lines in Figure 1.8, a polynomial of degree 2, or even of degree 3, may not be adequate. This is because the shapes of the two surfaces, the surface above the region R and the surface over the larger region, may be quite different. Over the larger rectangular region and particularly along its boundaries, the shape of the surface may require that additional terms be included in the model than are required to model the shape of the surface over the smaller region. When this is the case, Cornell and Montgomery (1996) propose fitting an interaction model that is a *reduced form* of a higher-than-degree-

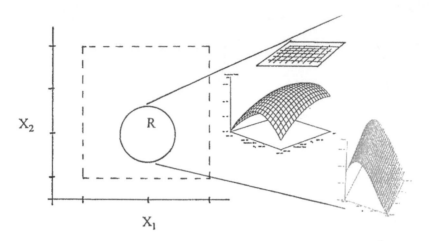

Figure 1.8   The localized neighborhood R and a larger rectangular (dashed lines) region.

3 polynomial. When compared to the form of the standard polynomial, interaction models contain a greater number of cross-product terms and therefore are useful in identifying terms that, when the fitted polynomial of degree d is found to be inadequate, can be added to the polynomial to improve the fit. Interaction models are introduced briefly in Section 2.7.1.

1.4   THE SEQUENTIAL NATURE OF A RESPONSE SURFACE INVESTIGATION

Applying the principles of the scientific method when attempting to enhance one's knowledge of the overall system can improve the chances for success. The scientific method is characterized by performing experiments in stages (or sequentially) and by using the information acquired from one set of experiments to plan the strategy for a follow-up set of experiments. This iterative or sequential pattern of experimentation was suggested by Box and Youle (1955).

Most response surface investigations are sequential. At first an idea or a conjecture is formed concerning which factors are important in terms of influencing some particular response of interest. This leads to planning or designing an experiment that can conceivably perform a dual role; to verify that the factors thought to be important are indeed influential, and to eliminate (weed out) factors that are unimportant. The experiment is then performed and the data are collected. The data are analyzed and the results lead to new ideas or conjectures.

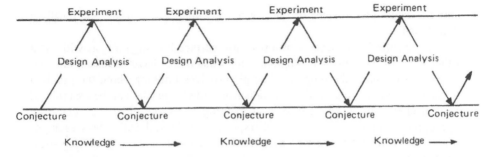

Figure 1.9 The sequential nature of obtaining knowledge of the system through experimentation.

The entire process of conjecture-design-experiment-analysis is repeated as shown in Figure 1.9. In the second set of experiments, additional knowledge is acquired about the relationship between the response and the important factors by learning *how* the factors influence the response. In the third and final stage of the learning process, a clearer understanding of the overall system is realized by answering *why* the factors are important.

Response surface methods are employed at the design and the analysis stages. The design stage is important because the design defines how the data are to be collected and how much data is to be collected. In the analysis of the data, the objective is to provide plausible explanations of the experimental evidence and to stimulate the process of conjecture on the part of the experimenter. Thus, the design of the experiment and the analysis of the data go hand in hand in helping an experimenter learn which factors are important, what role each factor plays in the system, and why these factors are important.

1.5 SOME COMMENTS ON AN EFFECTIVE STRATEGY FOR CARRYING OUT AN EXPERIMENTAL PROGRAM

The success of an experimental program depends largely on the commitment and cooperation among those involved in the program. Let us imagine, for example, that a decision is made to try and develop a new product or to improve an existing product, which of course, will require some experimental effort. Such a decision often originates with a principal investigator or a team of investigators, located typically in a research and development environment. Yet, other personnel will be involved in the project. These other personnel include subject- or process-oriented scientists (such as engineers and/or chemists), technicians or operators in the laboratory or pilot plant who usually have considerable insight into the actual performance or running of the experimental trials, and ideally a statistician.

Instrumental in the success of a response surface investigation, as with most other types of experimental programs, are what others have called the *art* and *science* of planning the program. The *art* involves thinking about the problem at hand and integrating all the nonstatistical knowledge that can be assimilated to deal effectively with issues such as defining goals and objectives, choosing appropriate experimental factors and response variables, deciding how each potential experimental factor will be treated (held fixed or varied), the logistical aspects of running the experiments, and so on. Several papers in the literature on this subject are Hahn (1977, 1984), Bishop, Petersen, and Trayser (1982), Hoadley and Kettenring (1990), and Coleman and Montgomery (1993).

The *science* of planning an experiment involves the statistical and mathematical principles that lead to the development of the experimental design as well as the methods of analyses of the resulting data. This book is directed at this scientific side of the strategy.

An effective strategy for carrying out a successful experimental program starts with planning the program. Montgomery (1991) presents a seven-step approach to take when planning and conducting experiments. The seven steps are listed in Table 1.2. These steps are also relevant when a response surface investigation is being planned. The first three steps constitute the preexperimental planning phase and are discussed extensively in Coleman and Montgomery (1993). Their discussion centers on the methodology used for evaluating potential experimental factors, determining the role each factor will play, understanding the various sources of experimental error, determining how randomization will be accomplished, and so on.

Much of the effort in a response surface investigation is empirical, requiring extensive use of experimentation. Thus the proper selection of the statistical design and the method of analysis can greatly increase the efficiency of a re-

Table 1.2 Seven Steps in the Planning and Execution of an Experimental Program

---
1. Statement of the problem
2. Selection of the response variable(s)
3. Choice of factors and their levels
4. Choice of experimental design
5. Execution of the experimental trials
6. Analysis of the data
7. Statement of conclusions and recommendations
---

sponse surface investigation as well as strengthen the conclusions reached. Planning the experiment carefully and selecting a reasonable design quite often lead to a straightforward analysis of the data, particularly with the extensive library of computing software that is available today.

As we prepare to embark on the scientific aspects of performing experiments of the response surface type in the remainder of this book, we would do well to remember that almost all experiments can be set up to be sequential, and that it is cost effective to think in this way. Generally, it is unwise to plan and execute a large, comprehensive experiment at the outset of the program, particularly if one is somewhat uncertain about which factors are important, or if one is uncertain about the ranges over which the important factors are to be varied, and so on. As we shall see, with most investigations, whether they are designed for confirming what is already known or whether they are set up for seeking new information, the recommended strategy is to begin on a small scale and to build the foundation of one's knowledge in stages. To this end, Box (1992) recommended investing only a fraction of the total experimental effort and budget, say 25–40%, for the initial set of experiments. Limiting expenditures in the initial set of experiments, which often is nothing more than a learning experience anyway, is a cost-effective approach, particularly when one suspects that, in all likelihood, additional experiments and resources will be needed in order to accomplish the final objectives of the experimental program.

REFERENCES AND BIBLIOGRAPHY

Regression Analysis

Allen, D. M. and F. B. Cady (1981). *Analyzing Experimental Data by Regression*, Belmont, CA.: Lifetime Learning Publications.

Belsley, D. A., E. Kuh, and R. E. Welsch (1980). *Regression Diagnostics: Identifying Influential Data and Sources of Collinearity*, New York: John Wiley.

Chatterjee, S. and B. Price (1977). *Regression Analysis by Example*, New York: John Wiley.

Cornell, J. A. and D. C. Montgomery (1996). Interaction Models as Alternatives to Low-Order Polynomials, *J. Quality Tech.*, **28**, (in press).

Draper, N. R. and H. Smith (1981). *Applied Regression Analysis*, 2nd ed., New York: John Wiley.

Gunst, R. F. and R. L. Mason (1980). *Regression Analysis and Its Application: A Data-Oriented Approach*, New York: Marcel Dekker.

Montgomery, D. C. and E. A. Peck (1992). *Introduction to Linear Regression Analysis*, 2nd ed., New York: John Wiley.

Seber, G. A. F. (1977). *Linear Regression Analysis*, New York: John Wiley.

Weisberg, S. (1980). *Applied Linear Regression*, New York: John Wiley.

Response Surface Methods

Box, G. E. P. and K. B. Wilson (1951). On the Experimental Attainment of Optimum Conditions (with discussion), *J. Roy. Statist. Soc.*, *B* **13**, 1–45.

Box, G. E. P. (1954). The Exploration and Exploitation of Response Surfaces: Some General Considerations and Examples, *Biometrics*, **10**, 16–60.

Box, G. E. P. and P. V. Youle (1955). The Exploration and Exploitation of Response Surfaces: An Example of the Link Between the Fitted Surface and the Basic Mechanism of the System, *Biometrics*, **11**, 287–323.

Box, G. E. P. and J. S. Hunter (1957). Multifactor Experimental Designs for Exploring Response Surfaces, *Ann. Math. Statist.*, **28**, 195–241.

Box, G. E. P. and N. R. Draper (1959). A Basis for the Selection of a Response Surface Design, *J. Amer. Statist. Assoc.*, **54**, 622–654.

Box, G. E. P., W. G. Hunter, and J. S. Hunter (1978). *Statistics for Experimenters: An Introduction to Design, Data Analysis, and Model Building*, New York: John Wiley.

Cochran, W. G. and G. M. Cox (1957). *Experimental Designs*, 2nd ed., New York: John Wiley.

Davies, O. L. (1956). *The Design and Analysis of Industrial Experiments*, 2nd ed., London: Oliver and Boyd.

Hill, W. J. and W. G. Hunter (1966). A Review of Response Surface Methodology: A Literature Survey, *Technometrics*, **8**, 571–590.

Mead, R. and D. J. Pike (1975). A Review of Response Surface Methodology from a Biometric Point of View, *Biometrics*, **31**, 803–851.

Montgomery, D. C. (1991). *Design and Analysis of Experiments*, 3rd ed., New York: John Wiley.

Myers, R. H. (1976). *Response Surface Methodology*, Blacksburg, VA: Author.

Myers, R. H., A. I. Khuri, and W. H. Carter, Jr. (1989). Response Surface Methodology: 1966–1988, *Technometrics*, **31**, 137–157.

Peng, K. C. (1967). *The Design and Analysis of Scientific Experiments*, Reading, MA: Addison-Wesley.

Planning Experiments and Experimental Design

Bishop, T., B. Petersen, and D. Trayser (1982). Another Look at the Statistician's Role in Experimental Planning and Design, *Amer. Statist.*, **36**, 387–389.

Box, G. E. P. (1992). Sequential Experimentation and Sequential Assembly of Designs, *Quality Eng.*, **5**, 321–330.

Coleman, D. E. and D. C. Montgomery (1993). A Systematic Approach to Planning for a Designed Industrial Experiment (with discussion), *Technometrics*, **35**, 1–27.

Hahn, G. J. (1977). Some Things Engineers Should Know About Experimental Design, *J. Quality Tech.*, **9**, 13–20.

Hahn, G. J. (1984). Experimental Design in a Complex World, *Technometrics*, **26**, 19–31.

Hoadley, A. and J. Kettenring (1990). Communications Between Statisticians and Engineers/Physical Scientists (with commentary), *Technometrics*, **32**, 243–274.

2

Matrix Algebra, Least Squares, the Analysis of Variance, and Principles of Experimental Design

2.1 INTRODUCTION

In this chapter, we shall review some basic definitions of the types of matrices that are used in RSM. Appendix 2A lists additional properties of matrices and determinants that are useful in the development of RSM. The method of least squares is reviewed along with the topic of analysis of variance. An introduction to the principles of experimental design and some of the more important properties of a response surface design are given in Section 2.8 of this chapter. Transformations on the response variable are presented in Section 2.9. Listed at the end of the chapter are sources for additional reading on these topics.

2.2 SOME FUNDAMENTAL MATRIX DEFINITIONS

Let us consider a linear transformation from the k variables $X_1, X_2, \ldots, X_k$ to the N variables $Y_1, Y_2, \ldots, Y_N$ expressed as

$$Y_1 = m_{11}X_1 + m_{12}X_2 + \cdots + m_{1k}X_k$$
$$Y_2 = m_{21}X_1 + m_{22}X_2 + \cdots + m_{2k}X_k$$
$$\vdots$$
$$Y_N = m_{N1}X_1 + m_{N2}X_2 + \cdots + m_{Nk}X_k$$

The rectangular (or square if $N = k$) array of coefficients $[m_{ij}]$ of the transformation is called the *matrix* of the linear transformation. The matrix $\mathbf{M}$ has N

rows and k columns

$$\underset{N \times k}{\mathbf{M}} = [m_{ij}] = \begin{bmatrix} m_{11} & m_{12} & \cdots & m_{1k} \\ m_{21} & m_{22} & \cdots & m_{2k} \\ \vdots & \vdots & \ddots & \vdots \\ m_{N1} & m_{N2} & \cdots & m_{Nk} \end{bmatrix}$$

and is, therefore, said to be of order $N \times k$. We shall denote the matrix $\mathbf{M}$ by square brackets $[m_{ij}]$ where m_{ij} is the element positioned in the ith row and jth column.

A matrix consisting of only a single column is called a *column vector*. A row vector is a matrix with just a single row. Notationally, matrices are denoted by uppercase boldface letters, and vectors are denoted by lowercase boldface letters. For example, the $N \times 3$ matrix $\mathbf{M}$ consists of the three $N \times 1$ column vectors $\mathbf{m}_1$, $\mathbf{m}_2$, and $\mathbf{m}_3$:

$$\mathbf{M} = [\mathbf{m}_1 : \mathbf{m}_2 : \mathbf{m}_3] = \begin{bmatrix} m_{11} & m_{12} & m_{13} \\ m_{21} & m_{22} & m_{23} \\ \vdots & \vdots & \vdots \\ m_{N1} & m_{N2} & m_{N3} \end{bmatrix} \tag{2.1}$$

Definition 2.1   A *square matrix* is a matrix in which the number of rows equals the number of columns. In Eq. 2.1 if $N = 3$, then $\mathbf{M}$ is a 3×3 square matrix.

Definition 2.2   The *transpose* of the $N \times k$ matrix $\mathbf{M}$ is the $k \times N$ matrix, $\mathbf{M}'$, obtained by interchanging rows with columns:

$$\mathbf{M}' = \begin{bmatrix} m_{11} & m_{21} & \cdots & m_{N1} \\ m_{12} & m_{22} & \cdots & m_{N2} \\ \vdots & \vdots & \cdots & \vdots \\ m_{1k} & m_{2k} & \cdots & m_{Nk} \end{bmatrix}$$

Definition 2.3   A *symmetric matrix* is equal to its transpose, that is, if $\mathbf{M} = \mathbf{M}'$, then $\mathbf{M}$ is symmetric.

Definition 2.4   A *diagonal matrix* is a square matrix whose off-diagonal elements, $m_{ij}, i \neq j$, are zero:

$$\mathbf{M} = \begin{bmatrix} m_{11} & & & 0 \\ & m_{22} & & \\ & & \ddots & \\ 0 & & & m_{kk} \end{bmatrix} = \mathrm{diag}(m_{11}, m_{22}, \ldots, m_{kk}).$$

Definition 2.5   A *scalar* is a single number (a 1×1 matrix).

Definition 2.6   The *identity* matrix $\mathbf{I}$ is a diagonal matrix with ones on the diagonal.

Definition 2.7   The *trace* of a square matrix **M** is the sum of the diagonal elements. The trace of the $k \times k$ matrix **M** is denoted by $\text{tr}(\mathbf{M}) = \Sigma_{i=1}^{k} m_{ii}$.

Definition 2.8   Associated with every square matrix **M** is a unique scalar called its *determinant*, denoted by $|\mathbf{M}|$. The determinants of the three smallest (in terms of order) square matrices are

$$|m_{11}| = m_{11}$$

$$\begin{vmatrix} m_{11} & m_{12} \\ m_{21} & m_{22} \end{vmatrix} = m_{11}m_{22} - m_{12}m_{21}$$

$$\begin{vmatrix} m_{11} & m_{12} & m_{13} \\ m_{21} & m_{22} & m_{23} \\ m_{31} & m_{32} & m_{33} \end{vmatrix} = m_{11}m_{22}m_{33} + m_{12}m_{23}m_{31} + m_{13}m_{21}m_{32}$$

$$- m_{13}m_{22}m_{31} - m_{11}m_{23}m_{32} - m_{12}m_{21}m_{33}$$

For larger matrices, let us define the minor of the element m_{ij} of **M** as the determinant of the matrix formed by deleting the ith row and jth column of **M**. The cofactor of m_{ij} is the minor multiplied by $(-1)^{i+j}$. If the cofactor of m_{ij} is written as $\text{Cof}(m_{ij})$, then the determinant of the $k \times k$ matrix **M** can be expressed as

$$|\mathbf{M}| = \sum_{j=1}^{k} m_{ij}\text{Cof}(m_{ij}) \qquad i = 1, 2, \ldots, k$$

$$= \sum_{i=1}^{k} m_{ij}\text{Cof}(m_{ij}) \qquad j = 1, 2, \ldots, k$$

Thus, the determinant of a 4×4 matrix can be expressed as a linear combination of 3×3 determinants; the determinant of a 5×5 matrix can be expressed as a linear combination of 4×4 determinants, and so forth.

Definition 2.9   A matrix **A** is a *submatrix* of a matrix **B** if **A** is obtained from **B** by deleting certain rows and/or columns.

Definition 2.10   If **A** is a matrix of order $m \times n$ and if all submatrices of order $(r + 1) \times (r + 1)$ have zero determinants while at least one submatrix of order $r \times r$ has a nonzero determinant, then **A** is said to be of *rank r*. Note that this definition necessarily implies that $r \leq m$, $r \leq n$. If $r = n$, then **A** is said to be of full column rank; if $r = m$, **A** is of full row rank.

Definition 2.11   The eigenvalues of the $k \times k$ matrix **M** are the solutions to the determinantal equation $|\mathbf{M} - \lambda\mathbf{I}| = 0$. The determinant is a kth degree polynomial in λ, therefore, **M** has k eigenvalues.

Definition 2.12   Associated with every eigenvalue, λ_i, of the square matrix $\mathbf{M}$ is a nonzero *eigenvector*, $\mathbf{v}_i$, whose elements satisfy the system of equations

$$(\mathbf{M} - \lambda_i \mathbf{I})\mathbf{v}_i = 0$$

Definition 2.13   A square matrix $\mathbf{M}$ is said to be *idempotent* if $\mathbf{MM} = \mathbf{M}^2 = \mathbf{M}$. We shall assume that all idempotent matrices are also symmetric.

Matrices that have the same order can be added together, element by element. They are said to be *conformable for addition*. For example,

$$\underset{2 \times 3}{\mathbf{M}} = \begin{bmatrix} m_{11} & m_{12} & m_{13} \\ m_{21} & m_{22} & m_{23} \end{bmatrix} \qquad \underset{2 \times 3}{\mathbf{N}} = \begin{bmatrix} n_{11} & n_{12} & n_{13} \\ n_{21} & n_{22} & n_{23} \end{bmatrix}$$

$$\text{and } \underset{2 \times 3}{\mathbf{M} + \mathbf{N}} = \begin{bmatrix} m_{11} + n_{11} & m_{11} + n_{12} & m_{13} + n_{13} \\ m_{21} + n_{21} & m_{22} + n_{22} & m_{23} + n_{23} \end{bmatrix}$$

The product of an $N \times k$ matrix $\mathbf{M}$ by a scalar c is the $N \times k$ matrix $c\mathbf{M} = [cm_{ij}]$. Two matrices are said to be *conformable for multiplication* if the number of columns of the matrix on the left is equal to the number of rows of the matrix on the right. For example, if $\mathbf{M} = [m_{ij}]$ is $N \times k$ and $\mathbf{T} = [t_{ij}]$ is $k \times r$, then their product $\mathbf{MT}$ is the matrix $\mathbf{S} = [s_{ij}]$, where $s_{ij} = \sum_{l=1}^{k} m_{il}t_{lj}$, which is of order $N \times r$. To illustrate,

$$\overset{\mathbf{M}}{\begin{bmatrix} m_{11} & m_{12} & m_{13} \\ m_{21} & m_{22} & m_{23} \\ m_{31} & m_{32} & m_{33} \end{bmatrix}} \overset{\mathbf{T}}{\begin{bmatrix} t_{11} & t_{12} \\ t_{21} & t_{22} \\ t_{31} & t_{32} \end{bmatrix}}$$
$$3 \times 3 \qquad 3 \times 2$$

$$\mathbf{S}$$
$$= \begin{bmatrix} (m_{11}t_{11} + m_{12}t_{21} + m_{13}t_{31}) = s_{11} & (m_{11}t_{12} + m_{12}t_{22} + m_{13}t_{32}) = s_{12} \\ (m_{21}t_{11} + m_{22}t_{21} + m_{23}t_{31}) = s_{21} & (m_{21}t_{12} + m_{22}t_{22} + m_{23}t_{32}) = s_{22} \\ (m_{31}t_{11} + m_{32}t_{21} + m_{33}t_{31}) = s_{31} & (m_{31}t_{12} + m_{32}t_{22} + m_{33}t_{32}) = s_{32} \end{bmatrix}$$
$$3 \times 2$$

Definition 2.14   The inverse of a square matrix $\mathbf{M}$ is a matrix denoted by $\mathbf{M}^{-1}$ such that $\mathbf{MM}^{-1} = \mathbf{M}^{-1}\mathbf{M} = \mathbf{I}$. A square matrix that has an inverse is said to be *nonsingular*. It is also said to be of full rank.

Definition 2.15   A square matrix $\mathbf{T}$ is said to be orthogonal if $\mathbf{T}' = \mathbf{T}^{-1}$, that is, the transpose of $\mathbf{T}$ is equal to its inverse. The transformation $\mathbf{w} = \mathbf{Tv}$ is called an *orthogonal* transformation.

We shall now review the method of least squares which is used for obtaining the estimates of the parameters in linear polynomial models. Additional definitions, theorems, and properties of matrices which are useful in the development of response surface techniques are presented in Appendix 2A.

2.3 A REVIEW OF THE METHOD OF LEAST SQUARES

Let us assume provisionally that N observations of the response are expressible by means of the first-order model

$$Y_u = \beta_0 + \beta_1 X_{u1} + \beta_2 X_{u2} + \cdots + \beta_k X_{uk} + \varepsilon_u \qquad u = 1, 2, \ldots, N \quad (2.2)$$

In Eq. 2.2, Y_u denotes the observed response for the uth trial, X_{ui} represents the level of factor i at the uth trial, β_0 and β_i, $i = 1, 2, \ldots, k$, are unknown parameters, and ε_u represents the random error in Y_u. The model $Y = f(X_1, \ldots, X_R) + \varepsilon$ of Eq. 2.2 is called a *linear model* because $f(X_1, \ldots, X_R)$ is a linear function of the parameters $\beta_0, \beta_1, \ldots, \beta_k$. Nonlinear models are presented in Chapter 10. Assumptions made about the errors are

1. Random errors ε_u have zero mean and common variance, σ^2.
2. Random errors ε_u are mutually independent in the statistical sense.

For tests of significance (t- and F-statistics), and confidence interval estimation procedures, an additional assumption must be satisfied:

3. Random errors ε_u are normally distributed.

The method of least squares selects as estimates for the unknown parameters in Eq. 2.2, those values, $b_0, b_1, \ldots, b_k$ respectively, which minimize the quantity

$$R(\beta_0, \beta_1, \ldots, \beta_k) = \sum_{u=1}^{N} (Y_u - \beta_0 - \beta_1 X_{u1} - \beta_2 X_{u2} - \cdots - \beta_k X_{uk})^2$$

The parameter estimates $b_0, b_1, \ldots, b_k$ which minimize $R(\beta_0, \beta_1, \ldots, \beta_k)$ are the solutions to the $k + 1$ normal equations,

$$b_0 N + b_1 \Sigma X_{u1} + b_2 \Sigma X_{u2} + \cdots + b_k \Sigma X_{uk} = \Sigma Y_u$$
$$b_0 \Sigma X_{u1} + b_1 \Sigma X_{u1}^2 + b_2 \Sigma X_{u1} X_{u2} + \cdots + b_k \Sigma X_{u1} X_{uk} = \Sigma X_{u1} Y_u$$
$$\cdots \qquad\qquad (2.3)$$
$$b_0 \Sigma X_{uk} + b_1 \Sigma X_{uk} X_{u1} + b_2 \Sigma X_{uk} X_{u2} + \cdots + b_k \Sigma X_{uk}^2 = \Sigma X_{uk} Y_u$$

where all summations are over $u = 1, 2, \ldots, N$.

Over N observations, the first-order model in Eq. 2.2 can be expressed, in matrix notation, as

$$\mathbf{Y} = \mathbf{X}\boldsymbol{\beta} + \boldsymbol{\varepsilon} \qquad\qquad (2.4)$$

where

$$Y = \begin{bmatrix} Y_1 \\ Y_2 \\ \vdots \\ Y_N \end{bmatrix} \qquad X = \begin{bmatrix} 1 & X_{11} & X_{12} & \cdots & X_{1k} \\ 1 & X_{21} & X_{22} & \cdots & X_{2k} \\ \vdots & \vdots & \vdots & & \vdots \\ 1 & X_{N1} & X_{N2} & \cdots & X_{Nk} \end{bmatrix}$$
$$\qquad N \times 1 \qquad\qquad\qquad N \times (k+1)$$

$$\beta = \begin{bmatrix} \beta_0 \\ \beta_1 \\ \vdots \\ \beta_k \end{bmatrix} \qquad \varepsilon = \begin{bmatrix} \varepsilon_1 \\ \varepsilon_2 \\ \vdots \\ \varepsilon_N \end{bmatrix}$$
$$(k+1) \times 1 \qquad N \times 1$$

The normal equations, Eq. 2.3, are written as

$$X'Xb = X'Y \tag{2.5}$$

The least squares estimates, **b**, of the elements of β in Eq. 2.4 are

$$b = (X'X)^{-1}X'Y \tag{2.6}$$

where $(X'X)^{-1}$ is the inverse of $X'X$. Since $X'X$ is symmetric, so is $(X'X)^{-1}$. Note that the matrix X in Eq. 2.4 is assumed to be of full column rank. The elements $b_0, b_1, \ldots, b_k$ of **b** in Eq. 2.6 are linear functions of the observations $Y_1, Y_2, \ldots, Y_N$ in **Y**.

2.3.1 Properties of the Coefficient Estimates

The statistical properties of the estimator **b** are easily defined once we recall the assumptions concerning the elements of ε. Let $E(\varepsilon)$ and $\text{Var}(\varepsilon)$ denote the vector of expectations or the mean vector and variance-covariance matrix, respectively, of the random vector ε. Then assumptions (1) and (2) in Section 2.3 are $E(\varepsilon) = 0$ and $\text{Var}(\varepsilon) = E(\varepsilon\varepsilon') = \sigma^2 I_N$, where I_N is the identity matrix of order $N \times N$. Thus the expectation or mean vector of **b** is

$$\begin{aligned} E(b) &= E[(X'X)^{-1}X'Y] \\ &= E[(X'X)^{-1}X'(X\beta + \varepsilon)] \\ &= \beta + E(X'X)^{-1}X'\varepsilon \\ &= \beta \end{aligned} \tag{2.7}$$

Simply stated, if the model $Y = X\beta + \varepsilon$ is correct, then **b** is an unbiased estimator of β.

The variance-covariance matrix of the vector of estimates, **b**, is

$$\begin{aligned} \text{Var}(b) &= \text{Var}[(X'X)^{-1}X'Y] \\ &= (X'X)^{-1}X' \, \text{Var}(Y)X(X'X)^{-1} \end{aligned}$$

Since $\mathrm{Var}(\mathbf{Y}) = \mathrm{Var}(\boldsymbol{\varepsilon}) = \sigma^2 \mathbf{I}_N$, then

$$\mathrm{Var}(\mathbf{b}) = (\mathbf{X}'\mathbf{X})^{-1}\sigma^2 \tag{2.8}$$

Along the main diagonal of the matrix $\mathbf{C} = (\mathbf{X}'\mathbf{X})^{-1}\sigma^2$, the iith element, $c_{ii}\sigma^2$, is the variance of b_i, the ith element of $\mathbf{b}$. The standard error of b_i is the positive square root of the variance of b_i, $\sqrt{c_{ii}\sigma^2}$. The ijth element of $\mathbf{C}$, $c_{ij}\sigma^2$, is the covariance between the elements b_i and b_j of $\mathbf{b}$. If the errors $\boldsymbol{\varepsilon}$ are jointly normally distributed, then $\mathbf{b}$ is said to be distributed as

$$\mathbf{b} \sim N[\boldsymbol{\beta}, (\mathbf{X}'\mathbf{X})^{-1}\sigma^2] \tag{2.9}$$

Additional properties of the estimator $\mathbf{b}$ are given in Section 2.3.4.

2.3.2 Predicted Response Values

One of the purposes in obtaining a fitted model is to use the model for predicting response values at points throughout the experimental region. Let $\mathbf{x}_p'$ denote a $1 \times p$ vector whose elements correspond to the elements of a row of the matrix $\mathbf{X}$ in Eq. 2.4, $p = k + 1$. If $\mathbf{b}$ represents the vector of estimates of the elements of $\boldsymbol{\beta}$, then an expression for the predicted value of the response, at any point $\mathbf{x} = (X_1, X_2, \ldots, X_k)'$ in the experimental region, is

$$\hat{Y}(\mathbf{x}) = \mathbf{x}_p'\mathbf{b} \tag{2.10}$$

The predicted value of the response corresponding to the uth observation is expressed as $\hat{Y}_u = \hat{Y}(\mathbf{x}_u) = \mathbf{x}_{up}'\mathbf{b}$, where $\mathbf{x}_{up}'$ is the uth row of $\mathbf{X}$. Hereafter we shall use the notation $\hat{Y}(\mathbf{x})$ to denote the predicted value of Y at the point $\mathbf{x}$.

A measure of the *precision* of the prediction $\hat{Y}(\mathbf{x})$, defined as the variance of $\hat{Y}(\mathbf{x})$, is expressed as

$$\begin{aligned} \mathrm{Var}[\hat{Y}(\mathbf{x})] &= \mathrm{Var}[\mathbf{x}_p'\mathbf{b}] \\ &= \mathbf{x}_p'\,\mathrm{Var}(\mathbf{b})\mathbf{x}_p \\ &= \mathbf{x}_p'(\mathbf{X}'\mathbf{X})^{-1}\mathbf{x}_p\sigma^2 \end{aligned} \tag{2.11}$$

The standard error of $\hat{Y}(\mathbf{x})$ is $\sqrt{\mathrm{Var}[\hat{Y}(\mathbf{x})]}$. Thus, the inverse matrix $(\mathbf{X}'\mathbf{X})^{-1}$ used for obtaining $\mathbf{b}$ in Eq. 2.6 is used also to determine the variances and covariances of the elements of $\mathbf{b}$ as well as the variance of $\hat{Y}(\mathbf{x})$.

2.3.3 Residuals

Residuals are defined as differences between the actual observed response values and those predicted for these response values using the fitted model. In other words, the difference between the observed value of the response at the uth trial and the value predicted with the fitted model for the uth trial is called the uth residual, $r_u = Y_u - \hat{Y}(\mathbf{x}_u)$, $u = 1, 2, \ldots, N$. If predicted values are obtained for each of the N trials, then the vector of residuals is $\mathbf{r} = (r_1, r_2, \ldots, r_N)' = \mathbf{Y} - \mathbf{Xb}$. It is important to distinguish between the error ε_u in Y_u with respect to the corresponding true value, $Y_u = \eta + \varepsilon_u$, and the residual r_u of Y_u with respect to the predicted value $\hat{Y}(\mathbf{x}_u)$. The individual residuals are used for checking how close the proposed model fits the data values as well as for checking the normality, independence, and constant variance assumptions of the errors.

Several properties of the residuals are worth mentioning. When the fitted model contains a constant term (b_0), then the sum of the N residuals equals zero, that is, $\Sigma_{u=1}^{N} r_u = 0$. Furthermore, the sum of the products $\Sigma_{u=1}^{N} r_u \hat{Y}(\mathbf{x}_u)$ equals zero, and the sum of the products $\Sigma_{u=1}^{N} r_u X_{ui}$ equals zero, for each $i = 1, 2, \ldots, k$. In matrix notation these properties are

1. $\mathbf{1}'\mathbf{r} = 0$, where $\mathbf{1}'$ is a $1 \times N$ vector of ones.
2. $\hat{\mathbf{Y}}(\mathbf{X})'\mathbf{r} = 0$, where $\hat{\mathbf{Y}}(\mathbf{X}) = \mathbf{Xb}$.
3. $\mathbf{X}'\mathbf{r} = \mathbf{0}$.

Proof of case 2. $\hat{\mathbf{Y}}(\mathbf{X})'\mathbf{r} = 0$. The $N \times 1$ vector of predicted values, $\hat{\mathbf{Y}}(\mathbf{X})$, is expressible as

$$\begin{aligned} \hat{\mathbf{Y}}(\mathbf{X}) &= \mathbf{Xb} \\ &= \mathbf{X}(\mathbf{X}'\mathbf{X})^{-1}\mathbf{X}'\mathbf{Y} = \mathbf{RY} \end{aligned}$$

where the $N \times N$ idempotent matrix $\mathbf{R}$ is called the *projection matrix*. The $N \times 1$ vector of residuals is also expressible in terms of $\mathbf{R}$:

$$\begin{aligned} \mathbf{r} &= \mathbf{Y} - \hat{\mathbf{Y}}(\mathbf{X}) \\ &= \mathbf{Y} - \mathbf{X}(\mathbf{X}'\mathbf{X})^{-1}\mathbf{X}'\mathbf{Y} \\ &= (\mathbf{I} - \mathbf{R})\mathbf{Y} \end{aligned}$$

Now the product, $\hat{\mathbf{Y}}(\mathbf{X})'\mathbf{r}$, equals

$$\begin{aligned} \hat{\mathbf{Y}}(\mathbf{X})'\mathbf{r} = \mathbf{Y}'\mathbf{R}'(\mathbf{I} - \mathbf{R})\mathbf{Y} &= \mathbf{Y}'\mathbf{R}(I - \mathbf{R})\mathbf{Y} \qquad \text{since } \mathbf{R}' = \mathbf{R} \\ &= \mathbf{Y}'(\mathbf{R} - \mathbf{R}^2)\mathbf{Y} \\ &= \mathbf{Y}'(\mathbf{R} - \mathbf{R})\mathbf{Y} \qquad \mathbf{R}^2 = \mathbf{R} \text{ since } \mathbf{R} \text{ is idempotent} \\ &= 0 \end{aligned}$$

The proofs of 1. and 3. are given as exercises.

2.3.4 Estimation of σ^2

In Eq. 2.8 for the variance-covariance matrix of $\mathbf{b}$, as well as in the formula Eq. 2.11 for the variance of $\hat{Y}(\mathbf{x})$, the variance of the errors, σ^2, was assumed known. This assumption is seldom true, however, and usually an estimate of

σ^2 is needed. The estimate is obtained in the analysis of the data values. For the general case where the fitted model contains p parameters and the total number of observations is $N(N > p)$, the estimate, s^2, is computed from

$$s^2 = \frac{1}{N - p} \sum_{u=1}^{N} r_u^2$$

$$= \frac{1}{N - p} (\mathbf{Y} - \mathbf{Xb})'(\mathbf{Y} - \mathbf{Xb})$$

$$= \frac{1}{N - p} \text{SSE} \qquad (2.12)$$

where SSE is the *sum of squared residuals*. The divisor $N - p$ is the degrees of freedom of the estimator s^2. When the true model is given by $\mathbf{Y} = \mathbf{X\beta} + \mathbf{\epsilon}$, then s^2 is an unbiased estimator of σ^2.

Some further remarks about s^2 and b. When the normality assumption concerning ϵ in Eq. 2.4 is true, the estimator s^2 in Eq. 2.12 is distributed independently of the vector $\mathbf{b}$ in Eq. 2.6. This is shown by using Theorem 4.17 of Graybill (1961), for example, which states that the $p(= k + 1)$ linear forms $\mathbf{b} = (\mathbf{X'X})^{-1}\mathbf{X'Y}$ are independent of the quadratic form $s^2 = [1/(N - p)]\mathbf{Y'(I -} \mathbf{X(X'X)^{-1}X')Y}$, if the product of the matrices for the linear forms and the quadratic form is the null matrix, as long as $\text{Var}(\mathbf{Y}) = \sigma^2\mathbf{I}$. In other words, $\mathbf{b}$ and s^2 are independent if $(\mathbf{X'X})^{-1}\mathbf{X'(I} - \mathbf{X(X'X)^{-1}X')} = \mathbf{0}$. Simply by performing the indicated multiplication, we can show that this is true.

A further desirable property of the least squares estimator $\mathbf{b}$ is the minimum variance property. Suppose we disregard the normality assumption of the errors and consider a class of linear estimators of the type $\mathbf{b^*} = \mathbf{AY}$ where $\mathbf{A} = [(\mathbf{X'X})^{-1}\mathbf{X'} + \mathbf{B}]$ and $\mathbf{BX} = \mathbf{0}$ so that $E(\mathbf{b^*}) = \mathbf{\beta}$. Then the least squares estimator, $\mathbf{b}$, produces minimum variance estimates of the elements of $\mathbf{\beta}$ (i.e., estimates that have minimum variances for a given sample size) in the class of all linear unbiased estimators of $\mathbf{\beta}$. As stated by the Gauss-Markoff theorem, Theorem 6.2 (Graybill, 1961), $\mathbf{b}$ in Eq. 2.6 is the best linear unbiased estimator (BLUE) of $\mathbf{\beta}$.

2.4 THE ANALYSIS-OF-VARIANCE TABLE

Following the program of experimentation, the data are analyzed and the results of the analysis are displayed in table form. The table is called an analysis-of-variance table. The entries in the table represent measures of information concerning the separate sources of variation in the data.

The total variation in a set of data is called the *total sum of squares* (SST). The quantity SST is computed by summing the squares of the deviations of the observed Y_u's about their average value, $\overline{Y} = (Y_1 + Y_2 + \cdots + Y_N)/N$,

$$SST = \sum_{u=1}^{N} (Y_u - \overline{Y})^2 \tag{2.13}$$

The quantity SST has associated with it $N - 1$ degrees of freedom since the sum of the deviations, $Y_u - \overline{Y}$, is equal to zero.

The total sum of squares can be partitioned into two parts, the sum of squares due to regression (or sum of squares explained by the fitted model) and the sum of squares unaccounted for by the fitted model. The formula for calculating the sum of squares due to regression (SSR) is

$$SSR = \sum_{u=1}^{N} (\hat{Y}(\mathbf{x}_u) - \overline{Y})^2 \tag{2.14}$$

The deviation $\hat{Y}(\mathbf{x}_u) - \overline{Y}$ is the difference between the value predicted by the fitted model for the uth observation and the overall average of the Y_u's. If the fitted model contains p parameters, then the number of degrees of freedom associated with SSR is $p - 1$. The sum of squares unaccounted for by the fitted model (SSE) is

$$SSE = \sum_{u=1}^{N} (Y_u - \hat{Y}(\mathbf{x}_u))^2 \tag{2.15}$$

The quantity SSE was called the sum of squares of the residuals in Eq. 2.12. The number of degrees of freedom for SSE was defined previously as $N - p$ which is the difference $(N - 1) - (p - 1) = N - p$.

Short-cut formulas for SST, SSR, and SSE are possible using matrix notation. Letting $\mathbf{1}'$ be a $1 \times N$ vector of ones, we have

$$SST = \mathbf{Y}'\mathbf{Y} - \frac{(\mathbf{1}'\mathbf{Y})^2}{N}$$

$$SSR = \mathbf{b}'\mathbf{X}'\mathbf{Y} - \frac{(\mathbf{1}'\mathbf{Y})^2}{N}$$

$$SSE = \mathbf{Y}'\mathbf{Y} - \mathbf{b}'\mathbf{X}'\mathbf{Y} \tag{2.16}$$

where $\mathbf{Y}$ is the $N \times 1$ vector of observations, $\mathbf{X}$ is the $N \times p$ matrix (of rank p) of the settings of the values of X_{ui}, and $\mathbf{b}$ is the $p \times 1$ vector of estimates of the coefficients in the model. The distributional properties of SSE and SSR are presented in Section 2.7 on hypothesis testing. The partitioning of the total sum of squares into SSR and SSE is displayed in the analysis-of-variance Table 2.1, where the fitted model in Eq. 2.10 is assumed to contain p parameters.

The usual test of the significance of the fitted regression equation is a test of the null hypothesis H_0: all values of β_i (excluding β_0) are zero. The alternative hypothesis is H_a: at least one value of β_i (excluding β_0) is not zero. Assuming normality of the errors, the test of H_0 involves first calculating the value of the F-statistic

$$F = \frac{\text{Mean Square Regression}}{\text{Mean Square Residual}} = \frac{\text{SSR}/(p-1)}{\text{SSE}/(N-p)} \tag{2.17}$$

If the null hypothesis is true, the F-statistic in Eq. 2.17 follows an F distribution with $p-1$ and $N-p$ degrees of freedom in the numerator and in the denominator, respectively. The second step of the test of H_0 is to compare the value of F in Eq. 2.17 to the table value, $F_{\alpha,p-1,N-p}$, which is the upper 100α percent point of the F distribution with $p-1$ and $N-p$ degrees of freedom, respectively. If the value of F in Eq. 2.17 exceeds $F_{\alpha,p-1,N-p}$, then the null hypothesis is rejected at the α level of significance and we infer that the variation accounted for by the model (through the values of b_i, $i \neq 0$) is significantly greater than the unexplained variation.

An accompanying statistic to the F-statistic of Eq. 2.17 is the coefficient of determination:

$$R^2 = \frac{\text{SSR}}{\text{SST}} \tag{2.18}$$

The value of R^2 is a measure of *the proportion of total variation of the values of Y_u about the mean $\bar{Y}$ explained by the fitted model*. It is often expressed as a percent by multiplying the ratio SSR/SST by 100.

Table 2.1 Analysis-of-Variance Table

Source of Variation	Degrees of Freedom (df)	Sum of Squares (SS)	Mean Square (MS)
Due to Regression (Fitted Model)	$p-1$	SSR	$\text{SSR}/(p-1)$
Residual (Error)	$N-p$	SSE	$\text{MSE} = \text{SSE}/(N-p)$
Total	$N-1$	SST	

A related statistic, called the adjusted R^2 statistic, is

$$R_A^2 = 1 - \frac{\text{SSE}/(N - p)}{\text{SST}/(N - 1)}$$

$$= 1 - (1 - R^2)\left(\frac{N - 1}{N - p}\right) \tag{2.19}$$

The adjustment is made by using the degrees of freedom corresponding to SSE and SST in Eq. 2.19. The R_A^2 statistic is a measure of the drop in the magnitude of the estimate of the error variance achieved by fitting a model other than $Y = \beta_0 + \varepsilon$ relative to the estimate of the error variance that would be obtained by fitting the model $Y = \beta_0 + \varepsilon$.

We shall now work through a numerical example to illustrate the calculating formulas for the sums of squares. Values of F, R^2, and R_A^2 for the fitted model are presented at the end of the example.

2.5 A NUMERICAL EXAMPLE: MODELING THE AMOUNT OF FOOD INGESTED BY LABORATORY RATS

Twelve laboratory rats of uniform size and age were deprived of food, except for one hour per day, for 10 days. On the eleventh day, each rat was inoculated with a hunger-reducing drug (the dosage levels were 0.3 and 0.7 mg/kg) and after a specific length of time (the time levels were 1, 5, and 9 hours), the rat was fed. Each of the six dosage by time combinations was applied to two rats. The weight (in grams) of the food ingested by each rat was measured. The purpose of the experiment was to determine if changing the dosage level of the drug as well as the length of time between inoculation and feeding had any effect on the rats in terms of the amount of food ingested. The data are presented in Table 2.2, where each entry represents the weight of food in grams ingested by one of the 12 rats. In Figure 2.1, the individual weight values are plotted for each combination of drug dosage and length of time between inoculation and feeding.

Table 2.2 Ingested Food Weights in Rat Feeding Experiment

Drug Dosage (mg/kg)	Length of Time Between Inoculation and Feeding (hours)			Totals
	1	5	9	
0.3	5.63, 6.42	11.57, 12.16	12.68, 13.31	61.77
0.7	1.38, 1.94	5.72, 4.69	8.28, 7.73	29.74
Totals	15.37	34.14	42.00	91.51

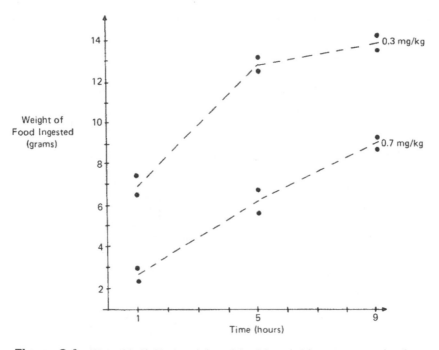

Figure 2.1 Plot of individual weights of food ingested by rats versus time between inoculation and feeding for each drug dosage.

The plot of the individual ingested food weights in Figure 2.1 suggests the higher dosage (0.7) suppresses hunger in rats better than the lower dosage level (0.3). This is because at each level of time, the average amount of food ingested at the 0.7 dosage level is lower than at the 0.3 dosage level. As time between inoculation and feeding increases from 1 hour to 9 hours, the average weight of food eaten increases. At the high dosage level, 0.7, the increase in average weight of food appears linear. At the lower dosage level, 0.3, the increase in average weight of food ingested is greater between 1 and 5 hours than between 5 and 9 hours. This difference in the increases of food ingested suggests a quadratic term in the time levels, $\beta_{22}X_2^2$, ought to be included in the model.

The proposed model form relating observed weight of food ingested per rat to dosage level of drug (X_1) and length of time between drug inoculation and feeding (X_2) is

$$Y_u = \beta_0 + \beta_1 X_{u1} + \beta_2 X_{u2} + \beta_{12} X_{u1} X_{u2} + \beta_{22} X_{u2}^2 + \varepsilon_u \qquad u = 1, 2, \ldots, 12$$
$$(2.20)$$

In matrix notation, the model in Eq. 2.20, expressed over the $N = 12$ observations, is

$$
\begin{array}{cccc}
Y & = & X & \beta & + & \varepsilon
\end{array}
$$

$$
\begin{array}{ccccc}
& 1 & X_1 & X_2 & X_1X_2 & X_2^2
\end{array}
$$

$$
\begin{bmatrix}
5.63 \\
6.42 \\
1.38 \\
1.94 \\
11.57 \\
12.16 \\
5.72 \\
4.69 \\
12.68 \\
13.31 \\
8.28 \\
7.73
\end{bmatrix}
=
\begin{bmatrix}
1 & 0.3 & 1 & 0.3 & 1 \\
1 & 0.3 & 1 & 0.3 & 1 \\
1 & 0.7 & 1 & 0.7 & 1 \\
1 & 0.7 & 1 & 0.7 & 1 \\
1 & 0.3 & 5 & 1.5 & 25 \\
1 & 0.3 & 5 & 1.5 & 25 \\
1 & 0.7 & 5 & 3.5 & 25 \\
1 & 0.7 & 5 & 3.5 & 25 \\
1 & 0.3 & 9 & 2.7 & 81 \\
1 & 0.3 & 9 & 2.7 & 81 \\
1 & 0.7 & 9 & 6.3 & 81 \\
1 & 0.7 & 9 & 6.3 & 81
\end{bmatrix}
\begin{bmatrix}
\beta_0 \\
\beta_1 \\
\beta_2 \\
\beta_{12} \\
\beta_{22}
\end{bmatrix}
+
\begin{bmatrix}
\varepsilon_1 \\
\varepsilon_2 \\
\varepsilon_3 \\
\varepsilon_4 \\
\varepsilon_5 \\
\varepsilon_6 \\
\varepsilon_7 \\
\varepsilon_8 \\
\varepsilon_9 \\
\varepsilon_{10} \\
\varepsilon_{11} \\
\varepsilon_{12}
\end{bmatrix}
$$

The normal equations are

$$
\begin{array}{ccc}
X'X & b & = & X'Y
\end{array}
$$

$$
\begin{bmatrix}
12 & 6 & 60 & 30 & 428 \\
 & 3.48 & 30 & 17.4 & 214 \\
 & & 428 & 214 & 3420 \\
 & \text{(Symmetric)} & & 124.12 & 1710 \\
 & & & & 28748
\end{bmatrix}
\begin{bmatrix}
b_0 \\
b_1 \\
b_2 \\
b_{12} \\
b_{22}
\end{bmatrix}
=
\begin{bmatrix}
91.51 \\
39.35 \\
564.07 \\
249.01 \\
4270.87
\end{bmatrix}
$$

and the solutions to the normal equations are

$$
\begin{array}{ccc}
b & = & (X'X)^{-1} & X'Y
\end{array}
$$

$$
\begin{bmatrix}
8.428 \\
-12.369 \\
1.782 \\
-0.195 \\
-0.085
\end{bmatrix}
=
\begin{bmatrix}
2.321 & -3.483 & -0.493 & 0.488 & 0.021 \\
 & 6.966 & 0.488 & -0.977 & 0.000 \\
 & & 0.203 & -0.098 & -0.015 \\
 & \text{(Symmetric)} & & 0.195 & 0.000 \\
 & & & & 0.001
\end{bmatrix}
\begin{bmatrix}
91.51 \\
39.35 \\
564.07 \\
249.01 \\
4270.87
\end{bmatrix}
$$

The fitted second-order model is

$$
\hat{Y}(x) = 8.428 - 12.369X_1 + 1.782X_2 - 0.195X_1X_2 - 0.085\,X_2^2 \qquad (2.21)
$$
$$
\;\;\;(1.172)\quad\;(2.031)\quad\;\;(0.347)\quad\;\;\;(0.340)\quad\;\;\;(0.029)
$$

The number in parentheses below the coefficient estimate is the estimated standard error (est. s.e.) of the coefficient estimate. The est. s.e.s are the positive square roots of the products of the diagonal elements of $(X'X)^{-1}$ and s^2, where $s^2 = 0.592$ is the residual or error mean square ($s^2 = $ MSE) in the analysis-of-variance table, Table 2.4 discussed below. The estimated standard error for $b_2 =$ 1.782, for example, is $\sqrt{c_{22}s^2} = \sqrt{(0.203)(0.592)} = 0.347$.

In Table 2.3 are listed the deviations, $Y_u - \bar{Y}$, $\hat{Y}(\mathbf{x}_u) - \bar{Y}$, and $Y_u - \hat{Y}(\mathbf{x}_u)$ corresponding to each of the 12 rats. The deviations, $Y_u - \bar{Y}$ and $\hat{Y}(\mathbf{x}_u) - \bar{Y}$, are listed to three decimal places, while the residuals, $Y_u - \hat{Y}(\mathbf{x}_u)$, are listed to two decimal places, the same number as the observed weight values. The sums of squares quantities in the analysis of variance, where $\bar{Y} = 91.51/12 = 7.626$, are

$$\text{SST} = \sum_{u=1}^{12} (Y_u - 7.626)^2 = 183.436 \text{ with } 12 - 1 = 11 \text{ degrees of freedom}$$

$$\text{SSR} = \sum_{u=1}^{12} (\hat{Y}(\mathbf{x}_u) - 7.626)^2 = 179.293 \text{ with } 5 - 1 = 4 \text{ degrees of freedom}$$

$$\text{SSE} = \sum_{u=1}^{12} (Y_u - \hat{Y}(\mathbf{x}_u))^2 = \text{SST} - \text{SSR} = 4.143 \text{ with } 12 - 5 = 7 \text{ degrees}$$

of freedom

The analysis of variance (ANOVA) is presented in Table 2.4. Note that in the ANOVA table, the source of variation other than regression and total has been labeled residual.

The F-ratio of Eq. 2.17 for testing the hypothesis H_0: $\beta_1 = \beta_2 = \beta_{12} = \beta_{22} = 0$ against the alternative H_a: at least one of the parameters of $\boldsymbol{\beta}$ in H_0 is not

Table 2.3 Predicted Response Values and Residuals Obtained with the Fitted Model (Eq. 2.21) for the Six Dosage and Time Combinations

Dosage (X_1)	Time (X_2)	Observed Weight Y_u	Deviation $Y_u - \bar{Y}$	Predicted Weight $\hat{Y}(\mathbf{x}_u)$	Deviation $\hat{Y}(\mathbf{x}_u) - \bar{Y}$	*Residual* $Y_u - \hat{Y}(\mathbf{x}_u)$
0.3	1	5.63	−1.996	6.36	−1.266	−0.73
		6.42	−1.206	6.36	−1.266	0.06
0.7	1	1.38	−6.246	1.33	−6.296	0.05
		1.94	−5.686	1.33	−6.296	0.61
0.3	5	11.57	3.944	11.20	3.574	0.37
		12.16	4.534	11.20	3.574	0.96
0.7	5	5.72	−1.906	5.87	−1.756	−0.15
		4.69	−2.936	5.87	−1.756	−1.18
0.3	9	12.68	5.054	13.33	5.704	−0.65
		13.31	5.684	13.33	5.704	−0.02
0.7	9	8.28	0.654	7.67	0.044	0.61
		7.73	0.104	7.67	0.044	0.06

Table 2.4   Analysis of Variance
for Model (Eq. 2.21) Fitted to the
Rat Feeding Experimental Data

Source	df	SS	MS
Regression	4	179.293	44.823
Residual	7	4.143	0.592
Total	11	183.436	

zero, is

$$F = \frac{179.293/4}{4.143/7}$$

$$= 75.73$$

The computed value of $F = 75.73$ exceeds the table value of $F_{0.01,4,7} = 7.85$ so that the hypothesis H_0 is rejected at the 0.01 level of significance. Having rejected H_0, we can only infer that at least one of the parameters (other than β_0) in the model of Eq. 2.20, based on the coefficient estimates in the fitted model, (Eq. 2.21), is not zero. The value of R^2 for the fitted model is, from Eq. 2.19, $R^2 = 0.9774$. This implies that approximately $100 \times R^2 = 97.74\%$ of the total variation in the ingested food weight values is explained by the fitted model of Eq. 2.21. A value of $R_A^2 = 0.9645$ means that MSE = 0.592 is only approximately 3.55% of the variance in the Y_i values.

2.5.1   Tests of Hypotheses Concerning the Individual Parameters in the Model

The F-statistic in Eq. 2.17 is a test of the hypothesis that all of the parameters (excluding β_0) in the proposed model are zero. In the rat feeding experiment, for example, this hypothesis is akin to saying neither the dosage of the drug nor the length of time between inoculation and feeding has an effect on the amount of food ingested by the rats. Yet, the analysis of the ingested food weight values produced a value of 75.73 for the F-statistic, which prompted us to reject the hypothesis that both factors, drug dosage and time, did not affect the response. We now look closer at the specific effects of the factors used in an experiment.

In general, tests of hypotheses concerning the individual parameters in the proposed model are performed by comparing the parameter estimates in the fitted model to their respective estimated standard errors. Let us denote the least squares estimate of β_i by b_i and the estimated standard error of b_i by est. s.e. (b_i). Then a test of the null hypothesis, $H_0: \beta_i = 0$, is performed by calculating the value of the test statistic

$$t = \frac{b_i}{\text{est. s.e.}(b_i)} \tag{2.22}$$

and comparing the value of t in Eq. 2.22 against a table value from the t-table. The choice of the table value depends on the alternative hypothesis, H_a, the level of significance, α, and the degrees of freedom for t in Eq. 2.22. If the alternative hypothesis is H_a: $\beta_i \neq 0$, the test is called a *two-sided test*, and the table value is $t_{\alpha/2}$, the upper $100(\alpha/2)$ percentage point of the t-distribution. If, on the other hand, the alternative hypothesis is H_a: $\beta_i > 0$ or H_a: $\beta_i < 0$, the test is a one-sided test, and the table value is t_α. The degrees of freedom for t in Eq. 2.22 are the degrees of freedom of s^2 used in est. s.e. (b_i).

To illustrate the test of Eq. 2.22, let us refer to the proposed model (Eq. 2.20) relating the weight of food ingested by the rats to drug dosage and time between inoculation and feeding. For the test of the null hypothesis H_0: $\beta_1 = 0$ (the linear effect on average weight of food ingested upon changing the drug dosage) against the alternative hypothesis, H_a: $\beta_1 < 0$, the value of the t-statistic is

$$t = \frac{b_1}{\text{est. s.e.}(b_1)} = \frac{-12.369}{2.031} = -6.09 \tag{2.23}$$

where the values of b_1 and est. s.e. (b_1) are taken from results of the fitted model (Eq. 2.21). Since the test is one-sided, owing to the alternative hypothesis, H_a: $\beta_1 < 0$, and if the level of significance, α, is set at $\alpha = 0.01$, then the table value, t_α, is $t_{0.01} = 2.998$. The t-statistic has 7 degrees of freedom, which is the number of degrees of freedom associated with the estimate, $s^2 = \text{MSE}$, taken from the ANOVA Table 2.4.

The absolute value of t in Eq. 2.23, $|-6.09| = 6.09$, exceeds the value of $t_{0.01} = 2.998$, and therefore the null hypothesis H_0: $\beta_1 = 0$ is rejected in favor of H_a: $\beta_1 < 0$. Similar tests of H_0: $\beta_2 = 0$ versus H_a: $\beta_2 > 0$, of H_0: $\beta_{12} = 0$ versus H_a: $\beta_{12} \neq 0$, and H_0: $\beta_{22} = 0$ versus H_a: $\beta_{22} < 0$ produced t-statistic values of 5.14, -0.57, and -2.93, respectively. The one-sided nature of the alternative hypotheses, H_a: $\beta_2 > 0$ and H_a: $\beta_{22} < 0$, were decided on from looking at the plot of the weight values versus time in Figure 2.1 and recalling that a parabola that increases initially followed by a leveling off or decrease is modeled as $\eta = \beta_0 + \beta_2 X_2 + \beta_{22} X_2^2$ (see Figure 1.2), where $\beta_2 > 0$ and $\beta_{22} < 0$.

Owing to the two-sided alternative, H_a: $\beta_{12} \neq 0$, the value of $t = -0.57$ is compared to the table value, $t_{\alpha/2}$. With α set equal to $\alpha = 0.01$, the table value is $t_{0.005} = 3.499$, and since $|-0.57| = 0.57 < 3.499$, we do not reject H_0: $\beta_{12} = 0$. However, at the $\alpha = 0.01$ level, H_0: $\beta_2 = 0$ is rejected in favor of $\beta_2 > 0$ since $|5.14| > 2.998$, but H_0: $\beta_{22} = 0$ is not rejected since $|-2.93| < 2.998$. At the $\alpha = 0.05$ level, H_0: $\beta_{22} = 0$ is rejected in favor of $\beta_{22} < 0$ since $|-2.93| >$

1.895. According to these separate tests then, we are inclined to infer that changing the drug dosage from 0.3 to 0.7 mg/kg reduces hunger in the rats as evidenced by a decrease in the average weight of food ingested. Increasing the time between inoculation and feeding from 1 to 9 hours increases the hunger since the average weight of food ingested increased. The increase in weight of food ingested is definitely linear (at the $\alpha = 0.01$), but there is some evidence ($\alpha = 0.05$) of a curvilinear effect of time. There is no evidence of interaction between drug dosage and time.

When the model contains more than one unknown parameter (in addition to β_0) and estimates of the coefficients of the terms, $\beta_i X_i$ and $\beta_j X_j$, in the model are not uncorrelated (i.e., $c_{ij}\sigma^2 \neq 0$), the tests of $H_0: \beta_i = 0$ and $H_0: \beta_j = 0$, using Eq. 2.22, are not independent. Consequently, when the null hypothesis is expressed as $H_0: \beta_i = 0$ versus $H_a: \beta_i \neq 0$, what is actually being tested is the hypothesis that the term $\beta_i X_i$ does not explain any additional amount of variation in the response values above that which is explained by the other terms in the model. In essence then, the test of $H_0: \beta_{12} = 0$ of the coefficient of the term, $\beta_{12}X_1X_2$, in model 2.20 is a test of the equivalence of the two models

$$Y = \beta_0 + \beta_1 X_1 + \beta_2 X_2 + \beta_{22} X_2^2 + \varepsilon$$

$$Y = \beta_0 + \beta_1 X_1 + \beta_2 X_2 + \beta_{12} X_1 X_2 + \beta_{22} X_2^2 + \varepsilon$$

while under the alternative hypothesis, the better model is

$$Y = \beta_0 + \beta_1 X_1 + \beta_2 X_2 + \beta_{12} X_1 X_2 + \beta_{22} X_2^2 + \varepsilon$$

Such a test is called a *partial F-test for* β_{12} (Draper and Smith, 1981: 101–102), since the value of $|t|$ in Eq. 2.22 equals $\sqrt{F}$-test statistic, which is calculated when comparing the two models that are equivalent under H_0. We shall discuss this F-test in greater detail in Section 2.7.

2.6   TESTING LACK OF FIT OF THE FITTED MODEL USING REPLICATED OBSERVATIONS

Before proceeding with tests of hypotheses concerning the form of the fitted model, let us discuss briefly one of the many procedures used for testing the adequacy of the fitted model. The F-test performed in the previous example involving the ratio of the mean squares for regression and residual was a test of the null hypothesis $H_0: \beta_1 = \beta_2 = \beta_{12} = \beta_{22} = 0$. Rejecting the hypothesis meant only that we could infer that at least one of the parameters in H_0 was not zero. This inference is not the same as saying the fitted model adequately describes the behavior of the response over the experimental ranges of the variables. A procedure for checking the adequacy of the fitted model is called *testing lack of fit of the fitted model*.

In general, to say the fitted model is inadequate or lacking in fit is to imply the proposed model does not contain a sufficient number of terms. This inadequacy of the model is due to either or both of the following causes:

1. Factors (other than those in the proposed model) that are omitted from the proposed model but which affect the response, and or,
2. The omission of higher-order terms involving the factors in the proposed model which are needed to adequately explain the behavior of the response.

In most modeling situations it is far easier from a design and analysis standpoint to upgrade (add terms to) the model in the factors already considered than to introduce new factors to the program. Thus we shall assume, upon detecting inadequacy of the fitted model, that the inadequacy is due to the omission of higher-order terms in the fitted model, case (2).

Detecting inadequacy of the fitted model is described in the following way. Let us write the proposed model as

$$Y = \mathbf{x}_p'\boldsymbol{\beta}_p + \varepsilon \tag{2.24}$$

where $\mathbf{x}_p'$ is a $1 \times p$ vector of terms in $1, X_1, \ldots, X_k$ and $\boldsymbol{\beta}_p$ is a $p \times 1$ vector of parameters associated with the terms in $\mathbf{x}_p$. For example, if the model in Eq. 2.24 is the equation of a hyperplane and $k = 3$ then $\mathbf{x}_{p=4}' = (1, X_1, X_2, X_3)$ and $\boldsymbol{\beta}_{p=4} = (\beta_0, \beta_1, \beta_2, \beta_3)'$. Consider the shape of the true surface to be more complicated than one that can be described with only the terms in the proposed model in Eq. 2.24. If the true surface shape can be represented by adding terms $\mathbf{x}_q'\boldsymbol{\beta}_q$ to the proposed model, then the true surface can be expressed as

$$E(Y) = \mathbf{x}_p'\boldsymbol{\beta}_p + \mathbf{x}_q'\boldsymbol{\beta}_q \tag{2.25}$$

In Eq. 2.25, $\mathbf{x}_q'$ is a $1 \times q$ vector of terms in $X_1, X_2, \ldots, X_k$ of higher order than those in $\mathbf{x}_p$ (or alternatively, $\mathbf{x}_q$ contains variables $X_{k+1}, \ldots, X_m$ not contained in $\mathbf{x}_p$) and $\boldsymbol{\beta}_q$ is a $q \times 1$ vector of parameters associated with the elements in $\mathbf{x}_q$. Given the expressions in Eq. 2.24 and 2.25 for the proposed and true surface models, respectively, the inability of the fitted model, $\hat{Y} = \mathbf{x}_p'\mathbf{b}_p$ to adequately account for the variation in the observed response values is reflected in the portion of the total variation that is called the *residual variation* or the *variation unaccounted for* by the fitted model. Isolating the *portion of the residual variation* that is directly attributed to underfitting the true surface (Eq. 2.25) with the fitted model, for example, is necessary in order to test for adequacy of the fitted model.

The test for adequacy (or zero lack of fit) of the fitted model requires two conditions be met regarding the collection (design) of the data values:

1. The number of distinct design points, n, must exceed the number of terms in the fitted model. If the fitted model contains p terms, then $n > p$.

2. An esimate of the experimental error variance that does not depend on the form of the fitted model is required. This can be achieved by collecting at least two replicate observations at one or more of the design points and calculating the variation among the replicates at each point. If an estimate of the error variance is available from previous experiments, then collecting replicate observations at one or more design points and calculating an estimate of the error variance from the replicates would enable one to check whether or not the magnitude of the error variance has changed since the time of the previous experiments. Such a test involves setting up the ratio of the two error variance estimates (the former and the present) in the form of an F-statistic.

In addition, we shall assume that the random errors are normal and independently distributed with a common variance σ^2.

When contributions (1) and (2) are met, the residual sum of squares, SSE, can be partitioned into two sources of variation; the variation among the replicates at those design points where replicates are collected, and the variation arising from the lack of fit of the fitted model. The sum of squares due to the replicate observations is called the *sum of squares for pure error* (abbreviated, SS_{PE}) and once calculated, it is then subtracted from the residual sum of squares to produce the sum of squares due to lack of fit (SS_{LOF}).

To illustrate the partitioning of the residual sum of squares, let us first give a formula for calculating the pure error sum of squares. Denote the uth observation at the lth design point by Y_{ul}, where $u = 1, 2, \ldots, r_l \geq 1, l = 1, 2, \ldots, n$. Define $\bar{Y}_l$ to be the average of the r_l observations at the lth design point. Then the sum of squares for pure error is calculated using

$$SS_{PE} = \sum_{l=1}^{n} \sum_{u=1}^{r_l} (Y_{ul} - \bar{Y}_l)^2 \tag{2.26}$$

The sum of squares due to lack of fit is found by subtraction

$$SS_{LOF} = SSE - SS_{PE}$$

or, alternatively, if $\hat{Y}_l$ is the predicted value of the response at the lth design point, then

$$SS_{LOF} = \sum_{l=1}^{n} r_l (\hat{Y}_l - \bar{Y}_l)^2 \tag{2.27}$$

Furthermore, the degrees of freedom associated with SS_{PE} in Eq. 2.26 is $\sum_{l=1}^{n} (r_l - 1) = N - n$, where N is the total number of observations, so that the degrees of freedom associated with SS_{LOF}, obtained by subtraction, is $(N - p) - (N - n) = n - p$. An expanded analysis-of-variance table that displays the partitioning of the residual sum of squares is presented in Table 2.5.

Table 2.5 Expanded Analysis-of-Variance Table Showing the Partitioning of the Residual Sum of Squares into Lack of Fit and Pure Error Sums of Squares

Source	df	SS	MS
Due to regression	$p - 1$	SSR	$\text{SSR}/(p - 1)$
Residual	$N - p$	$\sum_{l=1}^{n} \sum_{u=1}^{r_l} (Y_{ul} - \hat{Y}_l)^2 = \text{SSE}$	
Lack of fit	$n - p$	$\sum_{l=1}^{n} r_l(\hat{Y}_l - \bar{Y}_l)^2 = \text{SS}_{\text{LOF}}$	$\text{SS}_{\text{LOF}}/(n - p)$
Pure error	$N - n$	$\sum_{l=1}^{n} \sum_{u=1}^{r_l} (Y_{ul} - \bar{Y}_l)^2 = \text{SS}_{\text{PE}}$	$\text{SS}_{\text{PE}}/(N - n)$
Total	$N - 1$	$\sum_{l=1}^{n} \sum_{u=1}^{r_l} (Y_{ul} - \bar{Y})^2$	

The form of the test statistic to be used for testing lack of fit of the fitted model can now be defined. Let us write the form (Eq. 2.24) of the proposed model over N observations, in matrix notation, as

$$Y = X_p \beta_p + \varepsilon \qquad (2.28)$$

and write the true model in Eq. 2.25 as

$$E(Y) = X_p \beta_p + X_q \beta_q \qquad (2.29)$$

In the above equations, Y is an $N \times 1$ vector of observations, the matrix X_p is $N \times p$ and in Eq. 2.29, the matrix X_q is $N \times q$. The vector Y is assumed to be normally distributed with expectation $E(Y) = X_p \beta_p + X_q \beta_q$ and covariance matrix equal to $\sigma^2 I_N$.

The test of the null hypothesis of adequacy of fit (or lack of fit is zero) involves calculating the value of the F-ratio

$$F = \frac{\text{SS}_{\text{LOF}}/(n - p)}{\text{SS}_{\text{PE}}/(N - n)} \qquad (2.30)$$

and comparing the value in Eq. 2.30 with a table value of F. Lack of fit can be detected at the α level of significance if the value of F in Eq. 2.30 exceeds the table value, $F_{\alpha, n-p, N-n}$, where the latter quantity is the upper 100α percentage point of the central F-distribution.

The F-statistic in Eq. 2.30 possesses a central F-distribution when the model in Eq. 2.28 is the true model. The null hypothesis actually tested by F in Eq. 2.30 is H_0: $(X_q - X_p A)\beta_q = 0$, where $A = (X_p' X_p)^{-1} X_p' X_q$ is called the *alias matrix*, assuming $(X_p' X_p)^{-1}$ exists. On the other hand, if the true model is of the form in Eq. 2.29, then the F-statistic in Eq. 2.30 has a noncentral F-distribution. If in this latter case we write the alternative hypothesis as H_a: $(X_q - X_p A)\beta_q \neq 0$, then under H_a, F is distributed as $F_{n-p, N-n; \lambda_1}$ where λ_1 is the noncentrality parameter of the noncentral F-distribution. The noncentrality parameter, λ_1, can be written as $\lambda_1 = \boldsymbol{\beta}_q'(\mathbf{X}_q - \mathbf{X}_p\mathbf{A})'(\mathbf{X}_q - \mathbf{X}_p\mathbf{A})\boldsymbol{\beta}_q/2\sigma^2$. Also, under H_a, the expected values of the numerator and of the denominator of the F-statistic in Eq. 2.30 are written as $E[SS_{LOF}/(n - p)] = \sigma^2 + 2\sigma^2\lambda_1/(n - p)$ and $E[SS_{PE}/(N - n)] = \sigma^2$, respectively (Draper and Smith, 1981: 120; Draper and Herzberg, 1971).

When H_0 is rejected and we assume H_a is true, the next course of action is to upgrade the proposed model by adding higher-order terms in $X_1, X_2, \ldots, X_k$ to it. If extra design points are necessary in order to fit the augmented model form, they are added and data are then collected at the added points and the analysis is redone.

On the other hand, when the null hypothesis is not rejected, then there is no reason to doubt the adequacy of the fitted model. In this case, the values of SS_{LOF} and SS_{PE} are summed and their degrees of freedom are summed to obtain SSE and $N - p$, respectively. The residual mean square, $SSE/(N - p)$, is used to test the significance of the fitted model. This latter test is the same F-test that was defined in Eq. 2.17.

Let us illustrate the test for lack of fit of the second-order model of Eq. 2.21 that was obtained using the feed weight data from Table 2.2. Associated with the fitted model, Eq. 2.21, the sum of squares of the residuals is SSE = 4.143 with 7 degrees of freedom. Using the formula in Eq. 2.26 for calculating SS_{PE}, the value is

$$SS_{PE} = (5.63 - 6.025)^2 + (6.42 - 6.025)^2 + \cdots + (7.73 - 8.005)^2$$

$$= 1.523 \text{ with 6 degrees of freedom}$$

where, for example, 6.025 is the average of 5.63 and 6.42, which are the observations in the 0.3 dosage, 1 hour cell of Table 2.2. Subtracting SS_{PE} from SSE, we get

$$SS_{LOF} = 4.143 - 1.523 = 2.620 \text{ with } 7 - 6 = 1 \text{ degree of freedom}$$

The value of the F-statistic in Eq. 2.30 is

$$F = \frac{2.620/1}{1.523/6} = 10.32$$

Since $F = 10.32$ exceeds the table value $F_{0.05,1,6} = 5.99$, we infer that there is sufficient evidence to indicate the fitted model exhibits lack of fit at the $\alpha = 0.05$ level of significance. The presence of lack of fit is due possibly to the term $\beta_{122}X_1X_2^2$, which was omitted from the model in Eq. 2.20. Measuring the importance of a term such as $\beta_{122}X_1X_2^2$, by including it in a complete model and comparing the fit of the complete model relative to the fit of a reduced model without the term, is discussed in the next section.

2.7 TESTING HYPOTHESES ABOUT THE FORM OF THE MODEL: TESTS OF SIGNIFICANCE

The distributional properties of the parameter estimates shown in Eqs. 2.7, 2.8, and 2.9 depend on the assumptions made concerning the error term in the model. In general, the assumptions are $E(\varepsilon) = 0$, $E(\varepsilon\varepsilon') = \sigma^2 I_N$, and the errors are jointly normally distributed. The normality assumption on the errors allows us the additional flexibility of performing tests of significance on the individual parameters as well as the adequacy of the fitted model. This latter test uses entries from the analysis of variance table and the F-statistic in Eq. 2.30.

Occasionally, we may wish to test hypotheses on the parameters in the proposed model, $Y = X_p\beta_p + \varepsilon$, of the form, $H_0: C\beta_p = m$, where C is an $r \times p$ matrix of rank $r \leq p$ and m is an $r \times 1$ vector of constants. The elements in each row of C are coefficients in some linear combination of the elements of β_p so that the elements of $C\beta_p$ are linearly independent estimable functions, requiring rank $\begin{bmatrix} X \\ C \end{bmatrix}$ = rank (X) (see Searle, 1971: 180–188 for a discussion of estimable linear functions). For example, if in a model containing $p = 4$ terns, the elements of the parameter vector are $\beta_p = (\beta_0, \beta_1, \beta_2, \beta_{12})'$, and we wish to test the hypothesis that $\beta_1 = \beta_2$ and $\beta_{12} = 0$, then the matrix C and the vector m are

$$C = \begin{bmatrix} 0 & 1 & -1 & 0 \\ 0 & 0 & 0 & 1 \end{bmatrix} \qquad m = \begin{bmatrix} 0 \\ 0 \end{bmatrix}$$

As another example, suppose in the rat feeding experiment of Section 2.5 we wished to test the hypothesis, H_0: Drug dosage has zero effect. This is the same as hypothesizing, $H_0: \beta_1 = \beta_{12} = 0$ in the model in Eq. 2.20. Then the matrix C and the vector m are

$$C = \begin{bmatrix} 0 & 1 & 0 & 0 & 0 \\ 0 & 0 & 0 & 1 & 0 \end{bmatrix} \qquad m = \begin{bmatrix} 0 \\ 0 \end{bmatrix}$$

Let us now consider testing the hypothesis, $H_0: C\beta_p = m$. We shall refer to the model, $Y = X_p\beta_p + \varepsilon$, as *the complete model*. Let us define a reduced model, $Y = X_r\beta_r + \varepsilon$ where β_p is changed into β_r by the conditions of the hypothesis and X_r is the form of X_p corresponding to β_r. The sums of squares due to fitting the complete and reduced models are

$$\text{SSR}_{\text{complete}} = b_p'X_p'Y - \frac{(1'Y)^2}{N} \qquad \text{with } p - 1 \text{ degrees of freedom}$$

$$\text{SSR}_{\text{reduced}} = b_r'X_r'Y - \frac{(1'Y)^2}{N} \qquad \text{with } p - 1 - r \text{ degrees of freedom}$$

where $\mathbf{b}_p$ and $\mathbf{b}_r = \mathbf{b}_p - (\mathbf{X}_p'\mathbf{X}_p)^{-1}\mathbf{C}'[\mathbf{C}(\mathbf{X}_p'\mathbf{X}_p)^{-1}\mathbf{C}']^{-1}(\mathbf{Cb}_p - \mathbf{m})$ (Searle, 1971: 113), are the estimators of $\boldsymbol{\beta}_p$ and $\boldsymbol{\beta}_r$, respectively. The test of $\mathbf{C}\boldsymbol{\beta}_p = \mathbf{m}$ involves calculating the F-statistic

$$F = \frac{[\text{SSR}_{\text{complete}} - \text{SSR}_{\text{reduced}}]/r}{\text{SSE}_{\text{complete}}/(N - p)} \tag{2.31}$$

or

$$F = \frac{[\text{SSE}_{\text{reduced}} - \text{SSE}_{\text{complete}}]/r}{\text{SSE}_{\text{complete}}/(N - p)} \tag{2.32}$$

where $\text{SSE}_{\text{complete}} = \mathbf{Y}'\mathbf{Y} - \mathbf{b}_p'\mathbf{X}_p'\mathbf{Y}$ and $\text{SSE}_{\text{reduced}} = \mathbf{Y}'\mathbf{Y} - \mathbf{b}_r'\mathbf{X}_r'\mathbf{Y}$. Comparing the computed F-ratio value with the table value of $F_{\alpha, r, N-p}$ provides the test.

A more readily obtainable numerator for the F-ratio in Eq. 2.31 for testing $\mathbf{C}\boldsymbol{\beta}_p = \mathbf{m}$ is possible. In Searle (1971: 112), it is shown that the F-ratio for this test can be written as

$$F = \frac{(\mathbf{Cb}_p - \mathbf{m})'[\mathbf{C}(\mathbf{X}_p'\mathbf{X}_p)^{-1}\mathbf{C}']^{-1}(\mathbf{Cb}_p - \mathbf{m})}{r\{\text{SSE}_{\text{complete}}/(N - p)\}} \tag{2.33}$$

since in Eq. 2.32, $\text{SSE}_{\text{reduced}} = \text{SSE}_{\text{complete}} + (\mathbf{Cb}_p - \mathbf{m})'[\mathbf{C}(\mathbf{X}_p'\mathbf{X}_p)^{-1} \times \mathbf{C}']^{-1}(\mathbf{Cb}_p - \mathbf{m})$. Hence, once $\mathbf{C}$ and $\mathbf{m}$ are specified, only the complete model parameter estimates are necessary for the calculation of F in Eq. 2.33.

Note that $\text{SSR}_{\text{complete}}$ and $\text{SSR}_{\text{reduced}}$ can also be denoted by $\text{SSR}(\boldsymbol{\beta}_p)$ and $\text{SSR}(\boldsymbol{\beta}_r)$, respectively. Their difference, that is, $\text{SSR}(\boldsymbol{\beta}_p) - \text{SSR}(\boldsymbol{\beta}_r)$ is denoted by $\text{SSR}(\boldsymbol{\beta}_p|\boldsymbol{\beta}_r)$.

To illustrate the fitting of a complete model and the subsequent testing of the significance of the highest-degree term in the model, let us refer to the model in Eq. 2.20 in Section 2.5 and include the extra term $\beta_{122}X_1X_2^2$ in it so that this complete model contains six terms. Let us test the hypothesis $H_0: \beta_{122} = 0$ versus $H_a: \beta_{122} \neq 0$. Considering the data values in Table 2.2, the fitted model is

$$\begin{aligned}
\hat{Y}(\mathbf{x}) &= b_0 + b_1X_1 + b_2X_2 + b_{12}X_1X_2 + b_{22}X_2^2 + b_{122}X_1X_2^2 \\
&= \underset{(1.033)}{6.208} - \underset{(1.918)}{7.929X_1} + \underset{(0.533)}{3.331X_2} + \underset{(0.990)}{3.293X_1X_2} \\
&\quad - \underset{(0.052)}{0.240X_2^2} + \underset{(0.096)}{0.310X_1X_2^2}
\end{aligned} \tag{2.34}$$

where $\text{SSE}_{\text{complete}} = 1.523$ so that $s^2 = 1.523/6 = 0.2538$. The quantities in parentheses below the parameter estimates are the estimated standard errors. With the fitted model in Eq. 2.34, $R^2 = 0.9917$.

Corresponding to the null hypothesis $H_0: \beta_{122} = 0$, the 1×6 matrix $\mathbf{C}$ and 1×1 vector $\mathbf{m}$ are $\mathbf{C} = (0,0,0,0,0,1)$ and $\mathbf{m} = 0$. The numerator of the

F-statistic in Eq. 2.33 is

$$(\mathbf{Cb}_p - \mathbf{m})'[\mathbf{C}(\mathbf{X}_p'\mathbf{X}_p)^{-1}\mathbf{C}']^{-1}(\mathbf{Cb}_p - \mathbf{m})$$
$$= (b_{122} - 0)'0.03662^{-1}(b_{122} - 0)$$
$$= \frac{(0.310)^2}{0.03662} = 2.620$$

and therefore the value of the F-statistic is

$$F = \frac{2.620}{1(0.2538)}$$
$$= 10.32 \tag{2.35}$$

The value of $F = 10.32$ exceeds the table value $F_{0.05,1,6} = 5.99$ and $H_0: \beta_{122} = 0$ is rejected in favor of $\beta_{122} \neq 0$. Note the equality in the F values of Eq. 2.35 and the lack of fit test of the previous section, which in essence are testing the same hypothesis. The use of the residual sums of squares quantities for the complete and reduced models, where the reduced model is the fitted model, Eq. 2.21 from Section 2.5, produces the following value for Eq. 2.32

$$F = \frac{[SSE_{reduced} - SSE_{complete}]/r}{SSE_{complete}/(N - p)}$$
$$= \frac{[4.143 - 1.523]/1}{1.523/6}$$
$$= 10.32$$

which is identical to the value of F in Eq. 2.35.

2.7.1 Interaction Models

In Section 2.6 the significance of the lack-of-fit test of the fitted second-order model in Eq. 2.21 prompted us to suggest that the term $\beta_{122}X_1X_2^2$ be added to the model. Once the term was added, the complete fitted model in Eq. 2.34 was declared to be a more adequate model than the second-order model. A question that comes to mind is, "How did we know to suggest that the term $\beta_{122}X_1X_2^2$ be added to the second-order model?"

To answer this question, suppose we have two factors X_1 and X_2, whose separate effects on the mean response η are expressible as $f(X_1)$ and $g(X_2)$, respectively. A model that expresses the effects of X_1 and X_2 on η only *additively* is written as $\eta = \beta_0 + f(X_1) + g(X_2)$. If the effects of X_1 and X_2 on η are also interactive and the interactive effects can be written as $f(X_1) \times g(X_2)$, then a model that expresses both *additive* and *interactive* effects of X_1 and X_2 is

$$\eta = \beta_0 + f(X_1) + g(X_2) + f(X_1) \times g(X_2) \tag{2.36}$$

Models of the form shown in Eq. 2.36 are referred to in Cornell and Montgomery (1996) as *interaction models*.

Let us illustrate the form of the interaction model that could be fitted to the weight data in the rat feeding experiment of Section 2.5. We recall that the factors were dosage level of drug (X_1) at two levels and length of time between drug inoculation and feeding (X_2) at three levels. If the effects of X_1 and X_2 are expressible therefore as

$$f(X_1) = \beta_1 X_1 \qquad \text{and} \qquad g(X_2) = \beta_2 X_2 + \beta_{22} X_2^2$$

then the form of the interaction model of Eq. 2.36 is

$$\eta = \beta_0 + \beta_1 X_1 + \beta_2 X_2 + \beta_{22} X_2^2 + \beta_{12} X_1 X_2 + \beta_{122} X_1 X_2^2 \qquad (2.37)$$

Notationally in Eq. 2.37, $\beta_{12} X_1 X_2 = \beta_1 X_1 \times \beta_2 X_2$ and $\beta_{122} X_1 X_2^2 = \beta_1 X_1 \times \beta_{22} X_2^2$. Equation 2.37 is a reduced form of third-order polynomial.

The terms in the interaction portion, $f(X_1) \times g(X_2)$, of the models in Eqs. 2.36 and 2.37 contain information on how the effect of X_1, $f(X_1)$, is affected by factor X_2 and vice versa. If $f(X_1)$ in Eq. 2.36 contains terms up to and including degree d_1 while $g(X_2)$ contains terms up to and including degree d_2 (d_2 not necessarily equal to d_1), the terms in $f(X_1) \times g(X_2)$ will be of degree 2 up to and including degree $d_1 + d_2$. The number of terms in the interaction model of Eq. 2.36 is $(d_1 + 1)(d_2 + 1)$, whereas a polynomial of degree d contains at most $(d + 1)(d + 2)/2$ terms. Thus with two factors and when $d = d_1 = d_2$, the interaction model contains $d(d + 1)/2$ more cross-product terms than the polynomial. These additional cross-product terms in the interaction model can be very helpful in correctly describing the shape of the response surface at locations in the experimental region where the polynomial model fails. Two numerical examples are presented in Cornell and Montgomery (1996) illustrating the improved fit of an interaction model over the fit of a polynomial.

A final word about the terms in the interaction portion of the interaction model seems particularly germane as pertains to testing a fitted polynomial model of order d for *lack of fit*. Again let us consider only two factors where $d = d_1 = d_2$ and where the design supports the fit of the interaction model. If the test for lack of fit of the fitted polynomial model is significant, suggesting additional terms are needed, these additional terms of degree larger than d are among the extra $d(d + 1)/2$ cross-product terms in the interaction model that are not included in the polynomial model of order d. Realizing this, one can easily determine what the terms omitted from the polynomial are by writing the form of the interaction model. When the forms of the interaction model in Eq. 2.37 and the second-order model in Eq. 2.21 are compared, one notices immediately the presence of the extra term $\beta_{122} X_1 X_2^2$ in the interaction model whose absence in the fitted polynomial must have contributed to the significant lack of fit of Eq. 2.21.

2.8 SOME INTRODUCTORY COMMENTS ON THE DESIGN REGION, THE USE OF CODED VARIABLES, AND PROPERTIES OF A RESPONSE SURFACE DESIGN

In planning the program of experimentation, the experimenter is faced with (1) choosing the factors or input variables to be used in the experiment, and (2) selecting the range of values and the number of levels of each factor in order to adequately measure the effects of the factors on the response. Upon answering these questions, the experimental region is defined.

An example where the experimental region is a three-dimensional cube is the following. Suppose three controllable factors in a chemical reaction yield study are temperature of the reaction, length of time of the reaction, and amount of pressure exerted on the reactants. Let us suppose the experimenter desires only to measure the changes in the yield resulting from varying each factor over its range of values independently of the other factors as well as by varying the factors simultaneously. The levels chosen for temperature, time, and pressure are 250°F and 450°F, 5 sec and 10 sec, and 75 psi and 125 psi, respectively. If all eight temperature-time-pressure level combinations are to be performed, then the experimental region takes the form of a three-dimensional cube where each three-factor combination defines a vertex of the cube. The cube-shaped region arises from the use of coded (unitless) variables, which are defined now.

2.8.1 Defining Coded Variables

The use of coded variables in place of the input variables facilitates the construction of experimental designs. Coding removes the units of measurement of the input variables and as such, distances measured along the axes of the coded variables in a k-dimensional space are standardized (or defined in the same metric).

To illustrate the transformation in going from the system of the input variables to a unitless system in the coded variables, let us recall the chemical reaction process just discussed where the input variables (and their levels, X_i) are temperature ($X_1 = 250°F$ and $450°F$), time ($X_2 = 5$ sec and 10 sec), and pressure ($X_3 = 75$ psi and 125 psi). A convenient coding formula for defining the coded variable, x_i, is

$$x_i = \frac{2X_i - (X_{iL} + X_{iH})}{X_{iH} - X_{iL}} \qquad i = 1, 2, 3 \tag{2.38}$$

where X_{iL} and X_{iH} are the low and high levels of X_i, respectively. For example,

$$x_1 = \frac{2X_1 - 700}{200}$$

$$x_2 = \frac{2X_2 - 15}{5} \tag{2.39}$$

$$x_3 = \frac{2X_3 - 200}{50}$$

Thus, when an input variable has only two levels, the formula in Eq. 2.38 produces the familiar ± 1 notation for the levels of the coded variables associated with the two-level factorial arrangements. This is seen by noting that

$x_1 = -1$ when $X_1 = 250°F$, $x_2 = -1$ when $X_2 = 5$ sec,

$x_3 = -1$ when $X_3 = 75$ psi

$x_1 = +1$ when $X_1 = 450°F$, $x_2 = +1$ when $X_2 = 10$ sec,

$x_3 = +1$ when $X_3 = 125$ psi

An example of removing the units of measurement of drug dosage and time of feeding for the rat feeding experiment of Section 2.5 is illustrated in Figure 2.2.

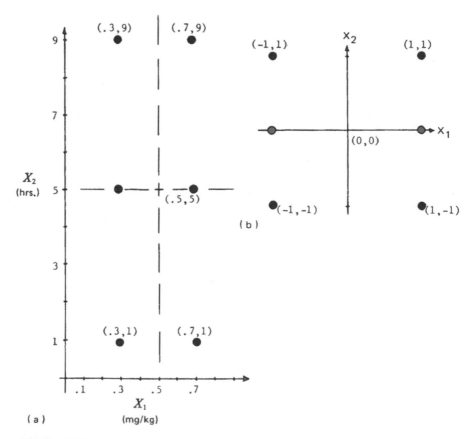

Figure 2.2 (a) Defining the point $(X_1, X_2) = (0.5, 5)$ as the origin $(0, 0)$ of a new system of variables; (b) defining standardized or unitless variables, $x_1 = (X_1 - 0.5)/0.2$ and $x_2 = (X_2 - 5)/4$.

The coded variables are defined as

$$x_1 = \frac{2X_1 - (0.3 + 0.7)}{0.4}$$

$$= \frac{X_1 - 0.5}{0.2}$$

$$x_2 = \frac{2X_2 - (1 + 9)}{8}$$

$$= \frac{X_2 - 5}{4}$$

The region defined by specifying the coded levels as ± 1 for k factors is a cuboidal region in a k-dimensional space. Geometrically, the cuboidal region has 2^k vertices where each vertex is defined by the coordinate (± 1) settings in (x_1, $x_2, \ldots, x_k$).

When a factor has three quantitative levels and the middle level is midway between the lower and upper levels, the coding formula, Eq. 2.38, produces the coded levels $x_i = -1, 0, +1$ associated with the low, middle, and high values of X_i, respectively. When all k factors have three levels so that all the values of x_i take the values of $-1, 0, +1, i = 1, 2, \ldots, k$, again the region in the coded variables is a k-dimensional cuboidal region. However, since the number of factor combinations is now 3^k, where 2^k of the combinations are the vertices of the k-dimensional cuboidal region, the remaining $3^k - 2^k$ combinations define the centroids of all of the lower dimensional boundaries of the k-dimensional cube along with the centroid $(0, 0, 0, \ldots, 0)$ of the cuboidal region. Figure 2.3 presents the cuboidal regions and the factor level designations for a 3^2 and 3^3 design, respectively.

Formulas, similar to Eq. 2.38, are easy to derive for defining coded variables with four or more levels. However, for the purpose of constructing design configurations that are not cuboidal in shape we shall introduce a design region (or region of interest) that is spherical in shape and centered at the origin (x_1, x_2, $\ldots, x_k$) = $(0, 0, \ldots, 0)$ of the coded variable region. For most cases where the design region is a sphere we shall adopt a coding convention in the values of x_i which forces the design points to fall within or on a sphere of radius ρ.

2.8.2 The Use of Coded Variables in the Fitted Model

There are several advantages to using coded variables rather than the original input variables when fitting polynomial models. Two of the more obvious advantages are:

1. Computational ease and increased accuracy in estimating the model coefficients
2. Enhanced interpretability of the coefficient estimates in the model

(a)

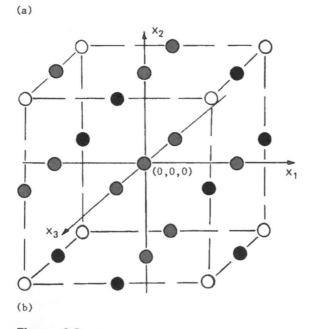

(b)

Figure 2.3 Experimental regions in (a) $k = 2$ and (b) $k = 3$ coded variables along with the coordinates (x_1, x_2) and (x_1, x_2, x_3) of the design points. The 2^k vertices are denoted by circles (O) and the remaining $3^k - 2^k$ points are denoted by dots (●).

Both of the advantages stem from the fact that the $\mathbf{X'X}$ matrix in the coded variables is usually of a simpler form (by a simpler form of $\mathbf{X'X}$ is meant there are many more zero-valued off-diagonal elements, $c_{ij} = 0$, $i \neq j$, where the simplest form is a diagonal matrix) than the $\mathbf{X'X}$ matrix in the original variables. In general, the simpler the form of the $\mathbf{X'X}$ matrix, the easier it is to invert $\mathbf{X'X}$, and as a result, the greater will be the computational accuracy of the model parameter estimates and the reduction in computing time (Snee, 1973). Furthermore, with the simpler form of $\mathbf{X'X}$, where some or all of the $c_{ij} = 0$, $i \neq j$, the corresponding parameter estimates are uncorrelated so that the resulting numerators of the test statistics associated with tests on the individual parameters are statistically independent.

Since the coding transformation in Eq. 2.38 is a one-to-one transformation, any linear polynomial equation in the values of x_i is expressible as (and equivalent to) a polynomial equation of the same degree in the values of X_i. To illustrate, let us recall the data in Table 2.2 from the rat feeding experiment of Section 2.5 and define the coded variables, using Eq. 2.38 as

$$x_1 = \frac{X_1 - 0.5}{0.2}$$

$$x_2 = \frac{X_2 - 5}{4}$$

(2.40)

The drug dosage levels 0.3 and 0.7 mg/kg yield x_1 values of -1 and $+1$, respectively, while the time levels 1, 5, and 9 hours yield x_2 values of -1, 0, and $+1$, respectively. If a model of the form in Eq. 2.20 is to be fitted in the coded variables, x_1 and x_2, then the coded variable settings and the vector of observations, when expressed in matrix notation, are

$$
\begin{array}{ccccc}
\mathbf{Y} & = & \mathbf{X} & \boldsymbol{\beta} & + & \boldsymbol{\varepsilon}
\end{array}
$$

$$
\begin{array}{ccccc}
 & 1 & X_1 & X_2 & X_1X_2 & X_2^2
\end{array}
$$

$$
\begin{bmatrix}
5.63 \\
6.42 \\
1.38 \\
1.94 \\
11.57 \\
12.16 \\
5.72 \\
4.69 \\
12.68 \\
13.31 \\
8.28 \\
7.73
\end{bmatrix}
=
\begin{bmatrix}
1 & -1 & -1 & 1 & 1 \\
1 & -1 & -1 & 1 & 1 \\
1 & 1 & -1 & -1 & 1 \\
1 & 1 & -1 & -1 & 1 \\
1 & -1 & 0 & 0 & 0 \\
1 & -1 & 0 & 0 & 0 \\
1 & 1 & 0 & 0 & 0 \\
1 & 1 & 0 & 0 & 0 \\
1 & -1 & 1 & -1 & 1 \\
1 & -1 & 1 & -1 & 1 \\
1 & 1 & 1 & 1 & 1 \\
1 & 1 & 1 & 1 & 1
\end{bmatrix}
\begin{bmatrix}
\beta_0 \\
\beta_1 \\
\beta_2 \\
\beta_{12} \\
\beta_{22}
\end{bmatrix}
+
\begin{bmatrix}
\varepsilon_1 \\
\varepsilon_2 \\
\varepsilon_3 \\
\varepsilon_4 \\
\varepsilon_5 \\
\varepsilon_6 \\
\varepsilon_7 \\
\varepsilon_8 \\
\varepsilon_9 \\
\varepsilon_{10} \\
\varepsilon_{11} \\
\varepsilon_{12}
\end{bmatrix}
$$

The normal equations in the coded variables are

$$
\underset{\mathbf{X'X}}{
\begin{bmatrix}
12 & 0 & 0 & 0 & 8 \\
 & 12 & 0 & 0 & 0 \\
 & & 8 & 0 & 0 \\
 & \text{(symmetric)} & & 8 & 0 \\
 & & & & 8
\end{bmatrix}
}
\underset{\mathbf{b}}{
\begin{bmatrix}
b_0 \\ b_1 \\ b_2 \\ b_{12} \\ b_{22}
\end{bmatrix}
}
=
\underset{\mathbf{X'Y}}{
\begin{bmatrix}
91.51 \\ -32.03 \\ 26.63 \\ -1.25 \\ 57.37
\end{bmatrix}
}
$$

and the fitted second-order model in the coded variables is

$$
\hat{Y}(\mathbf{x}) = \underset{(0.385)}{8.535} - \underset{(0.222)}{2.669x_1} + \underset{(0.272)}{3.329x_2} - \underset{(0.272)}{0.156x_1x_2} - \underset{(0.471)}{1.364x_2^2} \tag{2.41}
$$

The quantities in parentheses below the parameter estimates are the estimated standard errors, where $s^2 = 0.592$; the same value that was obtained previously and shown in Table 2.4.

Unlike the correlated coefficient estimates in the fitted model in Eq. 2.21 in the original input variables X_1 and X_2, the parameter estimates (except b_0 and b_{22}) in the coded variables are uncorrelated. [The matrix $(\mathbf{X'X})^{-1}$ is of the same form as the $\mathbf{X'X}$ matrix and thus only $\text{Cov}(b_0, b_{22}) \neq 0$.] It is interesting to notice the difference in the value of the t-statistic for testing the null hypothesis, H_0: $\beta_1 = 0$, with the model in Eq. 2.20 in X_1 and X_2 and the value of the t-statistic for testing the hypothesis, H_0: $\beta_1 = 0$ for the model

$$
Y = \beta_0 + \beta_1 x_1 + \beta_2 x_2 + \beta_{12} x_1 x_2 + \beta_{22} x_2^2 + \varepsilon \tag{2.42}
$$

From Eq. 2.23, the value of the t-statistic is -6.09, but for the model in Eq. 2.41, the value is

$$
t = \frac{-2.669}{0.222} = -12.02
$$

The different t-test values are the result of testing correlated estimates in the former model but uncorrelated estimates in the latter model. In other words, the null hypotheses corresponding to models in Eqs. 2.20 and 2.42, respectively, are different.

The fitted model in Eq. 2.41 in the coded variables can be rewritten in terms of the original input variables. The equivalent second-order model in X_1 and X_2 is

$$
\hat{Y}(\mathbf{x}) = 8.535 - 2.669 \left(\frac{X_1 - 0.5}{0.2} \right) + 3.329 \left(\frac{X_2 - 5}{4} \right)
$$

$$
- 0.156 \left(\frac{X_1 - 0.5}{0.2} \right) \left(\frac{X_2 - 5}{4} \right) - 1.364 \left(\frac{X_2 - 5}{4} \right)^2
$$

$$
= 8.428 - 12.369 X_1 + 1.782 X_2 - 0.195 X_1 X_2 - 0.085 X_2^2 \tag{2.43}
$$

which is identical to the previously fitted model in Eq. 2.21.

Since the fitted models in Eqs. 2.41 and 2.43 are equivalent, then at any point (x_1, x_2) in the experimental region, which of course can be decoded to the combination (X_1, X_2), the predicted value of the response is the same with the two models and the variance of prediction is the same value. Thus, later in Chapter 5, when discussing techniques for locating the maximum or minimum predicted response values, we shall work with the coded variables. Mathematically, the system in the coded variables is the more convenient system to work with.

2.8.3 Some Comments About the Properties of a Response Surface Design

Prior to performing the actual experimentation, quite often it is necessary to decide whether the main emphasis is in measuring the effects of the input variables or whether one is really more interested in modeling the response surface for prediction purposes. As we shall see, designs that are best suited to achieve these two objectives may be quite different in terms of their distribution of points in the experimental region. For the present we shall mainly concern ourselves with the properties of a design in which the primary emphasis is in modeling and exploring a response surface over the experimental region.

Box and Draper (1975) list 14 properties of a response surface design to be used when fitting a polynomial model to data collected at the design points. Some of the more important properties are:

1. The design should generate a satisfactory distribution of information throughout the region of interest.
2. The design should ensure that the fitted value at $\mathbf{x}$, $\hat{Y}(\mathbf{x})$, be as close as possible to the true value at $\mathbf{x}$, $\eta(\mathbf{x})$.
3. The design should give good detectability of model lack of fit.
4. The design should allow experiments to be performed in blocks.
5. The design should allow designs of increasing order to be built up sequentially.
6. The design should provide an internal estimate of the error variance σ^2.
7. The design should require a minimum number of experimental points.
8. The design should ensure simplicity of calculation of the model parameter estimates.

This less-than-exhaustive list of properties will serve our purposes in the next two chapters when selecting an appropriate grouping of points called the design.

Orthogonality and Rotatability. In addition to the properties listed above, there are times when we shall also want the design to possess the property of orthogonality and/or the property of rotatability. An orthogonal design is one in which the terms in the fitted model are uncorrelated with one another and thus the parameter estimates are uncorrelated. In this case, the variance of the predicted response at any point **x** in the experimental region, is expressible as a weighted sum of the variances of the parameter estimates in the model. A first-order design, for example, is orthogonal if and only if the corresponding **X′X** matrix is diagonal.

With a rotatable design, on the other hand, the variance of $\hat{Y}(\mathbf{x})$, which is known to depend on the location of the point **x**, is a function only of the distance from the point **x** to the center of the design. Thus, with a rotatable design, the prediction variance, $\text{Var}[\hat{Y}(\mathbf{x})]$, is the same at all points **x** that are equidistant from the design center. Consequently, in the space of the input variables, surfaces of constant prediction variance form concentric hyperspheres (circles and spheres in two-dimensional and three-dimensional Euclidean spaces, respectively). This constant variance property of the predictor, $\hat{Y}(\mathbf{x})$, is quite appealing when one is faced with determining in which direction to experiment next.

The concept of rotatability was first introduced by Box and Hunter (1957) and has since become an important design criterion. One of the desirable features of rotatability is that the quality of prediction, as measured by the magnitude of $\text{Var}[\hat{Y}(\mathbf{x})]$, is invariant to any rotation of the coordinate axes in the space of the input variables.

To learn more about the rotatability property of a design, we shall concentrate on certain parameters of the distribution of design points called the *design moments*. Suppose the model to be fitted is expressed in matrix notation as **Y** = **Xβ** + **ε**. If this model is of order d and is a function of k input variables (these variables are considered to have been coded), then a design moment of order δ ($\delta = 0, 1, \ldots, 2d$), denoted by $[1^{\delta_1} 2^{\delta_2} \ldots k^{\delta_k}]$, is equal to

$$[1^{\delta_1} 2^{\delta_2} \ldots k^{\delta_k}] = \frac{1}{N} \sum_{u=1}^{N} x_{u1}^{\delta_1} x_{u2}^{\delta_2} \ldots x_{uk}^{\delta_k} \tag{2.44}$$

where N is the total number of observations and $\delta_1, \delta_2, \ldots, \delta_k$ are nonnegative integers such that $\sum_{i=1}^{k} \delta_i = \delta$. These design moments are the elements of the matrix $N^{-1}\mathbf{X}'\mathbf{X}$ which is called the *moment matrix*. For example, suppose the proposed model is of the first order in the k coded variables $x_1, x_2, \ldots, x_k$, where

$$x_{ui} = \frac{X_{ui} - \bar{X}_i}{s_{X_i}}, \text{ and } s_{X_i} = \left\{ \sum_{u=1}^{N} \frac{(X_{ui} - \bar{X}_i)^2}{N} \right\}^{1/2} \quad \begin{array}{l} u = 1, 2, \ldots, N \\ \\ i = 1, 2, \ldots, k \end{array} \tag{2.45}$$

The denominator, s_{x_i}, is a measure of the spread of the design points in the direction of the X_i axis. The moment matrix is

$$
\frac{1}{N}\mathbf{X}'\mathbf{X} = \begin{array}{c} \\ x_1 \\ x_2 \\ x_3 \\ \vdots \\ x_k \end{array}
\begin{array}{ccccccc}
 & x_1 & x_2 & x_3 & \cdots & x_k \\
\end{array}
\left[\begin{array}{cccccc}
1 & [1] & [2] & [3] & \cdots & [k] \\
 & [11] & [12] & [13] & \cdots & [1k] \\
 & & [22] & [23] & \cdots & [2k] \\
 & & & [33] & \cdots & [3k] \\
 & \text{symmetric} & & & \ddots & \vdots \\
 & & & & & [kk]
\end{array}\right] \tag{2.46}
$$

where

$$
[i] = \frac{1}{N}\sum_{u=1}^{N} x_{ui} \quad [ii] = \frac{1}{N}\sum_{u=1}^{N} x_{ui}^2 \quad [ij] = \frac{1}{N}\sum_{u=1}^{N} x_{ui}x_{uj} \quad i, j = 1, 2, \ldots, k
$$

The moment $[i]$ is the average of the x_{ui} values over the N observations and is called the moment of the first order. The moment $[ii]$ is the average of the x_{ui}^2 values over all $u = 1, 2, \ldots, N$ and is called the pure second-order moment. This moment can also be written $[i^2]$. The moment $[ij]$, $i \neq j$, is called the mixed second-order moment. However, by the coding convention in Eq. 2.45, $\sum_{u=1}^{N} x_{ui} = 0$ and $\sum_{u=1}^{N} x_{ui}^2 = N$, so that in Eq. 2.46, $[i] = 0$ and $[ii] = 1$ for $i = 1, 2, \ldots, k$ and thus the moment matrix in Eq. 2.46 is simplified to

$$
\frac{1}{N}\mathbf{X}'\mathbf{X} = \begin{array}{c} \\ x_1 \\ x_2 \\ x_3 \\ \vdots \\ x_k \end{array}
\begin{array}{ccccccc}
 & x_1 & x_2 & x_3 & \cdots & x_k \\
\end{array}
\left[\begin{array}{cccccc}
1 & 0 & 0 & 0 & \cdots & 0 \\
 & 1 & [12] & [13] & \cdots & [1k] \\
 & & 1 & [23] & \cdots & [2k] \\
 & & & 1 & \cdots & [3k] \\
 & \text{symmetric} & & & \ddots & \vdots \\
 & & & & & 1
\end{array}\right] \tag{2.47}
$$

When the fitted model is of the second order, the matrix $N^{-1}\mathbf{X}'\mathbf{X}$ will contain moments up to order four. With two variables for example, and the model expressed as $Y = \beta_0 + \beta_1 x_1 + \beta_2 x_2 + \beta_{11}x_1^2 + \beta_{22}x_2^2 + \beta_{12}x_1 x_2 + \varepsilon$, the coding convention in Eq. 2.45 produces the moment matrix

$$
\frac{1}{N}\mathbf{X}'\mathbf{X} = \begin{array}{c} \\ x_1 \\ x_2 \\ x_1^2 \\ x_2^2 \\ x_1 x_2 \end{array}
\begin{array}{cccccc}
 & x_1 & x_2 & x_1^2 & x_2^2 & x_1 x_2 \\
\end{array}
\left[\begin{array}{cccccc}
1 & 0 & 0 & 1 & 1 & [12] \\
 & 1 & [12] & [111] & [122] & [112] \\
 & & 1 & [112] & [222] & [122] \\
 & & & [1111] & [1122] & [1112] \\
 & \text{symmetric} & & & [2222] & [1222] \\
 & & & & & [1122]
\end{array}\right] \tag{2.48}
$$

where, for example,

$$[111] = \frac{1}{N} \sum_{u=1}^{N} x_{u1}^3 \qquad [122] = \frac{1}{N} \sum_{u=1}^{N} x_{u1} x_{u2}^2$$

$$[1111] = \frac{1}{N} \sum_{u=1}^{N} x_{u1}^4 \qquad [1122] = \frac{1}{N} \sum_{u=1}^{N} x_{u1}^2 x_{u2}^2$$

These moments can also be denoted by $[1^3]$, $[12^2]$, $[1^4]$, and $[1^2 2^2]$, respectively.

A simple characterization of rotatability is given in terms of the elements of the $X'X$ matrix, or the moment matrix. The particular form of the moment matrix for a rotatable design is presented in Appendix 2B. From this form and when the coding scheme of Eq. 2.45 is used, we can deduce that a first-order rotatable design has zero-valued first-order moments $[i]$ and zero-valued mixed-order moments $[ij]$ while all pure second-order moments $[ii]$ are equal to 1. Thus with a first-order rotatable design, the distribution of the design points, when measured in the direction of each of the x_i-axes is centered at $(x_1, x_2, \ldots, x_k) = (0, 0, \ldots, 0)$ and has a variance equal to 1. For second-order rotatable designs the skewness and kurtosis properties of the distribution of the design points are considered. These properties are defined by the fourth-order moments $[iiii]$ and $[iijj]$ and take certain values as shown in Appendix 2B.

Any deviation of the moment matrix for a given response surface design from the form given in Appendix 2B, which characterizes rotatability, indicates that the design is nonrotatable. Loss of rotatability can be depicted in nonspherical-looking surfaces of constant prediction variance. The closer these surfaces are to being spherical, the higher the degree of rotatability. To quantify this degree of rotatability, Khuri (1988) introduced a measure of rotatability as a function of the moments of the design under consideration. The function is expressible as a percentage, and large values of this function indicate high degree of rotatability. The value 100 is attained if and only if the design is rotatable (see Section 12.5). The measure of rotatability can be useful in the following situations:

1. To compare designs on the basis of rotatability.
2. To assess the extent of departure from rotatability when an already rotatable design is deformed or modified. Quite often, design settings prescribed for rotatability may not be achieved exactly due to round-off errors, a malfunction in the measuring device, or because these settings are infeasible from the experimental point of view. Furthermore, rotatability may sometimes be purposefully deviated from by augmenting a rotatable design with additional design points in order to gain more information in certain areas of the experimental region, or to satisfy, or nearly satisfy, other design criteria. In all such cases, it might be of interest to determine the extent of "damage" done to rotatability.

3. To "repair" a nonrotatable design by the addition of design points so as to improve its percent rotatability.

2.9 TRANSFORMATIONS ON THE RESPONSE VARIABLE

Thus far in this chapter, we have discussed and reviewed

(i) The method of (unweighted) least squares for estimating the parameters in the proposed model
(ii) The properties (mean and variance) of the parameter estimates
(iii) The analysis-of-variance table for displaying the results of the analysis of the fitted model
(iv) Testing the fitted model for adequacy of fit

For each of these topics, especially the first, the proposed model $Y = X\beta + \varepsilon$ was assumed to be a linear function of the parameters and the random errors ε_u, $u = 1, 2, \ldots, N$ possessed the following properties:

1. The errors have zero mean and common variance, σ^2.
2. The errors are mutually independent in the statistical sense.
3. For (iv), the errors are normally distributed.

We went on to state in Section 2.3.4 that if the assumptions are true, the elements of **b** in Eq. 2.6 are the best linear unbiased estimates of the elements of β.

It is not uncommon to encounter data that appear to have been generated by a process for which some of the preceding assumptions do not hold. Two obvious cases are (1) the data were generated by a process in which, based on theoretical as well as practical laws, a nonlinear model is more appropriate than a linear model for describing the process and/or (2) the data have a probability distribution the variance of which is related to the mean. In either case, in order to use the procedures discussed in this chapter, it is necessary to *transform the response or data values* so that the behavior of the transformed data more closely conforms to the assumptions.

2.9.1 Transformations to Achieve Linearity of the Model

Data that are generated by a process driven by a nonlinear model are generally nonlinear or curvilinear in appearance when plotted. A simple example is the growth of bacteria, measured in CFUs (colony-forming units) over time. One model for the number n_t of CFUs at time t is $n_t = \beta_0 e^{\beta_1 t}$, where β_0 is the bacterial count at time $t = 0$ and β_1 is the rate of growth of CFUs per unit of time. When plotted over the range of time $t = 0, 1, 2, \ldots$, the values of n_t are exponential in appearance.

The nonlinear model $n_t = \beta_0 e^{\beta_1 t}$ is *intrinsically linear*—that is, it can be reexpressed as a linear model by a suitable transformation of the response variable n_t; otherwise it is *intrinsically nonlinear* (Section 10.1). To see this, if we take the natural logarithms of both sides of the equality sign, the result is a model in the units of $\ln(n_t)$ expressed as a linear function of time t

$$\ln(n_t) = \ln(\beta_0) + \beta_1 t \tag{2.49}$$
$$= \beta_0' + \beta_1 t$$

A transformation such as $\ln(n_t)$ can alter the structure and distribution of the error term associated with the transformed model, however (see Section 10.1). For example, if the errors in observing the counts n_t are multiplicative and log normally distributed so that $Y_t = n_t \varepsilon_t = \beta_0 e^{\beta_1 t} \varepsilon_t$, then

$$\ln(Y_t) = \beta_0' + \beta_1 t + \varepsilon_t' \tag{2.50}$$

where now the errors ε_t' are additive and normally distributed. On the other hand, if $Y_t = n_t + \varepsilon_t$, that is, the ε_t are additive and satisfy the assumptions of common variance and normality listed previously, then the model for the transformed value, denoted by y_t, is

$$y_t = \ln(Y_t) \tag{2.51}$$
$$= \beta_0' + \beta_1 t + \ln\left[1 + \frac{\varepsilon_t}{\beta_0 e^{\beta_1 t}}\right]$$

In Eq. 2.51 the errors $\ln[1 + \varepsilon_t/(\beta_0 e^{\beta_1 t})]$ have a nonnormal distribution with a variance that depends on t. We much prefer to operate under the assumptions and work with the model of Eq. 2.50.

Other types of linearizing transformations are listed in Table 2.6, where for simplicity we have expressed the response variable Y as a function of the single variable X. Graphs of these functions are presented in such textbooks as Chatterjee and Price (1977), Daniel and Wood (1980), and Montgomery and Peck (1992).

Table 2.6 Linearizable Functions with Corresponding Transformation

Function	Transformation	Linear form
$Y = \beta_0 X^{\beta_1}$	$y = \log(Y)$	$y = \log(\beta_0) + \beta_1 \log(X)$
$Y = \beta_0 e^{\beta_1 X}$	$y = \ln(Y)$	$y = \ln(\beta_0) + \beta_1 X$
$Y = \dfrac{X}{\beta_0 X + \beta_1}$	$y = Y^{-1}$	$y = \beta_0 + \beta_1 X^{-1}$
$Y = \dfrac{e^{\beta_0 + \beta_1 X}}{1 + e^{\beta_0 + \beta_1 X}}$	$y = \ln\left(\dfrac{Y}{1 - Y}\right)$	$y = \beta_0 + \beta_1 X$

2.9.2 Transformations to Stabilize the Variance

Assumption 1 of a common error variance is a basic requirement for relying on the (unweighted) least squares solution in Eq. 2.6 for the estimates of the parameters in the model. When the error variance is not constant over all the observations, the errors are said to be heteroscedastic. In some cases, heteroscedasticity can be removed by means of a suitable transformation.

Heteroscedasticity occurs when the response variable Y has a probability distribution the variance of which is a function of the mean of the distribution and the value of the mean, expressed in model form, depends on parameter values. The binomial and Poisson are two common probability distributions that have this characteristic. We know, for example, that a random variable that is distributed binomially with parameters n and π, has mean $n\pi$ and variance $n\pi(1 - \pi)$. The mean and variance are equal for a Poisson random variable. Therefore when the relationship between the mean and variance of a random variable is known, it is possible to find a suitable transformation of the variable that makes the variance approximately constant. Such a transformation is called *a variance-stabilizing transformation.*

Several commonly used variance-stabilizing transformations are listed in Table 2.7. These transformations also have the effect of creating the distribution of the transformed variable to be close to normal. Thus, these transformations serve the dual prupose of normalizing the data as well as making the estimated variance of the data functionally independent of the estimated mean. Finally, a useful class of transformations for correcting nonnormality and/or heteroscedasticity is the family of power transformations:

$$
y = \begin{cases} \dfrac{(Y^\lambda - 1)}{\lambda}, & \lambda \neq 0 \\ \ln(Y), & \lambda = 0 \end{cases} \tag{2.52}
$$

where the value of the parameter λ is estimated from the data. We shall not discuss the methods used to estimate λ since these methods are clearly explained in texts such as Draper and Smith (1981) and Montgomery and Peck (1992), we well as Box and Cox (1964).

Once a transformation is used to linearize the model or to stabilize the variance in the response, the resulting transformed data are then fitted using a linear model. All the usual procedures for testing the adequacy of the fitted model and, particularly, for testing the assumptions about the errors in the transformed data are performed on the residuals $r_u = y_u - \hat{y}(x_u)$, $u = 1, 2, \ldots, N$, generated by the fitted model. Some modelers prefer to work with standardized residuals, $r_u/\sqrt{\text{MSE}}$, where MSE is the estimated variance of the residuals, obtained from the analysis of variance, since standardized residuals are unitless. These procedures involve plotting the residuals versus $\hat{y}$ and/or values of the input variables

Table 2.7 Some Variance-Stabilizing Transformations

Probability distribution of variable Y	Variance of Y, σ^2, as a function of mean μ	Transformation y	Variance of y
Binomial	$\dfrac{\mu(1-\mu)}{n}$	$y = \sin^{-1}\sqrt{Y}$ (deg)	$\dfrac{821}{n}$
		$y = \sin^{-1}\sqrt{Y}$ (rad)	$\dfrac{1}{4n}$
Poisson	μ	$y = \sqrt{Y}$ or $\sqrt{Y + 0.375}$	$\dfrac{1}{4}$
Negative binomial	$\mu + \lambda^2 m^2$	$y = \lambda^{-1}\sinh^{-1}(\lambda\sqrt{Y})$	$\dfrac{1}{4}$
	$\propto \mu^2$	$y = \ln(Y)$ or $\log(Y)$	1
	$\propto \mu^3$	$y = Y^{-1/2}$	
	$\propto \mu^4$	$y = Y^{-1}$	

and checking for patterns in the plots. Heteroscedasticity, for example, is revealed when the values and thus the variation of the residuals tend to increase or decrease with increasing values of $\hat{y}$ or with increasing values of one or more of the input variables. Residual plots are also used for checking the normality assumption of the errors. The procedure entails plotting the ranked residuals against their cumulative probability $P_u = 100(u - 0.5)/N$, $u = 1, 2, \ldots, N$, on normal probability paper. If the errors are normally distributed, the plotted residuals form a straight line approximately. Montgomery and Peck (1992) present a lucid discussion of the plotting of residuals on normal probability paper and provide several plots that depict departures from normality that one might encounter in practice.

Appendix 2A OTHER DEFINITIONS AND THEOREMS CONCERNING MATRICES AND DETERMINANTS

2A.1 Products of Matrices

Theorem 1. The inverse of the product of several nonsingular matrices is equal to the product of the inverses in reverse order:

$$(\mathbf{MNP})^{-1} = \mathbf{P}^{-1}\mathbf{N}^{-1}\mathbf{M}^{-1}$$

Theorem 2. The transpose of the product of several matrices is equal to the product of their transposes taken in reverse order:

$$(\mathbf{MNP})' = \mathbf{P}'\mathbf{N}'\mathbf{M}'$$

Theorem 3. The inverse of the transpose of a square matrix is the transpose of the inverse of the matrix

$$(\mathbf{M}')^{-1} = (\mathbf{M}^{-1})'$$

2A.2 Determinants

Theorem 4. The determinant of a diagonal matrix is the product of the diagonal elements.

Theorem 5. If $\mathbf{M}$ is $k \times k$ and c is a scalar, then $|c\mathbf{M}| = c^k|\mathbf{M}|$.

Theorem 6. If $\mathbf{M}$ and $\mathbf{N}$ are $k \times k$ matrices, then $|\mathbf{MN}| = |\mathbf{M}||\mathbf{N}|$.

Theorem 7. If $\mathbf{M}$ is nonsingular, then $|\mathbf{M}^{-1}| = 1/|\mathbf{M}|$.

Theorem 8. The determinant of an orthogonal matrix is equal to 1 in absolute value.

2A.3 Eigenvalues and Eigenvectors

Theorem 9. The eigenvalues of a real-valued symmetric matrix are all real valued.

Theorem 10. The number of nonzero eigenvalues of a symmetric matrix $\mathbf{M}$ is equal to the rank of $\mathbf{M}$.

Theorem 11. The eigenvalues of a diagonal matrix are the diagonal elements.

Theorem 12. The nonzero eigenvalues of the product $\mathbf{MN}$ are equal to the nonzero eigenvalues of $\mathbf{NM}$.

Theorem 13. If λ_i and λ_j, $i \neq j$, are distinct eigenvalues of the symmetric matrix $\mathbf{M}$, their associated eigenvectors, $\mathbf{v}_i$ and $\mathbf{v}_j$, are orthogonal, that is $\mathbf{v}_i'\mathbf{v}_j = 0$.

Theorem 14. For every real symmetric matrix $\mathbf{M}$, there exists an orthogonal matrix $\mathbf{T}$ such that $\mathbf{T}'\mathbf{MT} = \mathbf{D}$ where $\mathbf{D}$ is a diagonal matrix containing the eigenvalues of $\mathbf{M}$. The columns of $\mathbf{T}$ are normalized eigenvectors of $\mathbf{M}$.

2A.4 Rank and Trace of a Matrix

Theorem 15. The rank of $\mathbf{A}$, denoted by $r(\mathbf{A})$, equals $r(\mathbf{A}') = r(\mathbf{AA}') = r(\mathbf{A}'\mathbf{A})$.

Theorem 16. $r(\mathbf{AB}) \leq \text{minimum}[r(\mathbf{A}), r(\mathbf{B})]$.

Theorem 17. $r(\mathbf{AB}) = r(\mathbf{A})$ if $\mathbf{B}$ is nonsingular.

Theorem 18. $tr(\mathbf{AB}) = tr(\mathbf{BA})$. Also, $tr(\mathbf{ABC}) = tr(\mathbf{BCA}) = tr(\mathbf{CAB})$.

Theorem 19. The rank of an idempotent matrix is equal to its trace.

2A.5 Quadratic Forms

Definition 1. A *quadratic form* in the k variables $x_1, x_2, \ldots, x_k$ is a scalar function of the form

$$Q = \sum_{i=1}^{k} \sum_{j=1}^{k} m_{ij} x_i x_j$$

where the m_{ij} are constants. In matrix notation, the quadratic form is expressed, using the vector $\mathbf{x} = (x_1, x_2, \ldots, x_k)'$ and the $k \times k$ matrix $\mathbf{M} = [m_{ij}]$, as

$$Q = \mathbf{x}'\mathbf{M}\mathbf{x}$$

The matrix $\mathbf{M}$ is called the matrix of the quadratic form Q and without loss of generality, we may consider $\mathbf{M}$ to be symmetric. This is because $Q = \mathbf{x}' \left(\dfrac{\mathbf{M} + \mathbf{M}'}{2} \right) \mathbf{x} = \mathbf{x}'\mathbf{M}_1\mathbf{x}$ and $\mathbf{M}_1$ is symmetric.

Definition 2. A quadratic form $Q = \mathbf{x}'\mathbf{M}\mathbf{x}$, which is positive for all values of $\mathbf{x}$ other than $\mathbf{x} = \mathbf{0}$, is said to be a *positive definite* quadratic form. The matrix $\mathbf{M}$ is called a positive definite matrix, and the eigenvalues of $\mathbf{M}$ are all greater than zero. If Q is nonnegative for all values of $\mathbf{x}$ and Q can be zero for some nonzero values of $\mathbf{x}$, then Q and the matrix $\mathbf{M}$ are said to be positive semi-definite. In this case, the eigenvalues of $\mathbf{M}$ are nonnegative.

Definition 3. If $\mathbf{M}$ is a positive definite matrix, the inequality $(\mathbf{x} - \mathbf{x}_0)' \times \mathbf{M}(\mathbf{x} - \mathbf{x}_0) \leq 1$ defines an ellipsoid with center at $\mathbf{x}_0$.

2A.6 Differentiation

Definition 4. Let $\mathbf{z} = (z_1, z_2, \ldots, z_k)'$ be a $k \times 1$ vector and $f(\mathbf{z})$ be some function of the elements of $\mathbf{z}$. Then the partial derivatives of $f(\mathbf{z})$ with respect to the elements of $\mathbf{z}$ is written as $\partial f(\mathbf{z})/\partial \mathbf{z}$ and is the $k \times 1$ vector

$$\frac{\partial f(\mathbf{z})}{\partial \mathbf{z}} = \begin{bmatrix} \partial f(\mathbf{z})/\partial z_1 \\ \partial f(\mathbf{z})/\partial z_2 \\ \vdots \\ \partial f(\mathbf{z})/\partial z_k \end{bmatrix}$$

Theorem 20. Given a $k \times 1$ vector $\mathbf{a} = (a_1, a_2, \ldots, a_k)'$ containing k constants and the $k \times 1$ vector $\mathbf{z}$. The derivative of the scalar $\mathbf{a}'\mathbf{z}$ with respect to $\mathbf{z}$ is

$$\frac{\partial \mathbf{a}'\mathbf{z}}{\partial \mathbf{z}} = \mathbf{a}$$

Theorem 21. The derivative of the quadratic form $\mathbf{z}'\mathbf{M}\mathbf{z}$ with respect to $\mathbf{z}$ is

$$\frac{\partial(\mathbf{z}'\mathbf{M}\mathbf{z})}{\partial \mathbf{z}} = \mathbf{M}\mathbf{z} + \mathbf{M}'\mathbf{z} = (\mathbf{M} + \mathbf{M}')\mathbf{z}$$

If $\mathbf{M}$ is symmetric, then $\mathbf{M} + \mathbf{M}' = 2\mathbf{M}$ and $\partial(\mathbf{z}'\mathbf{M}\mathbf{z})/\partial \mathbf{z} = 2\mathbf{M}\mathbf{z}$.

Appendix 2B   THE GENERAL FORM OF THE MOMENT MATRIX FOR ROTATABLE DESIGNS

To define the general form of the moment matrix for a rotatable design, we first recall that the design moment of order δ for a model of order d in k variables, coded as in Eq. 2.45, is

$$\frac{1}{N} \sum_{u=1}^{N} x_{u1}^{\delta_1} x_{u2}^{\delta_2} \ldots x_{uk}^{\delta_k} = [1^{\delta_1} 2^{\delta_2} \ldots k^{\delta_k}]$$

where the values of δ_i are nonnegative integers such that $\delta_1 + \delta_2 + \cdots + \delta_k = \delta$. For example, with $k = 3$, the mixed fourth-order moment, $[1123] = N^{-1} \sum_{u=1}^{N} x_{u1}^2 x_{u2} x_{u3}$, has $\delta_1 = 2$, $\delta_2 = 1$, and $\delta_3 = 1$ so that $\delta = 4$. For a model of order d, δ is at most $2d$.

A necessary and sufficient condition for a design to be rotatable is that the moment of order δ be of the form

$$[1^{\delta_1} 2^{\delta_2} \ldots k^{\delta_k}] = 0 \qquad \text{if any } \delta_i \text{ is odd}$$

$$= \frac{\lambda_\delta \prod_{i=1}^{k} (\delta_i)!}{2^{\delta/2} \prod_{i=1}^{k} \left(\frac{\delta_i}{2}\right)!} \qquad \text{if all } \delta_i \text{ are even} \qquad (2.B.1)$$

where λ_δ is some quantity which is a function of δ, $\prod_{i=1}^{k}$ is a product notation, and $(\delta_i)! = \delta_i(\delta_i - 1)(\delta_i - 2) \ldots 1$. The derivation of the expressions in Eq. 2.B.1 is given in Myers (1976: Appendix A.1), and in Box and Hunter (1957). By definition, a design moment of order δ is said to be odd if at least one δ_i in Eq. 2.B.1 is odd, and it is even if all the values of δ_i are even.

To illustrate the formula in Eq. 2.B.1, suppose we have the first-order model $Y_u = \beta_0 + \sum_{i=1}^{k} \beta_i x_{ui} + \varepsilon_u$, $u = 1, 2, \ldots, N$. Then all of the off-diagonal elements of the moment matrix will be of the form $[i]$ or $[ij]$ with $i \neq j$ and the δ_i are either zero or unity. The moments on the diagonal are $[ii]$ or $\delta_i = 2$, $\delta_j = 0$ for all $j \neq i$. From Eq. 2.B.1 these moments are expressed as $[ii] = \lambda_2 2!/2 = \lambda_2$ so that the moment matrix for a first-order rotatable design is

$$\frac{1}{N} X'X = \begin{bmatrix} 1 & 0' \\ 0 & \lambda_2 I_k \end{bmatrix} \tag{2.B.2}$$

where I_k is the identity matrix of order $k \times k$. Thus, a first-order rotatable design is an orthogonal design.

When fitting a second-order model in k variables, the form of the moment matrix for a second-order rotatable design is

$$\tag{2.B.3}$$

where J_k is a $k \times k$ matrix of ones and $k' = k(k-1)/2$. Aside from the moments $[ii] = \lambda_2$, $[iiii] = 3\lambda_4$ and $[iijj] = \lambda_4$, all of the other elements of the moment matrix in (2.B.3) are zero. Note that $\lambda_2 = 1$ under the coding convention in Eq. 2.45.

EXERCISES

2.1   In Section 2.3.1, the properties of the least squares estimator of β in the model $Y = X\beta + \varepsilon$ were discussed. It was shown that if $E(\varepsilon) = 0$, then $b = (X'X)^{-1}X'Y$ is an unbiased estimator of β.

   (a)   Suppose the true model is $Y = X\beta + \delta$, where $\delta = X_2\beta_2 + \varepsilon$, $E(\delta) = X_2\beta_2$ and $\text{Var}(\varepsilon) = \sigma^2 I$. If X is $N \times p$ and X_2 is $N \times p_2$, find $E(b)$ and $\text{Var}(b)$.

   (b)   The estimate of σ^2 when fitting the model $Y = X\beta + \varepsilon$ is $s^2 = Y'(I - X(X'X)^{-1}X')Y/(N - p)$. What is the expectation of s^2 if in fact the true model is $Y = X\beta + \delta$ as in (a)?

2.2 In a glass coating process, the percent of impurities remaining following burn-off of water is known to be affected by the temperature of the kiln. The following data represents percent impurities readings (Y) from 12 glass samples prepared at 5 temperatures (X).

Temperature (°F) X	Impurities (%) Y
200	18.4
200	17.6
200	18.0
210	11.7
210	10.3
220	7.7
220	8.3
230	6.5
230	6.7
240	6.6
240	7.2
240	6.7

(a) Fit the model $Y = \beta_0 + \beta_1 X + \varepsilon$ by calculating the least squares estimates of β_0 and β_1.

(b) Set of the ANOVA table and test $H_0: \beta_1 = 0$ versus $H_a: \beta_1 \neq 0$ at the $\alpha = 0.05$ level of significance

(c) Predict the mean percent impurities at a temperature of $X = 225°F$ and place 95% confidence limits on your estimate.

(d) Test the adequacy of the fitted first-order model at the 0.05 level of significance. If the model is not considered adequate, suggest another model and refit the new model to the data and retest.

(e) Recommend a kiln temperature for the process that you feel would not only produce low percent impurities, but also be cost effective assuming the cost for the process increases with increasing temperature.

2.3 Ten data values were recorded at five combinations of the levels of two variables denoted by V_1 and V_2 producing the table

$Y =$	6.5	7.0	9.3	9.9	9.2	11.4	13.2	12.7	13.2	12.6
$V_1 =$	2	2	3	3	3	4	4	5	6	6
$V_2 =$	1	1	2	2	2	1	1	2	2	2

Corresponding to the model forms, $Y_i = X_i\beta_i + \varepsilon_i$, $i = 1, 2, 3, 4$, the matrices and the respective inverses are

$$X_1 = [1:V_1]$$

$$X_2 = [1:V_2]$$

$$X_3 = [1:V_1:V_2]$$

$$X_4 = [1:V_1:V_2:V_1V_2]$$

$$(X_1'X_1)^{-1} = \begin{bmatrix} 0.83673 & -0.19388 \\ -0.19388 & 0.05102 \end{bmatrix}$$

$$(X_2'X_2)^{-1} = \begin{bmatrix} 1.16667 & -0.66667 \\ -0.66667 & 0.41667 \end{bmatrix}$$

$$(X_3'X_3)^{-1} = \begin{bmatrix} 1.34783 & -0.10870 & -0.52174 \\ & 0.06522 & -0.08696 \\ \text{symmetric} & & 0.53261 \end{bmatrix}$$

$$(X_4'X_4)^{-1} = \begin{bmatrix} 11.82353 & -3.38235 & -6.82353 & 1.88235 \\ & 1.08824 & 1.88235 & -0.58824 \\ \text{symmetric} & & 4.32353 & -1.13235 \\ & & & 0.33824 \end{bmatrix}$$

$$X_4'Y = \begin{bmatrix} 105.0 \\ 428.9 \\ 171.9 \\ 732.4 \end{bmatrix}$$

The following vectors of estimates were obtained:

$$b_1 = \begin{bmatrix} 4.7031 \\ 1.5255 \end{bmatrix}$$

$$b_2 = \begin{bmatrix} 7.9000 \\ 1.6250 \end{bmatrix}$$

$$b_3 = \begin{bmatrix} 5.2152 \\ 1.6109 \\ -0.5228 \end{bmatrix}$$

$$b_4 = \begin{bmatrix} -3.5500 \\ 4.3500 \\ 4.7500 \\ -1.5750 \end{bmatrix}$$

Writing the sum of squares for the fitted model, $\hat{Y} = b_0 + b_1 V_1$, as $SSR(\beta_1 | \beta_0)$, one obtains

$$SSR(\beta_1 | \beta_0) = 45.6128$$

$$SSR(\beta_2 | \beta_0) = 6.3375$$

$$SSR(\beta_1, \beta_2 | \beta_0) = 46.1260$$

$$SSR(\beta_1, \beta_2, \beta_{12} | \beta_0) = 53.4600$$

$$\sum_{i=1}^{10} (Y_i - \bar{Y})^2 = 56.3800$$

(a) Which variable, V_1 or V_2, by itself is more important in terms of explaining the variation in the Y values? Explain by testing $H_0: \beta_i = 0$ versus $H_a: \beta_i \neq 0$, $i = 1, 2$.

(b) Does either two-variable model, $Y = X_3\beta_3 + \varepsilon_3$ or $Y = X_4\beta_4 + \varepsilon_4$, fit the data better then a single-variable model, $Y = X_i\beta_i + \varepsilon$, $i = 1, 2$? Explain.

(c) The predicted values, $\hat{Y}_u$, and the residuals, $(Y_u - \hat{Y}_u)$, $u = 1, 2, \ldots, 10$ for the fitted models (i) $\hat{Y} = X_1 b_1$, (ii) $\hat{Y} = X_2 b_2$, (iii) $\hat{Y} = X_3 b_3$, and (iv) $\hat{Y} = X_4 b_4$ are

(i)	(ii)	(iii)	(iv)
7.75(−1.25)	9.53(−3.03)	7.91(−1.41)	6.75(−0.25)
7.75(−0.75)	9.53(−2.53)	7.91(−0.91)	6.75(0.25)
9.28(0.02)	11.15(−1.85)	9.00(0.30)	9.55(−0.25)
9.28(0.62)	11.15(−1.25)	9.00(0.90)	9.55(0.35)
9.28(−0.08)	11.15(−1.95)	9.00(0.20)	9.55(−0.35)
10.81(0.59)	9.53(1.88)	11.14(0.26)	12.30(−0.90)
10.81(2.39)	9.53(3.68)	11.14(2.06)	12.30(0.90)
12.33(0.37)	11.15(1.55)	12.22(0.48)	11.95(0.75)
13.86(−0.66)	11.15(2.05)	13.83(−0.63)	13.15(0.05)
13.86(−1.26)	11.15(1.45)	13.83(−1.23)	13.15(−0.55)

For each model (i)–(iv), plot the residuals against the values of $\hat{Y}_u$, V_{u1}, and V_{u2} separately.

2.4 Given the model $Y_u = \beta_0 + \beta_1 X_{u1} + \cdots + \beta_k X_{uk} + \varepsilon_u$, $u = 1, 2, \ldots, N$, prove the following properties of the residuals, $Y_u - \hat{Y}_u$:

(a) $\sum_{u=1}^{N} (Y_u - \hat{Y}_u) = 0$

(b) $\sum_{u=1}^{N} X_{ui}(Y_u - \hat{Y}_u) = 0$ $i = 1, 2, \ldots, k$.

2.5 When fitting the model $\mathbf{Y} = \mathbf{X}\boldsymbol{\beta} + \boldsymbol{\varepsilon}$, where $\mathbf{X}$ is $N \times p$ of full column rank and $\boldsymbol{\varepsilon} \sim N(0, \sigma^2\mathbf{I})$, the least squares estimates of $\boldsymbol{\beta}$ and σ^2, respectively, are

$$\mathbf{b} = (\mathbf{X}'\mathbf{X})^{-1}\mathbf{X}'\mathbf{Y}$$

and

$$s^2 = \frac{\mathbf{Y}'(\mathbf{I} - \mathbf{X}(\mathbf{X}'\mathbf{X})^{-1}\mathbf{X}')\mathbf{Y}}{N - p}$$

Show that $\mathbf{b}$ and s^2 are distributed independently of one another.

2.6 A certain resin is used as an undercoating to promote adhesion in a magnetic tape. Two additives (A_1 and A_2) used for enhancing adhesion are to be studied. Presently, it takes an average force of 24 lb/in to separate the tape from its bonding and an improvement of 5 lb/in would be very desirable. The data are

Y (lb/in of force)	A_1	A_2 (% of solution)
22.0	1	1
22.5	1	1.5
24.5	1	1.5
25.0	1	2
23.0	2	1
30.0	2	1.5
32.0	2	1.5
26.0	2	2

(a) Plot Y as a function of the levels of A_1 and A_2.

(b) Given the following matrices,

$$\mathbf{X}_1 = [\mathbf{1}:A_1]$$

$$\mathbf{X}_2 = [\mathbf{1}:A_2]$$

$$\mathbf{X}_3 = [\mathbf{1}:A_1:A_2]$$

$$\mathbf{X}_4 = [\mathbf{X}_3:A_1A_2:A_2^2:A_1A_2^2]$$

and the inverses,

$$(\mathbf{X}_1'\mathbf{X}_1)^{-1} = \begin{bmatrix} 1.25 & -0.75 \\ -0.75 & 0.50 \end{bmatrix}$$

$$(\mathbf{X}_2'\mathbf{X}_2)^{-1} = \begin{bmatrix} 2.375 & -1.5 \\ -1.5 & 1.0 \end{bmatrix}$$

$$(\mathbf{X}_3'\mathbf{X}_3)^{-1} = \begin{bmatrix} 3.5 & -0.75 & -1.5 \\ -0.75 & 0.5 & 0 \\ -1.5 & 0 & 1.0 \end{bmatrix}$$

$$(\mathbf{X}_4'\mathbf{X}_4)^{-1} = \begin{bmatrix} 385 & -231 & -525 & 315 & 170 & -102 \\ & 154 & 315 & -210 & -102 & 68 \\ & & 730 & -438 & -240 & 144 \\ & & & 292 & 144 & -96 \\ & \text{symmetric} & & & 80 & -48 \\ & & & & & 32 \end{bmatrix}$$

$$\mathbf{X}_4'\mathbf{Y} = \begin{bmatrix} 205 \\ 316 \\ 310.5 \\ 478.5 \\ 494.25 \\ 760.75 \end{bmatrix}$$

the vectors of estimates and the sums of squares quantities for the four fitted models, $\mathbf{Y}_i = \mathbf{X}_i\boldsymbol{\beta}_i + \boldsymbol{\varepsilon}_i$, $i = 1, 2, 3, 4$, are

$$\mathbf{b}_1 = \begin{bmatrix} 19.25 \\ 4.25 \end{bmatrix}$$

$$\mathbf{b}_2 = \begin{bmatrix} 21.125 \\ 3.000 \end{bmatrix}$$

$$\mathbf{b}_3 = \begin{bmatrix} 14.75 \\ 4.25 \\ 3.00 \end{bmatrix}$$

$$\mathbf{b}_4 = \begin{bmatrix} 70.0 \\ -51.0 \\ -75.0 \\ 78.0 \\ 26.0 \\ -26.0 \end{bmatrix}$$

$$\sum_{u=1}^{8} (Y_u - \bar{Y})^2 = 91.375$$

$$SSR(\beta_1 | \beta_0) = 36.125$$

$$SSR(\beta_2 | \beta_0) = 9.000$$

$$SSR(\beta_1, \beta_2 | \beta_0) = 45.125$$

$$SSR(\beta_1, \beta_2, \beta_{12}, \beta_{22}, \beta_{122} | \beta_0) = 87.375$$

The elements $c_{23} = c_{32} = 0$ of the matrix $(X_3'X_3)^{-1}$ suggest the estimates of the parameters for the terms $\beta_1 A_1$ and $\beta_2 A_2$ in the model, $Y = \beta_0 + \beta_1 A_1 + \beta_2 A_2 + \varepsilon$, are uncorrelated. Do the estimates in b_3, when compared to b_1 and b_2, reflect this property?

(c) Suggest a coding scheme for the levels of A_1 and A_2 that would make $(X_3'X_3)^{-1}$ diagonal where $X_3 = [1 : a_1 : a_2]$ and a_i is a coded variable for A_i, $i = 1, 2$. Use the coded variables a_1 and a_2 in the model $Y = \alpha_0 + \alpha_1 a_1 + \alpha_2 a_2 + \varepsilon$ and compare the estimates of α_0, α_1, and α_2 with the uncoded estimates in b_3.

(d) In fitting the model $Y_4 = X_4 \beta_4 + \varepsilon_4$, can you suggest a test statistic to test the null hypothesis, H_0: $\beta_{12} = \beta_{22} = \beta_{122} = 0$ against the alternative hypothesis, H_a: one or more equality signs is false. Perform the test at the 0.05 level of significance.

2.7 Let us refer to the rat feeding experimental data of Table 2.2 from which the fitted second-order model in Eq. 2.21 in the original variables, drug dosage (X_1) and time between inoculation and feeding (X_2), is

$$\hat{Y}(x) = 8.428 - 12.369X_1 + 1.782X_2 - 0.195X_1X_2 - 0.085X_2^2$$

Recall the coded variables, $x_1 = (X_1 - 0.5)/0.2$ and $x_2 = (X_2 - 5)/4$ from Eq. 2.40 in which an equivalent second-order fitted model is

$$\hat{Y}(x) = 8.535 - 2.669x_1 + 3.329x_2 - 0.156x_1x_2 - 1.364x_2^2$$

Compare the two fitted model forms and comment on which model is easier to work with in terms of interpreting the effects of drug dosage and time between inoculation and feeding on the amount of feed ingested. Explain by giving the meanings of the coefficient estimates in both fitted models.

REFERENCES AND BIBLIOGRAPHY

Box, G. E. P. and D. R. Cox (1964). An Analysis of Transformations (with discussion), *J. Roy. Statist. Soc.*, **B26**, 211–252.

Box, G. E. P. and N. R. Draper (1975). Robust Designs, *Biometrika*, **62**, 347–352.

Box, G. E. P. and J. S. Hunter (1957). Multifactor Experimental Designs for Exploring Response Surfaces, *Ann. Math. Statist.*, **28**, 195–242.

Box, G. E. P., W. G. Hunter, and J. S. Hunter (1978). *Statistics for Experimenters: An Introduction to Design, Data Analysis, and Model Building*, New York: John Wiley.

Chatterjee, S. and B. Price (1977). *Regression Analysis by Example*, New York: John Wiley.

Cornell, J. A. and D. C. Montgomery (1996). Interaction Models as Alternatives to Low-Order Polynomials, *J. Quality Tech.*, **28** (in press).

Daniel, C. and F. S. Wood (1980). *Fitting Equations to Data*, 2nd ed., New York: John Wiley.

Draper, N. R. and H. Smith (1981). *Applied Regression Analysis*, 2nd ed., New York: John Wiley.

Graybill, F. A. (1961). *An Introduction to Linear Statistical Models*, Vol. 1, New York: McGraw-Hill.

Khuri, A. I. (1988). A Measure of Rotatability for Response Surface Designs, *Technometrics*, **30**, 95–104.

Montgomery, D. C. and E. A. Peck (1992). *Introduction to Linear Regression Analysis*, 2nd ed., New York: John Wiley.

Myers, R. H. (1976). *Response Surface Methodology*, Blacksburg, VA: Author.

SAS Institute, Inc. (1989). *SAS/STAT User's Guide*, Vol. 2, Version 6, 4th ed., Cary, NC.

Searle, S. R. (1971). *Linear Models*, New York: John Wiley.

Shelton, J. T., A. I. Khuri, and J. A. Cornell (1983). Selecting Check Points for Testing Lack of Fit in Response Surface Models, *Technometrics*, **25**, 357–365.

Snee, R. D. (1973). Some Aspects of Nonorthogonal Data Analysis: Part I, Developing Prediction Equations, *J. Quality Tech.*, **5**, 67–79.

3

First-Order Models and Designs

3.1 INTRODUCTION

In Chapter 1 it was noted that in a response surface investigation, one of the objectives is the empirical determination of the functional relationship between the true mean response and a set of input variables. This relationship is represented by an unknown response function, which is assumed to be continuous in the input variables within some specified region of interest. Polynomial models are employed as approximating functions. In practice, low-degree polynomials are favored over high-degree polynomials because of their simpler form (fewer number of terms), particularly when one is studying the response over a small subregion, R, of the factor space. Over a larger region, for example, the operability region O or possibly the entire factor space, such low-degree polynomial representations may be inadequate and unrealistic due to the presence of lack of fit caused by higher-order terms in the true mean response model. In this case, a higher-degree polynomial is needed.

The most frequently used approximating polynomial models are of degrees one and two. This chapter is concerned with first-order models (i.e., models of degree one) and with designs that are set up for the purpose of collecting data for fitting such models. These designs are called *first-order designs*.

3.2 THE FIRST-ORDER MODEL

The general form of a first-order model in k input variables $X_1, X_2, \ldots, X_k$ is

$$Y = \beta_0 + \sum_{i=1}^{k} \beta_i X_i + \varepsilon \tag{3.1}$$

where Y is an observable response variable, $\beta_0, \beta_1, \ldots, \beta_k$ are unknown parameters, and ε is a random error term. If ε has a zero mean, then the non-random portion of the model in Eq. 3.1 represents the true mean response, η, that is,

$$\eta = \beta_0 + \sum_{i=1}^{k} \beta_i X_i \tag{3.2}$$

and ε in Eq. 3.1 is regarded as experimental error. If, however, the model in Eq. 3.2 is an inadequate representation of the true mean response (which happens when the response's relationship to the input variables is oversimplified), then ε contains, in addition to experimental error, a nonrandom (or systematic) error. The latter error is attributed to the absence of terms in $X_1, X_2, \ldots, X_k$ of degree higher than one (that are omitted from the model in Eq. 3.2) as well as to the absence of other variables which have some influence on the response. This additional error (excluding the experimental error) is called *lack of fit error* (see Section 2.6 in Chapter 2).

As an introduction to the construction of first-order designs, let us write the first-order model, over N observations, in matrix form as

$$\mathbf{Y} = \mathbf{X}\boldsymbol{\beta} + \boldsymbol{\varepsilon} \tag{3.3}$$

where $\mathbf{Y}$ is a vector of $N(N \geq k + 1)$ observations, $\boldsymbol{\beta} = (\beta_0, \beta_1, \ldots, \beta_k)'$ is a $(k + 1) \times 1$ vector of unknown parameters, $\boldsymbol{\varepsilon} = (\varepsilon_1, \varepsilon_2, \ldots, \varepsilon_N)'$ is an $N \times 1$ vector of errors, and $\mathbf{X}$ is an $N \times (k + 1)$ matrix of settings of the input variables. More specifically, the $\mathbf{X}$ matrix is of the form $\mathbf{X} = [\mathbf{1} : \mathbf{D}_X]$ where $\mathbf{1}$ is an $N \times 1$ column vector of ones and $\mathbf{D}_X$ is an $N \times k$ matrix whose (ui)th element, X_{ui}, is the value of the ith input variable used in the uth experimental run $(u = 1, 2, \ldots, N; i = 1, 2, \ldots, k)$. The matrix $\mathbf{D}_X$ will be referred to as the *design matrix* in the levels of the input variables. We assume that the random errors are independently distributed as normal variables with a zero mean and a common variance σ^2. As was noted in Chapter 2, when the matrix $\mathbf{X}$ is of full column rank, an estimator of $\boldsymbol{\beta}$ is the least squares estimator

$$\mathbf{b} = (\mathbf{X}'\mathbf{X})^{-1}\mathbf{X}'\mathbf{Y} \tag{3.4}$$

The estimator $\mathbf{b}$ is the best linear unbiased estimator (BLUE) of $\boldsymbol{\beta}$ as can be recalled from Section 2.3.4 in Chapter 2. The variance-covariance matrix of $\mathbf{b}$ is given by

$$\mathrm{Var}(\mathbf{b}) = (\mathbf{X}'\mathbf{X})^{-1}\sigma^2 \tag{3.5}$$

Throughout this chapter, we shall assume that the levels of the k input variables, X_i, are coded using the transformation,

$$x_{ui} = 2(X_{ui} - \bar{X}_i)/R_i \qquad u = 1, 2, \ldots, N; \quad i = 1, 2, \ldots, k \tag{3.6}$$

where $\bar{X}_i = \sum_{u=1}^{N} X_{ui}/N$ and R_i is the range between the lowest and highest settings. Then over the N observations, the values of the coded variables satisfy

$$\sum_{u=1}^{N} x_{ui} = 0 \qquad i = 1, 2, \ldots, k \tag{3.7}$$

To illustrate, let us recall the rat feeding experiment of Section 2.5 in Chapter 2 where the input variables were dosage of drug (X_1) with levels of 0.3 and 0.7 mg/kg and length of time between inoculation and feeding (X_2) with levels 1, 5, and 9 hours. Since the number of rats used in the experiment for each dosage-time combination was two, then $\bar{X}_1 = 0.5$ and $\bar{X}_2 = 5$, so that the coded variables in Eq. 3.6 are

$$x_{u1} = \frac{X_{u1} - 0.5}{0.2} \quad \text{and} \quad x_{u2} = \frac{X_{u2} - 5}{4} \qquad u = 1, 2, \ldots, 12$$

The transformation in Eq. 3.6 specifies the means, $\bar{X}_1, \bar{X}_2, \ldots, \bar{X}_k$, of the k input variables in the coded variables as the coordinates of the point $(x_1, x_2, \ldots, x_k) = (0, 0, \ldots, 0)$, which we shall designate the center of the design. Furthermore, if $\mathbf{D}$ is an $N \times k$ matrix whose uth row consists of the settings $x_{u1}, x_{u2}, \ldots, x_{uk}$, of the coded variables at the uth design point, and $\mathbf{1}'$ is a $1 \times N$ vector of ones, then Eq. 3.7 is expressible as $\mathbf{1}'\mathbf{D} = \mathbf{0}'$. Note that we have now defined two design matrices; $\mathbf{D}_X$ in the original k input variables and $\mathbf{D}$, the design matrix in the coded variables. Hereafter, when we refer to a design we shall be talking about the coordinate settings $x_{u1}, x_{u2}, \ldots, x_{uk}$ of the k coded variables rather than the respective levels of the k input factors even though the two sets of variables go hand in hand. The matrices $\mathbf{D}_X$ and $\mathbf{D}$ for the rat feeding experiment are

$$\mathbf{D}_X = \begin{array}{c} \begin{array}{cc} X_1 & X_2 \end{array} \\ \begin{bmatrix} 0.3 & 1 \\ 0.3 & 1 \\ 0.7 & 1 \\ 0.7 & 1 \\ 0.3 & 5 \\ 0.3 & 5 \\ 0.7 & 5 \\ 0.7 & 5 \\ 0.3 & 9 \\ 0.3 & 9 \\ 0.7 & 9 \\ 0.7 & 9 \end{bmatrix} \end{array} \qquad \mathbf{D} = \begin{array}{c} \begin{array}{cc} x_1 & x_2 \end{array} \\ \begin{bmatrix} -1 & -1 \\ -1 & -1 \\ 1 & -1 \\ 1 & -1 \\ -1 & 0 \\ -1 & 0 \\ 1 & 0 \\ 1 & 0 \\ -1 & 1 \\ -1 & 1 \\ 1 & 1 \\ 1 & 1 \end{bmatrix} \end{array}$$

3.3 DESIGNS FOR FITTING FIRST-ORDER MODELS

If the model in Eq. 3.2 is an adequate representation of the true mean response, then a design chosen for estimating the parameters must support the model to provide adequate prediction of the true response throughout the region of interest. More specifically, let $\hat{Y}(\mathbf{x})$ denote the predicted response value at a point $\mathbf{x} = (x_1, x_2, \ldots, x_k)'$ in a region of interest in the space of the coded variables which we denote by R (previously in Chapter 1, R was defined in the space of the values of X_i, but from now on it will be considered in the space of the values of x_i). Then $\hat{Y}(\mathbf{x})$ is written as

$$\hat{Y}(\mathbf{x}) = [1, \mathbf{x}']\mathbf{b} \tag{3.8}$$

The variance of the predicted response at the point $\mathbf{x}$ is expressed as

$$\text{Var}[\hat{Y}(\mathbf{x})] = [1, \mathbf{x}'][\text{Var}(\mathbf{b})][1, \mathbf{x}']' \tag{3.9}$$

From Eq. 3.5, and knowing that $\mathbf{1}'\mathbf{D} = \mathbf{0}'$, we can write Eq. 3.9 in the form

$$\text{Var}[\hat{Y}(\mathbf{x})] = [1, \mathbf{x}'] \,\text{diag}\left[\frac{1}{N}, (\mathbf{D}'\mathbf{D})^{-1}\right] [1, \mathbf{x}']'\sigma^2$$

or equivalently,

$$\text{Var}[\hat{Y}(\mathbf{x})] = \frac{\sigma^2}{N} + \mathbf{x}'(\mathbf{D}'\mathbf{D})^{-1}\mathbf{x}\sigma^2 \tag{3.10}$$

Equation 3.10 shows the dependence of $\text{Var}[\hat{Y}(\mathbf{x})]$ on the design matrix, $\mathbf{D}$.

A reasonable criterion for the choice of a first-order design is the minimization of the variance of $\hat{Y}(\mathbf{x})$. More specifically, we shall seek a design for which $\mathbf{x}'(\mathbf{D}'\mathbf{D})^{-1}\mathbf{x}$ is as small as possible within the region R, where the first-order model adequately represents the true mean response.

To accomplish this minimum variance property of $\hat{Y}(\mathbf{x})$, we note that

$$0 \le \mathbf{x}'(\mathbf{D}'\mathbf{D})^{-1}\mathbf{x} \le \|\mathbf{x}\|^2\|(\mathbf{D}'\mathbf{D})^{-1}\| \tag{3.11}$$

where $\|\cdot\|$ denotes the Euclidean norm defined as $\|\mathbf{x}\| = (\mathbf{x}'\mathbf{x})^{1/2}$ and $\|(\mathbf{D}'\mathbf{D})^{-1}\| = [\sum_{i=1}^{k} \sum_{j=1}^{k} (d^{ij})^2]^{1/2}$, where d^{ij} is the (ij)th element of $(\mathbf{D}'\mathbf{D})^{-1}$ $(i, j = 1, 2, \ldots, k)$. From Eq. 3.11 it follows that in order to minimize $\mathbf{x}'(\mathbf{D}'\mathbf{D})^{-1}\mathbf{x}$ for all $\mathbf{x} \in R$ we need to choose our design $\mathbf{D}$, that is, select the coordinate settings $x_{u1}, x_{u2}, \ldots, x_{uk}$ for $1 \le u \le N$, so that $\|(\mathbf{D}'\mathbf{D})^{-1}\|$ is as small as possible.

Let us denote the ith column of the matrix $\mathbf{D}$ by $\mathbf{d}_i$ $(i = 1, 2, \ldots, k)$ so that $\mathbf{D} = [\mathbf{d}_1:\mathbf{d}_2: \ldots :\mathbf{d}_k]$. Further, suppose that the region R places the following restrictions on the values of x_i:

$$\mathbf{d}_i'\mathbf{d}_i \le c_i^2 \qquad i = 1, 2, \ldots, k \tag{3.12}$$

where c_i is some fixed constant value ($i = 1, 2, \ldots, k$). Condition 3.12 means that the spread of the design in the direction of the ith coordinate axis is bounded by c_i^2. This is equivalent to

$$d_{ii} \leq c_i^2 \qquad i = 1, 2, \ldots, k \tag{3.13}$$

where d_{ii} is the ith diagonal element of $\mathbf{D}'\mathbf{D}$. In Appendix 3A it is shown that

$$d^{ii} \geq \frac{1}{d_{ii}} \qquad i = 1, 2, \ldots, k \tag{3.14}$$

where d^{ii} is the ith diagonal element of $(\mathbf{D}'\mathbf{D})^{-1}$. Thus, from Eqs. 3.13 and 3.14 we conclude that within the region R,

$$d^{ii} \geq \frac{1}{c_i^2} \qquad i = 1, 2, \ldots, k \tag{3.15}$$

Hence, Var(b_i) is at least equal to σ^2/c_i^2, since Var(b_i) $= \sigma^2 d^{ii}$. The equality in Eq. 3.14 holds when $d_{ij} = 0$, that is, when $\mathbf{d}_i'\mathbf{d}_j = 0$ for $i \neq j$. In this case, d^{ii} can assume the value $1/c_i^2$ in Eq. 3.15 when $d_{ii} = c_i^2$ ($i = 1, 2, \ldots, k$). Hence, if the design satisfies the conditions

$$\begin{aligned} \mathbf{d}_i'\mathbf{d}_j &= 0 \qquad i \neq j \\ \mathbf{d}_i'\mathbf{d}_i &= c_i^2 \qquad i = 1, 2, \ldots, k \end{aligned} \tag{3.16}$$

then d^{ii} assumes the minimum value $1/c_i^2$ ($i = 1, 2, \ldots, k$). It follows that the minimum value of $\|(\mathbf{D}'\mathbf{D})^{-1}\|$ over the region R can be attained when the conditions in Eq. 3.16 are satisfied.

A design satisfying the conditions in Eq. 3.16 causes the columns of the matrix $\mathbf{D}$ to be orthogonal to each other and the pure second-order moments

$$\frac{1}{N} \sum_{u=1}^{N} x_{ui}^2 = \frac{1}{N} d_{ii}$$

to be as large as possible within R for $i = 1, 2, \ldots, k$ (see Section 2.8.3 in Chapter 2 for a discussion on moments). A design satisfying the conditions in Eq. 3.16 thus produces minimum values equal to σ^2/c_i^2 for the variances of the values of b_i ($i = 1, 2, \ldots, k$).

Definition 3.1 A first-order design $\mathbf{D}$ for which $\mathbf{D}'\mathbf{D}$ is diagonal is called an *orthogonal design*.

Another consequence of using orthogonal designs is that the effects of the k input variables, as measured by the values of β_i ($i = 1, 2, \ldots, k$), can be assessed independently. This is because the off-diagonal elements of the variance-covariance matrix in Eq. 3.5 are zero, which implies that the elements of $\mathbf{b}$ are statistically independent.

Since orthogonal first-order designs are optimal in the sense of providing minimum variance estimates of the values of β_i ($i = 1, 2, \ldots, k$), we shall restrict our study of first-order designs to the class of orthogonal designs. This class includes

1. The 2^k factorial designs
2. Fractional replicates of the 2^k factorial designs
3. Simplex designs
4. Plackett-Burman designs

3.3.1 The 2^k Factorial Designs

In Section 2.8 of Chapter 2 an example was presented of a two-level factorial design. In a 2^k factorial design, each factor, or input variable, is measured at two levels which can be coded to take the value -1 when the factor is at its low level and $+1$ when at its high level (see Eq. 2.38 in Chapter 2). By considering all possible combinations of the levels of the k factors, we obtain a design matrix $\mathbf{D}$ in the coded variables consisting of 2^k rows. In the uth row, the elements are equal to $+1$ or -1 and represent the coordinates of the design point in the uth experimental run. Each row of $\mathbf{D}$ thus represents a combination of factor levels, or a treatment combination. In formula 2.38 it was shown that this coding convention can be arrived at by applying the transformation

$$x_{ui} = \frac{2(X_{ui} - \overline{X}_i)}{R_i} \qquad i = 1, 2, \ldots, k \tag{3.17}$$

where X_{ui} is the actual setting in the original units of the ith factor for the uth experimental run, $\overline{X}_i$ is the average of the low and high settings for the ith factor, and R_i is the range between the low and high settings. In coded form, the design matrix for a 2^3 factorial design, for example, has the form (hereafter we shall remove the "+" sign from +1)

$$\mathbf{D} = \begin{bmatrix} x_1 & x_2 & x_3 \\ -1 & -1 & -1 \\ 1 & -1 & -1 \\ -1 & 1 & -1 \\ 1 & 1 & -1 \\ -1 & -1 & 1 \\ 1 & -1 & 1 \\ -1 & 1 & 1 \\ 1 & 1 & 1 \end{bmatrix} = [\mathbf{d}_1 : \mathbf{d}_2 : \mathbf{d}_3]$$

The design points are displayed in Figure 3.1. As a check on the orthogonality property of the 2^3 factorial, we sum the cross-products of the elements in $\mathbf{d}_i$ and $\mathbf{d}_j$, $i \neq j$, to get $\mathbf{d}_i' \mathbf{d}_j = 0$. Furthermore, $\mathbf{d}_i' \mathbf{d}_i = 8$, $i = 1, 2, 3$, so that $\mathbf{D}'\mathbf{D} = \text{diag}$ (8,8,8). For the general case of k factors, $\mathbf{D}'\mathbf{D} = \text{diag}(2^k, 2^k, \ldots, 2^k) = 2^k \mathbf{I}_k$.

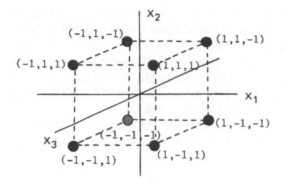

Figure 3.1 A 2^3 factorial design and coordinate settings (x_1, x_2, x_3) in the design variables x_1, x_2, and x_3.

Many texts (see, for example, Davies 1956) suggest the following notation for designating the treatment combinations. Let A, B, C denote the factors whose coded levels are denoted by x_1, x_2, and x_3, respectively. Then each row of **D** represents a treatment combination involving the levels of $A, B,$ and C. Notationally, we can represent a treatment combination by using the following scheme: If a high level of a factor is used, then the letter representing this factor in the treatment conbination is written in lowercase. Otherwise, if a low level is used, then that letter will not appear in the treatment combination. For example, ab denotes the treatment combination involving high levels of A and B and the low level of C. The symbol (1) denotes the treatment combination in which all factors are at their low levels. Using this scheme, the rows of the design matrix can be represented by the treatment combinations described below:

x_1	x_2	x_3	Treatment combination
-1	-1	-1	(1)
1	-1	-1	a
-1	1	-1	b
1	1	-1	ab
-1	-1	1	c
1	-1	1	ac
-1	1	1	bc
1	1	1	abc

In the remainder of this chapter, instead of using the above notation, we shall designate the treatment combinations in a 2^k factorial design in the following way. Let A_i denote the ith factor ($i = 1, 2, \ldots, k$). A treatment combination will be denoted by $a_1^{\alpha_1} a_2^{\alpha_2} \ldots a_k^{\alpha_k}$, where $a_i^{\alpha_i}$ denotes the level of the A_i factor used with α_i having the values $\alpha_i = 0$, for a low level of A_i, and $\alpha_i = 1$, for a high level of A_i ($i = 1, 2, \ldots, k$). Here again, (1) denotes the treatment combination in which all factors are at their low levels. For example, if $k = 4$, then $a_1^0 a_2^1 a_3^0 a_4^1$ would be written $a_2 a_4$, which means that factors A_2 and A_4 are at their high levels and factors A_1 and A_3 are at their low levels.

Without loss of generality we may consider that the values of X_i in Eq. 3.2 have been coded according to Eq. 3.17 so that the first-order model can be written as

$$\eta = \beta_0 + \sum_{i=1}^{k} \beta_i x_i$$

If a 2^k factorial design is used, then the least squares estimator **b** in Eq. 3.4 and the variance-covariance matrix in Eq. 3.5 become

$$\mathbf{b} = \frac{1}{2^k} \mathbf{X'Y}$$

$$\text{Var}(\mathbf{b}) = \frac{\sigma^2}{2^k} \mathbf{I}_{k+1}$$

The column of **X** corresponding to the ith input variable ($i = 1, 2, \ldots, k$) consists of positive and negative ones. The values -1 and $+1$ occur whenever the ith factor's levels used in the treatment combinations are low and high, respectively. Thus, $b_1, b_2, \ldots, b_k$ are contrasts in the observations obtained under the 2^k treatment combinations. These constrasts estimate the effects of the k factors. It follows that the sum of squares for the ith contrast, which is also the sum of squares (SS) for factor A_i, is given by

$$\text{SS}_{A_i} = 2^k b_i^2 \qquad i = 1, 2, \ldots, k$$

These sums of squares are statistically independent (since the b_i are statistically independent) and SS_{A_i} divided by the residual mean square MSE, which results from fitting the model to the 2^k observations, produces an F-statistic for testing the hypothesis $H_0: \beta_i = 0$. Under H_0, $F = \text{SS}_{A_i}/\text{MSE}$ has the F-distribution with 1 and $2^k - k - 1$ degrees of freedom. Large values of F are significant.

Example 3.1 In a study concerning coal conversion into oils, Fonseca et al. (1980) investigated the hydrogenolysis of a Canadian lignite using carbon monoxide and hydrogen mixtures as reducing agents. The input variables studied were $X_1 = $ temperature, $X_2 = $ CO/H$_2$ ratio, $X_3 = $ pressure, and $X_4 = $ contact time. One of the response variables under investigation was $Y = $ % lignite conversion. A 2^4 factorial design was used to fit a first-order model. The levels of the four factors were

A_1:	Reaction temperature (°C)	380	460
A_2:	Initial CO/H_2 ratio (molar ratio)	1/4	3/4
A_3:	Initial pressure (MPa)	7.1	11.1
A_4:	Contact time at reaction temperature (min)	10	50

The coded variables are defined as follows (see Eq. 3.17):

$$x_1 = \frac{X_1 - 420}{40}$$

$$x_2 = \frac{X_2 - 1/2}{1/4}$$

$$x_3 = \frac{X_3 - 9.1}{2.0}$$

$$x_4 = \frac{X_4 - 30}{20}$$

The treatment combinations and corresponding response values are given in Table 3.1

Using the data of Table 3.1 we obtain the fitted first-order model in the coded variables

$$\hat{Y} = 74.31 + 5.00x_1 + 3.15x_2 + 3.98x_3 + 1.86x_4$$
$$\phantom{\hat{Y} = 74.31 +} (1.41) \quad (1.41) \quad (1.41) \quad (1.41) \quad (1.41)$$

where the values inside parentheses are the estimated standard errors. These are equal to $(MSE/16)^{1/2}$, where $MSE = 31.71$ is the residual (or error) mean square. The corresponding ANOVA table is

Source	Degrees of Freedom	Sum of Squares	Mean Square	F
A_1	1	400.0	400.0	12.61[a]
A_2	1	158.76	158.76	5.01[c]
A_3	1	252.81	252.81	7.97[b]
A_4	1	55.50	55.50	1.75
Residual	11	348.825	31.71	
Total	15	1215.8975		

[a] Significant at 0.5%
[b] Significant at 2%
[c] Significant at 5%

Table 3.1 Treatment Combinations and Response
Values for the Coal Conversion Experiment

Treatment Combination	x_1	x_2	x_3	x_4	Y
(1)	−1	−1	−1	−1	53.3
a_1	1	−1	−1	−1	78.0
a_2	−1	1	−1	−1	62.4
a_1a_2	1	1	−1	−1	78.9
a_3	−1	−1	1	−1	75.9
a_1a_3	1	−1	1	−1	75.4
a_2a_3	−1	1	1	−1	71.3
$a_1a_2a_3$	1	1	1	−1	84.4
a_4	−1	−1	−1	1	64.5
a_1a_4	1	−1	−1	1	67.5
a_2a_4	−1	1	−1	1	72.8
$a_1a_2a_4$	1	1	−1	1	85.3
a_3a_4	−1	−1	1	1	71.4
$a_1a_3a_4$	1	−1	1	1	83.3
$a_2a_3a_4$	−1	1	1	1	82.9
$a_1a_2a_3a_4$	1	1	1	1	81.7

Source: R. Fonseca, E. Chornet, C. Roy, and M. Grandbois
(1980). Reproduced with permission of the Canadian Society
for Chemical Engineering.

From the ANOVA table we conclude that factors A_1, A_2, and A_3 are significant, whereas factor A_4 has a nonsignificant effect (the actual level of significance exceeds 20%). The value of R^2 for the fitted model is 0.713.

3.3.2 Fractional Replication of the 2^k Factorial Designs

Even though we have introduced the 2^k factorial design for fitting the first-order model in Eq. 3.1, we notice that when data are collected at all of the design points in a 2^k factorial design, it is possible in fact to estimate all of the coefficients in a model of the form

$$Y = \beta_0 + \sum_{i=1}^{k} \beta_i x_i + \sum \sum_{i=1<j}^{k} \beta_{ij} x_i x_j$$

$$+ \sum \sum \sum_{i=1<j<l}^{k} \beta_{ijl} x_i x_j x_l + \cdots + \beta_{12\ldots k} x_1 x_2 \cdots x_k + \varepsilon \qquad (3.18)$$

The additional higher-order terms, $\beta_{ij} x_i x_j$, $\beta_{ijl} x_i x_j x_l$, $\ldots$, $\beta_{12\ldots k} x_1 x_2 \ldots x_k$ in Eq. 3.18 are associated with two-factor, three-factor, $\ldots$, up to k-factor interactions among the k factors. Thus the number of design points, and hence the number of observations, along with the number of possible terms in the model in Eq. 3.18 with parameters that are estimable, increase rapidly with k, the number of factors. For example, with 10 factors at two levels each, the number of observations needed to fit the model in Eq. 3.18 is at least $2^{10} = 1024$, since 1024 is the number of parameters in the model. A cost-conscious practitioner cannot help wonder if it is necessary to run all the 2^k factorial combinations or whether some of them can be omitted when fitting only a first-order model of the form

$$Y = \beta_0 + \sum_{i=1}^{k} \beta_i x_i + \varepsilon$$

We shall now show that such a model can be fitted by using only a subset of the points (a fractional replication) of a 2^k factorial design provided that the number of points in the fraction is at least $k + 1$, which is the number of parameters to be estimated.

By definition, a fractional replication of a 2^k factorial design is a design consisting of a fraction of a complete 2^k factorial design. For example, a 2^{-1} (or 1/2) fractional replicate contains one-half the number of points of a 2^k design. This fraction is required to have properties similar to those of the complete factorial in the sense that it should be large enough to allow the estimation of the $k + 1$ parameters in a first-order model, in addition to being orthogonal. A fraction of a 2^k design consisting of only 2^{k-m} treatment combinations (m is a nonnegative integer such that $2^{k-m} \geq k + 1$) is called a 2^{-m}th fraction of a 2^k factorial design. Fractional factorial designs were first introduced by Finney (1945). The importance of applying fractional replication to research work was recognized by Davies and Hay (1950) and by Daniel (1959).

We now outline details of the construction of 2^k fractional factorial designs. Readers who are familiar with this subject may skip the construction and go directly to Example 3.2.

Construction. Let us denote the jth interaction among the k factors $A_1, A_2, \ldots, A_k$ by $A_1^{\alpha_{1j}} A_2^{\alpha_{2j}} \ldots A_k^{\alpha_{kj}}, j \leq 2^k - k - 1$, where α_{ij} is either 0 or 1 depending on the absence or presence of the factor A_i in the jth interaction. The total number of two-factor, three-factor, $\ldots$, k-factor interactions among the k factors is $2^k - k - 1$. With $k = 4$, for example, the three-factor interaction among the factors 1, 2, and 4 is $A_1 A_2 A_4$.

The construction of a 2^{-m}th fractional replicate is performed by first assuming that m high-order interactions are negligible. Such interactions will be sacrificed or confounded with one another as will soon be explained.

Let us begin with the simplest case, $m = 1$. To form a 2^{-1} fraction of the 2^k, we select a p-factor interaction, $p \leq k$, to be sacrificed and for simplicity let us write this interaction as containing the first p factors, that is, $A_1 A_2 \ldots A_p$. Of the 2^k treatment combinations, $2^k/2$ contain an even number of letters (e.g., a_1-

a_2) in common with $A_1A_2 \ldots A_p$ and the other $2^k/2$ contain an odd number of letters in common with $A_1A_2 \ldots A_p$. The choice of using the one half-fraction containing an even number of letters in common with $A_1A_2 \ldots A_p$ or the other that contains an odd number of letters in common with $A_1A_2 \ldots A_p$ is strictly arbitrary but many choose to use the principal fraction whose definition depends on whether p is an odd or even number. If p is odd, the principal fraction is the fraction containing those treatment combinations that have an odd number of letters in common with $A_1A_2 \ldots A_p$. If p is even, the other fraction is the principal fraction.

Let us continue with our discussion by constructing a 2^{-2} fraction, that is, $m = 2$. In this case, two high-order interactions are selected, say $A_1A_2 \ldots A_{p_1}$, and $A_1A_2 \ldots A_{p_2}$, $p_1, p_2 \leq k$. Now, of the 2^k treatment combinations, one-half contains an odd number of letters in common with $A_1A_2 \ldots A_{p_1}$ and the other half contains an even number of letters in common with $A_1A_2 \ldots A_{p_1}$. The same is true with the interaction $A_1A_2 \ldots A_{p_2}$. The number of letters then that the one-fourth fraction to be chosen has in common with the interactions $A_1A_2 \ldots A_{p_1}$ and $A_1A_2 \ldots A_{p_2}$ can take any one of the four cases: (odd, odd), (odd, even), (even, odd), and (even, even). The choice of the principal fraction will depend on the values of p_1 and p_2 as can be seen from the following table:

Number of Factors		Number of Letters in Common with	
p_1	p_2	$A_1A_2 \ldots A_{p_1}$	$A_1A_2 \ldots A_{p_2}$
Odd	Odd	Odd	Odd
Odd	Even	Odd	Even
Even	Odd	Even	Odd
Even	Even	Even	Even

That is, if p_1 and p_2 are odd, for example, then the principal fraction is the fraction consisting of all treatment combinations that have an odd number of letters in common with both $A_1A_2 \ldots A_{p_1}$ and $A_1A_2 \ldots A_{p_2}$.

In general, to construct a 2^{-m}th fractional replicate, we select m high-order interactions of the form $A_1^{\alpha_{1j}} A_2^{\alpha_{2j}} \ldots A_k^{\alpha_{kj}}$ ($j = 1, 2, \ldots, m$). Each one of these interactions is expressible as a contrast in the 2^k treatment combinations. If an interaction name consists of an odd (even) number of letters, then the treatment combinations having an odd (even) number of letters in common with that interaction's name are preceded by a $+$ sign in the corresponding contrast, and those having an even (odd) number of letters in common are preceded by a $-$ sign. Using this odd-even rule, the number of treatment combinations that are preceded by a $+$ sign in all the above mentioned m contrasts must be equal to 2^{k-m}. These 2^{k-m} treatment combinations form the principal 2^{-m}th fraction of the 2^k factorial design. A data set obtained from only this fraction does not permit the estimation of the above-mentioned m interaction effects since such effects cannot be distinguished from that of the overall mean in the model. We thus say that these interaction effects are confounded (aliased) with the identity effect, which we denote by I (i.e., the effect of the overall mean). This is written symbolically as

$$A_1^{\alpha_{1j}} A_2^{\alpha_{2j}} \ldots A_k^{\alpha_{kj}} = I \qquad j = 1, 2, \ldots, m \tag{3.19}$$

The confounding of m interaction effects with the identity effect amounts to a complete loss of information on these effects; they are thus completely sacrificed. The m relations defined in Eq. 3.19 are called *defining relations*. Thus, to obtain a principal 2^{-m}th fraction of a 2^k factorial design, we first introduce m defining relations such as Eq. 3.19. Next, from all 2^k treatment combinations we select those treatment combinations that have an odd (even) number of letters in common with the jth interaction in Eq. 3.19, if this interaction involves an odd (even) number of factors ($j = 1, 2, \ldots, m$). This amounts to selecting the treatment combinations that appear with a $+$ sign in the contrasts associated with the m interactions in Eq. 3.19.

Other 2^{-m}th fractions of the 2^k design can be obtained (there is a total of 2^m fractions in all) by considering all other variants of the odd-even rule. If, for example, treatment combinations having an odd (even) number of letters in common with the letters in an interaction's name involving an even (odd) number of factors are chosen, then this interaction will be confounded with $-I$ instead of I. These treatment combinations are preceded by a $-$ sign in the expression for the contrast associated with that interaction effect.

In addition to the interactions in Eq. 3.19, information on $2^m - m - 1$ other factorial effects is lost too. These effects are generalized interactions with each being the product of two interactions in the basic set of Eq. 3.19. In such products, the square of any letter is replaced by unity. For example, in our previous example where we generated the 2^{-2} fractional replicate using the interactions $A_1 A_2 \ldots A_{p_1}$ and $A_1 A_2 \ldots A_{p_2}$, suppose $k = 6$, and that these two interactions are $A_1 A_2 A_3$ and $A_1 A_2 A_3 A_4 A_5$, hence $p_1 = 3$ and $p_2 = 5$. Their generalized interaction is $(A_1 A_2 A_3)(A_1 A_2 A_3 A_4 A_5) = A_1^2 A_2^2 A_3^2 A_4 A_5 = A_4 A_5$, which is also confounded (or aliased) with $A_1 A_2 A_3$ and $A_1 A_2 A_3 A_4 A_5$. In general, with a 2^{-m}th fraction of a 2^k factorial design, a total of $2^m - 1$ effects will be sacrificed. Thus, care should be exercised in the choice of the defining relations in Eq. 3.19 in order to avoid sacrificing main effects or low-order interactions.

Information on effects other than the above mentioned $2^m - 1$ interaction effects is not completely lost. However, each effect will be aliased with $2^m - 1$ other effects. By definition, two effects are aliased if they are represented by the same contrast in the treatment combinations from the 2^{-m}th fraction used in the experiment. The aliases of any effect are obtained by taking its generalized interactions with the $2^m - 1$ interactions that were completely sacrificed. For example, with the defining relationships $A_1A_2A_3 = I$ and $A_1A_2A_3A_4A_5 = I$ and their generalized interaction A_4A_5, the aliases of the main effect of A_1 would be $A_1(A_1A_2A_3) = A_2A_3$, $A_1(A_1A_2A_3A_4A_5) = A_2A_3A_4A_5$ and $A_1(A_4A_5) = A_1A_4A_5$ so that the contrast among the treatment combinations in the principal fraction thought to estimate the effect of A_1 is in fact estimating the sum of the effects of A_1, A_2A_3, $A_2A_3A_4A_5$, and $A_1A_4A_5$. More details of fractional factorial designs are given in several experimental design books, such as those by Kempthorne (1952), Davies (1956), Hicks (1982), Box, Hunter, and Hunter (1978), Raktoe, Hedayat, and Federer (1981), Montgomery (1991), and McLean and Anderson (1984), as well as in the series of papers by Addelman (1961, 1963, 1969), Margolin (1969), and John (1961, 1962).

Example 3.2   Consider fitting a first-order model in five coded variables of the form

$$Y = \beta_0 + \sum_{i=1}^{5} \beta_i x_i + \varepsilon$$

Let us set up a one-fourth fraction of a 2^5 factorial design in which the five factors are denoted by A_1, A_2, A_3, A_4, and A_5. Suppose we choose to sacrifice information on the three-factor interactions, $A_1A_2A_3$ and $A_3A_4A_5$. The defining relations are thus $A_1A_2A_3 = A_3A_4A_5 = I$ with a generalized interaction, $A_1A_2A_3^2A_4A_5 = A_1A_2A_4A_5$. The alias structure can be determined on the basis of the defining relations

$$A_1A_2A_3 = A_3A_4A_5 = A_1A_2A_4A_5 = I \tag{3.20}$$

Since the number of letters in each of the three-factor interactions—$A_1A_2A_3$, $A_3A_4A_5$—is odd, the treatment combinations that have an odd number of letters in common with $A_1A_2A_3$ and $A_3A_4A_5$ make up the principal fraction. The remaining three one-fourth fractions can be obtained by selecting those treatment combinations whose numbers of letters in common with $A_1A_2A_3$ and $A_3A_4A_5$ are odd, even; even, odd; and even, even; respectively. The four fractions are displayed in Table 3.2.

Table 3.2   The Four One-Fourth Fractions of 2^5

$A_1A_2A_3$	$A_3A_4A_5$	Treatment Combinations
Odd	Odd	a_3, a_1a_4, a_2a_4, a_1a_5, a_2a_5, $a_1a_2a_3$, $a_3a_4a_5$, $a_1a_2a_3a_4a_5$
Odd	Even	a_1, a_2, a_3a_4, a_3a_5, $a_1a_4a_5$, $a_2a_4a_5$, $a_1a_2a_3a_4$, $a_1a_2a_3a_5$
Even	Odd	a_4, a_5, a_1a_3, a_2a_3, $a_1a_2a_4$, $a_1a_2a_5$, $a_1a_3a_4a_5$, $a_2a_3a_4a_5$
Even	Even	(1), a_1a_2, a_4a_5, $a_1a_3a_4$, $a_2a_3a_4$, $a_1a_3a_5$, $a_2a_3a_5$, $a_1a_2a_4a_5$

Table 3.3 Alias Structure for the Principal One-Fourth Fraction of the 2^5 Design with $A_1A_2A_3 = A_3A_4A_5 = A_1A_2A_4A_5 = I$

Effect		Aliases	
A_1	A_2A_3	$A_1A_3A_4A_5$	$A_2A_4A_5$
A_2	A_1A_3	$A_2A_3A_4A_5$	$A_1A_4A_5$
A_3	A_1A_2	A_4A_5	$A_1A_2A_3A_4A_5$
A_4	$A_1A_2A_3A_4$	A_3A_5	$A_1A_2A_5$
A_5	$A_1A_2A_3A_5$	A_3A_4	$A_1A_2A_4$
A_1A_4	$A_2A_3A_4$	$A_1A_3A_5$	A_2A_5
A_1A_5	$A_2A_3A_5$	$A_1A_3A_4$	A_2A_4

We note from Table 3.2 that the second, third, and fourth fractions can be derived from the principal fraction by multiplying each treatment combination in the principal fraction by a_4, a_1, and a_1a_4, respectively, and replacing the square of a lowercase letter by unity.

To find the aliases of any effect we obtain the generalized interaction of that effect with each interaction in Eq. 3.20. The complete alias structure is given in Table 3.3.

We note from Table 3.3 that every main effect is aliased with one or more two-factor interactions and higher-order interactions. The seven degrees of freedom for the treatment effects afforded by the use of the principal one-fourth fraction permit the estimation of the five main effects, A_1, A_2, A_3, A_4, and A_5 in addition to two two-factor interactions, such as A_1A_4 and A_1A_5, if the interactions aliased with these seven effects are negligible. The contrasts representing these seven effects can be obtained from Table 3.4. For example, the contrast asso-

Table 3.4 Contrasts Representing the Seven Effects in Table 3.3

	Treatment Combinations (Principal Fraction)							
Effect of	a_3	a_1a_4	a_2a_4	a_1a_5	a_2a_5	$a_1a_2a_3$	$a_3a_4a_5$	$a_1a_2a_3a_4a_5$
A_1	−	+	−	+	−	+	−	+
A_2	−	−	+	−	+	+	−	+
A_3	+	−	−	−	−	+	+	+
A_4	−	+	+	−	−	−	+	+
A_5	−	−	−	+	+	−	+	+
A_1A_4	+	+	−	−	+	−	−	+
A_1A_5	+	−	+	+	−	−	−	+

ciated with the effect of A_1 is

$$\phi_{A_1} = \tfrac{1}{4}(-a_3 + a_1a_4 - a_2a_4 + a_1a_5 - a_2a_5 + a_1a_2a_3 - a_3a_4a_5 + a_1a_2a_3a_4a_5)$$

In ϕ_{A_1} we have taken the liberty of denoting the response (or average response if replications are available) under a particular treatment combination by the same symbol used to denote that treatment combination. The divisor 4 in ϕ_{A_1} is used because there are four plus signs (also four minus signs) in that contrast. We note that the pattern of plus and minus signs in the contrasts for the main effects of A_1, A_2, A_3, A_4, and A_5 is such that if a treatment combination includes a high level of A_i ($i = 1, 2, 3, 4, 5$), then that treatment combination is preceded by a plus in the contrast for A_i, otherwise, the minus sign is used. The signs for an interaction contrast are obtained by multiplying the corresponding signs in the contrasts for the main effects involved in the interaction. For example, the signs of the contrast for the A_1A_4 interaction are obtained by multiplying the corresponding signs in the contrasts for A_1 and A_4.

The sum of squares for each effect in Table 3.4 can be obtained by computing the sum of squares for the associated contrast. For example, the sum of squares for factor A_1 is given by

$$SS_{A_1} = \frac{[\phi_{A_1}]^2}{8\left(\frac{1}{16}\right)} = 2[\phi_{A_1}]^2$$

Dividing this sum of squares by an independent estimate of the error variance (such as the residual mean square if a model with only the main effects is fitted) produces an F-statistic, which can be used to test the significance of the effect of factor A_1.

3.3.3 Biases of Parameter Estimators in Fractional Factorial Designs

A direct consequence to aliasing in fractional factorial designs is the introduction of bias in the least-squares estimators of the regression coefficients in the fitted first-order model when the true model contains interaction terms in addition to the main effects. This is because these estimators estimate the main effects of the factors in addition to the effects of the interactions with which they are aliased and which are present in the true model.

To show the bias in the estimates of the main effects, let us consider fitting the model of Eq. 3.3 using a 2^{-m}th fraction of a 2^k factorial design. Suppose that the true mean response, η, is of the form

$$\eta = \mathbf{X}\beta + \mathbf{X}_2\beta_2 \tag{3.21}$$

where the matrix X_2 contains as elements cross-product terms involving the elements of X and β_2 is a vector of parameters associated with the cross-product terms.

The expected value of b_i, the least squares estimator of the ith regression coefficient, β_i, in the model in Eq. 3.3 with coded variables as in Eq. 3.17 is given by

$$E(b_i) = 2^{-(k-m)}x_i'(X\beta + X_2\beta_2) \qquad i = 0, 1, \ldots, k, \qquad (3.22)$$

where x_i is the column in the X matrix associated with β_i. Now, x_i is orthogonal to all other columns of X, and is also orthogonal to all columns of X_2 except those columns which are associated with the effects that are aliased with the effect corresponding to β_i. The latter columns are identical to x_i, except possibly for a minus or a plus sign depending on the fraction used. It follows that the bias in b_i depends on those elements of β_2 which correspond to the effects that are aliased with the ith main effect.

To illustrate the bias, let us consider a 2^{3-1} fractional factorial in factors A_1, A_2, and A_3. Suppose we fit the model

$$Y = \beta_0 + \beta_1 x_1 + \beta_2 x_2 + \beta_3 x_3 + \varepsilon$$

The principal one-half fraction generated by the defining relation $A_1A_2A_3 = I$, using the odd-even rule described earlier, consists of the treatment combinations $a_1, a_2, a_3, a_1a_2a_3$. If we assume that the true mean response is of the form

$$\eta = \beta_0 + \beta_1 x_1 + \beta_2 x_2 + \beta_3 x_3 + \beta_{12}x_1x_2 + \beta_{13}x_1x_3 + \beta_{23}x_2x_3 + \beta_{123}x_1x_2x_3$$

then the X and X_2 matrices in Eq. 3.21 are

$$X = \begin{array}{c} \begin{array}{ccc} x_1 & x_2 & x_3 \end{array} \\ \begin{bmatrix} 1 & 1 & -1 & -1 \\ 1 & -1 & 1 & -1 \\ 1 & -1 & -1 & 1 \\ 1 & 1 & 1 & 1 \end{bmatrix} \end{array}$$

$$X_2 = \begin{array}{c} \begin{array}{cccc} x_1x_2 & x_1x_3 & x_2x_3 & x_1x_2x_3 \end{array} \\ \begin{bmatrix} -1 & -1 & 1 & 1 \\ -1 & 1 & -1 & 1 \\ 1 & -1 & -1 & 1 \\ 1 & 1 & 1 & 1 \end{bmatrix} \end{array}$$

From Eq. 3.22 it can be verified that

$$E(b_0) = \beta_0 + \beta_{123}$$
$$E(b_1) = \beta_1 + \beta_{23}$$
$$E(b_2) = \beta_2 + \beta_{13}$$
$$E(b_3) = \beta_3 + \beta_{12}$$

Thus, the biases in b_0, b_1, b_2, and b_3 are β_{123}, β_{23}, β_{13}, and β_{12}, respectively. This, of course, follows from the fact that $A_1A_2A_3$ is aliased with I, A_1 is aliased with A_2A_3, A_2 is aliased with A_1A_3, and A_3 is aliased with A_1A_2.

3.3.4   Design Resolution in Fractional Factorial Experiments

A 2^{-m}th fraction of a 2^k factorial design is said to be of resolution

III if main effects are confounded with two-factor interactions, but not with one another.

IV if main effects are not confounded with two-factor interactions; however, confounding can exist among the two-factor interactions.

V if main effects and all two-factor interactions are not confounded with one another; however, the two-factor interactions can be confounded with interactions of order three.

In general, if the system of confounding used in generating a 2^{-m}th fraction of a 2^k factorial design contains no interactions' names of length less than τ letters each, then this fractional design is of resolution τ. For example, a one-half fraction of a 2^3 factorial design generated by the defining relation $A_1A_2A_3 = I$ is of resolution III. A one-half fraction of a 2^6 factorial design generated by the defining relation $A_1A_2A_3A_4 = I$ is of resolution IV. Also, a one-fourth fraction of a 2^8 factorial design generated by the defining relations $A_3A_4A_5A_6 = I$ and $A_1A_2A_6A_7A_8 = I$ is of resolution IV. For a more detailed study of design resolution associated with two-level fractional factorial designs, the reader is referred to Box and Hunter (1961a,b), Addelman (1963, 1972) and Box, Hunter, and Hunter (1978).

3.3.5   The Simplex Design

The simplex design is an orthogonal design consisting of $N = k + 1$ design points, where k is the number of variables in the first-order model. These design points are located at the vertices of a k-dimensional regular-sided figure, or a simplex, and are characterized by the fact that the angle, θ, which any two points make with the origin is such that $\cos\theta = -1/k$ (see Box 1952). For $k = 2$, the simplex design points are the vertices of an equilateral triangle, and for $k = 3$, the design points are the vertices of a tetrahedron (see Figure 3.2).

To construct an N-point simplex design corresponding to $k = N - 1$ variables, we begin with an orthogonal matrix **P** of order $N \times N$ with equal elements in its first column. The design matrix for the simplex can then be made up of the last $N - 1$ columns of the matrix $\sqrt{N}$**P**. It is easy to verify that the **X** matrix

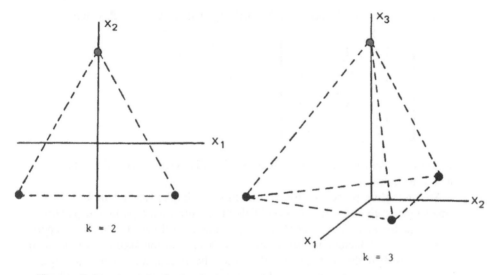

Figure 3.2 The simplex design in $k = 2$ and $k = 3$ dimensions.

corresponding to this design, which is equal to $\sqrt{N}\mathbf{P}$, satisfies the equation

$$\mathbf{X}'\mathbf{X} = N\mathbf{I}_N$$

The $\mathbf{P}'$ matrix can be obtained in the following way: select any nonsingular $N \times N$ matrix $\mathbf{Q}$ with equal first-column elements. Using the Gram-Schmidt orthonormalization technique, the columns of $\mathbf{Q}$ can be linearly transformed to another set of N columns which are orthogonal and have unit length where the first column consists of elements equal to $1/\sqrt{N}$. This linear transformation can be easily implemented by using the GSORTH subroutine in SAS (1990). The set of orthonormalized columns produces the matrix $\mathbf{P}$. For example, the following $\mathbf{X}$ matrix for a two-dimensional simplex:

$$\mathbf{X} = \begin{bmatrix} 1 & \sqrt{3/2} & \dfrac{1}{\sqrt{2}} \\ 1 & -\sqrt{3/2} & \dfrac{1}{\sqrt{2}} \\ 1 & 0 & \dfrac{-2}{\sqrt{2}} \end{bmatrix} \tag{3.23}$$

can be generated by the orthogonal matrix **P**, that is, $\mathbf{X} = \sqrt{3}\mathbf{P}$, where

$$\mathbf{P} = \begin{bmatrix} \dfrac{1}{\sqrt{3}} & \dfrac{1}{\sqrt{2}} & \dfrac{1}{\sqrt{6}} \\ \dfrac{1}{\sqrt{3}} & -\dfrac{1}{\sqrt{2}} & \dfrac{1}{\sqrt{6}} \\ \dfrac{1}{\sqrt{3}} & 0 & \dfrac{-2}{\sqrt{6}} \end{bmatrix}$$

It is easy to verify that $\mathbf{X'X} = 3\mathbf{P'P} = 3\mathbf{I}_3$. The vertices of this simplex are shown in Figure 3.3.

An alternative procedure for constructing a simplex design in $k = N - 1$ dimensions is given by Box (1952). This procedure uses a particular pattern of a "one factor at a time design" configuration as follows: In the first experimental run, all k factors are set at their low levels; in the second experimental run, the level of the first factor only would be raised, and in all subsequent experimental runs this factor would be held at the average of its first two levels. In the third experimental run, the level of the second factor would be raised, and in all subsequent experimental runs this factor would be held at the average of its first three levels. This procedure continues until the kth factor alone is subjected to a similar change. More explicitly, the coordinates of the design

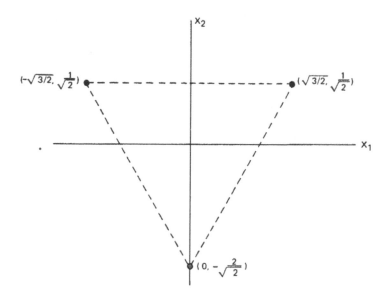

Figure 3.3 The vertices of the simplex design given in Eq. 3.23.

points for the simplex generated this way can have the form

$$
\begin{array}{cccccc}
-a_1 & -a_2 & -a_3 & \cdots & -a_i & \cdots & -a_k \\
a_1 & -a_2 & -a_3 & \cdots & -a_i & \cdots & -a_k \\
0 & 2a_2 & -a_3 & \cdots & -a_i & \cdots & -a_k \\
0 & 0 & 3a_3 & \cdots & -a_i & \cdots & -a_k \\
\vdots & \vdots & & & \vdots & & \vdots \\
& & & & ia_i & & \\
& & & & 0 & & \\
& & & & \vdots & & \\
0 & 0 & 0 & \cdots & 0 & \cdots & ka_k
\end{array}
$$

where $a_i = \{cN/i(i + 1)\}^{1/2}$, and c is a scaling constant to be selected.

The simplex design is a saturated design, that is, the number of design points is equal to the number of parameters in the model. Thus, the simplex design affords no degrees of freedom for estimating the experimental error variance or for testing lack of fit. Consequently, the usual tests of significance of the regression coefficients cannot be made unless replicate observations are taken in order to generate an estimate of the error variance, and lack of fit cannot be tested unless points are added to the design.

3.3.6 Plackett-Burman Designs

Plackett and Burman (1946) introduced fractional two-level factorial designs in k variables where the number of design points, N, is equal to $k + 1$. These designs are available only when N is a multiple of 4. They provide orthogonal designs alternative to the usual fractional factorial designs when N is not necessarily a power of 2 but is a multiple of 4.

The objective stated by Plackett and Burman was to obtain designs that can estimate all main effects with maximum precision possible for $N = k + 1$. If N is a power of 2 and $N > k + 1$, the effects of certain interactions among the factors may also be estimated with maximum precision. In this case (when N is a power of 2) Plackett-Burman designs are identical to the standard fractional two-level factorial designs. Plackett and Burman obtained design arrangements for $k = 3, 7, 11, \ldots, 99$ factors using $N = 4, 8, 12, \ldots, 100$ runs. Such designs may be used with qualitative or quantitative variables.

To construct a Plackett-Burman design in k variables, we select a row consisting of elements equal to $+1$ or to -1 such that the number of positive ones is $(k + 1)/2$ and the number of negative ones is $(k - 1)/2$ (note that $k + 1$ and $k - 1$ are both divisible by 2 since $k + 1$ is a multiple of 4). This row is chosen as the first row in the design. The next $k - 1$ rows are generated from the first row by shifting it cyclically one place to the right $k - 1$ times, that is, the second row is obtained from the first row by shifting it one place to the right, the third row is obtained from the second row by also shifting it one place, and so forth. Finally, a row of negative ones is added to the previous k rows producing a design with $N = k + 1$ rows. For example, for $k = 7$ we have the following design which is generated by the row $+1$ $+1$ $+1$ -1 $+1$ -1 -1:

x_1	x_2	x_3	x_4	x_5	x_6	x_7
$+1$	$+1$	$+1$	-1	$+1$	-1	-1
-1	$+1$	$+1$	$+1$	-1	$+1$	-1
-1	-1	$+1$	$+1$	$+1$	-1	$+1$
$+1$	-1	-1	$+1$	$+1$	$+1$	-1
-1	$+1$	-1	-1	$+1$	$+1$	$+1$
$+1$	-1	$+1$	-1	-1	$+1$	$+1$
$+1$	$+1$	-1	$+1$	-1	-1	$+1$
-1	-1	-1	-1	-1	-1	-1

3.3.7 An Example of a Plackett-Burman Design Used for Screening Variables for Reducing Oil/Fat Content of Potato Chips

Many deep fried foods, including potato chips, are prepared by frying raw vegetable materials in hot oil/fat. Deep frying causes the food products to absorb and retain relatively large amounts of oil/fat. Typically, regular potato chips contain approximately 30–40% (W/W) oil/fat, 5% fiber, 55–58% protein and carbohydrate, and 2% moisture.

The demand for snacks, such as potato chips, that are low in fat, and hence calories, has grown steadily over the past decade. Furthermore, the storage of high oil content snacks presents serious problems because of rancidity due to oxidative degradation of the oil/fat. Thus a laboratory-scale experiment was proposed to assess the technical feasibility of developing a commercial-scale-sized process for removing controlled amounts of oil/fat from potato chips and other food products.

Six process factors were identified as potentially important for the removal of oil/fat from unsalted and salted potato chips. The factors, each at two levels, were

Oil extraction: pressure at 30 and 50 MPa
temperature at 50 and 70°C
time at 45 and 75 min
Oil separation: pressure at 5 and 15 MPa
temperature at 25 and 40°C
volume of CO_2 at 600 and 800 kg/hr

An $N = 8$ run Plackett-Burman design in the six factors plus a seventh factor (X_1), that compares salted versus unsalted chips, is shown in Table 3.5. Also listed in Table 3.5 is the value of the response, percent of initial weight of oil removed, for each of the eight runs. For fitting a first-order model, coded variables for X_2 through X_7 are defined using Eq. 3.17 as $x_i = (2X_i - \text{(low + high)}/2)/(\text{high} - \text{low})$, $i = 2, \ldots, 7$.

A first-order or main-effects-only model in the coded variables x_1 through x_7 was fitted to the % wt of oil removed values in Table 3.5. The fitted model is

$$\hat{Y} = 50.41 + 1.44x_1 + 11.66x_2 + 5.09x_3 + 4.74x_4 + 4.31x_5$$
$$+ 3.56x_6 - 0.69x_7 \tag{3.24}$$

Unfortunately, no estimate of σ^2 is available from the analysis of the eight runs and therefore the magnitudes of the coefficient estimates in Eq. 3.24 cannot be tested to see which ones are different from zero.

To avoid this dilemma and obtain an estimate of σ^2, four additional runs were performed, two with salted chips and two with unsalted chips. The remaining six factors were set at their middle level, i.e., $X_2 = 40$, $X_3 = 60$, $X_4 = 60$, $X_5 = 10$, $X_6 = 32.5$, and $X_7 = 700$. With the salted chips the percent oil removed values were 38.6 and 48.0, and with the unsalted chips the response values were 45.2 and 53.3%. An estimate of σ^2, with 2 d.f., is

$$s^2 = \left[\frac{(38.6 - 48.0)^2}{2} + \frac{(45.2 - 53.3)^2}{2} \right] \div 2 = 38.49$$

Using the value of $s^2 = 38.49$, each coefficient estimate in the fitted model of Eq. 3.24 has an estimated standard error of $\sqrt{s^2/8} = 2.19$, so that now each

Table 3.5 Plackett-Burman Design in Seven Factors and Weight Percent of Oil Removed for Each Treatment Combination

Chip Type[a] X_1	Extraction			Separation		Vol. of CO$_2$ X_7	Percent wt of Oil Removed (Y)
	Pressure X_2	Temperature X_3	Time X_4	Pressure X_5	Temperature X_6		
Salted	50	70	45	15	25	600	65.3
Unsalted	50	70	75	5	40	600	70.4
Unsalted	30	70	75	15	25	800	47.2
Salted	30	50	75	15	40	600	48.4
Unsalted	50	50	45	15	40	800	58.0
Salted	30	70	45	5	40	800	39.1
Salted	50	50	75	5	25	800	54.6
Unsalted	30	50	45	5	25	600	20.3

[a] Type coded -1 for unsalted, $+1$ for salted chips.

coefficient estimate can be divided by 2.19 to produce a t value. When this is done the t values corresponding to $b_1 = 1.44$, $b_2 = 11.66$, . . . , $b_7 = -0.69$ are $t_1 = 0.66$, $t_2 = 5.32$, $t_3 = 2.32$, $t_4 = 2.16$, $t_5 = 1.97$, $t_6 = 1.63$, and $t_7 = -0.32$, respectively.

A ranking of the absolute values of the calculated t values (or of the absolute values of the coefficient estimates themselves) singled out the most important factors for the removal of oil from the potato chips as being extraction pressure, extraction temperature, and possibly extraction time. The estimates of the main effects ($=2b_i$) of extraction pressure, temperature, and time are 23.3%, 10.2%, and 9.5% of oil removed, respectively. So promising were the results from this initial experiment that a follow-up experiment was proposed and performed. In the second experiment, the levels of extraction pressure (X_2), temperature (X_3), and time (X_4) were again varied while the other factors X_5, X_6, and X_7 were set at their high, high, and low levels, respectively. Two new factors, CO_2 flow rate and chip feedstock, were introduced in the second experiment, where the data for the second experiment are listed in Exercise 3.7 at the end of this chapter.

3.4 SOME FURTHER COMMENTS ON LACK OF FIT OF FIRST-ORDER MODELS

One disadvantage of first-order designs is that they either provide no estimate of the experimental error variance, or, as in the case of saturated first-order designs, provide no means for testing lack of fit of the fitted first-order model. For example, if the true model contains pure quadratic effects in addition to the main effects, then their contribution to lack of fit cannot be detected since each factor is measured at only two levels. Lack of fit of a first-order model usually occurs when the model is fitted over a wide region where other terms of order higher than one become apparent and exert some influence on the response.

We may recall from Section 2.6 in Chapter 2 that the testing for lack of fit of a fitted model in general requires replicate observations be taken at some of the design points. The residual sum of squares, SSE, can then be partitioned into pure error sum of squares, SS_{PE}, due to the replicate observations, and lack of fit sum of squares, SS_{LOF}, which measures variation attributed to terms not included in the fitted model.

The addition of replicated points to a basic first-order design results in a new design with possibly different properties. If the basic first-order design is orthogonal, it would be desirable that this property remain invariant under such addition. If, for example, replicate observations are taken at all points of a factorial design, then an equal number of replications at all design points is needed in order to maintain the orthogonality property. This procedure, however, can be costly due to the amount of experimentation involved.

Another alternative is to augment the basic first-order orthogonal design with center point replications, that is, repeated observations taken at the design center $(0, 0, \ldots, 0)$. This enables us to obtain an independent estimate of the experimental error variance while maintaining the orthogonality property. If n_0 repeated observations are taken at the design center and if no other points are replicated, then the pure error sum of squares takes the form

$$SS_{PE} = Y'_0 \left(I_{n_0} - \frac{1_{n_0} 1'_{n_0}}{n_0} \right) Y_0$$

where Y_0 is the vector of center point observations, I_{n_0} and 1_{n_0} are the identity matrix of order $n_0 \times n_0$ and the vector of ones of dimension n_0, respectively. From Section 2.6 in Chapter 2, the lack of fit sum of squares can then be written as

$$SS_{LOF} = Y'[I_N - X(X'X)^{-1}X']Y - Y'_0 \left(I_{n_0} - \frac{1_{n_0} 1'_{n_0}}{n_0} \right) Y_0 \tag{3.25}$$

where Y is the vector of all N observations (including the n_0 center point replications). In Eq. 3.25 the X' matrix is of the form $X' = [X'_1 : X'_0]$, where X_1 is the $(N - n_0) \times (k + 1)$ matrix consisting of the column of ones along with the first-order orthogonal design, and X_0 is an $n_0 \times (k + 1)$ matrix which corresponds to the center point replications, that is, $X_0 = [1_{n_0} : 0]$, where 0 is a zero matrix of order $n_0 \times k$.

Draper and Herzberg (1971) partitioned the lack of fit sum of squares given in Eq. 3.25 into two portions as follows:

$$SS_{LOF} = L_1 + L_2$$

where

$$L_1 = Y'_1 (I_{N-n_0} - X_1 (X'_1 X_1)^{-1} X'_1) Y_1 \tag{3.26}$$

and

$$L_2 = \frac{n_0 (N - n_0)(\bar{Y}_1 - \bar{Y}_0)^2}{N}$$

In Eq. 3.26, I_{N-n_0} is the identity matrix of order $(N - n_0) \times (N - n_0)$ and Y_1 is the vector of observations taken at the points of the first-order orthogonal design. The quantities $\bar{Y}_0$ and $\bar{Y}_1$ are the averages of the center and noncenter point observations, respectively. The sums of squares L_1 and L_2 are independently distributed as $\sigma^2 \chi^2$ variables under certain null hypotheses (which will be mentioned shortly) with $(f - 1)$ and 1 degrees of freedom, respectively, where f is the lack of fit degrees of freedom and is given by $f = N - k - n_0$. If we assume that the true model is quadratic, that is, the true model is written as $E(\mathbf{Y}) = \mathbf{X}\boldsymbol{\beta} + \mathbf{X}_2\boldsymbol{\beta}_2$, where $\boldsymbol{\beta}_2$ has the form $\boldsymbol{\beta}_2 = (\beta_{11}, \beta_{22}, \ldots, \beta_{kk}, \beta_{12}, \beta_{13}, \ldots, \beta_{k-1,k})'$, then

$$E(L_1) = (f - 1)\sigma^2 + \boldsymbol{\beta}_2'\mathbf{X}_{12}'[\mathbf{I}_{N-n_0} - \mathbf{X}_1(\mathbf{X}_1'\mathbf{X}_1)^{-1}\mathbf{X}_1']\mathbf{X}_{12}\boldsymbol{\beta}_2 \tag{3.27}$$

and

$$E(L_2) = \sigma^2 + \frac{c^2 n_0 (\sum_{i=1}^{k} \beta_{ii})^2}{[N(N - n_0)]} \tag{3.28}$$

where $\mathbf{X}_{12}$ is the portion of the $\mathbf{X}_2$ matrix which corresponds to the $\mathbf{X}_1$ portion of $\mathbf{X}$, and $c = \sum_{i=1}^{N-n_0} x_{ij}^2$ ($j = 1, 2, \ldots, k$), assuming the same scaling for all the values of x. From Eq. 3.28 we note that L_2 can be used to test the hypothesis $H_0: \sum_{i=1}^{k} \beta_{ii} = 0$. The appropriate test statistic is of the form $F = L_2/\text{MS}_{\text{PE}}$ which, under H_0, has the central F-distribution with 1 and $(n_0 - 1)$ degrees of freedom, where $\text{MS}_{\text{PE}} = \text{SS}_{\text{PE}}/(n_0 - 1)$. We thus see that a lack of fit involving L_2 tests only for an overall measure of "pure curvature."

As for L_1, if the first-order orthogonal design has all odd moments of order four or less zero (this class of designs includes all fractional factorial designs of resolution greater than four with added center points), then $E(L_1)$ in Eq. 3.27 becomes

$$E(L_1) = (f - 1)\sigma^2 + \left(g - \frac{c^2}{N - n_0}\right) \sum_{i=1}^{k} \beta_{ii}^2$$
$$+ \left(h - \frac{c^2}{N - n_0}\right) \left(\sum_{i=1}^{k} \beta_{ii}\right)^2 + h \sum_{i=1}^{k-1} \sum_{j>i}^{k} \beta_{ij}^2$$

where $g = \sum_{l=1}^{N-n_0} x_{lj}^4$ ($j = 1, 2, \ldots, k$) and $h = \sum_{l=1}^{N-n_0} x_{li}^2 x_{lj}^2$ ($i \neq j$) are independent of i and j. If it can be assumed that all $\beta_{ii} = 0$, then L_1 provides a check on interaction terms by using the statistic $F = L_1/[(f - 1)\text{MS}_{\text{PE}}]$ which, under the hypothesis $H_0: \beta_{ij} = 0$, $j > i$, has the central F-distribution with $(f - 1)$ and $(n_0 - 1)$ degrees of freedom.

In an attempt to obtain a more detailed diagnosis of the lack of fit of the fitted model, Khuri and Cornell (1981) considered a further partitioning of the lack of fit sum of squares (which has f degrees of freedom) into f independent sums of squares each having one degree of freedom. The expected values of these single-degree-of-freedom sums of squares are used to identify at most f linearly independent causes for the lack of fit variation.

In summary, the addition of center point replications enables us to maintain the orthogonality property of the basic first-order design, and provides us with information concerning the existence of pure quadratic terms and of interaction terms when the actual response function is expressible in the form of a second-order model.

Example 3.3. Kizer et al. (1978) studied the behavior of a fluidized bed reactor for the catalytic oxidation of benzene to maleic anhydride. A 2^4 factorial design was used to determine the effects of X_1 = temperature in degrees centigrade (C°); X_2 = benzene/air ratio; X_3 = flow rate of reactants in m³/hr; and X_4 = height of the catalytic bed of particles at rest in cm on the conversion response Y, namely, the percentage of consumed benzene relative to the amount of moles of benzene introduced. The experimentation began at what was initially felt to be the best combination of $X_1 = 440°C$, $X_2 = 2$, $X_3 = 5$ m³/hr, and $X_4 = 5$ cm. At this point the experiment was repeated four times to provide an estimate of the error variance. The experiment was then continued around this central point following a 2^4 factorial arrangement. The original levels of factors A_1, A_2, A_3, and A_4 are given in Table 3.6. The variables in coded form, using Eq. 3.17, are

$$x_1 = \frac{X_1 - 440}{20}$$

$$x_2 = \frac{X_2 - 2}{0.5}$$

$$x_3 = \frac{X_3 - 5}{1}$$

$$x_4 = \frac{X_4 - 5}{1}$$

The design, in the coded variables, and the observed response values are given in Table 3.7. The fitted first-order model for the percentage of conversion

Table 3.6 Original Levels of the Input Variables

Factor	Input Variable	Low	High
A_1	Temperature, X_1 (°C)	420	460
A_2	Benzene/air ratio, X_2	1.5	2.5
A_3	Flow rate, X_3 (m³/hr)	4	6
A_4	Height, X_4 (cm)	4	6

Source: O. Kizer, C. Chavarie, C. Laguerie, and D. Cassimatis (1978). Reproduced with permission from the Canadian Society for Chemical Engineering.

Table 3.7 Experimental
Design and Percent
Conversion Values

x_1	x_2	x_3	x_4	Y
0	0	0	0	63.03
0	0	0	0	62.19
0	0	0	0	64.01
0	0	0	0	61.60
−1	−1	−1	−1	58.95
1	−1	−1	−1	78.34
−1	1	−1	−1	45.75
1	1	−1	−1	72.66
−1	−1	1	−1	46.36
1	−1	1	−1	68.62
−1	1	1	−1	35.16
1	1	1	−1	59.24
−1	−1	−1	1	71.62
1	−1	−1	1	84.01
−1	1	−1	1	61.18
1	1	−1	1	77.78
−1	−1	1	1	61.15
1	−1	1	1	74.83
−1	1	1	1	52.45
1	1	1	1	65.72

Source: O. Kizer, C. Chavarie, C.
Lagueir, and D. Cassimatis
(1978). Reproduced with permission of the Canadian Society for
Chemical Engineering.

is

$$\hat{Y} = 63.2325 + 9.2863x_1 - 4.6213x_2 - 5.4225x_3 + 5.2288x_4 \qquad (3.29)$$
$$\quad (0.614) \qquad (0.686) \qquad (0.686) \qquad (0.686) \qquad (0.686)$$

The numbers in parentheses in Eq. 3.29 are estimates of the standard errors of
the corresponding coefficient estimates of the model's parameters. The value of
R^2 is 0.959. The analysis of variance is given in Table 3.8. The 11 degrees of
freedom for cross-product terms represent the variation due to the six two-factor
interaction terms ($\beta_{ij}x_i x_j$), the four three-factor interaction terms ($\beta_{ijk}x_i x_j x_k$), and
the single four-factor interaction term ($\beta_{1234}x_1 x_2 x_3 x_4$). From Table 3.8 we note
that the cross-products portion is significant at the 5% level of significance

Table 3.8   Analysis of Variance Showing the Lack of Fit Partitioning for the Conversion Model

	Degrees of Freedom	Sum of Squares	Mean Square	F
A_1	1	1379.751	1379.751	
A_2	1	341.695	341.695	
A_3	1	470.456	470.456	
A_4	1	437.437	437.437	
Lack of fit	12	109.753	9.1461	8.328
Cross-products (L_1)	11	108.375	9.8523	8.97[a]
Pure quadratic (L_2)	1	1.378	1.378	1.25
Pure error	3	3.295	1.0983	
Total	19	2742.387		

[a] Significant at 5%

($F_{0.05,11,3}$ = 8.77). No significant presence of pure quadratic terms, however, can be detected ($F_{0.05,1,3}$ = 10.13).

A follow-up analysis indicated that the cross-product terms, $\beta_{12}x_1x_2$, and $\beta_{14}x_1x_4$ should be added to the model for a more complete analysis of the data. The new fitted model is of the form

$$\hat{Y} = 63.2325 + 9.2863x_1 - 4.6213x_2 - 5.4225x_3 + 5.2288x_4$$
$$(0.264) \qquad (0.295) \qquad (0.295) \qquad (0.295) \qquad (0.295)$$
$$+ \; 0.8213x_1x_2 - 2.2938x_1x_4 \tag{3.30}$$
$$(0.295) \qquad\quad (0.295)$$

The F-statistic value for the lack of fit test for the model in Eq. 3.30 is 1.35 (with 10 and 3 degrees of freedom), which is nonsignificant (the test is only significant at a level $\geq$ 0.4496). Thus, no significant lack of fit can be detected with this new model. The corresponding value of R^2 is 0.993. From the ANOVA table in Table 3.9 we note that all four main effects and two-factor interactions are significant.

Appendix 3A   A PROOF OF INEQUALITY (3.14)

To show that if d^{ii} and d_{ii} are the ith diagonal elements of $(\mathbf{D'D})^{-1}$ and $\mathbf{D'D}$, respectively, $i = 1, 2, \ldots , k$, then

$$d^{ii} \geq \frac{1}{d_{ii}}, \; i = 1, 2, \ldots , k \tag{3.A.1}$$

Table 3.9   Analysis of Variance for the Conversion Model in Eq. 3.30

Source	Degrees of Freedom	Sum of Squares	Mean Square	F
A_1	1	1379.751	1379.751	992.32[a]
A_2	1	341.695	341.695	245.75[a]
A_3	1	470.456	470.456	338.35[a]
A_4	1	437.437	437.437	314.61[a]
A_1A_2	1	10.791	10.791	7.76[b]
A_1A_4	1	84.181	84.181	60.54[a]
Residual	13	18.076	1.39	
Total	19	2742.387		

[a]Significant at a level ≥ 0.0001
[b]Significant at a level ≥ 0.0154

Proof. Let $D_{(-i)}$ be a matrix obtained from D by removing the ith column ($i = 1, 2, \ldots, k$). The cofactor of d_{ii} in $D'D$ is $|D'_{(-i)}D_{(-i)}|$, hence

$$d^{ii} = \frac{|D'_{(-i)}D_{(-i)}|}{|D'D|} \tag{3.A.2}$$

There exists a matrix E_i of order $k \times k$ whose determinant has an absolute value of one such that DE_i is a matrix whose first column is d_i, the ith column of D, and its remaining columns are the columns of $D_{(-i)}$, respectively, that is

$$DE_i = [d_i : D_{(-i)}] \qquad i = 1, 2, \ldots, k \tag{3.A.3}$$

From Eq. 3.A.3 it follows that

$$|D'D| = |E'_i D'DE_i| = |D'_{(-i)}D_{(-i)}| \times [d'_i d_i - d'_i D_{(-i)}(D'_{(-i)}D_{(-i)})^{-1}D'_{(-i)}d_i] \tag{3.A.4}$$

Thus, from Eqs. 3.A.2 and 3.A.4 we obtain

$$d^{ii} = \frac{1}{d'_i d_i - d'_i D_{(-i)}(D'_{(-i)}D_{(-i)})^{-1}D'_{(-i)}d_i} \qquad i = 1, 2, \ldots, k \tag{3.A.5}$$

Since $d'_i D_{(-i)}(D'_{(-i)}D_{(-i)})^{-1}D'_{(-i)}d_i \geq 0$, and $d'_i d_i = d_{ii}$, we conclude from Eq. 3.A.5 that

$$d^{ii} \geq \frac{1}{d_{ii}} \qquad i = 1, 2, \ldots, k \tag{3.A.6}$$

Equality in Eq. 3.A.6 is attained when $d'_i D_{(-i)} = 0'$, $i = 1, 2, \ldots, k$, that is, when the design is orthogonal.

EXERCISES

3.1 Consider the data of Example 3.1 given in Table 3.1.

  (a) Choose a one-half fraction of the 2^4 factorial design, then fit a first-order model in x_1, x_2, x_3, x_4 to the corresponding data.

  (b) Determine the alias structure on the basis of the fraction chosen in (a).

  (c) Obtain contrasts that can be used to estimate the main effects, then compute the corresponding sums of squares.

  (d) Obtain a complete ANOVA table based only on the fraction chosen in (a).

3.2 Construct a simplex design in

  (a) $k = 4$ dimensions
  (b) $k = 5$ dimensions

3.3 Construct a Plackett-Burman design for a first-order model involving $k = 11$ variables.

3.4 Consider fitting a first-order model of the form

$$\mathbf{Y} = \mathbf{X}\boldsymbol{\beta} + \boldsymbol{\varepsilon}$$

using a fractional factorial design with center point replications. The true mean response, however, is given by

$$\boldsymbol{\eta} = \mathbf{X}\boldsymbol{\beta} + \mathbf{X}_2\boldsymbol{\beta}_2$$

  (a) Show that the residual mean square from the analysis of the fitted model is a biased estimator of σ^2, the error variance. Give an expression for the amount of bias.

  (b) If, in particular, $\mathbf{X}'\mathbf{X}_2 = 0$, then

  (i) $\mathbf{b} = (\mathbf{X}'\mathbf{X})^{-1}\mathbf{X}'\mathbf{Y}$ is an unbiased estimator of $\boldsymbol{\beta}$.

  (ii) The bias in (a) is greater than the bias in the more general case, $\mathbf{X}'\mathbf{X}_2 \neq \mathbf{0}$. Deduce that in this special case it is easier to detect lack of fit of the fitted model

3.5 A 2^k factorial design with n_0 center point replications is used to fit a first-order model with k variables. The factors levels are coded as ± 1. The true mean response is quadratic of the form

$$\eta = \beta_0 + \sum_{i=1}^{k} \beta_i x_i + \sum_{i=1}^{k} \beta_{ii} x_i^2 + \sum_{i<j} \beta_{ij} x_i x_j$$

Let $\bar{Y}_0$ and $\bar{Y}_1$ be the averages of the center and noncenter point observations, respectively. Show that

  (a) $E(\bar{Y}_0) = \beta_0$

(b) $E(\bar{Y}_1) = \beta_0 + \Sigma_{i=1}^k \beta_{ii}$

(c)   Deduce Eq. 3.28 using (a) and (b).

3.6   Consider again the data of Example 3.1. The 2^4 factorial design was augmented with three center point replications. The percent lignite conversion values at the center were 82.7, 83.0, and 79.6.

(a)   Test for lack of fit of the model

$$Y = \beta_0 + \sum_{i=1}^{4} \beta_i x_i + \varepsilon$$

(b)   Obtain a partitioning of the lack of fit sum of squares into L_1 and L_2, then make significance tests concerning the pure quadratic terms and interaction terms assuming that the true mean response is of the second order.

3.7   The data for the seven-factor potato chip experiment are listed in Table 3.5 where the fitted main-effects-only model is Eq. 3.24. Four additional runs were performed in order to obtain an estimate of σ^2. The settings of X_1–X_7 for these runs were as shown in the following table

X_1	X_2	X_3	X_4	X_5	X_6	X_7	Y
Salted	40	60	60	10	32.5	700	38.6
Salted	40	60	60	10	32.5	700	48.0
Unsalted	40	60	60	10	32.5	700	45.2
Unsalted	40	60	60	10	32.5	700	53.3

(a)   Combine the preceding data with the data in Table 3.5 and refit the first-order model. How does this model differ from the fitted model of Eq. 3.24, if at all?

(b)   The residual mean square from the analysis in (a) is MSE = 43.64 with 4 degrees of freedom. If an estimate of σ^2, the error variance, is $s^2 = 38.49$, what is MSE = 43.64 an estimate of?

(c)   A follow-up experiment was performed using an $N = 8$ Plackett-Burman design. Factors $X_1 - X_4$ remained the same as before but two new factors $X_8 = CO_2$ flow rate (6 and 10 l/min) and $X_9 =$ chip feedstock (flat and round versus curled and elliptical) were introduced. The treatment combinations and % wt of oil removed values for this second set of experiments are as shown in the following table.

X_1	X_2	X_3	X_4	X_8	X_9	Y
Salted	50	70	45	10	Flat	61.4
Unsalted	50	70	75	6	Curled	79.8
Unsalted	30	70	75	10	Flat	50.1

Salted	30	50	75	10	Curled	48.2
Unsalted	50	50	45	10	Curled	57.6
Salted	30	70	45	6	Curled	46.0
Salted	50	50	75	6	Flat	65.3
Unsalted	30	50	45	6	Flat	40.2

Fit a first-order model to the % wt of oil removed values in the follow-up experiment. Combine the estimates of σ^2 in parts (b) and (c) to obtain an estimate of σ^2 with 3 degrees of freedom. Use the combined estimate of σ^2 to test $H_0: \beta_i = 0$ vs. $H_a: \beta_i \neq 0$, $i = 1, 2, \ldots, 7$ in Eq. 3.24 and $H_0: \beta_i = 0$ vs. $H_a: \beta_i \neq 0$, $i = 1, 2, 3, 4, 8, 9$ in the follow-up experiment model.

(d)   Based on the $N = 20$ experimental trials performed thus far, what can you conclude about the relative importance of the factors X_1, X_2, $\ldots, X_9$?

REFERENCES AND BIBLIOGRAPHY

Addelman, S. (1961). Irregular Fractions of the 2^n Factorial Experiments. *Technometrics,* **3,** 479–496.

Addelman, S. (1963). Techniques for Constructing Fractional Replicate Plans, *J. Amer. Statist. Assoc.,* **58,** 45–71.

Addelman, S. (1969). Sequences of Two-level Fractional Factorial Plans, *Technometrics,* **11,** 477–509.

Addelman, S. (1972). Recent Developments in the Design of Factorial Experiments, *J. Amer. Statist. Assoc.,* **67,** 103–111.

Box, G. E. P. (1952). Multi-Factor Designs of First Order, *Biometrika,* **39,** 49–57.

Box, G. E. P. and J. S. Hunter (1961a). The 2^{k-p} Fractional Factorial Designs, Part I, *Technometrics,* **3,** 311–351.

Box, G. E. P. and J. S. Hunter (1961b). The 2^{k-p} Fractional Factorial Designs, Part II, *Technometrics,* **3,** 449–458.

Box, G. E. P., W. G. Hunter, and J. S. Hunter (1978). *Statistics for Experimenters: An Introduction to Design, Data Analysis, and Model Building,* New York: John Wiley.

Daniel, C. (1959). Use of Half-Normal Plots in Interpreting Factorial Two-Level Experiments, *Technometrics,* **1,** 311–341.

Davies, O. L. (1956). *The Design and Analysis of Industrial Experiments,* 2nd ed., London: Oliver and Boyd.

Davies, O. L. and W. A. Hay (1950). The Construction and Uses of Fractional Factorial Designs in Industrial Research, *Biometrics,* **6,** 233–249.

Draper, N. R. and A. M. Herzberg (1971). On Lack of Fit, *Technometrics,* **13,** 231–241.

Finney, D. J. (1945). The Fractional Replication of Factorial Arrangements, *Annals of Eugenics,* **12,** 291–301.

Fonseca, R., E. Chornet, C. Roy, and M. Grandbois (1980). Statistical Study on the Batch Conversion of Estevan Lignite into Oils Using Carbon Monoxide and Hydrogen Mixtures, *Canadian J. of Chemical Engineering*, **58**, 458–465.

Hicks, C. R. (1982). *Fundamental Concepts of Design of Experiments*, 3rd ed., New York: Holt, Rinehart and Winston.

John, P. W. M. (1961). The Three-Quarter Replicates of 2^4 and 2^5 Designs, *Biometrics*, **17**, 319–321.

John, P. W. M. (1962). Three-Quarter Replicates of 2^n Designs, *Biometrics*, **18**, 172–184.

Kempthorne, O. (1952). *The Design and Analysis of Experiments*, New York: John Wiley.

Khuri, A. I. and J. A. Cornell (1981). Lack of Fit Revisited, Technical Report No. 167, Dept. of Statistics, Univ. of Florida, Gainesville, FL 32611.

Kizer, O., C. Chavarie, C. Laguerie, and D. Cassimatis (1978). Quadratic Model of the Behavior of a Fluidized Bed Reactor: Catalytic Oxidation of Benzene to Maleic Anhydride, *Canadian J. of Chemical Engineering*, **56**, 716–724.

Margolin, B. H. (1969). Results of Factorial Designs of Resolution IV for the 2^n and $2^n 3^m$ Series, *Technometrics*, **11**, 431–444.

McLean, R. A. and V. L. Anderson (1984). *Applied Factorial and Fractional Designs*, New York: Marcel Dekker.

Montgomery, D. C. (1991). *Design and Analysis of Experiments*, 3rd ed., New York: John Wiley.

Plackett, R. L. and J. P. Burman (1946). The Design of Optimum Multifactorial Experiments, *Biometrika*, **33**, 305–325.

Ratkoe, B. L., A. Hedayat, and W. T. Federer (1981). *Factorial Designs*, New York: John Wiley.

SAS Institute, Inc. (1990). SAS/IML Software, Version 6, 1st ed., Cary, NC.

4

Second-Order Models and Designs

4.1 INTRODUCTION

If knowledge concerning the shape of the true response surface is insufficient, an experimenter generally first attempts to approximate the shape by fitting a first-order model to the response values. When, however, the first-order model suffers from lack of fit arising from the existence of surface curvature, the first-order model is upgraded by adding higher-order terms to it. The next higher-order model is the second-order model

$$Y = \beta_0 + \sum_{i=1}^{k} \beta_i X_i + \sum_{i=1}^{k} \beta_{ii} X_i^2 + \sum_{\substack{i=1 \\ i<j}}^{k-1} \sum_{j=2}^{k} \beta_{ij} X_i X_j + \varepsilon \qquad (4.1)$$

where $X_1, X_2, \ldots, X_k$ are the input variables which influence the response Y; β_0, β_i $(i = 1, 2, \ldots, k)$, β_{ij} $(i = 1, 2, \ldots k; j = 1, 2, \ldots, k)$ are unknown parameters, and ε is a random error.

4.2 NOTATION

A design by means of which observed values of the response are collected for estimating the parameters in the second-order model (Eq. 4.1) is called a *second-order design*. The number of distinct design points must be at least $p = (k +$

1)$(k + 2)/2$, since p is the number of terms in the model in Eq. 4.1. For $k = 2$, 3, and 4, p is equal to 6, 10, and 15, respectively.

To begin our discussion concerning the choice of locations in the experimental region at which to assign the points of the design, let us define a set of standardized variables, x_i, whose levels for the uth trial $(u = 1, 2, \ldots, N)$ are given by

$$x_{ui} = \frac{X_{ui} - \bar{X}_i}{s_{X_i}} \qquad \begin{array}{l} u = 1, 2, \ldots, N \\ i = 1, 2, \ldots k \end{array} \tag{4.2}$$

where X_{ui} is the uth level of the ith input variable, $\bar{X}_i = \frac{1}{N} \sum_{u=1}^{N} X_{ui}$ is the mean of the X_{ui} values, $s_{X_i} = [\sum_{u=1}^{N} (X_{ui} - \bar{X}_i)^2/N]^{1/2}$ is a measure of the standard deviation, and N is the number of observations. Without loss of generality we may consider that the values of X_i in Eq. 4.1 have been replaced by the corresponding values of x_i $(i = 1, 2, \ldots, k)$. The observed value for the response for the uth trial can thus be represented as

$$Y_u = \beta_0 + \sum_{i=1}^{k} \beta_i x_{ui} + \sum_{i=1}^{k} \beta_{ii} x_{ui}^2 + \sum_{i=1}^{k-1} \sum_{\substack{j=2 \\ i<j}}^{k} \beta_{ij} x_{ui} x_{uj} + \varepsilon_u$$

$$u = 1, 2, \ldots, N \tag{4.3}$$

where ε_u is the experimental error in Y_u. We assume that the values of ε_u are independently distributed as random variables with zero means and variances σ^2 $(u = 1, 2, \ldots, N)$.

Model 4.3 can be written in the matrix form

$$\mathbf{Y} = \mathbf{X}\boldsymbol{\beta} + \boldsymbol{\varepsilon} \tag{4.4}$$

where $\mathbf{Y} = (Y_1, Y_2, \ldots, Y_N)'$, $\mathbf{X}$ is an $N \times p$ matrix whose uth row is a function of the design settings for the uth experimental trial $(u = 1, 2, \ldots, N)$; $p = (k + 1)(k + 2)/2$; $\boldsymbol{\beta}$ is a $p \times 1$ vector of unknown parameters and $\boldsymbol{\varepsilon} = (\varepsilon_1, \varepsilon_2, \ldots, \varepsilon_N)'$. The least squares estimator of $\boldsymbol{\beta}$ in Eq. 4.4 is given by

$$\mathbf{b} = (\mathbf{X}'\mathbf{X})^{-1}\mathbf{X}'\mathbf{Y} \tag{4.5}$$

The variance-covariance matrix of $\mathbf{b}$ is

$$\text{Var}(\mathbf{b}) = (\mathbf{X}'\mathbf{X})^{-1}\sigma^2 \tag{4.6}$$

The moment matrix, $N^{-1}\mathbf{X}'\mathbf{X}$, contains design moments of orders 0, 1, 2, 3, and 4. We recall from Appendix 2B that a design moment of order δ, $\delta \geq 0$, is of the form $\frac{1}{N} \sum_{u=1}^{N} x_{u1}^{\delta_1} x_{u2}^{\delta_2} \ldots x_{uk}^{\delta_k}$, where $\sum_{i=1}^{k} \delta_i = \delta$. Because of the scaling convention in Eq. 4.2, the first-order moments, $[i]$, are zero, and the second-order moments, $[ii]$, satisfy $[ii] = 1$ for $i = 1, 2, \ldots k$. For example, if $k = 2$

the moment matrix has the form

$$N^{-1}\mathbf{X'X} = \begin{array}{c} \\ 1 \\ x_1 \\ x_2 \\ x_1^2 \\ x_2^2 \\ x_1 x_2 \end{array} \begin{array}{cccccc} 1 & x_1 & x_2 & x_1^2 & x_2^2 & x_1 x_2 \\ \left[\begin{array}{cccccc} 1 & 0 & 0 & 1 & 1 & [12] \\ 0 & 1 & [12] & [111] & [122] & [112] \\ 0 & [12] & 1 & [112] & [222] & [122] \\ 1 & [111] & [112] & [1111] & [1122] & [1112] \\ 1 & [122] & [222] & [1122] & [2222] & [1222] \\ [12] & [112] & [122] & [1112] & [1222] & [1122] \end{array} \right] \end{array}$$

Under the scaling convention given in Eq. 4.2, the quantities $[iii]^2$ and $[iiii]$, are, respectively, measures of skewness and kurtosis of the distribution of the design points taken in the direction of the axis of the ith input ($i = 1, 2, \ldots,$ k) variable. If this distribution is symmetric about the origin, then $[iii]^2 = 0$, otherwise, the size of $[iii]^2$ measures departure from symmetry. The moment $[iiii]$ gives information concerning the spread of the design points in the ith direction, namely, whether the points tend to concentrate at the center and at the extremes of the ith range or are uniformly distributed.

4.3 ORTHOGONAL SECOND-ORDER DESIGNS

We have previously defined an orthogonal design as one for which the corresponding $\mathbf{X'X}$ matrix (or the moment matrix) is diagonal. For second-order designs a diagonal moment matrix is impossible to obtain. This is because the moments $[ii]$ and $[iijj]$ (which represent sums of products of 1 and x_i^2 and of x_i^2 and x_j^2, respectively) are necessarily positive.

Orthogonal second-order designs, however, can be obtained if we express the variables in the model in Eq. 4.3 in terms of orthogonal polynomials as shown below (see Box and Hunter, 1957: 200–201).

Let $P_m(x_{ui})$ be the orthogonal polynomial of degree m ($m \geq 0$) for the ith input variable x_i ($i = 1, 2, \ldots, k$). Then,

$$P_m(x_{ui}) = x_{ui}^m + \alpha_{m-1,m}x_{ui}^{m-1} + \cdots + \alpha_{1m}x_{ui} + \alpha_{0m} \tag{4.7}$$

where the values of the α's are chosen so that

$$\sum_{u=1}^{N} P_m(x_{ui})P_{m-j}(x_{ui}) = 0 \qquad P_0(x_{ui}) = 1 \qquad j = 1, 2, \ldots, m \tag{4.8}$$

The original second-order model can now be expressed in terms of these orthogonal polynomials as

$$\mathbf{Y} = (\mathbf{XP})(\mathbf{P}^{-1}\boldsymbol{\beta}) + \boldsymbol{\varepsilon} = \dot{\mathbf{X}}\dot{\boldsymbol{\beta}} + \boldsymbol{\varepsilon}$$

where $\mathbf{P}$ is the nonsingular matrix transforming the terms x_{ui}^m into $P_m(x_{ui})$, $\dot{\mathbf{X}} =$ $\mathbf{XP}$, and $\dot{\boldsymbol{\beta}} = \mathbf{P}^{-1}\boldsymbol{\beta}$. Thus, under the scaling convention given in Eq. 4.2 we have

$$P_1(x_{ui}) = x_{ui}, \quad P_2(x_{ui}) = x_{ui}^2 - [iii]x_{ui} - 1 \qquad i = 1, 2, \ldots, k \tag{4.9}$$

and Eq. 4.3 can be written as

$$Y_u = \left(\beta_0 + \sum_{i=1}^{k} \beta_{ii}\right) + \sum_{i=1}^{k} (\beta_i + [iii]\beta_{ii})P_1(x_{ui}) + \sum_{i=1}^{k} \beta_{ii}P_2(x_{ui})$$

$$+ \sum_{i=1}^{k-1}\sum_{j=2}^{k} \beta_{ij}P_1(x_{ui})P_1(x_{uj}) + \varepsilon_u \qquad u = 1, 2, \ldots, N \tag{4.10}$$

Since $\sum_{u=1}^{N} P_m(x_{ui}) = 0$ for $m = 1, 2$, and $i = 1, 2, \ldots, k$; and $\sum_{u=1}^{N} P_1(x_{ui})P_2(x_{ui}) = 0$ for $i = 1, 2, \ldots, k$, then the moment matrix $N^{-1}\mathbf{X}'\mathbf{X}$ is diagonal if

$[ij] = [ijj] = [iij] = 0$ for $i < j$

$[iijj] = 1$ for $i < j$

$[iijn] = [ijjn] = [ijnn] = 0$ for $i < j < n$ \hfill (4.11)

$[iiij] = [ijjj] = 0$ for $i < j$

$[ijns] = [ijn] = 0$ for $i < j < n < s$

Examples of such designs are the 3^k factorial designs and the orthogonal central composite designs which will be introduced later in this chapter.

4.4   ROTATABLE SECOND-ORDER DESIGNS

We recall from Chapter 2 that if a design is rotatable, the variance of the predicted response, $\hat{Y}$, remains constant at all points which are equidistant from the design center. The necessary and sufficient condition for rotatability was given in Appendix 2B. From this condition it can be deduced that in the case of a second-order model, all odd moments of order ≤ 4 must be zero and the remaining even moments must satisfy the equations, $[ii] = \lambda_2$; $[iijj] = \lambda_4$; and $[iiii] = 3\lambda_4$ $(i, j = 1, 2, \ldots, k; i < j)$. This implies that the pure fourth-order moment, $[iiii]$, must be three times as large as the mixed fourth-order moment, $[iijj]$. The value for the pure second-order moment, $[ii]$, is fixed at $\lambda_2 = 1$ by the scaling convention given by Eq. 4.2. Scaling, however, does not fix the value of λ_4, which can assume different values depending on other criteria that a rotatable second-order design may be required to have. For example, from Eq. 4.11, a rotatable second-order design is also orthogonal if $\lambda_4 = 1$.

Box and Hunter (1957: 213) showed that with any rotatable second-order design the variance of the predicted response at any point x in the experimental region is given by

$$\text{Var}[\hat{Y}(\mathbf{x})] = A\{2(k+2)\lambda_4^2 + 2\lambda_4(\lambda_4 - 1)(k+2)\rho^2 + [(k+1)\lambda_4 - (k-1)]\rho^4\}$$

(4.12)

where $\rho^2 = \mathbf{x}'\mathbf{x}$ and $A = \sigma^2\{2N\lambda_4[(k+2)\lambda_4 - k]\}^{-1}$. For the particular case of an orthogonal second-order rotatable design, that is, when $\lambda_4 = 1$, Eq. 4.12 becomes

$$\text{Var}[\hat{Y}(\mathbf{x})] = \frac{\sigma^2(k + 2 + \rho^4)}{2N}$$

In Figure 4.1 a plot of the precision function, $W(\rho) = \{N\,\text{Var}[\hat{Y}(\mathbf{x})]/\sigma^2\}^{-1}$, versus ρ is given for various values of λ_4 for the case of $k = 2$ where $\text{Var}[\hat{Y}(\mathbf{x})]$ is given by Eq. 4.12. We note that the precision, as measured by the values of $W(\rho)$, drops off rapidly as ρ exceeds unity. From Eq. 4.12, the variance of

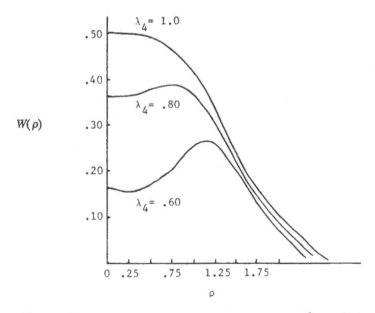

Figure 4.1   The measure of precision, $W(\rho) = \{N\,\text{Var}[\hat{Y}(\mathbf{x})]/\sigma^2\}^{-1}$ as a function of ρ and λ_4 for a second-order rotatable design when $k = 2$. [G. E. P. Box and J. S. Hunter (1957). Reproduced with permission of the Institute of Mathematical Statistics.]

$\hat{Y}(\mathbf{x})$ at the design center, that is, when $\rho = 0$, is

$$\text{Var}[\hat{Y}(\mathbf{0})] = \sigma^2 \lambda_4 \left\{ N \left(\lambda_4 - \frac{k}{k + 2} \right) \right\}^{-1} \tag{4.13}$$

It can be seen from Eq. 4.13 and Figure 4.1 that $\text{Var}[\hat{Y}(\mathbf{0})]$ is a decreasing function of λ_4 for $\lambda_4 > k/(k + 2)$. As λ_4 approaches or exceeds unity, the precision at the center of the design increases. In this case, however, the biases in the model's parameters in the presence of third-order terms (if, in reality, the true mean response is a polynomial of the third degree) will be high also (see Box and Hunter, 1957: 214, and Myers, 1976: 223).

For fixed N and $\rho \neq 0$, the quality of prediction as measured by the size of the precision function $W(\rho)$ depends on the rotatable design used through values of λ_4. The precision can decrease drastically as ρ moves away from zero as can be seen from Figure 4.1. Since the precision at the design center can be high for large values of λ_4, large variation in precision in the vicinity of the design center might be experienced. In order to maintain a somewhat uniform distribution of precision in the vicinity of the design center, one may choose the design so that the value of $\text{Var}[\hat{Y}(\mathbf{x})]$ at $\rho = 0$ is equal to the variance at $\rho = 1$. A rotatable second-order design satisfying this property is called a *uniform precision design*. This property provides approximately a uniform value of prediction variance inside a sphere of radius one. The purpose of this is to produce stability in the prediction variance in the vicinity of the design center. Values of λ_4 required for a rotatable uniform precision design are given in Table 4.1 for several values of k.

We note that when λ_4 approaches $k/(k + 2)$, the value of A in Eq. 4.12 becomes infinitely large, which renders the rotatable second-order design useless. Box and Hunter (1957) referred to this value of λ_4 (i.e., $\lambda_4 = k/(k + 2)$) as the "singular" value. This occurs when all the design points of a second-order design are equidistant from the design center. Such a design arrangement will be shown later (Section 4.5.5) to be rotatable. For such a set of points, $\rho^2 = \sum_{i=1}^{k} x_{ui}^2$ for $u = 1, 2, \ldots N$, hence

$$\rho^2 = N^{-1} \sum_{u=1}^{N} \sum_{i=1}^{k} x_{ui}^2 = \sum_{i=1}^{k} [ii] = k \tag{4.14}$$

since $[ii] = 1$ for $i = 1, 2, \ldots, k$. We also have

$$\rho^4 = N^{-1} \sum_{u=1}^{N} \left(\sum_{i=1}^{k} x_{ui}^2 \right)^2 = \sum_{i=1}^{k} [iiii] + \sum_{i=1}^{k} \sum_{j \neq i}^{k} [iijj] \tag{4.15}$$

Table 4.1 Values of λ_4 Required to Make the Variance of $\hat{Y}(\mathbf{x})$ at $\rho = 1$ Equal to That at $\rho = 0$ for a Rotatable Second-Order Design

k	2	3	4	5	6	7	8
λ_4	0.7844	0.8385	0.8704	0.8918	0.9070	0.9184	0.9274

Source: G. E. P. Box and J. S. Hunter (1957). Reproduced with permission of the Institute of Mathematical Statistics.

From Eqs. 4.14 and 4.15 we obtain

$$3k\lambda_4 + k(k - 1)\lambda_4 = k^2 \tag{4.16}$$

since $[iiii] = 3\lambda_4$ when the second-order design is rotatable. Equation 4.16 is equivalent to $\lambda_4 = k/(k + 2)$.

Methods for constructing rotatable second-order designs were introduced by Bose and Draper (1959) for the case $k = 3$. Draper (1960) presented a method for constructing a rotatable second-order design in k dimensions from a rotatable second-order design in $(k - 1)$ dimensions by adding a further coordinate to the coordinates of the $k - 1$ dimensional points followed by the addition of suitably chosen sets of points.

4.5 EXAMPLES OF SECOND-ORDER DESIGNS

Since model 4.1 contains pure quadratic terms, then any second-order design must involve at least three levels of each input variable. In this section several examples of second-order designs will be given along with the properties of the designs.

4.5.1 The 3^k Factorial Design

One possible second-order design is the 3^k factorial design, which requires that the response be observed at all possible combinations of the levels of k input variables which have three levels each. For example, for $k = 2$, if the three levels of the ith input variable, X_{ui}, are equally spaced and of the form $a_i - c_i$, a_i, $a_i + c_i$ $(i = 1, 2)$, then these levels can be coded so that they correspond to -1, 0, and 1, respectively, by using the transformation

$$x_{ui} = \frac{X_{ui} - a_i}{c_i} \qquad u = 1, 2, \ldots, 9, \quad i = 1, 2$$

Hence, in coded form, the design matrix for such a design has the form

$$
\mathbf{D} =
\begin{array}{cc}
x_1 & x_2 \\
\left[\begin{array}{cc}
-1 & -1 \\
-1 & 0 \\
-1 & 1 \\
0 & -1 \\
0 & 0 \\
0 & 1 \\
1 & -1 \\
1 & 0 \\
1 & 1
\end{array}\right]
\end{array}
$$

The design points are displayed in Figure 4.2.

In the case of the 3^k factorial design, the number, N, of experimental trials is $N = 3^k$ and can thus be excessively large, especially when a large number of input variables are under study. To reduce the total number of experimental design points, the use of fractional replications of these designs can be considered. A general procedure for constructing a 3^{-m}th fraction of the 3^k design for $m < k$ is given in Montgomery (1991: Chapter 12). Such a fraction is called a 3^{k-m} fractional factorial design. A thorough coverage of fractional 3^{k-m} designs is given in Connor and Zelen (1959) for $4 \le k \le 10$. These designs are also reproduced in Appendix 2 of McLean and Anderson (1984).

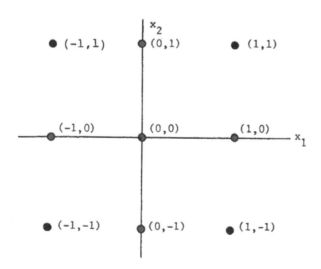

Figure 4.2 The 3^2 factorial design.

DeBaun (1959) introduced a number of three-factor, three-level designs. These designs consist of a combination of the following subsets of points taken from the 3^3 design:

1. The 3^3 factorial design
2. Cube (points at $(\pm 1, \pm 1, \pm 1)$)
3. Center point $(0, 0, 0)$
4. Octahedron (points at $(\pm 1, 0, 0; 0, \pm 1, 0; 0, 0, \pm 1)$)
5. Cuboctahedron (points at $(\pm 1, \pm 1, 0; \pm 1, 0, \pm 1; 0, \pm 1, \pm 1)$)

The designs investigated by DeBaun were:

1. The 3^3 factorial design
2. Cube + octahedron + n center points
3. Cube + 2 octahedra + n center points
4. Cuboctahedron + n center points
5. Cube + cuboctahedron + n center points
6. Cuboctahedron + octahedron + n center points

DeBaun compared the above six designs on the basis of the value $\{\text{Var}[\hat{Y}(\mathbf{x})]/\sigma^2\}^{-1}/N$ at $(0, 0, 0)$ in addition to its distribution on $\rho = (\mathbf{x}'\mathbf{x})^{1/2}$ along certain radii. He concluded that the 3^3 design is by no means the most efficient of the cases 1.–6.; he found that the 3^3 design is, in particular, excelled by the cube plus two octahedral designs with 22 and 24 points, respectively, and by the cuboctahedron design with 16 points.

If the three levels of each factor in a 3^k factorial design are evenly spaced, and if these levels are coded so that $[i] = 0$ and $[ii] = 1$ ($i = 1, 2, \ldots, k$), then it can be shown that this design is orthogonal and that $[iiii] = 3/2$ for $i = 1, 2, \ldots, k$. Since $[iijj] = 1$ and knowing that the design is orthogonal, we can easily conclude that the 3^k design is not rotatable since $[iiii] \neq 3 \, [iijj]$. Rotating this design affects the precision of the least squares estimators of the second-order model's parameters. For example, in a 3^2 factorial design with equally spaced levels for each factor, the levels can be scaled according to Eq. 4.2 to obtain the design matrix

$$
\mathbf{D} = \begin{array}{c} \begin{array}{cc} x_1 & x_2 \end{array} \\ \begin{bmatrix} -g & -g \\ -g & 0 \\ -g & g \\ 0 & -g \\ 0 & 0 \\ 0 & g \\ g & -g \\ g & 0 \\ g & g \end{bmatrix} \end{array}
$$

where $g = 3/\sqrt{6}$. It is easy to verify that $\lambda_4 = [1122] = 4g^4/9 = 1$ and that $[iiii] = 6g^4/9 = 3/2$ for $i = 1, 2$.

4.5.2   The Box-Behnken Designs

A class of three-level incomplete factorial designs for the estimation of the parameters in a second-order model was developed by Box and Behnken (1960). By definition, a three-level incomplete factorial design is a subset of the factorial combinations from the 3^k factorial design. The Box-Behnken designs are formed by combining two-level factorial designs with balanced incomplete block designs (BIBD) in a particular manner. The following example will help illustrate how a Box-Behnken design can be constructed.

Example 4.1.   Consider a BIBD involving four treatments and six blocks with each block containing two treatments. Each treatment appears three times in the design, once with each of the other treatments. If the treatments are denoted by asterisks, we obtain the design

		x_1	x_2	x_3	x_4
	1	*	*		
	2			*	*
Blocks	3	*			*
	4		*	*	
	5		*		*
	6	*		*	

We now combine the preceding BIBD with the 2^2 factorial design

$$
\begin{array}{cc}
x_i & x_j \\
\left[\begin{array}{rr}
-1 & -1 \\
1 & -1 \\
-1 & 1 \\
1 & 1
\end{array}\right]
\end{array}
$$

in the following manner: The two asterisks in every block are replaced by the two columns of the 2^2 design. A column of zeros is included when no asterisk appears. The design is augmented by the addition of center points, three being used in this example. The resulting three-level Box-Behnken design in four variables consists of the following 27 points:

x_1	x_2	x_3	x_4
−1	−1	0	0
1	−1	0	0
−1	1	0	0
1	1	0	0
0	0	−1	−1
0	0	1	−1
0	0	−1	1
0	0	1	1
0	0	0	0
. . . .			
−1	0	0	−1
1	0	0	−1
−1	0	0	1
1	0	0	1
0	−1	−1	0
0	1	−1	0
0	−1	1	0
0	1	1	0
0	0	0	0
. . . .			
0	−1	0	−1
0	1	0	−1
0	−1	0	1
0	1	0	1
−1	0	−1	0
1	0	−1	0
−1	0	1	0
1	0	1	0
0	0	0	0

The design obtained is rotatable and blocks orthogonally in three blocks indicated by the dotted lines (see Section 8.2 in Chapter 8 for a discussion on orthogonal blocking of second-order designs). In general, however, Box-Behnken designs are not always rotatable nor are they block orthogonal. Box and Behnken (1960) listed a number of second-order designs for $k = 3$, 4, 5, 6, 7, 9, 10, 11, 12, and 16 input variables.

Example 4.2. The following is a numerical example of a three-variable experiment using a Box-Behnken design: Tarwi (*Lupinus mutabilis*) is a legume native to the Andes in South America. Its seeds are similar in composition to

soybean. Removal of edible oil and alkaloids from Tarwi seed produces a source
of protein for inclusion in food products.

Supercritical carbon dioxide ($SC\text{-}CO_2$) extraction of oil from seeds has several advantages over the conventional use of hexane extraction (Rizvi, 1994).
CO_2 is nontoxic, nonexplosive, inexpensive, and easily removed from the residual extracted product. $SC\text{-}CO_2$ is a versatile solvent because extraction conditions (temperature and pressure) can be varied over wide ranges.

To measure the effects of temperature, pressure, and cosolvent or ethanol on
the extraction of oil and alkaloids from ground seeds, a three-factor 13-point
Box-Behnken design was chosen. The temperature levels, (35, 45, and 55°C)
were selected to prevent protein denaturation; the levels of pressure (10.4,
20.7, and 31.0 MPa) were chosen based on equipment restrictions, and cosolvent or ethanol levels (CO_2 containing 0, 5, and 10% ethanol) were chosen to
facilitate the extraction process. The design is displayed in Figure 4.3 where
the coordinates of the design points are defined in the coded variables x_1 =
(temperature − 45°C)/10°C, x_2 = (pressure − 20.7 MPa)/10.3 MPa, and x_3 =
(ethanol − 5%)/5%. Listed in Table 4.2 are the design settings in the original
and coded variables and the amounts of residual oil (g/100 g dry wt of seed)
remaining in the ground seed having an initial oil content of 22.3 g/100 g dry

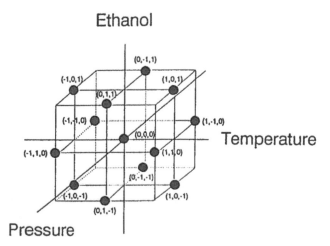

Figure 4.3 The three-variable Box-Behnken design. The coordinates of the design
points are expressed in the coded units for temperature (x_1), pressure (x_2), and ethanol
(x_3).

Table 4.2 Residual Oil Content After Supercritical Extraction

Run Order for First Replicate	Temperature (°C)	Pressure (MPa)	Ethanol (%)	Coded Design Settings x_1	x_2	x_3	Residual Oil (g/100 g dry wt)
1	35	20.7	0	−1	0	−1	20.79, 20.35, 21.03
2	45	20.7	5	0	0	0	10.68, 7.87, 6.10
3	45	10.4	0	0	−1	−1	22.08, 18.98, 20.08
4	55	10.4	5	1	−1	0	11.75, 12.50, 13.24
5	35	20.7	10	−1	0	1	3.25, 3.71, 4.16
6	45	31.0	0	0	1	−1	15.43, 12.83, 16.04
7	55	20.7	0	1	0	−1	17.84, 18.99, 18.76
8	45	20.7	5	0	0	0	8.22, 9.25, 7.49
9	35	10.4	5	−1	−1	0	10.31, 12.38, 14.45
10	35	31.0	5	−1	1	0	3.57, 2.25, 4.89
11	45	10.4	10	0	−1	1	1.93, 1.62, 1.30
12	45	31.0	10	0	1	1	7.16, 5.90, 6.53
13	45	20.7	5	0	0	0	10.31, 8.40, 6.15
14	55	31.0	5	1	1	0	0.34, 1.37, 0.86
15	55	20.7	10	1	0	1	3.25, 0.50, 1.76

wt of seed for three replicates of the entire set of 15 runs. Three replicates (runs 2, 8, and 13) of the center point settings were performed in each replicate of 15 runs. For this experiment, low values of the response (residual oil) were considered desirable.

A second-order polynomial in the coded variables was fitted to the 45 residual oil values listed in Table 4.2. The fitted model is

$$\hat{Y}(\mathbf{x}) = 8.27 - 0.83x_1 - 2.64x_2 - 7.59x_3 - 0.29x_1^2 - 0.66x_2^2$$
$$\quad\quad (0.51) \quad (0.51) \quad (0.51) \quad (0.75) \quad (0.75)$$

$$+ \; 3.21x_3^2 - 0.71x_1x_2 + 0.08x_1x_3 + 2.63x_2x_3 \quad\quad (4.17)$$
$$\quad (0.75) \quad\quad (0.72) \quad\quad (0.72) \quad\quad (0.72)$$

with $R^2 = $ SSR/SST $= 0.8921$ and $R_A^2 = 0.8643$. The quantities in parentheses below the coefficient estimates are the estimated standard errors of the coefficient estimates based on the residual mean square, MSE $= 6.157$. The corresponding analysis of variance table is as follows:

Source	Degrees of Freedom	Sum of Squares	Mean Square	F
Regression	9	1781.598	197.955	32.15
Residual	35	215.507	6.157	
Total	44	1997.105		

Prior to drawing inferences from the fitted model of Eq. 4.17 concerning the effects of temperature, pressure, and cosolvent on the extraction of residual oil from the seeds, the adequacy of the fitted model should be checked. From Section 2.6, and in particular Table 2.5, $SS_{PE} = 51.543$ with 32 degrees of freedom. Furthermore, $SS_{LOF} = 215.507 - 51.543 = 163.964$ with 3 degrees of freedom. The value of the F-statistic in Eq. 2.30 for testing zero lack of fit is

$$F = \frac{163.964/3}{51.543/32} = 33.93$$

Since $F = 33.93$ exceeds the table value $F_{0.01,3,32} = 9.15$, we conclude there is sufficient evidence to indicate that the fitted model of Eq. 4.17 exhibits lack of fit at the 0.01 level of significance.

An attempt to upgrade the fitted model of Eq. 4.17 by adding terms to it is facilitated by looking at the coefficient estimates in Eq. 4.17 and noticing those that are significantly different from zero when compared to their estimated standard errors. In particular, there is evidence of curvature ($b_{33} = 3.21$ and $t = 3.21/0.75 = 4.28$) in the residual oil values as the percent ethanol (x_3) is increased, but no evidence of curvature from changing the levels of temperature (x_1) or pressure (x_2). Hence, we might consider adding the terms $\beta_{133}x_1x_3^2$ and/or $\beta_{233}x_2x_3^2$ to the fitted model to improve the fit. We leave this as an exercise and direct the reader's attention to Exercise 4.6 at the end of this chapter.

4.5.3 The Central Composite Designs

Box and Wilson (1951) introduced an alternative class of designs to the 3^k factorial designs, namely, the class of central composite designs, (CCD). A central composite design consists of

1. A complete (or fraction of a) 2^k factorial design, where the factor levels are coded to the usual -1, $+1$ values. This is called the *factorial portion* of the design.
2. n_0 center points ($n_0 \geq 1$).
3. Two axial points on the axis of each design variable at a distance of α from the design center. This portion is called the axial portion of the design.

The total number of design points is thus $N = 2^k + 2k + n_0$. Values of n_0 and α are chosen appropriately as will be explained later. For example, a CCD in $k = 2$ variables with $n_0 = 1$ and $\alpha = \sqrt{2}$ is of the form

$$
\mathbf{D} = \begin{array}{cc}
x_1 & x_2 \\
\begin{bmatrix}
-1 & -1 \\
1 & -1 \\
-1 & 1 \\
1 & 1 \\
\sqrt{2} & 0 \\
-\sqrt{2} & 0 \\
0 & \sqrt{2} \\
0 & -\sqrt{2} \\
0 & 0
\end{bmatrix}
\end{array}
$$

The design points are represented graphically in Figure 4.4. The scaling convention given by Eq. 4.2 will be adopted for the CCD. In this case, it can be easily verified that the odd moments of a CCD through order 4 are zero, that is,

$$
\begin{aligned}
[i] &= 0 & i &= 1, 2, \ldots, k \\
[iii] &= 0 & i &= 1, 2, \ldots, k \\
[ij] &= 0 & i, j &= 1, 2, \ldots, k, \quad i \neq j \\
[iij] &= 0 & i, j &= 1, 2, \ldots, k, \quad i \neq j \\
[ijk] &= 0 & i &\neq j \neq k \\
[iiij] &= 0 & i, j &= 1, 2, \ldots k, \quad i \neq j \\
[iijk] &= 0 & i &\neq j \neq k
\end{aligned}
$$

The even moments, $[ii]$, $[iiii]$, and $[iijj]$ $(i, j = 1, 2, \ldots, k; i \neq j)$ are nonzero with $[ii] = 1$ by the scaling convention. The latter two moments are influenced by the choice of the number of center points, n_0, and by the value α of the axial points setting.

A CCD is rotatable if $[iiii] = 3 \, [iijj]$ for $i, j = 1, 2, \ldots, k; i \neq j$ (see Section 4.4). If g is a scale factor chosen so that $[ii] = 1$, then $g = [N/(F + 2\alpha^2)]^{1/2}$, where F is the number of points in the factorial portion of the design, and $N = F + 2k + n_0$ is the total number of points in the CCD. It follows that the condition of rotatability, $[iiii] = 3 \, [iijj]$, for a CCD can be written as

$$
Fg^4 + 2\alpha^4 g^4 = 3Fg^4
$$

or equivalently

$$
\alpha = F^{1/4} \tag{4.18}
$$

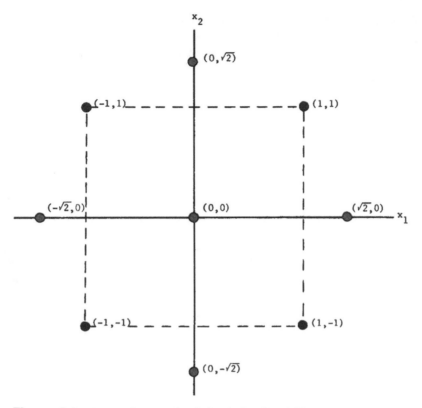

Figure 4.4   A central composite design in $k = 2$ variables.

A CCD is orthogonal if the mixed fourth-order moment, $[iijj]$, or equivalently, λ_4, is equal to unity (see Section 4.3). In terms of the scale factor g, $[iijj]$ is written

$$[iijj] = \frac{Fg^4}{N} = \frac{FN}{(F + 2\alpha^2)^2} \tag{4.19}$$

In order for $[iijj] = 1$ we must have

$$(F + 2\alpha^2)^2 = FN \tag{4.20}$$

Solving Eq. 4.20 for α we conclude that a CCD can be made orthogonal by choosing the value of the axial setting, α, to be equal to

$$\alpha = \left(\frac{(FN)^{1/2} - F}{2} \right)^{1/2} \tag{4.21}$$

Now, if it is desired for the CCD to be orthogonal as well as rotatable, it is possible to choose α and n_0 to achieve both properties. To do so, we only need to replace α^2 in Eq. 4.20 by $\sqrt{F}$ since $\alpha^4 = F$ is the condiiton for a CCD to be rotatable, to get

$$(F + 2\sqrt{F})^2 = F(F + 2k + n_0) \tag{4.22}$$

Solving Eq. 4.22 for n_0 we get

$$n_0 \simeq 4\sqrt{F} + 4 - 2k \tag{4.23}$$

In Eq. 4.23, n_0 is equal to the integer closest to the expression on the right-hand side. For example, if $k = 2$ and $F = 2^2 = 4$, then $n_0 = 8$ center point replications are needed for a rotatable CCD in two variables to be orthogonal. The number of center point replications can also be chosen to cause a rotatable CCD to have the uniform (or near uniform) precision property (see Section 4.4). Values of λ_4 required for a rotatable design to have the uniform precision property were given in Table 4.1. If we replace α^2 by $\sqrt{F}$ in Eq. 4.19 and solve for n_0, then it is possible to determine the number of center point replications that correspond to values of λ_4 for a uniform precision rotatable design according to the formula

$$n_0 \simeq \lambda_4(\sqrt{F} + 2)^2 - F - 2k \tag{4.24}$$

For example, for $k = 3$, $F = 2^3 = 8$, and $\lambda_4 = 0.8385$ from Table 4.1, a total of $n_0 = 6$ center point replications are needed to produce a rotatable CCD in three input variables with almost the uniform precision property.

When k is large, the factorial portion of a CCD can be replaced by a fraction of a 2^k factorial design. If a 2^{-m}th fraction of a 2^k factorial design is used in the factorial portion, then $2^m - 1$ effects will be sacrificed. The remaining $2^k - 2^m$ effects in a complete 2^k factorial experiment fall into $2^{k-m} - 1$ alias sets each of which contains 2^m effects which are aliased with one another. Hartley (1959) stated the following theorem:

Theorem 4.1. In any CCD in which no main effect is used as a defining relation of the 2^{-m}th fraction of a 2^k design, it is always possible to estimate the following parameters of the second-order model (Eq. 4.1): The constant β_0; all linear parameters, β_i ($i = 1, 2, \ldots, k$); all quadratic parameters, β_{ii} ($i = 1, 2, \ldots, k$); and one of the cross-product parameters, β_{ij} ($i, j = 1, 2, \ldots, k, i < j$) selected from each of the alias sets. It is not possible to estimate more than one of these cross-product parameters from each alias set.

Thus, in a CCD of the kind described in Theorem 4.1, the total number of parameters in model 4.3 that can be estimated is $2k + 1 + 2^{k-m} - 1$. These include β_0, each of the β_i ($i = 1, 2, \ldots, k$), each of the β_{ii} ($i = 1, 2, \ldots, k$) and $2^{k-m} - 1$ values of the β_{ij} ($i, j = 1, 2, \ldots, k, i < j$). Since the total number of points is $N = 2^{k-m} + 2k + n_0$, where n_0 is the number of center point replications, then the number of degrees of freedom for estimating the experimental error variance in this case is n_0.

Example 4.3: One-half Fraction of 2^4. Using the defining relation $A_1A_2A_3A_4 = I$, where A_i denotes the ith factor ($i = 1, 2, 3, 4$), all main and pure quadratic effects can be estimated. Of the seven alias sets, three consist of two-factor interactions aliased with other two-factor interactions. Therefore, only three two-factor interactions can be estimated. Thus, no fractions of a 2^4 factorial design should be used in the factorial portion of the CCD if the interest is in estimating all six two-factor interactions.

Example 4.4: One-half Fraction of 2^5. Using the defining relation $A_1A_2A_3A_4A_5 = I$, we note that in addition to the main and quadratic effects, all 10 two-factor interactions can be estimated since these are aliased with three-factor interactions.

Example 4.5: One-fourth Fraction of 2^5. Here the number of alias sets is equal to seven, therefore, we cannot estimate all the 10 two-factor interactions. The maximum number of $\beta_{ij}(i < j)$ which can be estimated is seven.

Example 4.6: One-fourth Fraction of 2^6. In this case the number of alias sets is 15 and the number of two-factor interactions is also 15; however, it is not always possible to choose defining relations such that each alias set contains at least one two-factor interaction. Hartley (1959) produced a fraction which permitted the estimation of all 15 values of β_{ij}.

Using Eqs. 4.23 and 4.24 and Table 4.1, values of n_0 required for a rotatable CCD in k input variables ($k = 2, 3, \ldots, 8$) to be either nearly orthogonal or nearly having the uniform precision (UP) property can be obtained. These values are given in Table 4.3, which also lists the corresponding values of F, N, and n_a, where n_a denotes the number of axial points.

Table 4.3 Uniform Precision or Orthogonal Rotatable CCD

k	2	3	4	5	$5(\frac{1}{2}\text{rep})$	6	$6(\frac{1}{2}\text{rep})$	7	$7(\frac{1}{2}\text{rep})$	8	$8(\frac{1}{2}\text{rep})$
F	4	8	16	32	16	64	32	128	64	256	128
n_a	4	6	8	10	10	12	12	14	14	16	16
n_0 (orth.) (Eq. 4.23)	8	9	12	17	10	24	15	35	22	52	33
n_0 (UP) (Eq. 4.24)	5	6	7	10	6	15	9	21	14	28	20
N (orth.)	16	23	36	59	36	100	59	177	100	324	177
N (UP)	13	20	31	52	32	91	53	163	92	300	164

Source: G. E. P. Box and J. S. Hunter (1957). Reproduced with permission of the Institute of Mathematical Statistics.

Draper (1982) listed several other criteria for choosing the number, n_0, of center point replications in a CCD in addition to the orthogonality and uniform precision criteria. Of these criteria, one is to minimize the function,

$$\phi_1 = \frac{Nr - p^2}{N^2} \tag{4.25}$$

with respect to n_0 for fixed values of α, F, and n_a. In Eq. 4.25, r is the sum of squares of the diagonal elements of $X(X'X)^{-1}X'$, and p is the number of parameters in the model. The criterion function (Eq. 4.25) was introduced previously by Box and Draper (1975), and by minimizing ϕ_1, we attempt to "even out" the pattern of variances of the predicted response at the design points by minimizing their spread.

A second criterion is concerned with the use of center point replications for testing lack of fit of the fitted second-order model. On the basis of this criterion alone and assuming no replicated points elsewhere in the design, Draper (1982) recommends that a minimum of four or five center point replications be taken in order to have a reasonably sensitive test of lack of fit. A third criterion listed by Draper (1982) is based on the integrated variance of the predicted response function normalized by N, σ^2, and the experimental region volume, that is,

$$\phi_2 = \frac{N}{\sigma^2} \int_R \text{Var}[\hat{Y}(x)] \, dx / \int_R dx$$

If the experimental region, R, is a hypersphere of radius ρ, then ϕ_2 is a function of ρ and n_0. Draper argued that a choice of $\rho = k^{1/2}$ is a reasonable one because in many CCD situations, the factorial and axial points lie on or within the same hypersphere with the above-mentioned radius. In this case, ϕ_2 reduces to a function of n_0 and the criterion here is to select the value of n_0 which minimizes ϕ_2. Draper noted that fewer center points were needed under the additional criteria than were initially recommended via the orthogonality and uniform precision criteria.

Design points, other than the center of a CCD, can be replicated for the purpose of providing a more general estimate of the experimental error and more reliable estimates of the second-order model's parameters. Dykstra (1960) presented and compared eight types of partially duplicated CCDs, all having second-order moments equal to unity. The comparisons among the designs were based on the following properties of the design: the biases in the second-order model's parameters due to the presence of third-order terms, the power of the test for lack of fit, and the precision of prediction. Dykstra noted that for all eight types of designs, the bias in the linear parameters due to third-order terms increased with n_0.

4.5.4 Efficiency of the Central Composite Design

As a measure of design efficiency, Box and Draper (1971) discussed the choice of design on the basis of maximizing the determinant of $\mathbf{X}'\mathbf{X}$. They showed that for a CCD consisting of 2^{k-m} factorial points with levels $\pm g$, $2k$ axial points with settings $\pm\alpha$, and n_0 center point replications, the determinant of $\mathbf{X}'\mathbf{X}$ is given by

$$|\mathbf{X}'\mathbf{X}|_{\mathrm{CCD}} = (2^{k-m}g^2 + 2\alpha^2)^k (2^{k-m}g^4)^{k(k-1)/2} (2\alpha^4)^{k-1} \times \{n_0(2\alpha^4 + 2^{k-m}kg^4)$$

$$+ 2^{k-m+1}(\alpha^2 - kg^2)^2\} \tag{4.26}$$

To maximize $|\mathbf{X}'\mathbf{X}|_{\mathrm{CCD}}$ for $k \geq 2$, Box and Draper indicated that the factorial and axial points should be positioned on the boundaries of the cuboidal region bounded by $x_i = \pm 1$, for $|g| \leq 1$. Lucas (1974) showed that $|\mathbf{X}'\mathbf{X}|_{\mathrm{CCD}}$ is an increasing function of α. Thus the axial points should be moved to the extremes of the experimental region.

The D-efficiency (D_{eff}) of a CCD is defined as

$$D_{\mathrm{eff}} = \left(\frac{|\mathbf{X}'\mathbf{X}|_{\mathrm{CCD}}}{N^p \max_\xi |\mathbf{M}(\xi)|}\right)^{1/p} \tag{4.27}$$

where p is the number of parameters in the second-order model, and $\mathbf{M}(\xi)$ is the information matrix for a design measure ξ. The maximum value of $|\mathbf{M}(\xi)|$ is attained when the design measure is D-optimal (see Chapter 12 for a more detailed discussion of D-optimality). Note that the information matrix for a discrete design measure, such as the one considered in this chapter, is equal to $\mathbf{X}'\mathbf{X}/N$. The G-efficiency of a CCD is defined as

$$G_{\mathrm{eff}} = p/d_{\max} \tag{4.28}$$

where $d_{\max}$ multiplied by σ^2/N is the maximum prediction variance over the experimental region. Lucas (1974, 1976, 1977) has examined the D-efficiency and G-efficiency of various designs including the CCD. In the case of the CCD, the D-efficiency and G-efficiency values are given in Table 4.4 for a hypercuboidal experimental region bounded by $x_i = \pm 1$ and in Table 4.5 for a hyperspherical experimental region of radius 1.

From Table 4.4 we note that the D-efficiency of a CCD is high for $k \leq 5$ (where the factorial portion consists of a full factorial array) and decreases as the number of factors increases. Table 4.5 indicates that the D-efficiency of a CCD is high for all values of k considered. Lucas (1974, 1976, 1977) indicated that an increase in the number of center points, while providing a better estimate of the experimental error variance, leads to a slight decrease in the D- and G-efficiencies.

Table 4.4 *D*- and *G*-Efficiencies of a CCD on a Hypercube[a]

Number of Factors	Number of Points[b]	*D*-Efficiency	*G*-Efficiency
2	9	0.974	0.828
3	15	0.942	0.836
4	25	0.911	0.780
5	43	0.885	0.602
5[c]	$27(I = A_1A_2A_3A_4A_5)$	0.842	0.749
6	77	0.853	0.417
6[c]	$45(I = A_1A_2A_3A_4A_5A_6)$	0.852	0.625
6[d]	$29(I = A_1A_2A_3 = A_4A_5A_6)$	0.485	0.074
7	143	0.813	0.264
7[c]	$79(I = A_1A_2A_3A_4A_5A_6A_7)$	0.845	0.442
7[d]	$47(I = A_1A_2A_3A_4A_5 = A_1A_6A_7)$	0.645	0.108
8	273	0.790	0.155
8[c]	$145(I = A_1A_2A_3A_4A_5A_6A_7A_8)$	0.821	0.283
8[d]	$81(I = A_1A_2A_3A_4A_5 = A_1A_2A_6A_7A_8)$	0.833	0.469

[a] The hypercube is bounded by $x_i = \pm 1$ ($i = 1, 2, \ldots, k$) and the axial portion of the CCD is determined by $\alpha = 1$.
[b] One center point is included.
[c] The factorial portion consists of a one-half fraction.
[d] The factorial portion consists of a one-fourth fraction.
Source: J. M. Lucas (1974). Reproduced with permission of the American Statistical Association.

Table 4.5 *D*- and *G*-Efficiencies of a CCD on a Hypersphere[a]

Number of Factors	Number of Points[b]	*D*-Efficiency	*G*-Efficiency
2	9	0.9862	0.6667
3	15	0.9914	0.6667
4	25	0.9923	0.6000
5	43	0.9860	0.4884
5[c]	$27(I = A_1A_2A_3A_4A_5)$	0.9843	0.7778
6[c]	$45(I = A_1A_2A_3A_4A_5A_6)$	0.9955	0.6222
7[c]	$79(I = A_1A_2A_3A_4A_5A_6A_7)$	0.9888	0.4557
8[c]	$145(I = A_1A_2A_3A_4A_5A_6A_7A_8)$	0.9640	0.3104
8[d]	$81(I = A_1A_2A_3A_4A_5 = A_1A_2A_6A_7A_8)$	0.9968	0.5556

[a] The hypersphere is centered at the origin and has radius one. The factorial portion has levels $\pm 1/\sqrt{k}$ and the axial portion is determined by $\alpha = 1$.
[b] One center point is included.
[c] The factorial portion consists of a one-half fraction.
[d] The factorial portion consists of a one-fourth fraction.
Source: J. M. Lucas (1976). Reproduced with permission of the American Statistical Association.

4.5.5 Equiradial Designs

An equiradial design is a design which consists of two or more sets of points where the points in each set are equidistant from the origin. Such sets of points are called *equiradial sets*. The rotatable CCD of the previous section is a member of the larger class of equiradial designs. For example, with $k = 2$ and an axial setting $\alpha = \sqrt{2}$, the four factorial points and the four axial points form one set of eight points on a circle of radius $\rho = \sqrt{2}$. The center point (or points) forms another set on a circle of radius zero. One single equiradial set cannot provide a design for fitting a second-order model. To see this, let ρ be the common distance of the points of an equiradial set from the origin (design center), so that $\sum_{i=1}^{k} x_{ui}^2 = \rho^2$ for $u = 1, 2, \ldots, N$. This implies that the column of ones and the columns corresponding to $x_1^2, x_2^2, \ldots x_k^2$ in the X-matrix for model 4.4 are linearly dependent which results in a singular $\mathbf{X'X}$ matrix.

An equiradial second-order design can be made rotatable. To show this, suppose the design consists of $s \geq 2$ equiradial sets of points. Let n_l and ρ_l denote the number of points and their common distance from the origin (design center), respectively, in the lth equiradial set ($l = 1, 2, \ldots, s$). Then

$$\sum_{u=1}^{N} \sum_{i=1}^{k} x_{ui}^2 = \sum_{l=1}^{s} n_l \rho_l^2 \tag{4.29}$$

Assuming that the design is scaled so that $[ii] = 1$ for $i = 1, 2, \ldots, k$, then from Eq. 4.29 we have

$$N^{-1} \sum_{l=1}^{s} n_l \rho_l^2 = k \tag{4.30}$$

Furthermore,

$$N^{-1} \sum_{l=1}^{s} n_l \rho_l^4 = N^{-1} \sum_{u=1}^{N} \left(\sum_{i=1}^{k} x_{ui}^2 \right)^2$$

$$= \sum_{i=1}^{k} [iiii] + \sum_{i=1}^{k} \sum_{\substack{j=1 \\ i \neq j}}^{k} [iijj] \tag{4.31}$$

From Eqs. 4.30 and 4.31 we conclude that

$$\frac{\sum_{l=1}^{s} n_l \rho_l^4}{\left(\sum_{l=1}^{s} n_l \rho_l^2 \right)^2} = \frac{\sum_{i=1}^{k} [iiii] + \sum_{i=1}^{k} \sum_{\substack{j=1 \\ i \neq j}}^{k} [iijj]}{Nk^2} \tag{4.32}$$

Now, in order for the second-order equiradial design to be rotatable, we must have $[iiii] = 3\lambda_4$ and $[iijj] = \lambda_4$ for $i, j = 1, 2, \ldots, k, i \neq j$. Hence, under rotatability, Eq. 4.32 becomes

$$\frac{\displaystyle\sum_{l=1}^{s} n_l \rho_l^4}{\left(\displaystyle\sum_{l=1}^{s} n_l \rho_l^2\right)^2} = \frac{3k\lambda_4 + k(k-1)\lambda_4}{Nk^2}$$

$$= \frac{\lambda_4(k+2)}{Nk} \tag{4.33}$$

Finally, from Eq. 4.33, λ_4 can be expressed as

$$\lambda_4 = \frac{Nk}{k+2} \frac{\displaystyle\sum_{l=1}^{s} n_l \rho_l^4}{\left(\displaystyle\sum_{l=1}^{s} n_l \rho_l^2\right)^2} \tag{4.34}$$

If one set of points is at the center, then its ρ_l value in Eq. 4.34 is zero. We recall that center point replications can be used to modify the value of λ_4. For example, if the design consists of n_0 center points plus a set of n_1 noncenter equiradial points, then Eq. 4.34 takes the form

$$\lambda_4 = \frac{k(n_0 + n_1)}{n_1(k+2)} \tag{4.35}$$

In general, the value $\lambda_4 = k/(k+2)$ should be avoided since for such a value the $\mathbf{X'X}$ matrix is singular. This occurs, for example, when the design consists of only one equiradial set with nonzero radius and no center points as can be seen from Eq. 4.35 with $n_0 = 0$.

Examples of second-order equiradial designs that are rotatable include those designs whose equiradial sets of points are characterized as being equally spaced points on a circle, a sphere, or a hypersphere and which thus form the vertices of a regular polygon, a polyhedron, or a polytope (see Box and Hunter 1957: Section 7). We consider first the case of two-dimensional designs.

Two-Dimensional Designs. For the case of two input variables, Box and Hunter (1957) showed that if a design consists of the vertices of a regular n-gon, all design moments up to order $n - 1$ are invariant under rotation. It follows that if the points of each equiradial set with a nonzero radius are equally spaced and their number is at least five, then the design moments through order four are invariant under rotation. This is sufficient to make the second-order equiradial design rotatable. Thus, by combining two or more concentric rings of equally spaced points with unequal radii and with each ring not of zero radius consisting of at least five points, we can obtain a class of two-dimensional second-order rotatable designs. For example, we can have a design consisting of $n_0 = 3$ center point replicates and one ring of nonzero radius which contains $n_1 = 5$ points (a pentagonal design with center points). A geometric configuration of this design is given in Figure 4.5. The corresponding value of λ_4 from Eq. 4.35 is equal to 0.8. In order for the pentagonal design to be orthogonal, that is, for $\lambda_4 = 1$, five center points are required. If we now consider two concentric rings with radii, $\rho_1 > 0$ and $\rho_2 > 0$, respectively, then the ratio of radii, ρ_2/ρ_1, can be chosen for given n_1 and n_2 from Eq. 4.34 in order for $\lambda_4 = 0.7844$, the value in Table 4.1 required for a uniform precision design, and for $\lambda_4 = 1$, the value required for orthogonality. Table 4.6 gives some such values of ρ_2/ρ_1.

The condition that each ring of nonzero radius, in a two-dimensional equiradial design consisting of s rings of equally spaced points, be composed of at

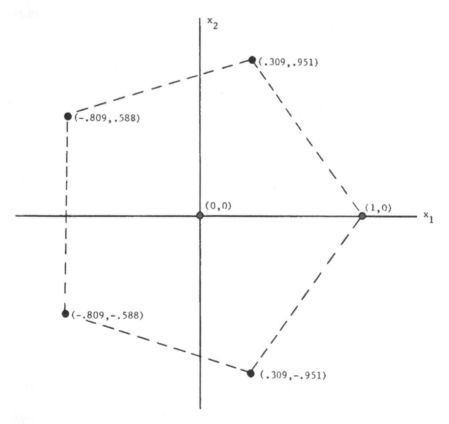

Figure 4.5 A pentagonal design with a center point.

Table 4.6 Ratio of Radii for Two Concentric Rings of a Rotatable Equiradial Design in Two Dimensions (No Center Points)

n_1	5	5	5	6	6	7
n_2	6	7	8	7	8	8
ρ_2/ρ_1 (uniform precision)	0.414	0.438	0.454	0.407	0.430	0.404
ρ_2/ρ_1 (orthogonality)	0.204	0.267	0.304	0.189	0.250	0.176

Source: G. E. P. Box and J. S. Hunter (1957). Reproduced with permission of the Institute of Mathematical Statistics.

least five points is *sufficient* but not necessary for second-order rotatability. In other words, if each ring of nonzero radius points consists of at least five equally spaced points, then the entire collection of s rings forms a second-order rotatable design. However, if the collection of s rings forms a second-order rotatable design, then it is not necessary that each ring should consist of at least five points.

Example. Let us consider s sets of points where each set consists of three points which form the vertices of an equilateral triangle with center coincident with the origin. Suppose that in the ith set the common distance of the three points from the origin is $\sqrt{2}\rho_i$. Suppose further that a line passing through the origin and extending to one of the points of the set makes an angle θ_i with the x_2 axis ($i = 1, 2, \ldots, s$). Box and Hunter (1957: 220) show that the second-order moment matrix for the collection of s such sets is

$$N^{-1}\mathbf{X}'\mathbf{X} = N^{-1}\sum_{i=1}^{s}\mathbf{X}_i'\mathbf{X}_i = \frac{1}{s}\sum_{i=1}^{s}\mathbf{W}_i$$

where $\mathbf{X}_i$ is the portion of the $\mathbf{X}$ matrix corresponding to the ith set and $\mathbf{W}_i$ is the matrix

$$\mathbf{W}_i = \begin{array}{c} \\ \\ \\ \\ \\ \\ \\ \end{array} \begin{array}{cccccc} 1 & x_1 & x_2 & x_1^2 & x_2^2 & x_1x_2 \\ \begin{bmatrix} 1 & 0 & 0 & \rho_i^2 & \rho_i^2 & 0 \\ 0 & \rho_i^2 & 0 & \rho_i^3 a_i & -\rho_i^3 a_i & -\rho_i^3 b_i \\ 0 & 0 & \rho_i^2 & -\rho_i^3 b_i & \rho_i^3 b_i & -\rho_i^3 a_i \\ \rho_i^2 & \rho_i^3 a_i & -\rho_i^3 b_i & \frac{3}{2}\rho_i^4 & \frac{1}{2}\rho_i^4 & 0 \\ \rho_i^2 & -\rho_i^3 a_i & \rho_i^3 b_i & \frac{1}{2}\rho_i^4 & \frac{3}{2}\rho_i^4 & 0 \\ 0 & -\rho_i^3 b_i & -\rho_i^3 a_i & 0 & 0 & \frac{1}{2}\rho_i^4 \end{bmatrix} \end{array} \; i = 1, 2, \ldots, s$$

(4.36)

where $a_i = 2^{-1/2}\sin 3\theta_i$ and $b_i = 2^{-1/2}\cos 3\theta_i$. The moment matrix, $N^{-1}\mathbf{X}'\mathbf{X}$, will have the form required for rotatability if

$$\sum_{i=1}^{s}\rho_i^3\sin 3\theta_i = 0 \qquad \sum_{i=1}^{s}\rho_i^3\cos 3\theta_i = 0$$

(4.37)

Since the design consisting of the s sets is rotatable, then from Eq. 4.34 the value of λ_4 for such a design is

$$\lambda_4 = \frac{s}{2} \frac{\sum_{i=1}^{s} \rho_i^4}{(\sum_{i=1}^{s} \rho_i^2)^2} \tag{4.38}$$

If $s > 2$, Eq. 4.37 can provide an infinite collection of rotatable equiradial designs, where each design consists of s sets of three points that have different ρ_i and θ_i values. For example, let $s = 3$ and let $\rho_1 = \rho$, $\theta_1 = 0$; $\rho_2 = 2\rho$, $\theta_2 = \pi/12$. Then, from Eq. 4.37 the values of ρ_3 and θ_3 must satisfy

$$\rho_3^3 \sin 3\theta_3 = -4\sqrt{2}\rho^3$$
$$\rho_3^3 \cos 3\theta_3 = -(1 + 4\sqrt{2})\rho^3 \tag{4.39}$$

A solution to Eq. 4.39 is given by $\rho_3 = 2.0595\rho$ and $\theta_3 = 0.4081\pi$, so that the three radii ρ_1, ρ_2, and ρ_3 are multiples of one another. The λ_4 value from Eq. 4.38 is equal to 0.6145, which is different from its singular value $\lambda_4 = 0.5$. If, however, $s = 2$, then to satisfy Eq. 4.37, ρ_1 must be equal to ρ_2 and the two sets of three points each form the vertices of a hexagon with λ_4 equal to its singular value 0.5, as can be seen from Eq. 4.38.

Equiradial Designs in More Than Two Dimensions. Box and Hunter (1957) list several examples of equiradial designs in three or more dimensions. In three dimensions, equiradial sets of points can be formed by the

1. Four vertices of the tetrahedron $(-1, -1, 1)$, $(1, -1, -1)$, $(-1, 1, -1)$, and $(1, 1, 1)$
2. Six vertices of the octahedron $(\pm\sqrt{3}, 0, 0)$, $(0, \pm\sqrt{3}, 0)$, $(0, 0, \pm\sqrt{3})$
3. Eight vertices of the cube $(\pm1, \pm1, \pm1)$
4. Twelve vertices of the icosahedron $(0, \pm a, \pm b)$, $(\pm b, 0, \pm a)$, $(\pm a, 0, \pm b)$
5. Twenty vertices of the dodecahedron $(0, \pm1/c, \pm c)$, $(\pm c, 0, \pm1/c)$, $(\pm1/c, \pm c, 0)$, $(\pm1, \pm1, \pm1)$

where $a = 1.473$, $b = 0.911$, and $c = 1.618$. The distance of each point from the origin $(0, 0, 0)$ in each of the above arrangements is $\rho = \sqrt{3}$. The vertices of the tetrahedron, octahedron, and cube do not individually support a rotatable design of order two. They do, however, support rotatable designs of order one. The vertices of the icosahedron and the dodecahedron individually form a singular rotatable design of order 2 with a singular value of $\lambda_4 = 3/5$. As in the two-dimensional case, we may consider augmenting the icosahedral and dodecahedral sets with either center points or with one another (if $\rho_1 \neq \rho_2$). We can also consider combining sets of points, which individually do not support a rotatable design of order 2, such as combining the eight vertices of a cube with the six vertices of the octahedron as long as $\rho_1 \neq \rho_2$ (see Figure 4.6).

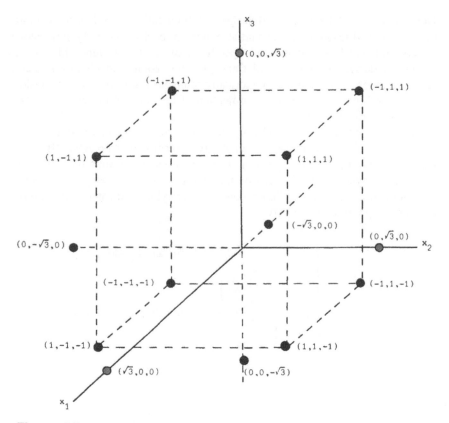

Figure 4.6 The eight vertices of a cube and the six vertices of an octahedron.

In more than three dimensions, equiradial sets can be formed by considering higher-dimensional analogues of the equiradial sets in three dimensions. These include the regular simplex (k-dimensional analogue of the tetrahedron with $k + 1$ vertices), the cross-polytope (k-dimensional analogue of the octahedron with $2k$ vertices), and the hypercube (k-dimensional analogue of the cube with 2^k vertices).

4.5.6 Cylindrically Rotatable Designs

Herzberg (1966) introduced this class of designs, which are less restrictive than rotatable designs and require fewer numbers of points while retaining some degree of rotatability.

Definition 4.1. A k-dimensional design is cylindrically rotatable if the variance of the predicted response is constant at points on the same $(k - 1)$-dimensional hypersphere which is centered on a specified axis. Such a design is identical to a rotatable design of the same order except in the required levels of one factor. An example of a second-order cylindrically rotatable design in $k = 3$ dimensions is shown in Figure 4.7, where the projection of the points onto the x_1, x_2 plane produces a rotatable design.

General conditions of cylindrical rotatability for a model of order d in k variables are similar to those of rotatability described in Appendix 2B. If the variance of the predicted response is to be kept equal at points on a $(k - 1)$-dimensional hypersphere centered on the axis $x_1 = \cdots = x_{i-1} = x_{i+1} = \cdots = x_k = 0$, then the necessary and sufficient condition for cylindrical rotatability requires that the design moments satisfy

$$[1^{\delta_1} 2^{\delta_2} \ldots k^{\delta_k}] = \begin{cases} 0 & \text{for any } \delta_i \text{ odd}, j \neq i \\[4ex] \dfrac{\lambda_{\delta,\delta_i} \left(\prod\limits_{j=1}^{k} \delta_j! \right)}{2^{\delta/2} \prod\limits_{\substack{j=1 \\ j \neq i}}^{k} \left(\dfrac{1}{2} \delta_j \right)!} & \text{for all } \delta_j \text{ even}, j \neq i \end{cases}$$

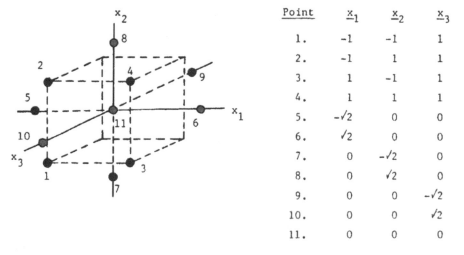

Point	x_1	x_2	x_3
1.	-1	-1	1
2.	-1	1	1
3.	1	-1	1
4.	1	1	1
5.	$-\sqrt{2}$	0	0
6.	$\sqrt{2}$	0	0
7.	0	$-\sqrt{2}$	0
8.	0	$\sqrt{2}$	0
9.	0	0	$-\sqrt{2}$
10.	0	0	$\sqrt{2}$
11.	0	0	0

Figure 4.7 A cylindrically rotatable design in $k = 3$ dimensions with respect to the x_3 axis.

where $\delta = \sum_{j=1}^{k} \delta_j$ is the order of the moment, $0 \le \delta \le 2d$, and $\lambda_{\delta_1 \delta_i}$ is a constant which depends on δ and δ_i and can be chosen so that the cylindrically rotatable design satisfies other design criteria.

4.5.7 Asymmetric Rotatable Designs

A symmetric design is one in which all the factors have the same number of levels. By contrast, an asymmetric design is one in which the number of levels is not the same for all the factors. Rotatable designs that have been discussed thus far are of the symmetric type. Rotatable designs that are asymmetric are useful in situations where it is desirable to maintain rotatability but it is difficult or infeasible to have the same number of levels for all the factors.

Mehta and Das (1968) showed that a second-order asymmetric rotatable design can be obtained from a second-order symmetric rotatable design by means of an orthogonal transformation. This transformation permits the factors to have different numbers of levels while preserving rotatability. It is not always possible, however, to achieve any specified number of levels for each factor.

As an example, let us consider the second-order rotatable central composite design in four factors $(\pm a, \pm a, \pm a, \pm a)$, $(\pm 2a, 0, 0, 0)$, $(0, \pm 2a, 0, 0)$, $(0, 0, \pm 2a, 0)$, $(0, 0, 0, \pm 2a)$, $(0, 0, 0, 0)$, where a is a nonzero constant. Let us consider transforming these design points using the transformation

$$\mathbf{z}'_u = \mathbf{x}'_u \mathbf{B}' \qquad u = 1, 2, \ldots, 25$$

where $\mathbf{x}'_u$ denotes the coordinates of the uth design point in the original settings and $\mathbf{z}'_u$ denotes the coordinates of the new design point $(u = 1, 2, \ldots, 25)$. The matrix $\mathbf{B}$ is an orthogonal matrix of the form

$$\mathbf{B} = \begin{bmatrix} \frac{1}{\sqrt{2}} & \frac{-1}{\sqrt{2}} & 0 & 0 \\ \frac{1}{\sqrt{2}} & \frac{1}{\sqrt{2}} & 0 & 0 \\ 0 & 0 & \frac{1}{\sqrt{5}} & \frac{2}{\sqrt{5}} \\ 0 & 0 & \frac{-2}{\sqrt{5}} & \frac{1}{\sqrt{5}} \end{bmatrix}$$

This transformation produces a rotatable design in four factors in which the first two factors each has the three levels 0 and $\pm \sqrt{2}a$ while the remaining two factors each has the seven levels 0, $\pm a/\sqrt{5}$, $\pm 2a/\sqrt{5}$, and $\pm 3a/\sqrt{5}$. A number of second-order asymmetric rotatable designs are given by Mehta and Das (1968).

4.5.8 Other Second-Order Designs

Box-Draper Saturated Designs. A design and its corresponding model are called saturated when the number of design points is exactly equal to the number of terms in the model to be fitted. Box and Draper (1971) list several saturated second-order designs for k variables (that is; the number of design points is equal to $p = \frac{1}{2}(k + 1)(k + 2)$), and suggest optimal settings of the design points by using the criterion of maximizing the determinant $|\mathbf{X}'\mathbf{X}|$ for the cases of $k = 2$ and 3 variables, where $\mathbf{X}$ is the matrix given in Eq. 4.4. In these designs, the design variables are coded so that $-1 \leq x_i \leq 1$ for $i = 1, 2, \ldots, k$. For $k = 2$, the six design points of Box and Draper are $(-1, -1)$, $(1, -1)$, $(-1, 1)$, $(-\delta, -\delta)$, $(1, 3\delta)$, and $(3\delta, 1)$, where $\delta = (4 - \sqrt{13})/3 = 0.1315$ (see Figure 4.8a). For $k = 3$, the 10 design points are $(-1, -1, -1)$, $(1, -1, -1)$, $(-1, 1, -1)$, $(-1, -1, 1)$, $(-1, \lambda, \lambda)$, $(\lambda, -1, \lambda)$, $(\lambda, \lambda, -1)$, $(\mu, 1, 1)$, $(1, \mu, 1)$, $(1, 1, \mu)$, where $\lambda = 0.1925$ and $\mu = -0.2912$ (see Figure 4.8b).

Box and Draper (1974) generalized these designs for $k \geq 4$. The generalized designs, however, are not optimal in the sense of maximizing $|\mathbf{X}'\mathbf{X}|$ for $k = 4$ and $k \geq 7$ while for $k = 5$ and 6, the question of their optimality remains unresolved. Box and Draper pointed out that Dubova and Federov (1972) had found a better design for $k = 4$ and that Jack Kiefer had established (in an unpublished correspondence) that their generalized designs cannot be optimal for $k \geq 7$. Box and Draper argued that since no better designs (better in the sense of producing a larger value of $|\mathbf{X}'\mathbf{X}|$ but containing the same number of points) are known for $k \geq 5$, then their generalization of the designs for $k = 2$ and 3 variables provides the only saturated designs with reasonably high D-efficiency when the number of variables exceeds four.

Uniform Shell Designs. Uniform shell designs were developed by Doehlert (1970) and Klee (Doehlert and Klee 1972). They consist of points uniformly spaced on concentric spherical shells. For this reason they are called *uniform shell designs*. The requirement that the design points be uniformly spaced was motivated by Scheffé (1963), who presented good arguments for using designs with an equally spaced distribution of points. Many levels of each factor are generally needed for these designs.

Doehlert (1970) described a general method for generating these designs. We demonstrate this method for $k = 2$ and 3 factors. For $k = 2$, we consider the equilateral triangle whose vertices have the coordinates $(0, 0)$, $(1.0, 0)$, $(0.5, 0.866)$. If we subtract each point from the other points we obtain the following four points: $(-1.0, 0)$, $(-0.5, -0.866)$, $(-0.5, 0.866)$, $(0.5, -0.866)$. The seven points thus obtained form a regular hexagon with a center point as in Figure 4.9. The initial points of the equilateral triangle are labelled with diamonds in this figure.

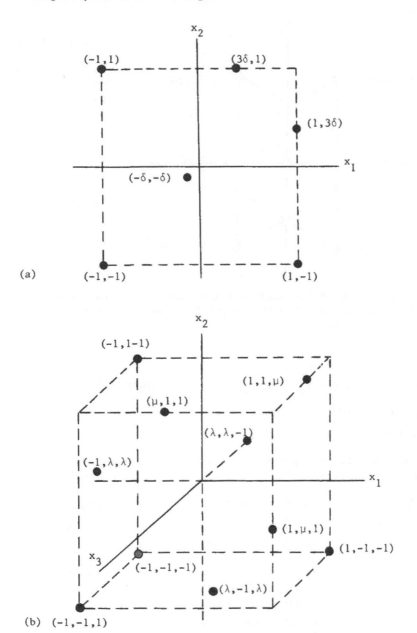

Figure 4.8 Box-Draper saturated designs for a second-order model in k variables ($\mu = -0.2912$, $\lambda = 0.1925$, $\delta = 0.1315$); (a) $k = 2$; (b) $k = 3$.

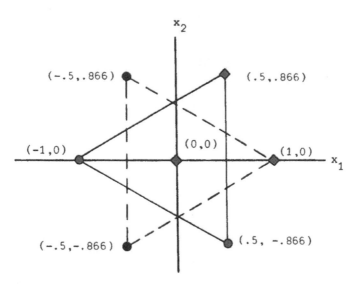

Figure 4.9 A uniform shell design in two dimensions. The full design is a double-simplex with a center point. [D. H. Doehlert (1970). Reproduced with permission of the Royal Statistical Society.]

For $k = 3$, we consider the tetrahedron with the vertices

(0, 0, 0)	(0.5, 0.866, 0)
(1.0, 0, 0)	(0.5, 0.289, 0.816)

Again, if each of these four points is subtracted from each other point, we obtain the following nine points

(−1.0, 0, 0)	(0.5, −0.866, 0)	(0.5, −0.289, −0.816)
(−0.5, −0.866, 0)	(−0.5, −0.289, −0.816)	(0, −0.577, 0.816)
(−0.5, 0.866, 0)	(−0.5, 0.289, 0.816)	(0, 0.577, −0.816)

The collection of 13 points form a cuboctahedron with a center point.

In general, the uniform shell designs can be generated from the points of a regular simplex by taking differences among its points. Designs for two to 10 factors are given in Doehlert (1970). If the number of factors is k, then in addition to the center point, a total of $k^2 + k$ design points lie on a hypersphere of radius one. Thus, a single shell of points plus the center point (with possible replications at the center point) would suffice to estimate the parameters of the second-order model.

Hoke Designs. Hoke (1974) presented economical second-order designs for $k \geq 3$ factors at 3 levels. These designs are based on saturated irregular fractions of the 3^k factorial that are partially balanced. Hoke (1974) compared his best design with Box and Behnken's (1960) and Hartley's (1959) designs that were of comparable size. The comparison was made by using the $\text{tr}[(\mathbf{X'X})^{-1}]$ and $|\mathbf{X'X}|$ criteria. On the basis of this comparison Hoke concluded that his designs compared favorably with the latter designs.

Hybrid Designs. Hybrid designs, introduced by Roquemore (1976), are designs that satisfy (or nearly satisfy) several desirable criteria. A hybrid design for k factors can be constructed using a central composite design (with a center point) for $k - 1$ factors augmented with a column for factor k and, possibly, another row or two. These designs can achieve the same degree of orthogonality as a central composite design while being near saturated, and near rotatable. Eight hybrid designs were presented by Roquemore (1976) for $k = 3$, 4, and 6 factors. A hybrid design in 3 factors is shown in Figure 4.10.

Lucas (1976) compared the performance of several types of second-order designs, namely, central composite designs, Box-Behnken designs, uniform shell designs, Hoke designs, and Box-Draper saturated designs, on the basis of their D- and G-efficiencies. He concluded that all of the compared designs had high efficiencies, and that while more efficient designs were possible, they either remained undiscovered or required significantly more design points. It was also

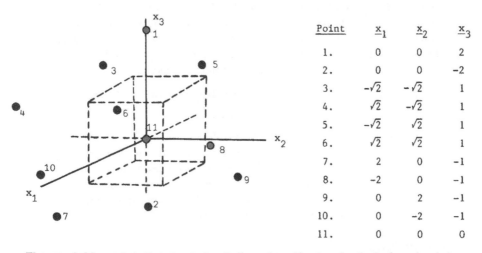

Point	x_1	x_2	x_3
1.	0	0	2
2.	0	0	-2
3.	$-\sqrt{2}$	$-\sqrt{2}$	1
4.	$\sqrt{2}$	$-\sqrt{2}$	1
5.	$-\sqrt{2}$	$\sqrt{2}$	1
6.	$\sqrt{2}$	$\sqrt{2}$	1
7.	2	0	-1
8.	-2	0	-1
9.	0	2	-1
10.	0	-2	-1
11.	0	0	0

Figure 4.10 A hybrid design in $k = 3$ dimensions (the dotted cube is the unit cube). [K. G. Roquemore (1976). Reproduced with permission of the American Statistical Association.]

noted that some of the Hoke designs fared well in comparison to the central composite designs and required fewer points. However, as the number of factors increased, the relative efficiency of the Hoke designs decreased. Furthermore, the former designs had higher D-efficiency values than the Box-Draper designs for $k \geq 4$. In a hyperspherical region of radius one, the study by Lucas (1976) indicates that the three designs; central composite, Box-Behnken, and uniform shell; have high D-efficiency and moderate-to-high G-efficiency. The uniform shell design, however, is not as efficient as the other two designs. The addition of center point replications leads to a slight decrease in the D- and G-efficiencies, but is very beneficial in providing a good estimate of the error variance, and in checking the adequacy of the fitted second-order model.

4.6 A NUMERICAL EXAMPLE: THE EFFECT OF HEAT AND OTHER FACTORS UPON THE FOAMING PROPERTIES OF WHEY PROTEIN CONCENTRATES

Richert et al. (1974) investigated the effects of heating temperature (X_1), pH level (X_2), Redox potential (X_3), sodium oxalate (X_4), and sodium lauryl sulfate (X_5) on foaming properties of whey protein concentrates. Several response variables were studied, one of which was the percent of undenatured protein, Y. A central composite design with six center points was employed in order to fit a second-order model. The design in coded form and the observed data values are listed in Table 4.7. The original and coded levels of the input variables are listed in Table 4.8. Table 4.9 contains the least squares estimates of the second-order model's parameters, their corresponding standard errors, the coefficient of determination, R^2, and the error mean square, MSE. An examination of the parameter estimates and their standard errors reveals that the terms containing x_4 and x_5 in the model are nonsignificant. By fitting a reduced model not containing these variables to the same data set we obtain the model

$$\hat{Y}(\mathbf{x}) = 77.17 - 10.12x_1 - 8.68x_2 - 0.10x_3 - 4.10x_1^2 - 4.72x_2^2 - 2.30x_3^2$$
$$- 6.21x_1x_2 + 2.77x_1x_3 - 1.68x_2x_3$$

For this model, MSE = 18.23 and $R^2 = 0.9395$.

The factorial portion of the central composite design used in this experiment consists of a one-half fraction of 2^5. The composite design is rotatable since the axial setting, α, is the fourth root of the number of factorial points, that is, $\alpha = \sqrt[4]{16}$. We also note that the design has the uniform precision property since the number of center point replications is six (see Table 4.3).

EXERCISES

4.1 Consider a 3^2 factorial design with each factor having the levels $-g$, 0, g, where $g = \sqrt{3/2}$. This design is not rotatable since $\kappa = [iiii]/[1122]$ is not

Table 4.7 Experimental Design (Coded) and Response Values

x_1	x_2	x_3	x_4	x_5	Y (%)
-1	-1	-1	-1	1	80.6
1	-1	-1	-1	-1	67.9
-1	1	-1	-1	-1	83.1
1	1	-1	-1	1	38.1
-1	-1	1	-1	-1	79.7
1	-1	1	-1	1	74.7
-1	1	1	-1	1	71.2
1	1	1	-1	-1	36.8
-1	-1	-1	1	-1	81.7
1	-1	-1	1	1	66.8
-1	1	-1	1	1	73.0
1	1	-1	1	-1	40.5
-1	-1	1	1	1	74.9
1	-1	1	1	-1	74.2
-1	1	1	1	-1	63.5
1	1	1	1	1	42.8
-2	0	0	0	0	80.9
2	0	0	0	0	42.4
0	-2	0	0	0	73.4
0	2	0	0	0	45.0
0	0	-2	0	0	66.0
0	0	2	0	0	71.7
0	0	0	-2	0	77.5
0	0	0	2	0	76.3
0	0	0	0	-2	67.4
0	0	0	0	2	86.5
0	0	0	0	0	77.4
0	0	0	0	0	74.6
0	0	0	0	0	79.8
0	0	0	0	0	78.3
0	0	0	0	0	74.8
0	0	0	0	0	80.9

Source: S. H. Richert, C. V. Morr, and C. M. Cooney (1974). Reprinted from Journal of Food Science. Copyright © by Institute of Food Technologies.

Table 4.8 Original and Coded Levels of Independent Variables

Variable	Original	Coded				
		-2	-1	0	1	2
Heating temperature	X_1 (°C/30 min)	65.0	70.0	75.0	80.0	85.0
pH	X_2	4.0	5.0	6.0	7.0	8.0
Redox potential	X_3 (volt)	-0.025	0.075	0.175	0.275	0.375
Sodium oxalate	X_4 (Molar)	0.0	0.0125	0.025	0.0375	0.05
Sodium lauryl sulfate	X_5 (% of solids)	0.0	0.05	0.10	0.15	0.20

Source: S. H. Richert, C. V. Morr, and C. M. Cooney (1974). Reprinted from Journal of Food Science. Copyright © by Institute of Food Technologies.

Table 4.9 The Least Squares Fit and Parameter Estimates

Model Term	Estimate
Intercept	77.79 (1.86)[a]
x_1	-10.12 (0.95)
x_2	-8.68 (0.95)
x_3	-0.10 (0.95)
x_4	-0.71 (0.95)
x_5	1.37 (0.95)
x_1^2	-4.16 (0.86)
x_2^2	-4.77 (0.86)
x_3^2	-2.36 (0.86)
x_4^2	-0.34 (0.86)
x_5^2	-0.33 (0.86)
$x_1 x_2$	-6.21 (1.16)
$x_1 x_3$	2.77 (1.16)
$x_1 x_4$	1.77 (1.16)
$x_1 x_5$	0.71 (1.16)
$x_2 x_3$	-1.68 (1.16)
$x_2 x_4$	-0.26 (1.16)
$x_2 x_5$	0.48 (1.16)
$x_3 x_4$	0.04 (1.16)
$x_3 x_5$	1.51 (1.16)
$x_4 x_5$	0.03 (1.16)
MSE	21.69
R^2	0.964

[a]The number in parentheses is the standard error.

equal to 3 for $i = 1$, 2. In this case, κ can be considered as a measure of departure from rotatability. Augment this design with the four axial points $(\pm h, 0)$, $(0, \pm h)$.

(a) Compute κ for the augmented design.

(b) Compare contour plots of $\text{Var}[\hat{Y}(\mathbf{x})]/\sigma^2$ for the original 3^2 design and the augmented design using the value $h = 1.6$. What conclusion can you draw from this comparison?

(c) Can h be chosen so that the augmented design is rotatable?

4.2 Construct a Box-Behnken design for fitting a second order model in $k = 5$ input variables. Is this design rotatable?

4.3 Consider a CCD for $k = 3$ input variables where each design point in its factorial portion is replicated two times.

(a) Determine the value of the axial setting α so that the design is rotatable.

(b) Determine the number of center points needed to make this CCD rotatable as well as orthogonal.

4.4 Suppose that the fitted model is of the second order in $k = 2$ input variables and that the true model is of the third order.

(a) Obtain an expression for the bias in the linear parameters of the fitted model given that the design used is a CCD.

(b) Show that the bias in (a) increases with the number of center points.

4.5 Consider the CCD used in the example described in Section 4.6 along with the data given in Table 4.7.

Suppose the design is augmented with one more center point (thus, the total number of center points is now 7), and let the corresponding response value be equal to 76.7. Conduct an analysis of variance including the sources of variation due to:

(i) Linear terms in the second-order model.

(ii) Quadratic terms in the second-order model.

(iii) Lack of fit.

4.6 In the analysis of the residual oil values listed in Table 4.2, the adequacy of the fitted second-order model of Eq. 4.17 was tested and found to suffer from lack of fit at the 0.01 level of significance. It was then suggested that the terms $\beta_{133}x_1x_3^2$ and $\beta_{233}x_2x_3^2$ be added to the model and the augmented model fitted. Fit the augmented model and comment on your findings (i.e., comment on whether or not the augmented model is any better than Eq. 4.17 and why). Suggest a final model form for prediction purposes by including first-degree and only significant ($P < 0.05$) second- and higher-degree terms in the model and explain the pros and cons of using a reduced higher-order (≥ 3) model form for prediction purposes as well as for interpreting the effects of the factors included in the model.

REFERENCES AND BIBLIOGRAPHY

Bose, R. C. and N. R. Draper (1959). Second Order Rotatable Designs in Three Dimensions, *Ann. Math. Statist.*, **30**, 1097–1112.

Box, G. E. P. and D. W. Behnken (1960). Some New Three Level Designs for the Study of Quantitative Variables, *Technometrics*, **2**, 455–475.

Box, G. E. P. and N. R. Draper (1975). Robust Designs, *Biometrika*, **62**, 347–352.

Box, G. E. P. and J. S. Hunter (1957). Multi-Factor Experimental Designs for Exploring Response Surfaces, *Ann. Math. Statist.*, **28**, 195–241.

Box, G. E. P. and K. B. Wilson (1951). On the Experimental Attainment of Optimum Conditions (with discussion), *J. Roy. Statist. Soc.*, **B13**, 1–45.

Box, M. J. and N. R. Draper (1971). Factorial Designs, the $|X'X|$ Criterion and Some Related Matters, *Technometrics*, **13**, 731–742.

Box, M. J. and N. R. Draper (1974). On Minimum-Point Second-Order Designs, *Technometrics*, **16**, 613–616.

Connor, W. S. and M. Zelen (1959). Fractional Factorial Experimental Designs for Factors at Three Levels. National Bureau of Standards, Washington, D.C., Applied Mathematics Series, No. 54.

DeBaun, R. M. (1959). Response Surface Designs for Three Factors at Three Levels, *Technometrics*, **1**, 1–8.

Doehlert, D. H. (1970). Uniform Shell Designs, *J. Roy. Statist. Soc.*, **C19**, 231–239.

Doehlert, D. H. and V. L. Klee (1972). Experimental Designs Through Level Reduction of the *d*-Dimensional Cuboctahedron, *Discrete Mathematics*, **2**, 309–334.

Draper, N. R. (1960). Second Order Rotatable Designs in Four or More Dimensions, *Ann. Math. Statist.*, **31**, 23–33.

Draper, N. R. (1982). Center Points in Second-Order Response Surface Designs, *Technometrics*, **24**, 127–133.

Dubova, I. S. and V. V. Fedorov (1972). Tables of Optimum Designs II (Saturated *D*-Optimal Designs on a Cube). Preprint No. 40, in Russian. Issued by Interfaculty Laboratory of Statistical Methods, Moscow University.

Dykstra, O. (1960). Partial Duplication of Response Surface Designs, *Technometrics*, **2**, 185–195.

Hartley, H. O. (1959). Smallest Composite Designs for Quadratic Response Surfaces, *Biometrics*, **15**, 611–624.

Herzberg, A. M. (1966). Cylindrically Rotatable Designs. *Ann. Math. Statist.*, **37**, 242–247.

Hoke, A. T. (1974). Economical Second-Order Designs Based on Irregular Fractions of the 3^n Factorial, *Technometrics*, **16**, 375–384.

Lucas, J. M. (1974). Optimum Composite Designs, *Technometrics*, **16**, 561–567.

Lucas, J. M. (1976). Which Response Surface Design Is Best, *Technometrics*, **18**, 411–417.

Lucas, J. M. (1977). Design Efficiencies for Varying Numbers of Center Points, *Biometrika*, **64**, 145–147.

McLean, R. A. and V. L. Anderson (1984). *Applied Factorial and Fractional Designs*, New York: Marcel Dekker.

Mehta, J. S. and M. N. Das (1968). Asymmetric Rotatable Designs and Orthogonal Transformations, *Technometrics*, **10**, 313–322.

Montgomery, D. C. (1991). *Design and Analysis of Experiments*, 3rd ed., New York: John Wiley.

Myers, R. H. (1976). *Response Surface Methodology*, Blacksburg, VA: Author.

Richert, S. H., C. V. Morr, and C. M. Cooney (1974). Effect of Heat and Other Factors Upon Foaming Properties of Whey Protein Concentrates, *J. Food Science*, **39**, 42–48.

Rizvi, S. S. H. (1994). *Supercritical Fluid Processing of Food and Biomaterials*, Blackie Academic & Professional, New York: Chapman & Hall.

Roquemore, K. G. (1976). Hybrid Designs for Quadratic Response Surfaces, *Technometrics*, **18**, 419–423.

Scheffé, H. (1963). The Simplex-Centroid Design for Experiments With Mixtures, *J. Roy. Statist. Soc.*, **B25**, 235–263.

5

Determining Optimum Conditions

5.1 INTRODUCTION

In this chapter we present several different procedures that can be used to find the settings of the input variables which produce the most desirable response values. These response values may be the maximum yield or the highest level of quality coming off the production line. Similarly, we may seek the variables settings that minimize the cost of making the product. In any case, the set of values of the input variables which result in the most desirable response values is called *the set of optimum conditions*.

The first step in the process of seeking optimum conditions is to identify the input variables that have the greatest influence on the response. Generally, the fewer the number of variables that have an effect on the response, the easier it is to identify them. Once the important variables are discovered, the next step is to postulate a model which expresses the response of interest as a function of the variables. If nothing, or even if very little, is known of the relationship between the response variable and the important input variables, then the simplest form of model equation (usually a first-degree polynomial) is postulated. The first-order model provides the basis for performing an initial set of experiments, which, upon completion, may suggest the fitting of a different model form along with performing further experimentation. If at any time in the model developing process it is discovered that further experimentation appears uneconomical, the procedure is terminated. The sequence of fitting and testing the model forms and the eventual selection of a model are the prelude to the determination of the optimum operating conditions for a process.

The strategy in developing an empirical model through a sequential program of experimentation is as follows:

1.   The simplest polynomial model is fitted (a first-order model) to a set of data collected at the points of a first-order design. If extra points are included from which data are collected and an estimate of the error variance is available, the model is tested for adequacy of fit.

2.   If the fitted first-order model is adequate, the information provided by the fitted model is used to locate areas in the experimental region, or outside the experimental region, but within the boundaries of the operability region, where more desirable values of the response are suspected to be.

3.   In the new region, the cycle is repeated in that the first-order model is fitted and tested for adequacy of fit. If nonplanarity in the surface shape is detected through the test for lack of fit of the first-order model, the model is upgraded by adding cross-product terms and/or pure quadratic terms to it. The first-order design is likewise augmented with points to support the fitting of the upgraded model.

4.   If curvature of the surface is detected and a fitted second-order model is found to be appropriate, the second-order model is used to map or describe the shape of the surface, through a contour plot, in the experimental region. If the optimal or most desirable response values are achieved within the boundaries of the experimental region, then locating the best values as well as the settings of the input variables that produce the best response values is the next order of business.

5.   Finally, in the region where the most desirable response values are suspected to be found, additional experiments are performed to verify that this is so. Once the location of the most desirable response values is determined, the shape of the response surface in the immediate neighborhood of the optimum is described.

Sections 5.2 through 5.5 illustrate the steps of the sequential procedure just described. In Section 5.2, we introduce a set of artificial data and present the analysis surrounding the fitting of a first-order model. Additional data values are then provided in Sections 5.3 and 5.4 to enable us to end up with a fitted second-order model. For simplicity of presentation we shall assume there is only one response variable to be studied although in practice there can be several response variables that are under investigation simultaneously. In Chapter 7, methods are presented for analyzing multiresponse data.

We shall now review the analysis of a first-order surface which is represented by the equation of a hyperplane in k coded variables

$$\eta = \beta_0 + \beta_1 x_1 + \beta_2 x_2 + \cdots + \beta_k x_k \tag{5.1}$$

As mentioned in Chapter 3, the model, $Y = \eta + \varepsilon$, is appropriate when

1. The measured response values are collected over a limited region of the overall space of the k input variables and the assumption that the surface can be approximated reasonably well by the hyperplane is valid, or
2. When nothing is known about the shape of the surface in the region of interest and, therefore, the hyperplane represents an inexpensive first-order approximation.

Designs for fitting a first-order model of the form in Eq. 5.1 were discussed in Chapter 3.

5.2 THE ANALYSIS OF A FITTED FIRST-ORDER SURFACE

We begin by introducing the following hypothetical example. In a particular chemical reaction setting, the temperature, X_1, of the reaction and the length of time, X_2, of the reaction are known to affect the reaction rate and thus the percent yield. An experimenter, interested in determining if an increase in the percent yield is possible, decides to perform a set of experiments by varying the reaction temperature and reaction time while holding all other factors fixed. The initial set of experiments consists of looking at two levels of temperature (70° and 90°C) and two levels of time (30 sec and 90 sec). The response of interest is the percent yield, which is recorded in terms of the amount of residual material burned off during the reaction resulting in a measure of the purity of the end product. The process currently operates in a range of percent purity between 55% and 75%, but it is felt that a higher percent yield is possible.

For the initial set of experiments, the two-variable model to be fitted is

$$\%\text{Yield} = \alpha_0 + \alpha_1 X_1 + \alpha_2 X_2$$

Each of the four temperature-time settings, 70°–30 sec, 70°–90 sec, 90°–30 sec, 90°–90 sec, is replicated twice and the percent yield recorded for each of the eight trials. The measured yield values associated with each temperature-time combination are listed in Table 5.1 along with the values of the coded variables x_1 and x_2, which are defined as

$$x_1 = \frac{X_1 - 80}{10} \qquad x_2 = \frac{X_2 - 60}{30} \tag{5.2}$$

The location of the four-point design, drawn relative to the overall region of possible temperature-time combinations, is shown in Figure 5.1. Also displayed in Figure 5.1 are the contours (dashed curves) of the true shape of the percent yield surface. The shape of the surface is unknown to the experimenter.

Table 5.1 Percent Yield Values Observed at Each of the Four Temperature-Time Combinations Including Settings of the Coded Variables Defined in Eq. 5.2

Original Variables		Coded Variables		Percent Yield
Temperature (°C)	Time (sec)	x_1	x_2	Y
70	30	-1	-1	49.8, 48.1
90	30	$+1$	-1	57.3, 52.3
70	90	-1	$+1$	65.7, 69.4
90	90	$+1$	$+1$	73.1, 77.8

Figure 5.1 The initial four temperature and time settings and the observed percent yield values overlaying the contours (dashed curves) of hypothetical percent yield surface.

Expressed in terms of the coded variables in Eq. 5.2, the observed percent yield values are modeled as

$$Y = \beta_0 + \beta_1 x_1 + \beta_2 x_2 + \varepsilon \tag{5.3}$$

where the parameters β_0, β_1, and β_2 are functions of α_0, α_1 and α_2 in Eq. 5.1, respectively. In particular,

$$\beta_0 = \alpha_0 + 80\alpha_1 + 60\alpha_2 \qquad \beta_1 = 10\alpha_1 \qquad \beta_2 = 30\alpha_2$$

The remaining term, ε, in Eq. 5.3 represents random error in the yield values. The eight observed percent yield values, when expressed as functions of the levels of the coded variables, in matrix notation, are

$$
\begin{array}{cccc}
\mathbf{Y} & = & \mathbf{X} & \boldsymbol{\beta} + \boldsymbol{\varepsilon}
\end{array}
$$

$$
\begin{bmatrix} 49.8 \\ 48.1 \\ 57.3 \\ 52.3 \\ 65.7 \\ 69.4 \\ 73.1 \\ 77.8 \end{bmatrix}
=
\begin{bmatrix}
1 & -1 & -1 \\
1 & -1 & -1 \\
1 & +1 & -1 \\
1 & +1 & -1 \\
1 & -1 & +1 \\
1 & -1 & +1 \\
1 & +1 & +1 \\
1 & +1 & +1
\end{bmatrix}
\begin{bmatrix} \beta_0 \\ \beta_1 \\ \beta_2 \end{bmatrix}
+
\begin{bmatrix} \varepsilon_1 \\ \varepsilon_7 \\ \varepsilon_4 \\ \varepsilon_6 \\ \varepsilon_2 \\ \varepsilon_8 \\ \varepsilon_3 \\ \varepsilon_5 \end{bmatrix}
$$

The estimates of the coefficients in the first-order model in Eq. 5.3 are found by solving the normal equations

$$
\begin{array}{ccc}
\mathbf{X'X} & \mathbf{b} & = \quad \mathbf{X'Y}
\end{array}
$$

$$
\begin{bmatrix} 8 & 0 & 0 \\ 0 & 8 & 0 \\ 0 & 0 & 8 \end{bmatrix}
\begin{bmatrix} b_0 \\ b_1 \\ b_2 \end{bmatrix}
=
\begin{bmatrix} 493.5 \\ 27.5 \\ 78.5 \end{bmatrix}
$$

The estimates are

$$
\mathbf{b} = (\mathbf{X'X})^{-1} \quad \mathbf{X'Y}
$$

$$
\begin{bmatrix} b_0 \\ b_1 \\ b_2 \end{bmatrix}
=
\begin{bmatrix} \frac{1}{8} & 0 & 0 \\ 0 & \frac{1}{8} & 0 \\ 0 & 0 & \frac{1}{8} \end{bmatrix}
\begin{bmatrix} 493.5 \\ 27.5 \\ 78.5 \end{bmatrix}
=
\begin{bmatrix} 61.6875 \\ 3.4375 \\ 9.8125 \end{bmatrix}
$$

where $(\mathbf{X'X})^{-1}$ is the inverse of the matrix $\mathbf{X'X}$. Substituting the estimates b_0, b_1, and b_2 into Eq. 5.3, the fitted first-order model in the coded variables is

$$\hat{Y}(\mathbf{x}) = 61.6875 + 3.4375x_1 + 9.8125x_2 \tag{5.4}$$

The equivalent first-order model in the original variables, temperature (X_1) and time (X_2), is

$$\hat{Y}(\mathbf{x}) = 14.5625 + 0.34375X_1 + 0.32708X_2 \tag{5.5}$$

The separate sums of squares (S.S.) quantities, rounded off to four decimal places, in the corresponding analysis of variance are

$$\text{Total S.S.} = \mathbf{Y'Y} - \frac{(\mathbf{1'Y})^2}{8} = 898.7488 \text{ with } 8 - 1 = 7 \text{ d.f.}$$

$$\text{Regression S.S.} = \mathbf{b'X'Y} - \frac{(\mathbf{1'Y})^2}{8} = 864.8125 \text{ with } 3 - 1 = 2 \text{ d.f.}$$

$$\text{Residual S.S.} = \mathbf{Y'Y} - \mathbf{b'X'Y} = 33.9363 \text{ with } 8 - 3 = 5 \text{ d.f.}$$

$$\text{Pure Error S.S.} = (\tfrac{1}{2})[(49.8 - 48.1)^2 + (57.3 - 52.3)^2$$
$$+ (65.7 - 69.4)^2 + (73.1 - 77.8)^2]$$
$$= (\tfrac{1}{2})(63.67) = 31.8350 \text{ with } 4 \text{ d.f.}$$

$$\text{Lack of Fit S.S.} = \text{Residual S.S.} - \text{Pure Error S.S.}$$
$$= 33.9363 - 31.8350$$
$$= 2.1013 \text{ with } 5 - 4 = 1 \text{ d.f.}$$

The analysis of variance (ANOVA) is displayed in Table 5.2.

To perform a test on the adequacy of the fitted model in Eq. 5.4, the errors in the observed percent yield values are assumed to be distributed normally with mean zero and variance σ^2. The value of the lack of fit test statistic is

$$F = \frac{\text{Lack of Fit Mean Square}}{\text{Pure Error Mean Square}}$$

$$= \frac{2.1013}{7.9588} = 0.264$$

Since the value $F = 0.264$ does not exceed the table value $F_{0.05,1,4} = 7.71$, we do not have sufficient evidence to doubt the adequacy of the fitted model.

Not having discovered evidence of lack of fit, next the fitted model in Eq. 5.4 is tested to see if it explains a significant amount of the variation in the observed percent yield values. This test is equivalent to testing the null hypoth-

Table 5.2 ANOVA Table for the First-Order Model in Eq. 5.4
Fitted to the Percent Yield Values Listed in Table 5.1

Source	Degrees of Freedom	Sum of Squares	Mean Square	F
Regression	2	864.8125	432.4063	63.71
Residual	5	33.9363	6.7873	
Lack of Fit	1	2.1013	2.1013	0.264
Pure Error	4	31.8350	7.9588	

esis, H_0: $\beta_1 = \beta_2 = 0$, or that both temperature and time have zero or no effect on percent yield. The test is highly significant since

$$F = \frac{\text{Regression Mean Square}}{\text{Residual Mean Square}}$$

$$= \frac{432.4063}{6.7873}$$

$$= 63.71 > F_{0.01,2,5} = 13.27$$

Hence, one or both of the parameters, β_1 and β_2, in Eq. 5.3 are nonzero.

At this point in the model development exercise, tests are performed on the magnitudes of the separate effects of temperature and time on percent yield to see if both terms, $b_1 x_1$ and $b_2 x_2$, are needed in the fitted model. These tests were discussed in Section 2.5.1 and proceed as follows: The residual mean square (MSE = 6.7873), taken from the ANOVA Table 5.2, is used as an estimate of σ^2. The diagonal elements, 1/8, of the $(\mathbf{X'X})^{-1}$ matrix, when multiplied by $MSE = 6.7873$, produce the estimates of the variances of the parameter estimates, that is, $\widehat{\text{Var}}(b_i) = (1/8)6.7873 = 0.8484$, $i = 0, 1, 2$. The estimated standard error of each b_i is $(0.8484)^{1/2} = 0.9211$. Assuming the errors in Eq. 5.3 are independent and normally distributed with mean zero and variance σ^2, the test of the magnitudes of the effects of temperature and time on percent yield are

Test of H_0: $\beta_1 = 0$ Test of H_0: $\beta_2 = 0$

$$t = \frac{b_1 - 0}{\sqrt{\widehat{\text{Var}}(b_1)}} \qquad t = \frac{b_2 - 0}{\sqrt{\widehat{\text{Var}}(b_2)}}$$

$$= \frac{3.4375}{0.9211} \qquad\qquad = \frac{9.8125}{0.9211}$$

$$= 3.73 \qquad\qquad\qquad = 10.65 \tag{5.6}$$

Each of the null hypotheses H_0: $\beta_1 = 0$ and H_0: $\beta_2 = 0$ is rejected at the $\alpha = 0.05$ level of significance owing to the calculated values, 3.73 and 10.65, being greater in absolute value than the tabled value, $t_{0.025,5} = 2.571$. We infer, therefore, that both temperature and time have an effect on percent yield. Furthermore, since both b_1 and b_2 are positive ($b_i > 0$, $i = 1, 2$), the effects are positive. Thus, by raising either the temperature or the time of reaction, this produced a significant increase in percent yield.

From the results of the tests in Eq. 5.6 performed on the parameter estimates and in view of the objective of the experiment, which is to find the temperature and time settings that maximize the percent yield, the experimenter quite naturally might ask, "If additional experiments can be performed, at what settings of temperature and time should the additional experiments be run in order to find higher values of percent yield?'' To answer this question, we enter the second stage of our sequential program of experimentation.

5.2.1 Using the Fitted First-Order Model to Map the Surface in the Experimental Region

The fitted model described in Eq. 5.4 can now be used to map values of the estimated response surface over the experimental region. The mapping takes the form of a contour plot (see Section 1.3.7 in Chapter 1) of the estimated surface as shown in Figure 5.2. The contour lines are drawn by connecting two points (coordinate settings of x_1 and x_2) in the experimental region that produce the same value of $\hat{Y}$.

To illustrate the construction of surface contour lines using the fitted plane in Eq. 5.4, suppose the value $\hat{Y}(\mathbf{x}) = 60.0$ is substituted in the fitted Eq. 5.4.

Figure 5.2 Contours of the predicted percent yield planar surface in Eq. 5.5 for yield values of 55, 60, 65, and 70%. The arrow drawn perpendicular to the contour lines indicates the direction of steepest ascent.

The result is the equation of a straight line, that is,

$$0 = 1.6875 + 3.4375x_1 + 9.8125x_2$$

Setting $x_1 = 0$ and solving for x_2, we obtain

$$x_2 = \frac{-1.6875}{9.8125} = -0.172$$

Now setting $x_2 = 0$ and solving for x_1 produces

$$x_1 = \frac{-1.6875}{3.4375} = -0.495$$

The contour line for $\hat{Y} = 60.0$ passes through the coordinate settings $(x_1, x_2) = (0, -0.172)$ and $(-0.495, 0)$, and is drawn to extend slightly beyond the boundaries of the experimental region defined by $x_1 = \pm 1$ and $x_2 = \pm 1$. In Figure 5.2 are shown the contour lines of the estimated planar surface for percent yield corresponding to values of $\hat{Y}(\mathbf{x}) = 55$, 60, 65, and 70%.

The direction of tilt of the estimated percent yield planar surface is indicated by the direction of the arrow which is drawn perpendicular to the surface contour lines in Figure 5.2. The arrow points upward and to the right indicating that higher values of the response (i.e., higher percent yield values) are expected by increasing the values of x_1 and x_2 each above $+1$. This action corresponds to increasing the temperature of the reaction above 90°C and increasing the time of reaction above 90 sec. These recommendations comprise the beginning steps in a series of single experiments to be performed along the path of steepest ascent up the surface.

Thus far we have fitted a planar surface in the region of an initial set of conditions and have learned from the fitted model that a further improvement, in terms of higher values of the response, is possible. We now discuss a strategy which will move us in a direction perpendicular to the contours (see arrow in Figure 5.2) and ''up'' the surface. Such a plan is known as performing experiments along the path of steepest ascent.

5.3 PERFORMING EXPERIMENTS ALONG THE PATH OF STEEPEST ASCENT

The steepest ascent procedure consists of performing a sequence of experiments along the path of maximum increase in response. (Reminder: the direction is dependent on the scale of the coded variables.) The procedure begins by approximating the response surface using an equation of a hyperplane. The information provided by the estimated hyperplane is used to determine a direction toward which one may expect to observe increasing values of the response. As one moves up the surface of increasing response values and approaches a region where curvature in the surface is present, the increase in the response values will eventually level off at the highest point of the surface in the particular direction. If one continues in this direction and the surface height decreases, a new set of experiments is performed and again the first-order model is fitted. A new direction toward increasing values of the response is determined from which another sequence of experiments along the path toward increasing response values is performed. This sequence of trials continues until it becomes evident that little or no additional increase in response can be achieved from the method.

To describe the method of steepest ascent mathematically, we begin by assuming the true response surface can be approximated locally with an equation of a hyperplane $\eta = \beta_0 + \Sigma_{i=1}^k \beta_i x_i$. Data are collected from the points of a first-order design and the data are used to calculate the coefficient estimates to obtain the fitted first-order model

$$\hat{Y}(\mathbf{x}) = b_0 + \sum_{i=1}^k b_i x_i \tag{5.7}$$

The next step is to move away from the center of the design, a distance of r units, say, in the direction of the maximum increase in the response. By choosing the center of the design in the coded variable $x_1, x_2, \ldots, x_k$ to be denoted by $(0, 0, \ldots, 0)$ then movement from the center r units away is equivalent to finding the values of $(x_1, x_2, \ldots, x_k)$ which maximize $b_0 + \Sigma_{i=1}^k b_i x_i$ subject to the constraint $\Sigma_{i=1}^k x_i^2 = r^2$.

Maximization of the response function is performed using Lagrange multipliers. Let

$$Q(x_1, x_2, \ldots, x_k) = b_0 + \sum_{i=1}^k b_i x_i - \mu \left(\sum_{i=1}^k x_i^2 - r^2 \right) \tag{5.8}$$

where μ is the Lagrange multiplier. To maximize $\hat{Y}(\mathbf{x})$ in Eq. 5.7 subject to the above-mentioned constraint, first we set equal to zero the partial derivatives

$$\frac{\partial Q(x_1, x_2, \ldots, x_k)}{\partial x_i} \qquad i = 1, 2, \ldots, k$$

and

$$\frac{\partial Q(x_1, x_2, \ldots, x_k)}{\partial \mu}$$

Setting the partial derivatives equal to zero produces

$$\frac{\partial Q(\mathbf{x})}{\partial x_i} = b_i - 2\mu x_i = 0 \qquad i = 1, 2, \ldots, k \tag{5.9}$$

and

$$\frac{\partial Q(\mathbf{x})}{\partial \mu} = -\sum_{i=1}^{k} x_i^2 + r^2 = 0 \tag{5.10}$$

The solutions to Eqs. 5.9 and 5.10 are the values of x_i satisfying

$$x_i = \frac{b_i}{2\mu} \qquad i = 1, 2, \ldots, k \tag{5.11}$$

where the value of μ is yet to be determined. Note from Eq. 5.11 that the proposed next value of x_i is directly proportional to the value of b_i. Since this value depends on the range of x_i, and the range of x_i depends on the units of measurement of the original input variables through the transformation $x_i = (X_i - \overline{X}_i)/s_{X_i}$, then, the direction for the next value of the original variable X_i, proposed by Eq. 5.11 for x_i, also depends on s_i (for simplicity, we shall write the scale constant as s_i instead of s_{X_i}).

Let us pause briefly in the discussion of the method of steepest ascent to illustrate the dependency of direction in Eq. 5.11 on the scale of x_i relative to X_i. Suppose in the chemical reaction example, that instead of choosing the levels ± 1 for x_2 which resulted from using the value $s_2 = 30$ sec in Eq. 5.2, we had chosen to use the values $\pm \theta$, $(\theta \neq 1)$ for x_2, so that in Eq. 5.2 $s_2 = 30/\theta$. Then the coefficient estimate, $b_{2\theta}$ say, in the fitted model of Eq. 5.4 would equal b_2 divided by θ, that is, $b_{2\theta} = 9.8125/\theta$. For a fixed value of μ in Eq. 5.11, say $\mu = 1/2$, the value of x_2 in Eq. 5.11 would be $b_{2\theta} = b_2/\theta$ so that the corresponding change in X_2 (sec) is $X_2 = (s_2 b_2/\theta^2) + \overline{X}_2$, which differs from $X_2 = (s_2 b_2 + \overline{X}_2)$ when $\theta \neq 1$. In practice this means that the method of steepest ascent can only be applied in a subjective manner, using the experimenter's assumed knowledge of the amount of variation in the response to be expected while moving in the direction of increasing values of the original factors. In other words, the method of steepest ascent, or descent, should never be applied in an automatic fashion. For further discussion on what effect the size of the scale units has on the "direction of steepest ascent," the reader is directed to the discussion of the paper by Box and Wilson (1951), in particular, the comments by N. L. Johnson.

Keeping in mind the dependency of the direction of steepest ascent on the scaled units of the x_i, let us continue with the development of the direction that led to Eq. 5.11. To calculate the value of x_i in Eq. 5.11, first a value of μ must be obtained. To do this, one method is to isolate one of the variables, X_j, say, which if the value of X_j were changed by an amount Δ_j, would result in an increase in the value of the response. Based on the incremental change, Δ_j, we then calculate $x_j = \Delta_j/s_j$, where s_j is that scale factor considered in the denominator of $x_j = (X_j - \overline{X}_j)/s_j$. Substituting the value of Δ_j/s_j for x_j in Eq. 5.11, the value of μ becomes $\mu = b_j/2x_j = b_j s_j/2\Delta_j$. This value of μ produces the settings of the remaining values of x_i, using Eq. 5.11 and the resulting values of $x_1, x_2 , \ldots , x_k$ define the first point on the path of steepest ascent.

Let us illustrate the procedure with the first-order model of Eq. 5.4 that was fitted to the percent yield values of Table 5.1,

$$\hat{Y}(\mathbf{x}) = 61.6875 + 3.4375x_1 + 9.8125x_2$$

Since b_2 is positive, we know that the surface increases in height with increasing values of x_2 (or for values of time greater than 60 sec). Now, let us find the value of μ that corresponds to an arbitrary change in time (X_2) of $\Delta_2 = 45$ sec. In the coded variable x_2, the change of 45 sec in X_2 corresponds to a change of $(45/30) = 1.5$ units so that from Eq. 5.11,

$$\mu = \frac{b_2}{2x_2} = \frac{9.8125}{2(1.5)} = 3.27$$

The value of x_1, corresponding to a change of 1.5 units in x_2, is

$$x_1 = \frac{b_1}{2\mu} = \frac{3.4375}{2(3.27)} = 0.526 \text{ or } 0.53$$

and the incremental change in temperature is $\Delta_1 = 0.53(10) = 5.3°C$.

The first point on the path of steepest ascent, therefore, is located at the coordinates $(x_1, x_2) = (0.53, 1.5)$, which corresponds to the settings in the original variables of $(X_1, X_2) = (85.3, 105)$. To check if an increase in the response value is predicted at this point, the estimate of the response at $(x_1, x_2) = (0.53, 1.5)$, using Eq. 5.4, is compared to the estimate of the response at the center of the design, that is, $\hat{Y}(0.53, 1.5) - \hat{Y}(0, 0) = 16.5406$. Although the estimate $\hat{Y}$ (0.53, 1.5) is extrapolated outside the region, we feel that because the difference is positive, additional higher response values will be observed if one runs extra experiments in the direction specified in Eq. 5.11.

Additional experiments are now performed along the path of steepest ascent at points corresponding to the increments of distances $1.5\Delta_i, 2\Delta_i, 3\Delta_i,$ and $4\Delta_i$ ($i = 1, 2$). Table 5.3 lists the coordinates of the points along the path of steepest ascent up to the point corresponding to the increment $4\Delta_i$, and the corresponding percent yield values. The percent yield values increase to a value of 84.7% at the settings in X_1 and X_2 of 95.9°C and 195 sec, respectively, and then the value drops to 80.1% at $X_1 = 101.2°C$ and $X_2 = 240$ sec. Our thinking at this moment is that either the temperature of 101.2°C is too high or the length of time of 240 sec is too long and therefore additional experimentation along the path at higher values of X_1 and X_2 would not be useful. [Myers and Khuri (1979) present a stopping rule procedure that takes into account the random error variation in

Table 5.3 Points Along the Path of Steepest Ascent and Observed Percent Yield Values at the Points

	Temperature (°C)	Time (sec)	Percent Yield
Base	80.0	60	
Δ_i	5.3	45	
Base + Δ_i	85.3	105	74.3
Base + 1.5Δ_i	87.95	127.5	78.6
Base + 2Δ_i	90.6	150	83.2
Base + 3Δ_i	95.9	195	84.7
Base + 4Δ_i	101.2	240	80.1

the observed response values. The procedure protects against taking too many observations along the path of steepest ascent when in fact the unknown true mean response is decreasing in value and also protects against stoppimg prematurely when the unknown true mean response is increasing.]

Having observed a decrease in the percent yield value at $X_1 = 101.2°C$ and $X_2 = 240$ sec (center + 4Δ_i) when compared to the observed yield value of 84.7% at $X_1 = 95.9°C$ and $X_2 = 195$ sec, the decision is made to conduct a second group of experiments and again fit a first-order model. The new design is set up with the point $X_1 = 95.9°C$ and $X_2 = 195$ sec as its center. For this design, the spread of temperature and time settings (levels) is kept the same as with the initial set of experiments so that for the second group of experiments the coded variables are defined as

$$x_1 = \frac{\text{Temperature } (X_1) - 95.9}{10} \qquad x_2 = \frac{\text{time } (X_2) - 195}{30} \qquad (5.12)$$

When $x_1 = -1$ and $x_1 = +1$, the temperature settings are 85.9°C and 105.9°C, and $x_2 = -1$, +1 correspond to time settings of 165 sec and 225 sec, respectively. The four new design settings, the center point, and the measured percent yield values are listed in Table 5.4. Two replicate yield values were collected at each of the four factorial combinations along with a second replicated observation at the center point.

The fitted model for the second group of experiments whose percent yield values are listed in Table 5.4 is

$$\hat{Y}(\mathbf{x}) = 82.70 + 3.575x_1 - 2.750x_2 \qquad (5.13)$$

Table 5.4 Sequence of Experimental Trials Performed in Moving to a Region of High Percent Yield Values

Design Two		x_1	x_2	Temp. (T)	Time (t)	% Yield	Fitted Model
		−1	−1	85.9	165	82.9, 81.4	
		+1	−1	105.9	165	87.4, 89.5	(5.13)
		−1	+1	85.9	225	74.6, 77.0	Considered
		+1	+1	105.9	225	84.5, 83.1	Adequate
		0	0	95.9	195	84.7, 81.9	
Step. Direction Two							
1.	Center + Δ_i	+1	−0.77	105.9	171.9	89.0	
2.	Center + $2\Delta_i$	+2	−1.54	115.9	148.8	90.2	
3.	Center + $3\Delta_i$	+3	−2.31	125.9	125.7	87.4	
4.	Center + $4\Delta_i$	+4	−3.08	135.9	102.6	82.6	
	Retreat to Center + $2\Delta_i$ and Proceed in Direction Three						
5.		+2	−1.54	115.9	148.8	91.0	
6.		+3	−0.77	125.9	171.9	93.6	
7.		+4	0	135.9	195	96.2	
8.		+5	0.77	145.9	218.1	92.9	
Set up Design Three Using Points of Steps 6, 7, and 8 along with the Following Points. Center of Design is $(T, t) = (135.9, 195)$.							
9.		+3	0.77	125.9	218.1	91.7	
10.		+5	−0.77	145.9	171.9	92.5	
11.	(Replicated 7)	+4	0	135.9	195	97.0	
Fitted Model Using Percent Yield Values in Steps 6–11							
$\hat{Y}(\mathbf{x}) = 93.983 + 0.025x_1 - 0.375x_2$							

The analysis of variance for the 10 yield values is

Source	d.f.	SS	MS	F
Regression	2	162.745	81.372	42.34
Residual	7	13.455	1.922	
Lack of Fit	2	2.345	1.173	0.53
Pure Error	5	11.110	2.222	
Total	9	176.200		

The test for adequacy of fit of the model 5.13 produced an F value of $F = 1.173/2.222 = 0.53$, which is not significant. The test for significance of the fitted model produced a highly significant $F = 42.34$ value. Thus, the information obtained from the fitted model 5.13 is used to obtain a new direction in which to perform additional experiments in seeking higher percent yield values.

Table 5.4 lists the sequence of experimental trials that were performed along with the measured yield values beginning with the direction of steepest ascent determined by using the coefficient estimates in Eq. 5.13. Figure 5.3 shows the sequence of experiments performed by numbering the points which are listed as steps 1–11 in Table 5.4, where steps 1–4 represent the experimental trials taken along the second direction of steepest ascent. Step 5 denotes a return to the point in step 2 and replicating the experiment to validate the previously high percent yield value. Having verified the percent yield at the settings of steps 2

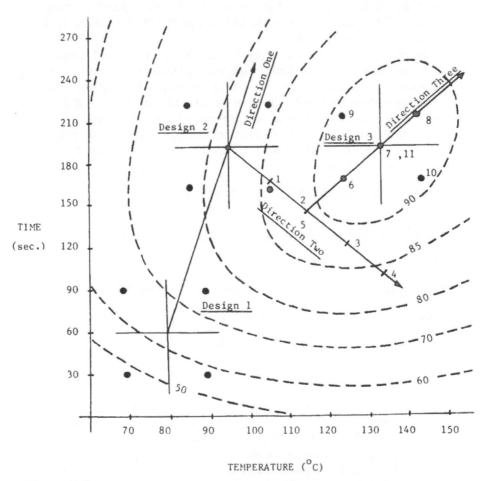

Figure 5.3 Sequence of experiments performed (steps 1–4) along the second path of steepest ascent. Points 5–8 represent experiments run in a third direction. Points 6–11 comprise design three.

and 5, step 6 represents an experiment performed at a setting where a higher yield value was expected since both temperature and time were increased relative to their values in step 5. In fact, at step 6 the coded values of the temperature and time settings are +1, +1, respectively. Comprising a 3/4 replicate of a 2^2 factorial are the points at steps 1, 3, and 6, with the point at step 5 as the center. Upon observing a higher percent yield at step 6 than at step 5, steps 7 and 8 represent additional experiments performed along a third direction defined by the line joining the points of steps 5 and 6. The choice of this third direction represents a deviation from the conventional steepest ascent approach and was undertaken in an attempt to reduce the amount of work required in setting up a complete 2^2 factorial experiment with center at point 5 and the subsequent fitting of another first-order model.

The percent yield observed at step 8 was lower than observed at step 7, and so it was decided to go no further than step 8. Design three was set up using the point at step 7 as its center. The settings of temperature and time that comprise design three are listed in the temperature and time columns of Table 5.4, steps 6–11 , where step 11 denotes a replicate of the center point.

For the third design (steps 6–11), the settings of temperature range from 125.9°C to 145.9°C and the settings of time from 171.9 sec to 218.1 sec. If we redefine the coded variables

$$x_1 = \frac{\text{Temperature} - 135.9}{10} \qquad x_2 = \frac{\text{time} - 195}{23.1}$$

then the first-order model fitted to the percent yield values of steps 6–11 is

$$\hat{Y}(\mathbf{x}) = 93.983 + 0.025x_1 - 0.375x_2 \tag{5.14}$$

The analysis of variance table for the fitted model in Eq. 5.14 is

Source	d.f.	SS	MS	F
Regression	2	0.5650	0.2825	0.04
Residual	3	22.1833	7.3944	
Total	5	22.7483		

It is obvious from the ANOVA table that the fitted model in Eq. 5.14 does not explain a significant amount of the overall variation in the percent yield values. In fact, a plot of the residuals $(Y_u - \hat{Y}_u)$ at the points of the design, shown in Figure 5.4, indicates that the fitted model in Eq. 5.14 underestimates the percent yield at the center but overestimates the percent yield at the four factorial settings. This is indicative of a less than adequate attempt by a plane to fit a curved surface, in particular, a mound. We shall now present a test for surface curvature. The test was introduced previously in Section 3.4 in Chapter 3.

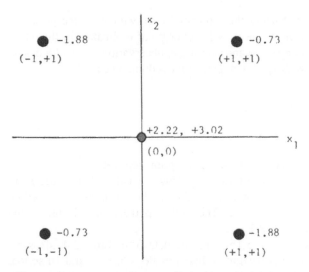

Figure 5.4   Residuals, $Y_u - \hat{Y}_u$, at the points (x_1, x_2) of design three.

5.3.1   The Single Degree of Freedom Test for Surface Curvature

Let n_1 observations be collected at the n points of a first-order orthogonal design where $n_1 \geq n$. We assume that whenever $n_1 > n$, an equal number of replications, r_1, is taken at each design point, that is, $n_1 = nr_1$. Denote the average of the n_1 observations by $\bar{Y}_1$. Suppose further that $n_0 > 1$ observations are collected at the center of the design, where the center point is defined by the coordinate settings of the coded variables, $(x_1, x_2, \ldots, x_k) = (0, 0, \ldots, 0)$. Denote the average of the center point observations by $\bar{Y}_0$. If the true surface is represented by the second-order model

$$\eta = \beta_0 + \sum_{i=1}^{k} \beta_i x_i + \sum_{i=1}^{k} \beta_{ii} x_i^2 + \sum_{i<j} \sum \beta_{ij} x_i x_j \tag{5.15}$$

then the difference, $\bar{Y}_1 - \bar{Y}_0$, is an unbiased estimator of the sum of the quadratic coefficients, $\sum_{i=1}^{k} \beta_{ii}$. A test of the null hypothesis $H_0 : \sum_{i=1}^{k} \beta_{ii} = 0$ versus the alternative hypothesis $H_a : \sum_{i=1}^{k} \beta_{ii} \neq 0$ is

$$F = \frac{(\bar{Y}_1 - \bar{Y}_0)^2}{s^2 \left(\frac{1}{n_1} + \frac{1}{n_0}\right)} \tag{5.16}$$

where s^2 is an estimate of the error variance, σ^2. If $n_1 = n$, then s^2 is calculated from the $n_0 > 1$ center point replicates and has $n_0 - 1$ degrees of freedom. If

$n_1 > n$, then s^2 is calculated using the replicated observations at the points of the first-order design as well as the $n_0 > 1$ center point replicates. In this latter case, let s_l^2 represent the sample variance among r_1 observations at the lth design point ($l = 1, 2, \ldots, n$) with $r_1 - 1$ degrees of freedom. Then

$$s^2 = \frac{\sum_{l=1}^{n} (r_1 - 1)s_l^2 + (n_0 - 1)s_0^2}{n_1 - n + n_0 - 1} \tag{5.17}$$

where s_0^2 is the sample variance of the n_0 center point replicates.

The test of $H_0: \sum_{i=1}^{k} \beta_{ii} = 0$ involves comparing the calculated value, F, in Eq. 5.16 to the table value of $F_{\alpha,1,\nu}$ where ν is the number of degrees of freedom in s^2, namely, $\nu = n_1 + n_0 - n - 1$. The null hypothesis is rejected at the α-level of significance when $F > F_{\alpha,1,\nu}$.

Let us illustrate the test in Eq. 5.16 using the data from Table 5.4, steps 6–11. The points of the first-order design are listed in steps 6, 8, 9, and 10 so that $\overline{Y}_1 = (93.6 + 92.9 + 91.7 + 92.5)/4 = 92.675$. Steps 7 and 11 represent two replicates of the center point so that $\overline{Y}_0 = (96.2 + 97.0)/2 = 96.6$. Since $r_1 = 1$, and $n_0 = 2$, then only the observations at the center point are used to calculate

$$s^2 = \frac{(96.2 - 96.6)^2 + (97.0 - 96.6)2}{2 - 1} = 0.32$$

with $2 - 1 = 1$ degree of freedom. The value of F in Eq. 5.16 is

$$F = \frac{(92.675 - 96.6000)^2}{0.32(\frac{1}{4} + \frac{1}{2})}$$

$$= 64.19$$

Since $F = 64.19 > F_{0.10,1,1} = 39.9$, we reject $H_0: \sum_{i=1}^{2} \beta_{ii} = 0$ at the 10% level of significance and infer that one or both, β_{11} or β_{22}, are nonzero. The next course of action would be to set up a design enabling us to fit a second-order (quadratic) model in x_1 and x_2 to account for the surface curvature in the immediate region of interest.

To this point, we have discovered that not only does the fitted first-order model in Eq. 5.14 not adequately fit the observed response values as well as another model might, but we suspect that there is curvature in the surface that may be picked up with the addition of terms to the model like $\beta_{11}x_1^2$ and $\beta_{22}x_2^2$. The fitting of the full second-order model, however, would require that additional points be added to the basic design so that all the coefficients associated with the second-order terms in the model can be estimated. We shall now discuss the fitting of a second-order model.

5.4 FITTING A SECOND-ORDER MODEL: LOCAL EXPLORATION OF THE FITTED RESPONSE SURFACE

In this section our goals will be to:

1. Obtain an adequate model representation of the estimated second-order response surface.

2. Use the model in 1. to locate the coordinates $(x_{10}, x_{20}, \ldots, x_{k0})$ of the stationary point which is where the partial derivatives of the estimated response function in 1. are equal to zero. If the stationary point is found to be inside the experimental region, then we proceed to goal 3. If the stationary point is not inside the region, then further experimentation taken in the direction of the stationary point is necessary.

3. Describe the nature of the stationary point. Is it a maximum, a minimum, or a saddle point (minimax point)?

Once these goals have been attained, we shall then

4. Describe the shape of the response surface in the vicinity of the stationary point.

Before embarking on our plan of attack to reach the goals listed above, we remark that with the two-variable (temperature and time of reaction) example that we have used up to this point there are basically only two types of response surfaces that can be modeled with a second-order model (Eq. 5.15). One is where the stationary point is a maximum (it could be a minimum) and the second type of surface is one where the stationary point is a minimax point (also called a *saddle point*). With a minimax point, the height of the estimated surface drops off as we move away from the stationary point in certain directions, but the height increases as we move away from the point in some other directions. Often the presence of a minimax point denotes the existence of two distinct regions containing maxima, which implies the existence of two distinct fundamental and different mechanisms.

5.4.1 Fitting a Second-Order Response Surface

Let us consider the fitting of a second-order model in k variables of the form

$$Y = \beta_0 + \sum_{i=1}^{k} \beta_i x_i + \sum_{i=1}^{k} \beta_{ii} x_i^2 + \sum_{i<j}^{k} \beta_{ij} x_i x_j + \varepsilon \tag{5.18}$$

The number of terms in Eq. 5.18 is $p = (k + 1)(k + 2)/2$; for example, when $k = 2$, then $p = 6$.

Designs that are used for collecting observed values of the response for estimating the coefficients in a second-order model of the form in Eq. 5.18 were presented in Chapter 4. For our purposes, let us assume that observed response values are collected at the points of a second-order design and the fitted second-order model is

$$\hat{Y}(\mathbf{x}) = b_0 + \sum_{i=1}^{k} b_i x_i + \sum_{i=1}^{k} b_{ii} x_i^2 + \sum \sum_{i<j}^{k} b_{ij} x_i x_j \tag{5.19}$$

After the fitted model in Eq. 5.19 is checked for adequacy of fit in the region defined by the coordinates of the design and is found to be adequate, the model is then used to locate the coordinates of the stationary point and to perform a more detailed analysis of the response system.

Let us return to the chemical reaction example of the previous section where the observed percent yield values at the points and at the center of design three, listed in Table 5.4, are presented in Figure 5.5. Suppose that four additional experiments are performed, one at each of the axial settings $(x_1, x_2) = (\pm\sqrt{2}, 0)$ and $(x_1, x_2) = (0, \pm\sqrt{2})$. These four design settings along with the four factorial settings $(x_1, x_2) = (\pm1, \pm1)$ and center point comprise a central

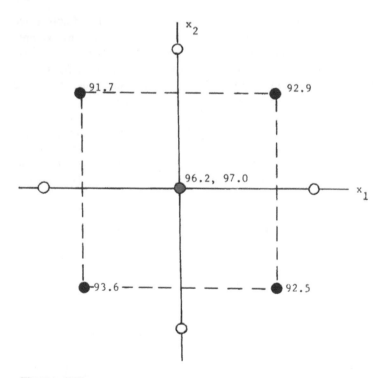

Figure 5.5 Percent yields at the four outer points ($\bullet$) and center ($\bullet$) of design three. The circles ($\circ$) represent the axial settings that together with points ($\bullet$) form a central composite rotatable design.

composite rotatable design (Section 4.5.3 in Chapter 4). The percent yield values and the corresponding nine design settings are listed in Table 5.5.

Using the percent yield values listed in Table 5.5 along with settings of the coded variables, the fitted second-order model is

$$\hat{Y}(\mathbf{x}) = 96.60 + 0.03x_1 - 0.31x_2 - 1.98x_1^2 - 1.83x_2^2 + 0.58x_1x_2 \qquad (5.20)$$

A computer printout, using the SAS (1989) system of statistical software (PROC RSREG), of the analysis of variance table is listed in Table 5.6.

The test for adequacy of fit of the fitted model in Eq. 5.20 produced an F value (lack of fit mean square/pure error mean square) less than unity, which is clearly not significant. The pure quadratic coefficient estimates are each highly significant ($P < 0.001$), which indicates that surface curvature is present in the observed percent yield values.

With the fitted second-order model in Eq. 5.20, we can predict percent yield values for values of x_1 and x_2 inside the region of experimentation. Several values of x_1 and x_2 were selected arbitrarily, and the predicted or estimated percent yield values are listed in Table 5.7. Also listed in Table 5.7 is the estimated variance of the predicted percent yield value at each point, where an estimate of the error variance is MSE = residual mean square = 0.113.

A contour plot of the estimated percent yield surface, over the x_1 and x_2 region, is shown in Figure 5.6. The contours represent predicted yield values of $\hat{Y}(\mathbf{x}) = 95.0$ to $\hat{Y}(\mathbf{x}) = 96.5$ percent in steps of 0.5 percent. The units of the original variables, reaction temperature, and reaction time are listed also. The contours are elliptical and centered at the point $(x_1, x_2) = (-0.0048, -0.0857)$ or $(T, t) = (135.85°C, 193.02 \text{ sec})$. The coordinates of this centroid point are

Table 5.5 Percent Yield Values at the Nine Points of a Central Composite Rotatable Design

Coded Variables		Percent Yield	Original Variables	
x_1	x_2	Y	Temp. (°C)	Time (sec)
-1	-1	93.6	125.9	171.9
-1	1	91.7	125.9	218.1
1	-1	92.5	145.9	171.9
1	1	92.9	145.9	218.1
0	0	96.2, 97.0	135.9	195.0
$-\sqrt{2}$	0	92.7	121.75	195.0
$\sqrt{2}$	0	92.8	150.04	195.0
0	$-\sqrt{2}$	93.4	135.9	162.3
0	$\sqrt{2}$	92.7	135.9	227.7

Table 5.6 SAS/RSREG Computer-Generated Analysis of Variance for the Percent Yield Second-Order Fitted Model

Response Surface for Variable Y

Response Mean	93.55
Root MSE	0.3368665
R-Square	0.9824777
Coef of Variation	0.3600925

Regression	d.f.	Type I SS	R-Square	F-Ratio	Prob $> F$
Linear	2	0.78226810	0.0302	3.45	0.1348
Quadratic	2	23.34631580	0.9012	102.87	0.0004
Cross-Product	1	1.32250000	0.0511	11.65	0.0269
Total Regression	5	25.45108391	0.9825	44.86	0.0013

Residual	d.f.	SS	Mean Square	F-Ratio	Prob $> F$
Lack of Fit	3	0.13391609	0.04463870	0.139	0.9248
Pure Error	1	0.32000000	0.32000000		
Total Error	4	0.45391609	0.11347902		

Parameter	d.f.	Estimate	Std Dev	T-Ratio	Prob $> |T|$
Intercept	1	96.60000431	0.23820057	405.54	0.0001
$x1$	1	0.03017779	0.11910086	0.25	0.8125
$x2$	1	-0.31124548	0.11910086	-2.61	0.0592
$x1 * x1$	1	-1.98127242	0.15755681	-12.57	0.0002
$x2 * x1$	1	0.57500000	0.16843324	3.41	0.0269
$x2 * x2$	1	-1.83126954	0.15755681	-11.62	0.0003

Factor	d.f.	SS	Mean Square	F-Ratio	Prob $> F$
$x1$	3	19.27421	6.424738	56.62	0.0010
$x2$	3	17.42761	5.809203	51.19	0.0012

called *the coordinates of the stationary point*, and the method of determining the values of the coordinates of the stationary point and the estimate of the mean response at the stationary point are described in the next section.

From the contour plot in Figure 5.6, we see that as one moves away from the point where $T = 135.85°$ and $t = 193.02$ sec by increasing or decreasing the values of either temperature or time, the predicted percent yield value decreases. We discuss in Section 5.5.1 how to measure the rate of change of the height of the surface as one moves away from the stationary point. This is made possible by expressing the estimated surface in its canonical form.

Table 5.7 Predicted Percent Yield Values and Their Variances

Coded Variables		Temperature (°C)	Time (sec)	$\hat{Y}(x)$	$\widehat{Var}[\hat{Y}(x)]$
x_1	x_2				
0.5	0.5	140.9	206.55	95.65	0.0415
−0.5	0.5	130.9	206.55	95.33	0.0415
−0.5	−0.5	130.9	183.45	95.93	0.0415
0.5	−0.5	140.9	183.45	95.67	0.0415
1	0	145.9	195.0	94.65	0.0388
0	1	135.9	218.1	94.46	0.0388
−1	0	125.9	195.0	94.59	0.0388
0	−1	135.9	171.9	95.08	0.0388
0.3	0.3	138.9	201.93	96.22	0.0497
−0.3	0.3	132.9	201.93	96.10	0.0497
−0.3	−0.3	132.9	188.07	96.39	0.0497
0.3	−0.3	138.9	188.07	96.31	0.0497

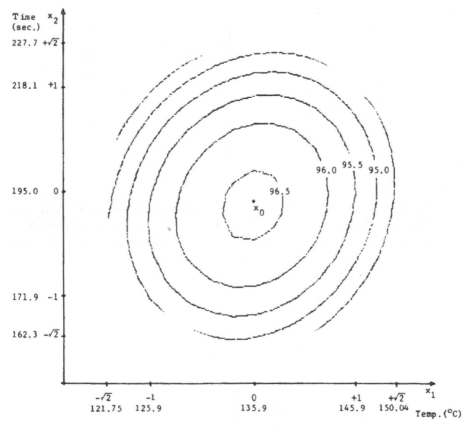

Figure 5.6 Computer-generated contour plot of the estimated percent yield surface. The stationary point is located at $T = 135.85°C$ and $t = 193.02$ sec.

5.5 DETERMINING THE COORDINATES OF THE STATIONARY POINT OF A RESPONSE SYSTEM

A near stationary region is defined as a region where the surface slopes (or gradients along the variables axes) are small compared to the estimate of experimental error. The *stationary point* of a near stationary region is the point at which the slope of the response surface is zero when taken in all directions. The coordinates $\mathbf{x}_0 = (x_{10}, x_{20}, \ldots, x_{k0})'$ of the stationary point are calculated by differentiating the estimated response equation with respect to each x_i, equating these derivatives to zero, and solving the resulting k equations simultaneously.

To obtain the coordinates of the stationary point, let us write the fitted second-order model in k variables, using matrix notation, as

$$\hat{Y}(\mathbf{x}) = b_0 + \mathbf{x}'\mathbf{b} + \mathbf{x}'\mathbf{B}\mathbf{x} \tag{5.21}$$

where

$$
\mathbf{x} = \begin{bmatrix} x_1 \\ x_2 \\ \vdots \\ x_k \end{bmatrix}
\qquad
\mathbf{b} = \begin{bmatrix} b_1 \\ b_2 \\ \vdots \\ b_k \end{bmatrix}
\qquad
\mathbf{B} = \begin{bmatrix}
b_{11} & \dfrac{b_{12}}{2} & \cdots & \dfrac{b_{1k}}{2} \\[2mm]
 & b_{22} & \cdots & \dfrac{b_{2k}}{2} \\[2mm]
 & & \ddots & \vdots \\[2mm]
 & & & \dfrac{b_{k-1,k}}{2} \\[2mm]
\text{symmetric} & & & b_{kk}
\end{bmatrix}
$$

The elements of the $k \times 1$ vector $\mathbf{b}$ are the estimated coefficients of the first-order terms in Eq. 5.21, and the elements of the $k \times k$ symmetric matrix $\mathbf{B}$ are the estimated coefficients of the second-order terms in Eq. 5.21. The partial derivatives of $\hat{Y}(\mathbf{x})$ with respect to $x_1, x_2, \ldots, x_k$ are

$$
\left.
\begin{aligned}
\frac{\partial \hat{Y}(\mathbf{x})}{\partial x_1} &= b_1 + 2b_{11}x_1 + \sum_{j=2}^{k} b_{1j}x_j \\[2mm]
\frac{\partial \hat{Y}(\mathbf{x})}{\partial x_2} &= b_2 + 2b_{22}x_2 + \sum_{j\neq2}^{k} b_{2j}x_j \\
&\ \ \vdots \\
\frac{\partial \hat{Y}(\mathbf{x})}{\partial x_k} &= b_k + 2b_{kk}x_k + \sum_{j=1}^{k-1} b_{kj}x_j
\end{aligned}
\right\} = \mathbf{b} + 2\mathbf{B}\mathbf{x}
$$

Setting each of the k derivatives equal to zero and solving for the values of the x_i, we find that the coordinates of the stationary point are the values of the elements of the $k \times 1$ vector $\mathbf{x}_0$ given by

$$\mathbf{x}_0 = -\frac{\mathbf{B}^{-1}\mathbf{b}}{2} \tag{5.22}$$

where $\mathbf{B}^{-1}$ is the inverse of the matrix $\mathbf{B}$ in Eq. 5.21.

For the percent yield surface represented by the second-order fitted model in Eq. 5.20 we have

$$\mathbf{B} = \begin{bmatrix} -1.98127 & 0.28750 \\ 0.28750 & -1.83127 \end{bmatrix}$$

$$\mathbf{B}^{-1} = \begin{bmatrix} -0.51650 & -0.08109 \\ -0.08109 & -0.55880 \end{bmatrix}$$

$$\mathbf{b} = \begin{bmatrix} 0.03018 \\ -0.31125 \end{bmatrix}$$

so that the stationary point is located at

$$\mathbf{x}_0 = -\frac{1}{2}\mathbf{B}^{-1}\mathbf{b} = \begin{bmatrix} -0.00486 \\ -0.08568 \end{bmatrix} \tag{5.23}$$

In the original variables, temperature and time, the settings at the stationary point are $T = 135.85°C$ and $t = 193.02$ sec, respectively.

The elements of the vector $\mathbf{x}_0$ in Eq. 5.22 do not tell us anything about the nature of the surface at the stationary point. For example, when $k = 2$, the elements of $\mathbf{x}_0 = \begin{bmatrix} x_{10} \\ x_{20} \end{bmatrix}$ might represent the point at which the fitted surface attains a maximum (see Figure 5.7a), or the point at which the fitted surface attains a minimum (Figure 5.7a), or where the fitted surface comes together in the form of a minimax point (Figure 5.7b). For each of these cases, we are assuming that the stationary point is located inside the experimental region. When, on the other hand, the stationary point is outside the region, then we might have encountered a rising ridge system or a falling ridge system as drawn in Figure 5.7c, or possibly a stationary ridge as shown in Figure 5.7d. Actually, these latter two ridge systems are special cases of the types of systems shown in Figure 5.7a where the optimum (maximum or minimum) is located outside the boundary of the basic system.

A procedure for constructing a $(1 - \alpha) \times 100\%$ confidence region for the true location of the stationary point was first presented by Box and Hunter

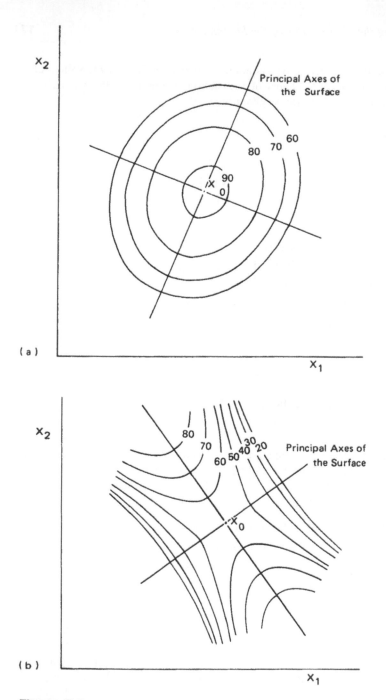

Figure 5.7 (a) A maximum point of the surface; increasing response values moving away from x_0 would indicate the point x_0 is a minimum. (b) Saddle point or minimax of the surface. (c) Rising ridge surface; a falling ridge surface would have decreasing values of the response as one moves toward x_0. (d) A stationary ridge.

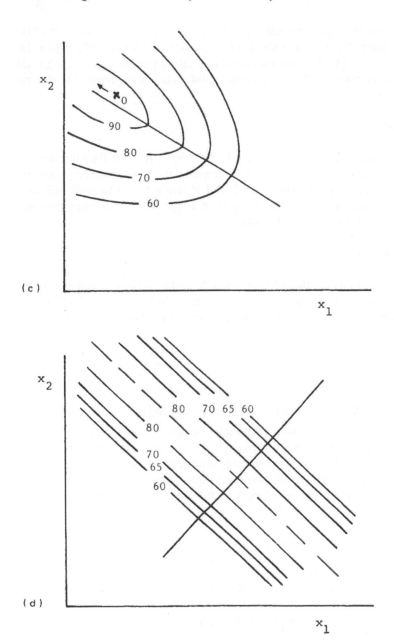

(1954). Normally such a procedure is carried out only when the location of the stationary point (Eq. 5.22) is estimated to be inside the experimental region. Let $\xi = (\xi_1, \xi_2, \ldots, \xi_k)'$ represent the coordinate settings of $(x_1, x_2, \ldots, x_k)$ at the true stationary point. If the second-order model (Eq. 5.18) represents the true surface, then

$$\beta + 2B_2\xi = 0$$

'where $\beta = (\beta_1, \beta_2, \ldots, \beta_k)'$ and the $k \times k$ matrix B_2 is of the same form as the matrix B in Eq. 5.21 except that B_2 contains the true quadratic parameters $\beta_{11}, \beta_{12}, \ldots, \beta_{kk}$. If the elements of β and B_2 are replaced by their estimates, b and B, respectively, where the elements in b and B are distributed multinormally, we can define the $k \times 1$ vector

$$\delta = b + 2B\xi$$

so that δ is distributed $N(0, \sigma^2 V)$, where $\sigma^2 V$ is the variance-covariance matrix of the elements in δ. Furthermore, if an estimate, MSE, of σ^2 is distributed as a chi-square independently of the estimates b and B, then a $(1 - \alpha) \times 100\%$ confidence region for ξ is

$$\frac{\delta' V^{-1} \delta}{k \text{MSE}} \leq F_{\alpha, k, \nu} \tag{5.24}$$

where ν is the degrees of freedom associated with MSE.

Let us illustrate the calculations used in obtaining a 95% (and 99%) confidence region for the two-variable percent yield example where the second-order fitted model is Eq. 5.20 and the estimated stationary point settings are given in Eq. 5.23. Associated with the nine-point central composite rotatable design (Table 5.5), the variances and covariances of the coefficient estimates in the fitted model (Eq. 5.20) are

$$\text{Var}(b_0) = 0.5\sigma^2 \qquad \text{Var}(b_1) = \text{Var}(b_2) = 0.125\sigma^2$$
$$\text{Var}(b_{11}) = \text{Var}(b_{22}) = 0.21875\sigma^2 \qquad \text{Var}(b_{12}) = 0.25\sigma^2$$
$$\text{Cov}(b_0, b_{11}) = \text{Cov}(b_0, b_{22}) = -0.25\sigma^2 \qquad \text{Cov}(b_{11}, b_{22}) = 0.09375\sigma^2$$

An estimate of σ^2, taken from the ANOVA Table 5.6, is MSE = 0.113479. Define the 2×1 vector

$$\begin{bmatrix} \delta_1 \\ \delta_2 \end{bmatrix} = \begin{bmatrix} b_1 + 2b_{11}\xi_1 + b_{12}\xi_2 \\ b_2 + b_{12}\xi_1 + 2b_{22}\xi_2 \end{bmatrix}$$

Then

$$
\text{Var} \begin{pmatrix} \delta_1 \\ \delta_2 \end{pmatrix}
$$

$$
= \begin{bmatrix} 0.125 + 4(0.21875)\xi_1^2 + 0.25\xi_2^2 & 0.25\xi_1\xi_2 + 4(0.09375)\xi_1\xi_2 \\ 0.25\xi_1\xi_2 + 4(0.09375)\xi_1\xi_2 & 0.125 + 0.25\xi_1^2 + 4(0.21875)\xi_2^2 \end{bmatrix} \sigma^2
$$

$$
= \mathbf{V}\sigma^2
$$

The 95% confidence region consists of all values (ξ_1, ξ_2) satisfying the inequality 5.24, that is,

$$
\frac{(\delta_1, \delta_2)\mathbf{V}^{-1}\begin{pmatrix} \delta_1 \\ \delta_2 \end{pmatrix}}{2(0.113479)} \leq F_{0.05,2,4} = 6.94
$$

or

$$
(\delta_1, \delta_2)\mathbf{V}^{-1}\begin{pmatrix} \delta_1 \\ \delta_2 \end{pmatrix} - 1.575 \leq 0
$$

In Figure 5.8 are shown the 95% (smaller ellipse) and 99% (larger ellipse) confidence regions for (ξ_1, ξ_2).

When the estimated second-order response function in Eq. 5.19 is subject to a set of m constraints $g_i(\mathbf{x})$, $i = 1, 2, \ldots, m$, where each $g_i(\mathbf{x})$ is a second-order equation, Stablein, Carter, and Wampler (1983) present a procedure for constructing a confidence region for ξ subject to the m constraints. Their procedure is an extension of the work by Myers and Carter (1973), who considered the problem of optimizing a second-order primary-response function in the presence of a second-order secondary-response function.

At the stationary point, the predicted response value, denoted by $\hat{Y}_0$, is obtained by substituting $\mathbf{x}_0$ for $\mathbf{x}$ in $\hat{Y}(\mathbf{x})$ of Eq. 5.21,

$$
\hat{Y}_0 = b_0 + \mathbf{x}_0'\mathbf{b} + \mathbf{x}_0'\mathbf{B}\mathbf{x}_0
$$

$$
= b_0 + \frac{\mathbf{x}_0'\mathbf{b}}{2} \tag{5.25}
$$

For the percent yield example, the predicted yield at the stationary point, whose coordinates are given in Eq. 5.23, is $\hat{Y}_0 = 96.60 + 0.013 = 96.613$.

Thus far we have presented a method for obtaining the coordinates of the stationary point and used these coordinates to obtain an estimate of the mean response at the stationary point (assuming the stationary point is inside the region of experimentation). The next step is to turn our attention to expressing the response system in canonical form so as to be able to describe in greater

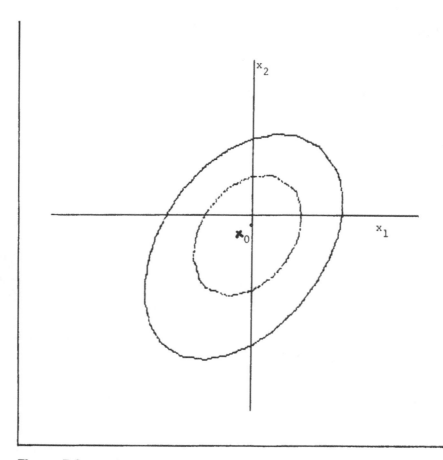

Figure 5.8 Confidence regions for the coordinates of the true stationary point. The smaller (inner) region is the 95% confidence region, and the larger (outer) region is the 99% confidence region.

detail the nature of the response system in the neighborhood of the stationary point.

5.5.1 The Canonical Equation of a Second-Order Response System

The first step in developing the canonical equation for a k-variable system is to translate the origin of the system from the center of the design to the stationary point, that is, to move from $(x_1, x_2, \ldots, x_k) = (0, 0, \ldots, 0)$ to $\mathbf{x}_0$. This is done by defining the intermediate variables $(z_1, z_2, \ldots, z_k)' = (x_1 - x_{10}, x_2 -$

$x_{20}, \ldots, x_k - x_{k0})'$ or $\mathbf{z} = \mathbf{x} - \mathbf{x}_0$. Then the second-order response equation (Eq. 5.21), expressed in terms of the values of z_i, is

$$\hat{Y}(\mathbf{z}) = b_0 + (\mathbf{z} + \mathbf{x}_0)'\mathbf{b} + (\mathbf{z} + \mathbf{x}_0)'\mathbf{B}(\mathbf{z} + \mathbf{x}_0)$$

$$= \hat{Y}_0 + \mathbf{z}'\mathbf{Bz} \qquad (5.26)$$

In the intermediate variables, the predicted response is a linear function of the estimate of the response at the stationary point, $\hat{Y}_0$, plus a quadratic form in the values of z_i. The axes of the values of z_i are aligned with the corresponding axes of the values of x_i, as shown in Figure 5.9a for the case where $k = 2$.

Now, to obtain the canonical form of the predicted response equation, let us define a set of variables $W_1, W_2, \ldots, W_k$ such that $\mathbf{W} = (W_1, W_2, \ldots, W_k)'$ is given by

$$\mathbf{W} = \mathbf{M}'\mathbf{z} \qquad (5.27)$$

where $\mathbf{M}$ is a $k \times k$ orthogonal matrix whose columns are eigenvectors of the matrix $\mathbf{B}$. The matrix $\mathbf{M}$ has the effect of diagonalizing $\mathbf{B}$, that is, $\mathbf{M}'\mathbf{BM} = \text{diag}(\lambda_1, \lambda_2, \ldots, \lambda_k)$ by Theorem 14 in Appendix 2A of Chapter 2, where λ_1, $\lambda_2, \ldots, \lambda_k$ are the corresponding eigenvalues of $\mathbf{B}$. The axes associated with the variables $W_1, W_2, \ldots, W_k$ are called the *principal axes* of the response system. The transformation in Eq. 5.27 is a rotation of the z_i axes to form the W_i axes (see Figure 5.9b for $k = 2$). Furthermore, if we write $\mathbf{W} = \mathbf{M}'\mathbf{z} = \mathbf{M}'(\mathbf{x} - \mathbf{x}_0)$, then the coefficients of the x_i are the direction cosines of the W_i axes with respect to the x_i axes and a coefficient of x_i equal to unity would indicate that the W_i axis is parallel to the x_i axis.

To express Eq. 5.26 in the W_i variables, write the quadratic form, $\mathbf{z}'\mathbf{Bz}$, as

$$\mathbf{z}'\mathbf{Bz} = \mathbf{W}'\mathbf{M}'\mathbf{BMW}$$

$$= \lambda_1 W_1^2 + \cdots + \lambda_k W_k^2 \qquad (5.28)$$

The eigenvalues λ_i are real-valued (since the matrix $\mathbf{B}$ is a real-valued, symmetric matrix) and represent the coeffiicents of the W_i^2 terms in the canonical equation

$$\hat{Y} = \hat{Y}_0 + \sum_{i=1}^{k} \lambda_i W_i^2 \qquad (5.29)$$

The canonical equation for the percent yield surface is

$$\hat{Y} = 96.613 - 1.6091 W_1^2 - 2.2034 W_2^2 \qquad (5.30)$$

The canonical equation (Eq. 5.29) of the second-order response surface performs the same function for the second-order approximating polynomial (Eq. 5.21) as does the direction of steepest ascent (Eq. 5.11) for the first-order model of Eq. 5.1. It is easy to see that if $\lambda_1, \lambda_2, \ldots, \lambda_k$ are

Figure 5.9 (a) Defining the origin and axes of the (z_1, z_2) system at the stationary point. (b) Rotation of the z_1 and z_2 axes to form the W_1 and W_2 axes. The W_1 and W_2 axes are the principal axes of the response surface.

1. All negative, as in (5.30), then at x_0 the surface is a maximum (see Figure 5.7a).
2. All positive, then at x_0 the surface is a minimum.
3. Of mixed signs, that is, some are positive and others are negative, then x_0 is a saddle point of the fitted surface (see Figure 5.7b).

Furthermore, some of the λ_i values may be zero (or very close to zero), meaning the response does not change in value as one moves away from x_0 in the direction of the W_i-axis. This is the case, when for $k = 2$, the fitted surface is a stationary ridge system, as in Figure 5.7d, where $\lambda_2 = 0$ and $\lambda_1 < 0$. The surface drops off in the W_1 direction but does not change in the W_2 direction. In this case, if the experimenter is seeking a maximum response, then a range of values exist (or possible operating conditions) along the W_2 axis, all of which give approximately the same estimated optimum response value.

The magnitudes of the individual values of the λ_i tell how quickly the surface height changes along the W_i axes as one moves away from x_0. For example, if with $k = 2$, $|\lambda_2| > |\lambda_1|$, then the height of the response surface changes more rapidly when moving in the direction of the W_2 axis than when moving in the direction of the W_1 axis. The drop in the height of the surface represented by Eq. 5.30, for example, is approximately 37% = 100% (2.2034 − 1.6091)/1.6091 greater in the direction of the W_2 axis than in the direction of the W_1 axis (at equal distances along the axes from the stationary point), since $\lambda_1 = -1.6091$ and $\lambda_2 = -2.2034$.

Today there are computer software packages available that perform the steps of locating the coordinates of the stationary point, predict the response at the stationary point, and compute the eigenvalues, $\lambda_1, \lambda_2, \ldots, \lambda_k$ and the corresponding eigenvectors. For example, Table 5.8, lists the solution for the optimum response generated from PROC RSREG in SAS (1989) for the chemical reaction data of Table 5.5.

The final stage in expressing the response surface in its canonical form is that of expressing the values of W_i as functions of the values of x_i. This is done by noting that in the orthogonal transformation (Eq. 5.27) we stated that the k columns of the matrix $\mathbf{M}$ are the eigenvectors associated with the k eigenvalues, λ_i, of $\mathbf{B}$. Now, suppose we partition the $k \times k$ matrix $\mathbf{M}$ as

$$\mathbf{M} = [\mathbf{m}_1 : \mathbf{m}_2 : \ldots : \mathbf{m}_k]$$

where $\mathbf{m}_i$ is the ith column of $\mathbf{M}$. The elements of $\mathbf{m}_i$ are obtained by finding an eigenvector associated with λ_i and then normalizing it so that the sum of squares of the elements in $\mathbf{m}_i$ is unity, that is, $\mathbf{m}_i'\mathbf{m}_i = 1$. In other words, the elements of $\mathbf{m}_i$ are obtained by solving

$$(\mathbf{B} - \lambda_i \mathbf{I}_k)\mathbf{m}_i = 0 \qquad\qquad (5.31)$$

Table 5.8 SAS/RSREG Computer Printout of
the Location of the Stationary Point x_0, the
Predicted Percent Yield at x_0, and the Coefficients
λ_1 and λ_2 of the Canonical Eq. 5.30

Canonical Analysis of Response Surface		
	Factor	Critical Value
	$x1$	-0.004825623
	$x2$	-0.08573840
Predicted Value at Stationary Point	96.61327	
Eigenvalues	Eigenvectors	
	$x1$	$x2$
-1.60915	0.611381	0.7913363
-2.20339	0.7913363	-0.611381
Stationary point is a maximum		

for which $m_i'm_i = 1$. For two distinct $\lambda_i \neq \lambda_j$, the eigenvectors of **B** must satisfy the property that $m_i'm_j = 0$, $i \neq j$.

Once the values of the elements of the matrix **M** are known and the values of λ_i are found, a description of the estimated response surface can be made through the use of the canonical Eq. 5.29. Plots of the surface contours (when $k = 2$ or $k = 3$) are easily drawn to pictorially represent the changes in the response as one moves away from the stationary point. Since the canonical equation, which is used to represent the estimated surface, is expressed in the system of W_i, $i = 1, 2, \ldots, k$, we should like to express W_i in terms of x_i. Such a relationship is, from Eq. 5.27,

$$\mathbf{W} = \mathbf{M'z}$$
$$= \mathbf{M'(x - x_0)} \tag{5.32}$$

Before we present a numerical example in two variables to illustrate the steps in performing the canonical analysis of a second-order response system, let us add some further comments on typical surfaces generated by second-order equations when $k = 3$. With three factors, the contour surfaces can be built up from contour lines of two dimensions by superposition. The contour surfaces are again revealed by the signs of λ_1, λ_2, and λ_3, in the canonical equation

$$\hat{Y} = \hat{Y}_0 + \lambda_1 W_1^2 + \lambda_2 W_2^2 + \lambda_3 W_3^2$$

For example, if the signs are $(-, -, -,)$ the contour surfaces are ellipsoids. The following is a list of some typical contour surfaces generated by quadratic equations in three variables where examples of the surfaces are drawn in Figure 5.10.

1. Ellipsoids $(-, -, -)$
2. Hyperbolic paraboloid $(+, -, 0)$
3. Hyperboloids of one sheet $(+, +, -)$
4. Hyperboloids of two sheets $(+, -, -)$

5.5.2 The Canonical Analysis of the Percent Yield Surface

Let us recall the fitted second-order model in Eq. 5.20 which expresses percent yield as a function of the coded variables, $x_1 =$ (temp. $- 135.9)/10$ and $x_2 =$ (time $- 195)/23.1$. Carrying the coefficient estimates to four decimal places, the

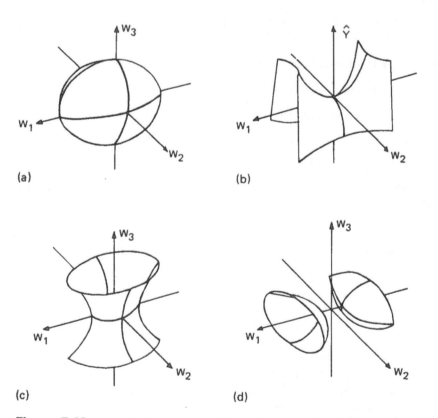

Figure 5.10 Some higher-dimensional contour surfaces (a, c, d), and a hyperbolic paraboloid response surface (b) for $k = 3$. (a) Ellipsoid; (b) hyperbolic paraboloid; (c) hyperboloid of one sheet; (d) hyperboloid of two sheets.

model is

$$\hat{Y}(x) = 96.6000 + 0.0302x_1 - 0.3112x_2 - 1.9813x_1^2 - 1.8313x_2^2 + 0.5750x_1x_2$$

$$= 96.6000 + \begin{bmatrix} x_1 \\ x_2 \end{bmatrix}' \begin{bmatrix} 0.0302 \\ -0.3112 \end{bmatrix} + \begin{bmatrix} x_1 \\ x_2 \end{bmatrix}' \begin{bmatrix} -1.9813 & 0.2875 \\ 0.2875 & -1.8313 \end{bmatrix} \begin{bmatrix} x_1 \\ x_2 \end{bmatrix}$$

$$= b_0 + \mathbf{x'b} + \mathbf{x'Bx}$$

The coordinates, $\mathbf{x}_0$, of the stationary point were obtained in Eq. 5.23, and the predicted percent yield at the stationary point was found to be $\hat{Y}_0 = 96.613$. The canonical equation for the percent yield surface was given as Eq. 5.30.

To obtain the form in Eq. 5.30 of the canonical equation, the eigenvalues (λ_1 and λ_2) of the matrix $\mathbf{B}$ are found by solving the determinantal equation $|\mathbf{B} - \lambda\mathbf{I}| = 0$, that is,

$$\begin{vmatrix} (-1.9813 - \lambda) & 0.2875 \\ 0.2875 & (-1.8313 - \lambda) \end{vmatrix} = 0$$

Expanding the determinant, we find

$$\lambda^2 + 3.8125\lambda + 3.5455 = 0$$

The roots of this equation are

$$\frac{-3.8125 \pm \sqrt{(3.8125)^2 - 4(1)(3.5455)}}{2(1)} = \frac{-3.8125 \pm 0.5942}{2}$$

resulting in $\lambda_1 = -1.6091$ and $\lambda_2 = -2.2034$. The canonical equation is, carrying the estimates to four decimal places,

$$\hat{Y} = \hat{Y}_0 + \lambda_1 W_1^2 + \lambda_2 W_2^2$$

$$= 96.6133 - 1.6091 W_1^2 - 2.2034 W_2^2$$
(5.33)

Since both λ_1 and λ_2 are negative, the surface is maximum at the stationary point. The estimated surface drops off faster as we move away from the point $\mathbf{x}_0$ in the direction of the W_2 axis (the minor axis) than when we move in the direction of the W_1 axis (the major axis).

In order to draw the surface contours (as shown in Figure 5.6), one needs to express the axes of the W_1 and W_2 system in terms of the axes in the x_1 and x_2 system. From Eq. 5.32, $\mathbf{W} = \mathbf{M'}(\mathbf{x} - \mathbf{x}_0)$, and therefore to find $\mathbf{M'}$ one needs the elements of $\mathbf{M}$. Since $(\mathbf{B} - \lambda_i\mathbf{I})\mathbf{m}_i = \mathbf{0}$, $i = 1, 2$, where $\mathbf{M} = [\mathbf{m}_1 : \mathbf{m}_2]$, then corresponding to $\lambda_1 = -1.6091$,

$$\begin{bmatrix} (-1.9813 + 1.6091) & 0.2875 \\ 0.2875 & (-1.8313 + 1.6091) \end{bmatrix} \begin{bmatrix} m_{11} \\ m_{21} \end{bmatrix} = \begin{bmatrix} 0 \\ 0 \end{bmatrix}$$

or

$$-0.3722m_{11} + 0.2875m_{21} = 0 \qquad (5.34)$$
$$0.2875m_{11} - 0.2222m_{21} = 0$$

To solve Eq. 5.34 for m_{11} and m_{21} subject to the normalizing condition that $m_{11}^2 + m_{21}^2 = 1$, we initially set $m_{11} = 1$ and solve for m_{21} in the first equation $-0.3722(1) + 0.2875m_{21} = 0$. The value of m_{21} is $m_{21} = 1.2945$. The normalized values of m_{11} and m_{21} are

$$m_{11} = \frac{1}{\sqrt{1^2 + (1.2945)^2}} = 0.6113$$

$$m_{21} = \frac{1.2945}{\sqrt{1^2 + (1.2945)^2}} = 0.7914$$

where as a check $(0.6113)^2 + (0.7914)^2 = 1.0000$, so that

$$\mathbf{m}_1 = \begin{bmatrix} 0.6113 \\ 0.7914 \end{bmatrix}$$

Corresponding to $\lambda_2 = -2.20335$, the system of equations to solve simultaneously is

$$0.2221m_{12} + 0.2875m_{22} = 0$$
$$0.2875m_{12} + 0.3721m_{22} = 0$$

and the normalized solutions are

$$m_{12} = \frac{-1.2945}{\sqrt{(-1.2945)^2 + 1^2}} = -0.7914 \qquad m_{22} = 0.6113$$

The matrix $\mathbf{M}'$ and the relationship between the W_i variables and the x_i variables, from Eq. 5.32, is

$$\begin{bmatrix} W_1 \\ W_2 \end{bmatrix} = \begin{bmatrix} 0.6113 & -0.7914 \\ 0.7914 & 0.6113 \end{bmatrix}' \begin{bmatrix} x_1 + 0.0049 \\ x_2 + 0.0857 \end{bmatrix}$$

so that

$$W_1 = 0.6113x_1 + 0.7914x_2 + 0.0708$$

and

$$W_2 = -0.7914x_1 + 0.6113x_2 + 0.0485$$

A plot of the W_1 and W_2 axes, relative to the x_1 and x_2 axes, is drawn in Figure 5.11.

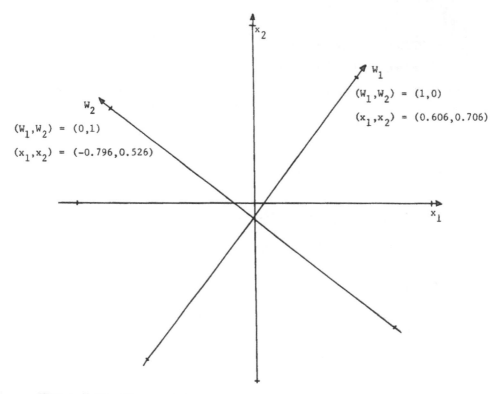

Figure 5.11   The principal axes of the percent yield surface defined in W_1 and W_2 relative to the coded temperature (x_1) and coded time (x_2) axes.

5.6   COMPUTER ANALYSIS OF A THREE-VARIABLE SNAP BEAN YIELD SURFACE

A central composite rotatable design (Section 4.5.3 in Chapter 4) was set up to investigate the effects of three fertilizer ingredients on the yield of snap beans under field conditions. The fertilizer ingredients and actual amounts applied were nitrogen (N), from 0.94 to 6.29 lb/plot; phosphoric acid (P_2O_5), from 0.59 to 2.97 lb/plot; and, potash (K_2O), from 0.60 to 4.22 lb/plot. The response of interest is the average yield in pounds per plot of snap beans.

The levels of nitrogen, phosphoric acid, and potash are coded, and the coded variables are defined as

$$x_1 = \frac{N - 3.62}{1.59} \qquad x_2 = \frac{P_2O_5 - 1.78}{0.71} \qquad x_3 = \frac{K_2O - 2.42}{1.07}$$

The values 3.62, 1.78, and 2.42 lb/plot represent the centers of the values for nitrogen, phosphoric acid, and potash, respectively. Five levels of each variable are used in the experimental design. The coded and measured levels for the variables are listed as

	Level of x_i				
	-1.682	-1.000	0.000	$+1.000$	$+1.682$
N	0.94	2.03	3.62	5.21	6.29
P_2O_5	0.59	1.07	1.78	2.49	2.97
K_2O	0.60	1.35	2.42	3.49	4.22

Six center point replications were run in order to obtain an estimate of the experimental error variance. Table 5.9 lists the design settings of x_1, x_2, and x_3 and the observed yield values at the 15 design points.

Table 5.9 Central Composite Rotatable Design Settings in the Coded Variables x_1, x_2, and x_3, the Original Variables N, P_2O_5, and K_2O, and the Average Yield of Snap Beans at Each Setting

x_1	x_2	x_3	N	P_2O_5	K_2O	Yield (Y)
-1	-1	-1	2.03	1.07	1.35	11.28
1	-1	-1	5.21	1.07	1.35	8.44
-1	1	-1	2.03	2.49	1.35	13.19
1	1	-1	5.21	2.49	1.35	7.71
-1	-1	1	2.03	1.07	3.49	8.94
1	-1	1	5.21	1.07	3.49	10.90
-1	1	1	2.03	2.49	3.49	11.85
1	1	1	5.21	2.49	3.49	11.03
-1.682	0	0	0.94	1.78	2.42	8.26
1.682	0	0	6.29	1.78	2.42	7.87
0	-1.682	0	3.62	0.59	2.42	12.08
0	1.682	0	3.62	2.97	2.42	11.06
0	0	-1.682	3.62	1.78	0.60	7.98
0	0	1.682	3.62	1.78	4.22	10.43
0	0	0	3.62	1.78	2.42	10.14
0	0	0	3.62	1.78	2.42	10.22
0	0	0	3.62	1.78	2.42	10.53
0	0	0	3.62	1.78	2.42	9.50
0	0	0	3.62	1.78	2.42	11.53
0	0	0	3.62	1.78	2.42	11.02

The complete second-order model to be fitted to the yield values is

$$Y = \beta_0 + \sum_{i=1}^{3} \beta_i x_i + \sum_{i=1}^{3} \beta_{ii} x_i^2 + \sum_{i<j}\sum \beta_{ij} x_i x_j + \varepsilon$$

The analysis of the data was performed using PROC RSREG in SAS (1989). The fitted second-order model (carrying the estimates to four decimals) is

$$\hat{Y}(\mathbf{x}) = 10.4624 - 0.5737x_1 + 0.1834x_2 + 0.4555x_3 - 0.6764x_1^2 + 0.5625x_2^2$$

$$- 0.2734x_3^2 - 0.6775x_1x_2 + 1.1825x_1x_3 + 0.2325x_2x_3 \qquad (5.35)$$

The test of the lack of fit hypothesis is not significant at the $\alpha = 0.10$ level, and so we proceed with the fitted model given in Eq. 5.35 for studying the shape characteristics of the yield surface. The coordinates of the stationary point are found to be

$$\mathbf{x}_0 = -\frac{1}{2} \begin{bmatrix} -0.6764 & -0.3387 & 0.5912 \\ -0.3387 & 0.5625 & 0.1162 \\ 0.5912 & 0.1162 & -0.2734 \end{bmatrix}^{-1} \begin{bmatrix} -0.5737 \\ 0.1834 \\ 0.4555 \end{bmatrix}$$

$$= \begin{bmatrix} -0.394 \\ -0.364 \\ -0.175 \end{bmatrix}$$

Expressed in units of pounds per plot, at $\mathbf{x}_0$ the levels of the original factors are N = 2.99, P_2O_5 = 1.52, and K_2O = 2.23. The point $\mathbf{x}_0$ is located inside the experimental region since $\mathbf{x}_0'\mathbf{x}_0 \leq 3$. The estimated yield at $\mathbf{x}_0$ is $\hat{Y}_0 = 10.4624 + \mathbf{x}_0'\mathbf{b}/2 = 10.5024$. These values are displayed on the computer printout of Table 5.10.

To obtain the canonical equation of the fitted surface, the eigenvalues of the matrix $\mathbf{B}$ are taken from Table 5.10 and are

$$\lambda_1 = 1.8413 \qquad \lambda_2 = 0.3673 \qquad \lambda_3 = -3.3040$$

The canonical equation of the snap bean yield surface is given by

$$\hat{Y} = 10.50 + 1.8413W_1^2 + 0.3673W_2^2 - 3.3040W_3^2 \qquad (5.36)$$

Since λ_1 and λ_2 are positive while λ_3 is negative the stationary point is a saddle point, or minimax point, of the surface. The magnitudes of the λ_i indicate the height of the surface changes faster when moving along the W_3 axis (where it decreases in value) while along the W_2 axis the response increases less rapidly than along the W_1 axis.

To obtain the relationship between the W_i variables and the x_i variables, the elements of the eigenvectors corresponding to λ_1, λ_2, and λ_3, respectively, are listed in Table 5.10. The eigenvectors, $\mathbf{m}_1$, $\mathbf{m}_2$, and $\mathbf{m}_3$ are the columns of the

Table 5.10 SAS/RSREG Computer Printout of the Analysis of the Snap Bean Yield Surface, with the Stationary Point a Saddle Point or Minimax Point

Response Surface for Variable Y

Response Mean		10.198	
Root MSE		0.995974	
R-Square		0.786146	
Coef of Variation		9.7664	

Regression	d.f.	Type I SS	R-Square	F-Ratio	Prob > F
Linear	3	7.78826086	0.1679	2.62	0.1088
Quadratic	3	13.38626756	0.2886	4.50	0.0303
Cross-Product	3	15.29095000	0.3297	5.14	0.0209
Total Regression	9	36.46547843	0.7861	4.08	0.0193

Residual	d.f.	SS	Mean Square	F-Ratio	Prob > F
Lack of Fit	5	7.38004157	1.47600831	2.906	0.1333
Pure Error	5	2.53960000	0.50792000		
Total Error	10	9.91964157	0.99196416		

Parameter	d.f.	Estimate	Std Error	T-Ratio	Prob > $|T|$
Intercept	1	10.46243541	0.40620958	25.76	0.0001
$x1$	1	−0.57371780	0.26949486	−2.13	0.0591
$x2$	1	0.18335880	0.26949486	0.68	0.5117
$x3$	1	0.45546837	0.26949486	1.69	0.1219
$x1 * x1$	1	−0.67635605	0.26231107	−2.58	0.0275
$x2 * x1$	1	−0.67750000	0.35212998	−1.92	0.0833
$x2 * x2$	1	0.56254334	0.26231107	2.14	0.0576
$x3 * x1$	1	1.18250000	0.35212990	3.36	0.0073
$x3 * x2$	1	0.23250000	0.35212998	0.66	0.5240
$x3 * x3$	1	−0.27340446	0.26231107	−1.04	0.3218

Factor	d.f.	SS	Mean Square	F-Ratio	Prob > F
$x1$	4	25.94912	6.48728	6.54	0.0075
$x2$	4	9.125901	2.281475	2.30	0.1302
$x3$	4	15.52996	3.882491	3.91	0.0364

Canonical Analysis of Response Surface

Factor	Critical Value
$x1$	−0.39427410
$x2$	−0.36431730
$x3$	−0.17458479

Predicted Value at Stationary Point 10.50238

Eigenvalues	Eigenvectors		
	$x1$	$x2$	$x3$
1.841262	−0.268043	0.9620844	−0.0504618
0.367301	0.5273195	0.1903476	0.8280712
−3.304048	0.8062797	0.1953494	−0.558347

Stationary Point is a Saddle Point

matrix $\mathbf{M}$ where $\mathbf{W} = \mathbf{M}'(\mathbf{x} - \mathbf{x}_0)$, that is,

$$\begin{bmatrix} W_1 \\ W_2 \\ W_3 \end{bmatrix} = \begin{bmatrix} -0.2680 & 0.9621 & -0.0505 \\ 0.5273 & 0.1903 & 0.8281 \\ 0.8063 & 0.1953 & -0.5583 \end{bmatrix} \begin{bmatrix} x_1 + 0.3943 \\ x_2 + 0.3643 \\ x_3 + 0.1746 \end{bmatrix}$$

Examination of the canonical Eq. 5.36 reveals that an increase in estimated yields of snap beans occurs upon moving away from the stationary point along the W_1 and W_2 axes while a decrease in yield results when moving away from $\mathbf{x}_0$ along the W_3 axis. It would seem reasonable that an experimenter might use either an equation in the original variables N, P_2O_5, and K_2O (Eq. 5.36) or Eq. 5.35 in the coded x_i variables, to find possible combinations of N, P_2O_5 and K_2O that produce high estimated yields. In Table 5.11 are listed 51 combinations of the levels of N, P_2O_5, and K_2O that produced estimated yields of $\hat{Y} = 11.5$, $\hat{Y} = 11.0$, and $\hat{Y} = 10.0$ lb/plot, respectively.

5.7 THE METHOD OF RIDGE ANALYSIS

During the analysis of a fitted response surface one may discover that the location of the stationary point is not inside the experimental region, and yet one may still wish to find the optimal response value within the boundaries of the region. As an example, suppose for $k = 2$ the boundary of the experimental region is represented by the perimeter of the largest circle in Figure 5.12. If the surface is a rising ridge system, then one may want to calculate the absolute maximum (minimum) value of the estimated response, $\hat{Y}$, over the experimental region. Such a search for the optimal value of $\hat{Y}$ is possible by the method of ridge analysis.

In general, this method is used for finding the absolute maximum (or minimum) of $\hat{Y}$ on concentric spheres of varying radii, R_l ($l = 1, 2, \ldots$) which are centered at $(x_1, x_2, \ldots, x_k) = (0, 0, \ldots, 0)$ and are contained within the experimental region. In Figure 5.12, three circles ($l = 1, 2, 3$) are drawn with different radii. The objective is to find the maximum value of $\hat{Y}$ as well as the location of the maximum value of $\hat{Y}$ on the perimeter of each circle.

Let us assume the fitted model over the region of the k coded variables x_1, $x_2, \ldots, x_k$ is of the second order and is expressible in the form

$$\hat{Y} = b_0 + \mathbf{x}'\mathbf{b} + \mathbf{x}'\mathbf{B}\mathbf{x} \tag{5.37}$$

Suppose we restrict the search of the stationary point to that of a point lying on the boundary of a sphere of radius R, that is, restrict our search to finding the coordinates of the variables which maximize $\hat{Y}$ subject to the condition

$$\sum_{i=1}^{k} x_i^2 = R^2 \tag{5.38}$$

Table 5.11   Combinations of Nitrogen (N), Phosphoric Acid (P_2O_5) and Potash (K_2O) Estimated to Produce Snap Beans Yields of 11.5, 11.0, and 10.0 Pounds per Plot (Based on 200 Plants per Plot)

Combination No.	Estimated Yield	N	P_2O_5	K_2O	Combination No.	Estimated Yield	N	P_2O_5	K_2O
1	11.5	3.52	0.67	2.30	37	10.0	3.83	1.61	1.84
2		2.47	2.37	2.17	38		2.15	1.43	2.63
3		4.79	1.06	3.73	39		5.17	1.82	2.97
4		3.94	2.42	3.63	40		3.12	1.59	3.93
5		1.20	1.99	0.74	41		3.28	1.52	1.17
6		2.05	0.62	0.84	42		1.48	1.33	2.00
7		4.07	0.79	2.90	43		6.17	1.96	3.64
8		3.05	2.43	2.77	44		5.09	2.39	2.82
9		5.60	1.16	3.83	45		5.77	1.30	2.90
10		4.74	2.53	3.72	46		3.21	1.02	4.10
11		3.75	2.42	4.19	47		2.53	2.11	4.01
12		3.11	2.47	3.60	48		4.51	1.72	2.47
13		4.13	0.83	3.72	49		3.85	2.35	1.58
14		1.86	2.22	0.75	50		4.71	0.97	1.69
15		1.92	2.39	2.40	51		1.28	2.07	2.78
16		4.06	0.66	2.07					
17		4.11	2.58	3.13					
18		5.13	0.94	3.26					
19	11.0	3.37	0.92	2.28					
20		2.62	2.12	2.19					
21		4.63	1.78	3.96					
22		4.55	1.40	3.68					
23		1.44	1.65	0.79					
24		4.12	2.08	3.63					
25		1.87	0.97	0.84					
26		2.67	0.87	1.58					
27		3.32	2.17	2.89					
28		3.88	0.89	2.07					
29		2.10	2.16	2.40					
30		2.96	0.79	2.50					
31		4.29	0.79	1.93					
32		1.70	2.26	2.54					
33		3.27	2.43	1.81					
34		4.86	1.81	3.92					
35		5.61	1.91	4.08					
36		4.48	1.77	4.10					

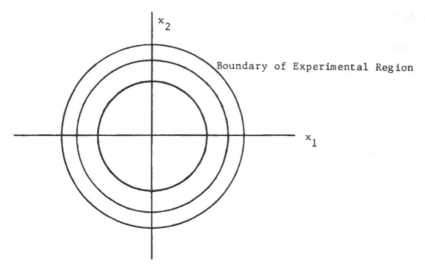

Figure 5.12   Concentric circles on which to locate maximum values of $\hat{Y}$, $k = 2$.

Once these coordinates or settings are found for a particular value of R, one can then change the value of R and repeat the procedure. Repeating the procedure by choosing different values for R and plotting the R values against the appropriate coordinates $x_1, x_2, \ldots, x_k$, and $\hat{Y}$, produces plots of the values of the maximum $\hat{Y}$ for various distances from the design center.

To maximize $\hat{Y}$ in Eq. 5.37 subject to the constraint of Eq. 5.38, we consider the function

$$F = \hat{Y} - \mu(\mathbf{x}'\mathbf{x} - R^2) \tag{5.39}$$

where μ is a Lagrangian multiplier and $\mathbf{x} = (x_1, x_2, \ldots, x_k)'$. Differentiating Eq. 5.39 with respect to each x_i, we have

$$\frac{\partial F}{\partial \mathbf{x}} = \mathbf{b} + 2\mathbf{B}\mathbf{x} - 2\mu\mathbf{x} \tag{5.40}$$

Equating Eq. 5.40 to zero, we obtain

$$(\mathbf{B} - \mu\mathbf{I})\mathbf{x} = \frac{-\mathbf{b}}{2} \tag{5.41}$$

Values for $x_1, x_2, \ldots, x_k$ are found by substituting a value for μ that is not an eigenvalue of $\mathbf{B}$ in Eq. 5.41 after which a value of R and a value of $\hat{Y}$ are calculated using Eqs. 5.38 and 5.37, respectively.

The choice of the value of μ selected for Eq. 5.41 can have an effect on the nature of the stationary point that is found. That is to say, some values of μ will produce stationary points of maximum estimated response while other values of μ will produce points of minimum estimated response. Furthermore, it may happen that different values of μ used in Eq. 5.41 may lead to the same R value in Eq. 5.38. The selection of the value of μ for generating a particular type of stationary point is now discussed in greater detail.

The choice of the value of μ for generating a particular type of stationary point. A well-known method (Kaplan, 1952: 128) of obtaining the stationary values of a function $f(x_1, x_2, \ldots, x_k) = f(\mathbf{x})$ of k variables subject to the n constraints

$$h_j(x_1, x_2, \ldots, x_k) = 0 \qquad j = 1, 2, \ldots, n \tag{5.42}$$

is the following. Form the function

$$F = f(\mathbf{x}) - \sum_{j=1}^{n} \mu_j h_j(\mathbf{x}) \tag{5.43}$$

where the μ_j are Lagrangian multipliers. Differentiating Eq. 5.43 partially with respect to each x_i and setting the results equal to zero produces the k equations

$$\frac{\partial F}{\partial x_i} = \frac{\partial f(\mathbf{x})}{\partial x_i} - \sum_{j=1}^{n} \mu_j \frac{\partial h_j(\mathbf{x})}{\partial x_i} = 0 \qquad i = 1, 2, \ldots, k \tag{5.44}$$

The $n + k$ equations, Eqs. 5.42 and 5.44, can be solved for the values of $x_1, x_2, \ldots, x_k$ and $\mu_1, \mu_2, \ldots, \mu_n$. Actually, the values of $\mu_1, \mu_2, \ldots, \mu_n$ are not of interest except to use them for calculating the values of $x_1, x_2, \ldots, x_k$. In fact, what is usually done is to eliminate the values of μ_j rather than calculate values for them and concentrate on solving for the values of $x_1, x_2, \ldots, x_k$ as follows.

Consider our ridge analysis problem where $n = 1$ and the single constraint is Eq. 5.38. Suppose $(x_1, x_2, \ldots, x_k) = (a_1, a_2, \ldots, a_k)$ is a solution of Eqs. 5.38 and 5.41 after the elimination of μ. Let

$$\mathbf{M}(\mathbf{x}) = \begin{bmatrix} \dfrac{\partial^2 F}{\partial x_1^2} & \dfrac{\partial^2 F}{\partial x_1 \partial x_2} & \cdots & \dfrac{\partial^2 F}{\partial x_1 \partial x_k} \\[2mm] \dfrac{\partial^2 F}{\partial x_2 \partial x_1} & \dfrac{\partial^2 F}{\partial x_2^2} & \cdots & \dfrac{\partial^2 F}{\partial x_2 \partial x_k} \\[2mm] \vdots & & \ddots & \vdots \\[2mm] \dfrac{\partial^2 F}{\partial x_k \partial x_1} & \dfrac{\partial^2 F}{\partial x_k \partial x_2} & \cdots & \dfrac{\partial^2 F}{\partial x_k^2} \end{bmatrix} = 2(\mathbf{B} - \mu \mathbf{I}_k) \tag{5.45}$$

be the symmetric $k \times k$ matrix of second-order partial derivatives and define $M(a)$ to be $M(x)$ evaluated at $x_1 = a_1, x_2 = a_2, \ldots, x_k = a_k$. Then the nature of the stationary point depends on the properties of $M(a)$. In particular,

1. If $M(a)$ is positive definite, that is, if $d'M(a)d > 0$, for all d where d is any nonzero $k \times 1$ real vector, then $\hat{Y}$ achieves a local minimum at $x = a$.
2. If $M(a)$ is negative definite, that is, if $d'M(a)d < 0$, for all $d \neq 0$, then $\hat{Y}$ achieves a local maximum at $x = a$.
3. If $M(a)$ is indefinite, further investigation of the mean response near the point $x = a$ is required to determine what sort of stationary point has been obtained.

Note from Eq. 5.45 that the property of the matrix $M(x)$ depends on the nature of the real symmetric matrix B of second-order regression coefficients and the particular choice of the value of μ. Several results regarding the value of μ and the corresponding values of $x_1, x_2, \ldots, x_k$, of R and $\hat{Y}$ are stated now. For proofs of these results, see Myers (1976) and Draper (1963).

Result 1. Consider two solutions of Eq. 5.41, call them x_1 and x_2 corresponding to the two values μ_1 and μ_2, respectively. If with x_1 and x_2 the estimates are $\hat{Y}_1$ and $\hat{Y}_2$ on spheres of radii R_1 and R_2, respectively, then if $R_1 = R_2$ and $\mu_1 > \mu_2$, then $\hat{Y}_1 > \hat{Y}_2$. This result says that for two stationary points that are the same distance from the design center, the response estimate will be larger for that stationary point corresponding to the larger value of μ.

Result 2. If $R_1 = R_2$, and $M(x_1)$ is positive definite but $M(x_2)$ is indefinite, then $\hat{Y}_1 < \hat{Y}_2$.

Result 3. Let μ_1 be a chosen value of μ in Eq. 5.41, and let x_1 be the resulting solution, with R_1 being the corresponding radius. If $\mu_1 > \lambda_i$ (all i) where λ_i is the ith eigenvalue of B, then x_1 is a point at which $\hat{Y}$ attains a local maximum on R_1. If, on the other hand, $\mu_1 < \lambda_i$ (all i), then x_1 is a point at which $\hat{Y}$ attains a local minimum on R_1.

Result 4. Suppose, as R increases, the values of the coordinates of the corresponding stationary points are obtained along with the changing values of $\hat{Y}$. Then, as R increases, the value of $\hat{Y}$ changes in one of the following ways (assuming the response surface is quadratic):

1. $\hat{Y}$ decreases monotonically.
2. $\hat{Y}$ increases monotonically.
3. $\hat{Y}$ passes through a maximum (or minimum) and decreases (or increases) monotonically.

The proof of condition 3. is presented in Draper (1963).

The above four results have the following implication. In order for the experimenter to be assured that stationary points used for the ridge analysis plot are local maxima, the value of μ used in Eq. 5.41 should be larger than the largest eigenvalue of the matrix **B** (as a matter of fact, the resulting values of x obtained will be points of absolute maxima, not merely local maxima). When searching for local minima, the values of μ selected should be smaller than the smallest eigenvalue of **B** (the resulting values of x will give the absolute minima of $\hat{Y}$). A plot of R against values of μ over the range of values $-\infty < \mu < +\infty$ is provided in Myers (1976) and Draper (1963).

The results of ridge analysis optimization can be easily obtained using PROC RSREG in SAS (1989) by simply adding the "Ridge Max(Min)" statement to the RSREG SAS code. For example, let us again consider the snap bean example in Section 5.6. We recall that the stationary point, x_0, was a saddle point. In this case, the maximum predicted response over the experimental region is attained on the boundary of the region. Applying "Ridge Max" to this example produces the following information: Maximum estimated yield = 12.886, which is attained at the point $x_1 = -0.5443$, $x_2 = 1.5890$, $x_3 = 0.0892$. Expressed in units of pounds per plot, the corresponding levels of the original factors are N = 2.755, P_2O_5 = 2.9082, and K_2O = 2.515.

For the interested reader, the complete SAS code used for this example is the following:

```
DATA;
INPUT X1 X2 X3 Y;
CARDS;
(insert here values of x1, x2, x3, and Y from Table 5.9)
PROC SORT;
BY X1-X3;
RUN;
PROC RSREG;
MODEL Y = X1-X3/LACKFIT;
RIDGE MAX;
RUN;
```

Note that the two statements, "PROC SORT" and "BY X1–X3," are needed to perform a lack-of-fit test (see Table 5.10). The data are sorted by the variables x_1, x_2, and x_3 so that the replicated observations at the center are grouped together (see SAS, 1989, p. 1463).

5.8 MODIFIED RIDGE ANALYSIS

In the method described in Section 5.7, no consideration was given to the size of the prediction variance. It should be noted, however, that the determination of the optimum value of $\hat{Y}$ on a sphere S_R of radius R is meaningful provided that $\text{Var}(\hat{Y})$ is constant on S_R, which occurs only when the design is rotatable. Furthermore, small values of $\text{Var}(\hat{Y})$ are desirable for a reliable estimate of the optimum. It is therefore important that the prediction variance be given serious consideration while searching for the optimum. Otherwise, the usefulness of the method of ridge analysis can be severely impaired, especially if $\text{Var}(\hat{Y})$ is highly dispersed on S_R. This can happen, for example, in poorly designed experiments where multicollinearity exists among some of the columns of the $\mathbf{X}$ matrix for the fitted model. Thus the procedure given in Section 5.7 should be modified so as to account for the prediction variance. The steps taken for determining the optimum value of $\hat{Y}$ on a sphere of radius R, subject to certain constraints on the value of $\text{Var}(\hat{Y})$, are described now.

Let us write the second-order model of Eq. 5.18 as

$$Y = \mathbf{f}'(\mathbf{x})\gamma + \varepsilon \qquad (5.46)$$

where

$$\mathbf{f}(\mathbf{x}) = (1, x_1, \ldots, x_k, x_1^2, \ldots, x_k^2, x_1 x_2, x_1 x_3, \ldots, x_{k-1} x_k)'$$

$$\gamma = (\beta_0, \beta_1, \ldots, \beta_k, \beta_{11}, \ldots, \beta_{kk}, \beta_{12}, \beta_{13}, \ldots, \beta_{k-1 k})'$$

Equation 5.37 can then be expressed as

$$\hat{Y} = \mathbf{f}'(\mathbf{x})(\mathbf{X}'\mathbf{X})^{-1}\mathbf{X}'\mathbf{Y} \qquad (5.47)$$

where $\mathbf{X} = [\mathbf{f}(\mathbf{x}_1) : \mathbf{f}(\mathbf{x}_2) : \ldots : \mathbf{f}(\mathbf{x}_N)]'$ with $\mathbf{x}_u$ being the vector of design settings at the uth experimental run ($u = 1, 2, \ldots, N$), and $\mathbf{Y}$ is the corresponding vector of observations. The prediction variance is given by

$$\text{Var}(\hat{Y}) = \mathbf{f}'(\mathbf{x})(\mathbf{X}'\mathbf{X})^{-1}\mathbf{f}(\mathbf{x})\sigma^2 \qquad (5.48)$$

where σ^2 is the error variance.

Next, let us write $\text{Var}(\hat{Y})$ as a weighted sum of quadratic forms with the weights being the eigenvalues of $(\mathbf{X}'\mathbf{X})^{-1}$. Let $\nu_1, \nu_2, \ldots, \nu_p$ denote the eigenvalues of $\mathbf{X}'\mathbf{X}$, where $p = (k + 1)(k + 2)/2$ is the number of elements of γ. If the matrix $\mathbf{X}$ is of full column rank, then $\nu_i > 0$ for all i. From Eq. 5.48, the following bounds on $\text{Var}(\hat{Y})$ can be obtained:

$$\frac{\sigma^2 \mathbf{f}'(\mathbf{x})\mathbf{f}(\mathbf{x})}{\nu_{\max}} \leq \text{Var}(\hat{Y}) \leq \frac{\sigma^2 \mathbf{f}'(\mathbf{x})\mathbf{f}(\mathbf{x})}{\nu_{\min}} \qquad (5.49)$$

where $\nu_{\min}$ and $\nu_{\max}$ are, respectively, the smallest and largest of the ν_i's. These bounds result from applying, for example, Theorem 2.3.16 in Khuri (1993) and the fact that $1/\nu_1, 1/\nu_2, \ldots, 1/\nu_p$ are the eigenvalues of $(\mathbf{X}'\mathbf{X})^{-1}$. It is easy to see that when $\nu_{\min}$ is small, large values of $\text{Var}(\hat{Y})$ can be expected at some points inside the experimental region. This occurs when the columns of $\mathbf{X}$ are multicollinear resulting in an ill-conditioned $\mathbf{X'X}$ matrix. However, since $\mathbf{X'X}$ is a symmetric matrix, by Theorem 14 in Appendix 2A (Chapter 2), there exists an orthogonal matrix $\mathbf{V}$ of order $p \times p$ such that

$$\mathbf{X'X} = \mathbf{V}\text{Diag}(\nu_1, \nu_2, \ldots, \nu_p)\mathbf{V'} \tag{5.50}$$

The columns of $\mathbf{V}$, denoted by $\mathbf{v}_1, \mathbf{v}_2, \ldots, \mathbf{v}_p$, are orthonormal eigenvectors of $\mathbf{X'X}$. Substituting Eq. 5.50 into Eq. 5.48 we obtain

$$\text{Var}(\hat{Y}) = \mathbf{f'}(\mathbf{x})\mathbf{V}\text{Diag}\left(\frac{1}{\nu_1}, \frac{1}{\nu_2}, \ldots, \frac{1}{\nu_p}\right)\mathbf{V'}\mathbf{f}(\mathbf{x})\sigma^2$$

$$= \sigma^2 \sum_{i=1}^{p} \frac{[\mathbf{f'}(\mathbf{x})\mathbf{v}_i]^2}{\nu_i} \tag{5.51}$$

Suppose that the elements of $\mathbf{v}_i$ are denoted in a manner corresponding to that of $\mathbf{f}(\mathbf{x})$, that is,

$$\mathbf{v}_i = (v_{0i}, v_{1i}, \ldots, v_{ki}, v_{11i}, \ldots, v_{kki}, v_{12i}, v_{13i}, \ldots, v_{k-1\,ki})',$$
$$i = 1, 2, \ldots, p$$

Then, $\mathbf{f'}(\mathbf{x})\mathbf{v}_i$ can be expressed as a quadratic form, namely,

$$\mathbf{f'}(\mathbf{x})\mathbf{v}_i = v_{0i} + \mathbf{x'}\boldsymbol{\tau}_i + \mathbf{x'}\mathbf{T}_i\mathbf{x}, \qquad i = 1, 2, \ldots, p \tag{5.52}$$

where $\boldsymbol{\tau}_i = (v_{1i}, v_{2i}, \ldots, v_{ki})'$ and

$$\mathbf{T}_i = \begin{bmatrix} v_{11i} & \frac{1}{2}v_{12i} & \frac{1}{2}v_{13i} & \cdots & \frac{1}{2}v_{1ki} \\ & v_{22i} & \frac{1}{2}v_{23i} & \cdots & \frac{1}{2}v_{2ki} \\ & & \cdot & & \cdot \\ & & & \cdot & \cdot \\ & & & & \cdot \\ & & & & \frac{1}{2}v_{k-1\,ki} \\ \text{symmetric} & & & & v_{kki} \end{bmatrix}, \qquad i = 1, 2, \ldots, p$$

From Eqs. 5.51 and 5.52, we then have

$$\text{Var}(\hat{Y}) = \sigma^2 \sum_{i=1}^{p} \frac{(v_{0i} + \mathbf{x'}\boldsymbol{\tau}_i + \mathbf{x'}\mathbf{T}_i\mathbf{x})^2}{\nu_i} \tag{5.53}$$

where v_{0m}, $\boldsymbol{\tau}_m$, and $\mathbf{T}_m$ are the values of v_{0i}, $\boldsymbol{\tau}_i$, and $\mathbf{T}_i$, respectively, that correspond to $\nu_{\min}$, the smallest eigenvalue of $\mathbf{X'X}$. As was noted earlier, a small value of $\nu_{\min}$ can lead to inflated prediction variance values at some points inside the region where optimization of $\hat{Y}$ is to take place. Such points should be avoided during the process of optimization on spheres.

To reduce the size of the prediction variance within the region explored by ridge analysis, we should consider placing constraints on the absolute value of

$v_{0m} + x'\tau_m + x'T_m x$. This is needed in order to offset the smallness of v_{min}. Let us therefore consider optimizing $\hat{Y}$ under the following constraints:

$$x'x = R^2 \tag{5.54}$$

$$|v_{0m} + x'\tau_m + x'T_m x| \le q \tag{5.55}$$

where q is a positive constant, small enough to minimize the effect of v_{min}. As a rule of thumb, q can be chosen to have an order of magnitude comparable to that of the largest value taken by $|v_{0m} + x'\tau_m + x'T_m x|$ at the design points. The rationale behind this rule is that the prediction variance is smaller at the design points than at other locations inside the experimental region.

Inequality 5.55 represents an added constraint to the usual procedure of ridge analysis, which, if we recall, uses Eq. 5.54 as the only constraint. With this added constraint, a better control on the size of the prediction variance can be maintained.

Optimization of $\hat{Y}$ under the constraints given in Eqs. 5.54 and 5.55 can be carried out using the so-called dual response technique (see Section 7.6.1 and also Myers and Carter, 1973). As in ridge analysis, this technique uses the method of Lagrange multipliers. Instead of the function F in Eq. 5.39, we now consider the function

$$\bar{F} = \hat{Y} - \lambda(v_{0m} + x'\tau_m + x'T_m x - \omega) - \mu(x'x - R^2)$$

where $\hat{Y}$ is expressed as in Eq. 5.37, λ and μ are Lagrange multipliers, and ω represents a value of $v_{0m} + x'\tau_m + x'T_m x$ such that $|\omega| \le q$. Differentiating $\bar{F}$ with respect to $x_1, x_2, \ldots, x_k$ and equating the partial derivatives to zero, we obtain

$$(B - \lambda T_m - \mu I_k)x = \tfrac{1}{2}(\lambda \tau_m - b) \tag{5.56}$$

Following the recommendation given in Myers and Carter (1973), values of λ and μ are appropriately chosen to ensure that the solution of Eq. 5.56 yields the desired optimum (maximum or minimum). Thus for a given value of λ, the $k \times k$ matrix of second-order partial derivatives of $\bar{F}$, namely $2(B - \lambda T_m - \mu I_k)$, is made positive definite, and hence an absolute minimum of $\hat{Y}$ is achieved on a sphere of radius R, if μ is chosen smaller than the smallest eigenvalue of $B - \lambda T_m$. For an absolute maximum, μ is chosen larger than the largest eigenvalue of the same matrix. Solving Eq. 5.56 under such chosen values of λ and μ will result in an optimum of $\hat{Y}$ subject to a fixed $R = (x'x)^{1/2}$ and a fixed value of $v_{0m} + x'\tau_m + x'T_m x$. By selecting several values of λ and μ in this manner, we can identify operating conditions that produce optimal response values with pleasingly low values of $Var(\hat{Y})$ within the region of experimentation.

5.8.1 Numerical Example

An experiment was conducted for the purpose of determining conditions on the concentration of three basic substances that maximize a certain mechanical modular property of a solid propellant. It was initially planned to use a central composite design (CCD) for the three components in the system. However, because of experimental difficulties, it was not possible to do so. Instead, the design settings actually used in the experiment constituted a "deformed" CCD, that is, a design not having the exact settings of those of a CCD. This deformation led to problems with multicollinearity as far as the fitting of a second-order model is concerned. The design settings and corresponding response values are given in Table 5.12. We note that most of the design points are at a distance of 2 or less from the design center.

The smallest eigenvalue of $X'X$ is $\nu_{\min} = 0.0321$. The corresponding values of ν_{0m}, τ_m, and T_m (see inequality 5.55) are $\nu_{0m} = -0.2935$, $\tau_m = (0.0469, 0.4081, 0.4071)'$, and

$$T_m = \begin{bmatrix} 0.1129 & 0.0095 & 0.2709 \\ 0.0095 & -0.1382 & -0.0148 \\ 0.2709 & -0.0148 & 0.6453 \end{bmatrix}$$

Table 5.12 Design Settings and Response Values for Example 5.8.1

x_1	x_2	x_3	Y
−1.020	−1.402	−0.998	13.5977
0.900	0.478	−0.818	12.7838
0.870	−1.282	0.882	16.2780
−0.950	0.458	0.972	14.1678
−0.930	−1.242	−0.868	9.2461
0.750	0.498	−0.618	17.0167
0.830	−1.092	0.732	13.4253
−0.950	0.378	0.832	16.0967
1.950	−0.462	0.002	14.5438
−2.150	−0.402	−0.038	20.9534
−0.550	0.058	−0.518	11.0411
−0.450	1.378	0.182	21.2088
0.150	1.208	0.082	25.5514
0.100	1.768	−0.008	33.3793
1.450	−0.342	0.182	15.4341

Source: A. I. Khuri and R. H. Myers (1979). Reproduced with permission of the American Statistical Association.

The value of $|v_{0m} + x'\tau_m + x'T_m x|$ was computed at each of the 15 design points in Table 5.12. The largest value was found to be 0.087. For inequality 5.55, q is chosen as less than twice this value. We should therefore consider maximizing $\hat{Y}$ subject to the constraints

$$x'x \leq 4.0 \qquad\qquad (5.57)$$
$$|v_{0m} + x'\tau_m + x'T_m x| \leq 0.17$$

Table 5.13 gives maximum values of $\hat{Y}$ on concentric spheres of varying radii, $R \leq 2.0$. The results of applying the standard procedure of ridge analysis, that is, without the additional constraint of Eq. 5.57, are displayed in Table 5.14. This table also gives values of $\omega^2/v_{\min}$, which should be compared with the corresponding values of $\text{Var}(\hat{Y})/\sigma^2$. We recall from Eq. 5.53 that $\omega^2/v_{\min}$ is a portion of $\text{Var}(\hat{Y})/\sigma^2$. The closer the former is to the latter, the greater the effect of the additional constraint in Eq. 5.57 on reducing the prediction variance.

From Tables 5.13 and 5.14 it is easy to see that the precision of $\hat{Y}$ at the estimated maxima in Table 5.13 is markedly better than in Table 5.14, especially for large values of R. We note that the highest value of $\hat{Y}$ in Table 5.13 is 46.26 with $\text{Var}(\hat{Y})/\sigma^2 = 5.336$. The corresponding values in Table 5.14 are 57.176 and 145.907, respectively. Even though the former value of $\hat{Y}$ is smaller than the latter, the big discrepancy in their corresponding variances makes the value 46.26 a more reliable estimate of the maximum mean response. The user should therefore consider a tradeoff between a high response value and a pleasingly low prediction variance.

For more details concerning the role of the prediction variance in ridge analysis, see Khuri and Myers (1979).

5.9   OTHER FUNCTION OPTIMIZATION TECHNIQUES

Previously in this chapter we showed how to determine the location of the stationary point of a response surface where the surface could be expressed in

Table 5.13   Results of Modified Ridge Analysis

R	0.848	1.162	1.530	1.623	1.795	1.850	1.904	1.935	2.000
$|\omega|$	0.006	0.074	0.136	0.139	0.048	0.086	0.126	0.146	0.165
$\frac{1}{\sigma^2}\text{Var}(\hat{Y})$	1.170	1.635	2.922	3.147	1.305	2.330	3.510	4.177	5.336
x_1	0.410	0.563	0.773	0.785	0.405	0.601	0.750	0.820	0.965
x_2	0.737	1.015	1.320	1.422	1.752	1.751	1.750	1.752	1.752
x_3	0.097	0.063	0.019	0.011	0.000	0.000	0.012	0.015	−0.030
$\hat{Y}$	22.420	27.780	35.242	37.190	37.042	40.222	42.830	44.110	46.260

Source: A. I. Khuri and R. H. Myers (1979). Reproduced with permission of the American Statistical Association.

Table 5.14   Results of Standard Ridge Analysis

R	0.140	0.379	0.698	0.938	1.146	1.394	1.484	1.744	1.944	1.975	2.000
$|\omega|$	0.241	0.124	0.104	0.337	0.587	0.942	1.085	1.553	1.958	2.025	2.080
$\omega^2/\nu_{\min}$	1.809	0.479	0.337	3.538	10.734	27.644	36.674	75.134	119.432	127.745	134.779
$\frac{1}{\sigma^2}\mathrm{Var}(\hat{Y})$	2.592	1.554	2.104	6.138	14.305	32.787	42.475	83.380	129.834	138.668	145.907
x_1	0.037	0.152	0.352	0.515	0.660	0.835	0.899	1.085	1.227	1.249	1.265
x_2	0.103	0.255	0.422	0.531	0.618	0.716	0.749	0.845	0.916	0.927	0.936
x_3	0.087	0.235	0.431	0.577	0.705	0.858	0.912	1.074	1.197	1.217	1.232
$\hat{Y}$	12.796	16.021	21.365	26.229	31.086	37.640	40.197	48.332	55.147	56.272	57.176

Source: A. I. Khuri and R. H. Myers (1979). Reproduced with permission of the American Statistical Association.

the form of a second-order model in k variables. When the location of the stationary point was inside the experimental region, we went on to show how to express the surface using a canonical form of the model in a set of variables whose axes represent the principal axes of the response system.

When the location of the stationary point was outside the region of interest, we could choose to remain within the boundaries of the experimental region and search for the optimal response value on concentric spheres of varying radii. Such a search was described as performing ridge analysis.

5.9.1 A Derivative-Free Optimization Technique: The Nelder-Mead Simplex Method

The Nelder-Mead simplex method for function minimization is a "direct" method that requires no derivatives of the function to be minimized. An objective function in k variables is evaluated at the $k + 1$ vertices of a general simplex, and minimization of the function involves moving away from the vertex, with the highest value of the function, to another point. The simplex procedure adapts itself to the local landscape and continues to search for the final minimum point. The generality of the method has been illustrated by Olsson and Nelson (1975) in solving such problems as the direct maximization of the logarithms of a likelihood function (maximization is accomplished by minimizing the negative of the function), the solution of simultaneous equations, the maximization of a quadratic function that is subject to a quadratic constraint, the fitting of a line by minimizing the sum of squares of perpendicular distances from the points to the line, nonlinear least squares, and the fitting of approximations to tabular data.

To describe the simplex method, let us consider the minimization of a function of k variables where $Y = f(x_1, x_2, \ldots, x_k)$ is the observed value of the function at the settings $x_1, x_2, \ldots, x_k$. Let $\mathbf{P}_0, \mathbf{P}_1, \ldots, \mathbf{P}_k$ be the $k + 1$ vertices of a k-dimensional simplex. Let us denote the value of the function at the point

P_i by $Y_i = f(P_i)$ and define the maximum (high) and minimum (low) values of the function as

$$Y_H = \max_{0 \le i \le k} (Y_i) \qquad Y_L = \min_{0 \le i \le k} (Y_i)$$

that is,

$$f(\mathbf{P}_H) \ge f(\mathbf{P}_i) \qquad \text{and} \qquad f(\mathbf{P}_L) \le f(\mathbf{P}_i) \qquad i = 0, 1, \ldots, k$$

Further, we define $\overline{\mathbf{P}}$ as the centroid of the points other than $\mathbf{P}_H$ and write $[\mathbf{P}_i\mathbf{P}_j]$ for the distance from $\mathbf{P}_i$ to $\mathbf{P}_j$. At each stage in the minimization process, $\mathbf{P}_H$, the point at which Y is maximum, is replaced by a new point according to one of three operations that could be performed—reflection, contraction, and expansion. These actions are defined as follows:

Definition 5.1. *Reflection*: The reflection of $\mathbf{P}_H$ is the point $\mathbf{P}^*$ defined by the relation

$$\mathbf{P}^* = (1 + \alpha)\overline{\mathbf{P}} - \alpha\mathbf{P}_H$$

where $\alpha = [\mathbf{P}^*\overline{\mathbf{P}}]/[\mathbf{P}_H\overline{\mathbf{P}}] > 0$ is the reflection coefficient. In Figure 5.13, $\mathbf{P}^*$ is on the line joining $\mathbf{P}_H$ and $\overline{\mathbf{P}}$ but lies on the opposite of $\overline{\mathbf{P}}$ from $\mathbf{P}_H$ so that $[\mathbf{P}^*\overline{\mathbf{P}}] = \alpha[\mathbf{P}_H\overline{\mathbf{P}}]$. If $Y_L < Y^* < Y_H$ where $Y^* = f(\mathbf{P}^*)$, then $\mathbf{P}_H$ is replaced by $\mathbf{P}^*$, and we start the process over again with the simplex whose vertices are $\mathbf{P}^*$, $\mathbf{P}_1$, and $\mathbf{P}_2$.

Definition 5.2. *Expansion*: If $Y^* < Y_L$, then $\mathbf{P}^*$ is expanded to $\mathbf{P}_E^{**}$ by the relation

$$\mathbf{P}_E^{**} = \gamma\mathbf{P}^* + (1 - \gamma)\overline{\mathbf{P}}$$

where $\gamma = [\mathbf{P}_E^{**}\overline{\mathbf{P}}]/[\mathbf{P}^*\overline{\mathbf{P}}] > 1$ is the expansion coefficient. If $Y_E^{**} = f(\mathbf{P}_E^{**}) < Y_L$, replace $\mathbf{P}_H$ by $\mathbf{P}_E^{**}$ and restart the process. However, if $Y_E^{**} > Y_L$, then replace $\mathbf{P}_H$ by $\mathbf{P}^*$ and restart.

Definition 5.3. *Contraction*: If upon reflecting $\mathbf{P}_H$ to $\mathbf{P}^*$ we find $Y^* > Y_i$ for all $i \ne H$, then define a new $\mathbf{P}_H$ to be either the old $\mathbf{P}_H$ or $\mathbf{P}^*$, whichever has the lower Y value. Now form

$$\mathbf{P}_C^{**} = \beta\mathbf{P}_H + (1 - \beta)\overline{\mathbf{P}}$$

where $\beta = [\mathbf{P}_C^{**}\overline{\mathbf{P}}]/[\mathbf{P}_H\overline{\mathbf{P}}] < 1$ and substitute $\mathbf{P}_C^{**}$ for $\mathbf{P}_H$ and restart, unless $Y_C^{**} > \min(Y_H, Y^*)$, that is, the contracted point is worse than the better of $\mathbf{P}_H$ or $\mathbf{P}^*$. When such a contraction fails, we replace all of the values of $\mathbf{P}_i$ (i.e., $\mathbf{P}_0$, $\mathbf{P}_1$, and $\mathbf{P}_2$ in Figure 5.13) by $(\mathbf{P}_i + \mathbf{P}_L)/2$ and restart the process.

As an aid to illustrating the step-by-step procedure, a flow diagram is drawn in Figure 5.14. The form of the flow diagram is similar to a flow diagram presented

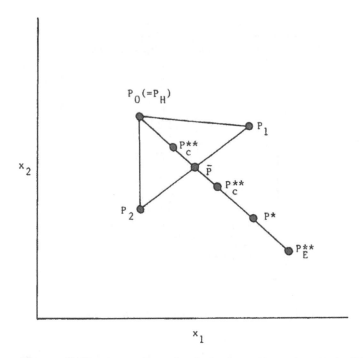

Figure 5.13 A two-dimensional simplex with vertices P_0 (= P_H), P_1, and P_2 and possible subsequent points.

by Nelder and Mead (1965: 309). Figure 5.14 lists the explanations of steps 1 through 6.

The criterion used to stop the search is one of variation in the Y_i values. At each step the standard error of the Y values, $s = \{\Sigma_{i=1}^{n} (Y_i - \overline{Y})^2/(n - 1)\}^{1/2}$ is calculated and compared against some preselected value, S say. The search is stopped when s is less than S. The reasoning behind this criterion is that when curvature near the minimum is slight, the variation in the Y values will not be due to curvature, but rather to experimental error. If this is large, there is no sense in trying to pinpoint the minimum very accurately. If the curvature is marked, then there is justification for pinning down the minimum more exactly.

Nelder and Mead (1965) discuss the strategy they used, in terms of values of α, β, and γ and step-length, in minimizing each of the functions:

1. $Y = 100(x_2 - x_1^2)^2 + (1 - x_1)^2$; Rosenbrock's (1960) parabolic valley.
2. $Y = (x_1 + 10x_2)^2 + 5(x_3 - x_4)^2 + (x_2 - 2x_3)^4 + 10(x_1 - x_4)^4$; Powell's (1962) quadratic function.

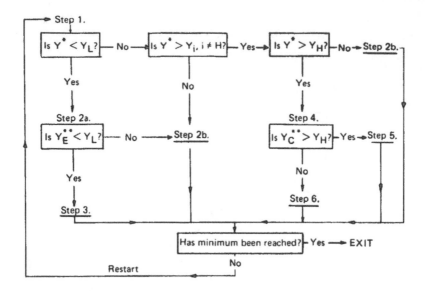

Select initial values of P_i, $i = 0, 1, 2, \ldots, k$ and calculate the values of Y_i at the P_i.

Step 1.   Determine H and L and calculate $\bar{P} = \sum_{i \neq H}^{k} P_i/k$. Select $\alpha > 0$, say $\alpha = 1/2$, 2/3, or 1 and find $P^* = (1 + \alpha)\bar{P} - \alpha P_H$ and calculate Y^*.

Step 2.   (a) Calculate $P_E^{**} = \gamma P^* + (1 - \gamma)\bar{P}$ by choosing $\gamma > 1$, say $\gamma = 1.5$, and calculate Y_E^{**}. (b) Replace P_H with P^*.

Step 3.   Replace P_H with P_E^{**}.

Step 4.   Calculate $P_C^{**} = \beta P_H + (1 - \beta)\bar{P}$ by choosing $0 < \beta < 1$ and calculate Y_C^{**}.

Step 5.   Replace all values of P_i with $(P_i + P_L)/2$.

Step 6.   Replace P_H with P_C^{**}.

Figure 5.14   Flow diagram of simplex procedure.

3. $Y = 100[x_3 - 10\theta(x_1, x_2)]^2 + [\sqrt{x_1^2 + x_2^2} - 1]^2 + x_3^2$, where $2\pi\theta(x_1, x_2) = \arctan(x_2/x_1)$, $x_1 > 0$ and $2\pi\theta(x_1, x_2) = \pi + \arctan(x_2/x_1)$, $x_1 < 0$; Fletcher and Powell's (1963) helical valley.

In their paper they compare the simplex method with a minimization technique suggested by Powell (1962). Using the mean number of evaluations required for convergence to the minimum for step-lengths ranging from 0.1 to 3.0, the simplex method averaged 144 evaluations for function (1), 216 for function (2), and 228 for function (3). Powell's technique similarly required 150 for function

(1), 235 for function (2), and somewhere between 177 and 375 for function (3) depending on the initial step-length. While there appears to be little difference between the two methods as far as the number of evaluations required to reach the minimum, the simplex method appears to establish an initial advantage over Powell's technique up to say 200 evaluations after which Powell's method apparently rapidly closes the gap. A table of values for functions (1), (2), and (3) for 20, 40, 60, . . . , 220 evaluations (i.e., in increments of 20 evaluations) using the two methods is presented by Nelder and Mead (1965).

EXERCISES

5.1 The equation of a surface is expressed as $\eta = \beta_0 + \beta_1 X_1 + \beta_2 X_2 + \beta_{12} X_1 X_2$ where X_1 and X_2 are the respective levels of two input variables.

  (a) What type of surface is η?
  (b) Show how as one moves along the X_i axis, $i = 1, 2$, the shape or height of the surface above the $X_1 X_2$ plane depends on the level X_j, $j = 1, 2, j \neq i$.

5.2 Using the percent yield values in Table 5.1, fit the model in the coded variables,

$$Y = \beta_0 + \beta_1 x_1 + \beta_2 x_2 + \beta_{12} x_1 x_2 + \varepsilon$$

Show that the test of $H_0: \beta_{12} = 0$ versus $H_a: \beta_{12} \neq 0$ is equivalent to the lack of fit test for the model $Y = \beta_0 + \beta_1 x_1 + \beta_2 x_2 + \varepsilon$ at the 0.05 level.

5.3 In Section 5.3.1, it was stated that the difference, $\bar{Y}_1 - \bar{Y}_0$, where $\bar{Y}_1$ is the average of n_1 responses at the points of a first-order orthogonal design and $\bar{Y}_0$ is the average of n_0 center point replicates, is an unbiased estimator of the sum of the quadratic coefficients, $\sum_{i=1}^{k} \beta_{ii}$, in the second-order model (Eq. 5.15). Show this.

5.4 Table 5.6 presents the computer printout of the analysis of variance for the percent yield second-order fitted model. The reciprocal of the coefficient of variation (1/0.0036) is an estimate of the signal-to-noise ratio for this data set. Explain what is meant by the signal-to-noise ratio.

5.5 Derive the expression, $\hat{Y}_0 = b_0 + x_0' b / 2$, for the estimate of the response at the stationary point of a second-order surface where the coordinates of the stationary point are given by Eq. 5.22.

5.6 Two types of fertilizers, one a standard N-P-K combination and the other a nutritional supplement, were applied to experimental plots to assess their effects on the yield of peanuts measured in pounds per plot. The level of the amount (lb/plot) of each fertilizer applied to a plot was determined by the coordinate settings of a central composite rotatable design. The data below represent the harvested yields of peanuts taken from two replicates of each experimental combination.

Fertilizer 1	Fertilizer 2	x_1	x_2	Yields Rep. 1	(lb/plot) Rep. 2
50	15	-1	-1	7.52	8.12
120	15	$+1$	-1	12.37	11.84
50	25	-1	$+1$	13.55	12.35
120	25	$+1$	$+1$	16.48	15.32
35.5	20	$-\sqrt{2}$	0	8.63	9.44
134.5	20	$+\sqrt{2}$	0	14.22	12.57
85	12.9	0	$-\sqrt{2}$	7.90	7.33
85	27.1	0	$+\sqrt{2}$	16.49	17.40
85	20	0	0	15.73	17.00

(a) Fit a second-order model to the peanut yields in the coded variables x_1 and x_2 and construct the corresponding ANOVA table.

(b) Fit a second-order model to the peanut yields in the coded variables $x_1 = (F_1 - 85)/49.5$ and $x_2 = (F_2 - 20)/7.1$ and construct the corresponding ANOVA table.

(c) Test the fitted models in (a) and (b) for lack of fit at the 0.05 level of significance.

(d) If the fitted models in (a) and (b) are adequate, locate the coordinates of the stationary point associated with each fitted model in (a) and (b). What is the estimated yield at the stationary point?

(e) What is the nature of the surface at the stationary point? What are the settings of fertilizers 1 and 2 at the stationary point?

REFERENCES AND BIBLIOGRAPHY

Box, B. E. P. and N. R. Draper (1987). *Empirical Model-Building and Response Surfaces*, New York: John Wiley.

Box, G. E. P. and J. S. Hunter (1954). A Confidence Region for the Solution of a Set of Simultaneous Equations with an Application to Experimental Design, *Biometrika*, **41**, 190–199.

Box, G. E. P., W. G. Hunter, and J. S. Hunter (1978). *Statistics for Experimenters: An Introduction to Design, Data Analysis, and Model Building*, New York: John Wiley.

Box, G. E. P. and K. B. Wilson (1951). On the Experimental Attainment of Optimum Conditions (with discussion), *J. R. Statist. Soc.*, **B13**, 1–45.

Bradley, R. A. (1958). Determination of Optimum Operating Conditions by Experimental Methods: Part I, Mathematics and Statistics Fundamental to the Fitting of Response Surfaces, *Industrial Quality Control*, **15**, 16–20.

Cornell, J. A. (1990). How to Apply Response Surface Methodology, *The ASQC Basic References in Quality Control: Statistical Techniques*, Vol. 8, revised ed., Milwaukee: ASQC.

Davies, O. L. (1956). *The Design and Analysis of Industrial Experiments*, 2nd ed., London: Oliver and Boyd.

Draper, N. R. (1963). "Ridge Analysis" of Response Surfaces, *Technometrics*, **5**, 469–479.

Fletcher, R. and M. J. D. Powell (1963). A Rapidly Convergent Descent Method for Minimization, *The Computer J.*, **6**, 163–168.

Hunter, J. S. (1958). Determination of Optimum Operating Conditions by Experimental Methods: Part II–1. Models and Methods, *Industrial Quality Control*, **15**, 16–24.

Hunter, J. S. (1959). Part II–2, Models and Methods, *Industrial Quality Control*, **16**, 7–15.

Hunter, J. S. (1959). Part II–3, Models and Methods, *Industrial Quality Control*, **16**, 6–14.

Kaplan, W. (1952). *Advanced Calculus*, Reading, Mass.: Addison-Wesley.

Khuri, A. I. (1993). *Advanced Calculus with Applications in Statistics*, New York: John Wiley.

Khuri, A. I. and R. H. Myers (1979). Modified Ridge Analysis, *Technometrics*, **21**, 467–473.

Montgomery, D. C. (1991). *Design and Analysis of Experiments*, 3rd ed., New York: John Wiley.

Myers, R. H. (1976). *Response Surface Methodology*, Blacksburg, VA: Author.

Myers, R. H. and W. H. Carter, Jr. (1973). Response Surface Techniques for Dual Response Systems, *Technometrics*, **15**, 301–317.

Myers, R. H. and A. I. Khuri (1979). A New Procedure for Steepest Ascent, *Commun. Statist. Theor. Meth. A.*, **8**, 1359–1376.

Nelder, J. A. and R. Mead (1965). A Simplex Method for Function Minimization, *The Computer J.*, **7**, 308–313.

Olsson, D. M. (1974). A Sequential Simplex Program for Solving Minimization Problems, *J. Quality Tech.*, **6**, 53–57.

Olsson, D. M. and L. S. Nelson (1975). The Nelder-Mead Simplex Procedure for Function Minimization, *Technometrics*, **17**, 45–51.

Powell, M. J. D. (1962). An Iterative Method for Finding Stationary Value of a Function of Several Variables, *The Computer J.*, **5**, 147–151.

Rosenbrock, H. (1960). An Automatic Method for Finding the Greatest or Least Value of a Function, *The Computer J.*, **3**, 175–184.

SAS Institute, Inc. (1989). *SAS/STAT User's Guide*, Vol. 2, Version 6, 4th ed., Cary, NC.

Spendley, W., G. R. Hext, and F. R. Himsworth (1962). Sequential Application of Simplex Designs in Optimization and Evolutionary Operation, *Technometrics*, **4**, 441–461.

Stablein, D. M., W. H. Carter, Jr., and G. L. Wampler (1983). Confidence Regions for Constrained Optima in Response-Surface Experiments, *Biometrics*, **39**, 759–763.

6

Methods of Estimating Response Surfaces That Rival Least Squares Based on the Integrated Mean Squared Error Criterion

6.1 INTRODUCTION

In Section 2.8.3 in Chapter 2, eight properties of a response surface design were listed. Each of the separate properties plays a distinct role in the overall strategy of sequentially fitting polynomial models for the purpose of exploring a response surface over the experimental region. Two other design properties, orthogonality and rotatability, were likewise defined. We went on to show how these latter two properties are obtainable by concentrating on the values of the moments of the distribution of design points in the design region.

In this chapter, we concentrate on the actual specification of the coordinates of the design points in the experimental region as dictated by the integrated mean squared error (IMSE) of $\hat{Y}(\mathbf{x})$ design criterion, abbreviated IMSE $[\hat{Y}(\mathbf{x})]$. The IMSE $[\hat{Y}(\mathbf{x})]$ criterion involves two properties of the fitted model, the variance and the bias. We shall discuss two rival methods (i.e., they rival standard least squares) of estimating a response surface. These alternative approaches differ from standard least squares by placing a greater emphasis on minimizing the bias or the variance of the predicted response in the region of experimentation.

In much of the material which follows, the design criterion of interest will be defined within the context of fitting the general linear model

$$Y_u = \mathbf{f}'(\mathbf{x}_u)\boldsymbol{\beta} + \varepsilon_u \qquad (6.1)$$

where $\mathbf{x}_u$ is a $k \times 1$ vector of predictor variables whose values are defined for the uth experimental trial $(1 \le u \le N)$, $\mathbf{f}'(\mathbf{x}_u)$ is a vector of $p(\ge k + 1)$ functions that model how the response depends on $\mathbf{x}_u$, $\boldsymbol{\beta}$ is the $p \times 1$ vector of unknown parameters, and ε_u is the random error associated with the uth observation, Y_u. Expressed in matrix notation over N observations, the model is

$$\mathbf{Y} = \mathbf{X}\boldsymbol{\beta} + \boldsymbol{\varepsilon} \tag{6.2}$$

In Eq. 6.2, $\mathbf{Y}$ is an $N \times 1$ vector of observations, $\mathbf{X}$ is an $N \times p$ matrix, with the uth row being $\mathbf{f}'(\mathbf{x}_u)$, and $\boldsymbol{\varepsilon}$ is the $N \times 1$ vector of independently and identically distributed random errors, with mean zero and variance σ^2. The standard least squares estimators of the elements of $\boldsymbol{\beta}$ are given by

$$\mathbf{b} = (\mathbf{X}'\mathbf{X})^{-1}\mathbf{X}'\mathbf{Y}$$

with a variance-covariance structure

$$\mathrm{Var}(\mathbf{b}) = (\mathbf{X}'\mathbf{X})^{-1}\sigma^2 \tag{6.3}$$

At some point $\mathbf{x}$ inside the experimental region, the predicted value of the response is

$$\hat{Y}(\mathbf{x}) = \mathbf{f}'(\mathbf{x})\mathbf{b}$$

and the predicted value $\hat{Y}(\mathbf{x})$ has a variance of

$$\mathrm{Var}[\hat{Y}(\mathbf{x})] = \mathbf{f}'(\mathbf{x})(\mathbf{X}'\mathbf{X})^{-1}\mathbf{f}(\mathbf{x})\sigma^2 \tag{6.4}$$

where $\mathbf{f}'(\mathbf{x})$ assumes the same role as $\mathbf{f}'(\mathbf{x}_u)$ in Eq. 6.1 except that the elements of $\mathbf{f}'(\mathbf{x})$ are evaluated at $\mathbf{x}$.

When fitting the model in Eq. 6.2, the design problem consists of selecting the N vectors $\mathbf{x}_u$, $u = 1, 2, \ldots, N$, so that the design is, in some sense, optimal. Some optimality criteria (or design criteria) are defined using the properties in Eqs. 6.3 or 6.4 since in both cases what is sought is a design that makes the $p \times p$ matrix $(\mathbf{X}'\mathbf{X})^{-1}$, or in the case of Eq. 6.4, some real-valued function, $\mathbf{f}'(\mathbf{x})(\mathbf{X}'\mathbf{X})^{-1}\mathbf{f}(\mathbf{x})$, small in some sense. For example, in Section 3.3 in Chapter 3, it was shown that for fitting a first-order model, the value of Eq. 6.4 is minimized by using an orthogonal design. In Section 4.5.4 in Chapter 4, the D- and G-effeciency values of a central composite design on a hypercube and on a hypersphere for fitting a second-order model were listed in Tables 4.4 and 4.5, respectively. These criteria are variance minimizing criteria and will be discussed in more detail in Chapter 12.

Equally important to considering designs that minimize the variance of $\hat{Y}(\mathbf{x})$ is that of seeking designs that minimize the bias of $\hat{Y}(\mathbf{x})$, particularly when the fitted model is of low order (i.e., first- or second-order). In the framework of modeling a response surface, the bias of $\hat{Y}(\mathbf{x})$ is defined as the difference between the expected value of the predicted response at $\mathbf{x}$, which is obtained with the fitted model, and the value of the true response at **x**. In most cases the difference, if present, is assumed to be the result of underestimating the shape of the true surface with the low-order fitted model. Testing for lack of fit of the fitted model was discussed in Section 2.6 in Chapter 2 and Section 3.4 in Chapter 3. The design criterion that we address now is therefore one that considers the magnitude of the variance of $\hat{Y}(\mathbf{x})$ as well as establishing some protection against the possibility that the fitted model is an inadequate representation (i.e., an underestimate) of the true surface. This criterion is called *the average or integrated mean squared error of* $\hat{Y}(\mathbf{x})$, IMSE $[\hat{Y}(\mathbf{x})]$.

6.2 THE IMSE CRITERION

Let us write the model to be fitted in some specified region of interest, R, as $\mathbf{Y} = \mathbf{X}_1\boldsymbol{\beta}_1 + \boldsymbol{\varepsilon}$, where the order of the model is d_1. If $d_1 = 1$, then the number of terms in the model or the number of columns of the matrix, $\mathbf{X}_1$ is $p = k + 1$. If the fitted model is of order two ($d_1 = 2$), then $p = (k + 1)(k + 2)/2$. Suppose further that over the operability region, O, the true surface, η, is expressible by a model of order d_2, where $d_2 > d_1$. This true model is written as $\eta(\mathbf{x}) = \mathbf{X}_1\boldsymbol{\beta}_1 + \mathbf{X}_2\boldsymbol{\beta}_2$. Then, at each point **x** in R, the mean squared error (MSE) of $\hat{Y}(\mathbf{x})$, is

$$\text{MSE}[\hat{Y}(\mathbf{x})] = E[\hat{Y}(\mathbf{x}) - \eta(\mathbf{x})]^2$$

where $\hat{Y}(\mathbf{x})$ is the estimate of $\eta(\mathbf{x})$ obtained with the fitted model. If we average this quantity over the region R, we obtain

$$\Omega \int_R E[\hat{Y}(\mathbf{x}) - \eta(\mathbf{x})]^2 d\mathbf{x} \tag{6.5}$$

where $\Omega^{-1} = \int_R d\mathbf{x}$ and $d\mathbf{x}$ means $dx_1\, dx_2 \ldots dx_k$. Now, in order to be able to center our attention on designs which may or may not contain a fixed number of points and also to remove the dependence of the mean squared error of $\hat{Y}(\mathbf{x})$ on the value of the variance σ^2, our criterion, the average or IMSE of $\hat{Y}(\mathbf{x})$, is put on a "per observation" basis and is expressed as

$$J = \frac{N\Omega}{\sigma^2} \int_R E[\hat{Y}(\mathbf{x}) - \eta(\mathbf{x})]^2 d\mathbf{x} \tag{6.6}$$

In Eq. 6.6, the difference $\hat{Y}(\mathbf{x}) - \eta(\mathbf{x})$ can be partitioned into

$$\hat{Y}(\mathbf{x}) - \eta(\mathbf{x}) = \{\hat{Y}(\mathbf{x}) - E[\hat{Y}(\mathbf{x})]\} + \{E[\hat{Y}(\mathbf{x})] - \eta(\mathbf{x})\}$$

This partitioning enables us to separate J into two parts, that is,

$$J = \frac{N\Omega}{\sigma^2} \int_R \text{Var}[\hat{Y}(\mathbf{x})] d\mathbf{x} + \frac{N\Omega}{\sigma^2} \int_R \{E[\hat{Y}(\mathbf{x})] - \eta(\mathbf{x})\}^2 d\mathbf{x}$$

$$= V + B \tag{6.7}$$

The quantities V and B in Eq. 6.7 are called the average variance of $\hat{Y}(\mathbf{x})$ and the average squared bias of $\hat{Y}(\mathbf{x})$, respectively, where average means averaged over the region R.

To the casual observer, the use of J as a design criterion is appealing for two reasons: (1) it takes into consideration the quality of $\hat{Y}(\mathbf{x})$ as an estimator of $\eta(\mathbf{x})$ over the entire region R, that is, it provides a measure of the "goodness" of $\hat{Y}(\mathbf{x})$ overall, and (2) it simultaneously considers the variance and the bias of $\hat{Y}(\mathbf{x})$. It has been shown in Ott and Cornell (1974) that the averaging in J may in fact mask a poor performance by $\hat{Y}(\mathbf{x})$ for certain locations of $\mathbf{x}$ in R. Since both the variance and the bias of $\hat{Y}(\mathbf{x})$ are affected by the particular choice of design, we shall consider the following three cases separately: a design that

1. Minimizes V while ignoring the value of B
2. Minimizes B while ignoring the value of V
3. Minimizes the sum, $V + B$

Each of these three cases will be covered in greater detail in the next three sections. Before doing this however, let us illustrate the use of the J criterion (Eq. 6.6) for the simple case of fitting a first-order model in one independent variable ($k = 1$).

6.2.1 Fitting a Single-Variable First-Order Model When the True Response Is Quadratic

Let us assume the experimenter is interested in fitting the simple linear regression model

$$Y_u = \mathbf{f}'(x_u)\boldsymbol{\beta} + \varepsilon_u = \beta_0 + \beta_1 x_u + \varepsilon_u \qquad u = 1, 2, \ldots, N$$

where the values of x_u are coded so that the experimental region, R, is the interval $[-1, +1]$. The objective is to construct a design for collecting data to obtain the fitted model

$$\hat{Y}(x) = b_0 + b_1 x \tag{6.8}$$

where b_0 and b_1 are respectively the least-squares estimates of β_0 and β_1. Now, if indeed the shape of the true surface (which is unknown to us) is better described by the quadratic equation

$$E[Y] = \eta(x) = \beta_0 + \beta_1 x + \beta_2 x^2 \tag{6.9}$$

then the design question to be answered is: For what values of x_u will the value of $J = V + B$ for the fitted model (Eq. 6.8) be minimized over the region R?

To minimize J in Eq. 6.7, it is necessary first to obtain separate expressions for V and B. To this end, we note that by using the values of the coded variable

x_u to obtain the fitted model in Eq. 6.8, the following restriction is placed on the N settings of x_u at the design points, $\sum_{u=1}^{N} x_u = 0$. Associated with the fitted model in Eq. 6.8, the restriction $\sum_{u=1}^{N} x_u = 0$ creates the following expression for $\text{Var}[\hat{Y}(\mathbf{x})]$,

$$\text{Var}[\hat{Y}(\mathbf{x})] = \text{Var}(b_0) + x^2\text{Var}(b_1)$$

so that with $E(Y)$ expressed as in Eq. 6.9, J can be written as

$$J = \frac{N\Omega}{\sigma^2}\left(\int_{-1}^{1}[\text{Var}(b_0) + x^2\text{Var}(b_1)]dx\right.$$
$$\left. + \int_{-1}^{1}\{E[\hat{Y}(x)] - \beta_0 - \beta_1x - \beta_2x^2\}^2dx\right) \tag{6.10}$$

where $\Omega^{-1} = \int_{-1}^{1}dx = 2$.

In the average variance portion of J in Eq. 6.10,

$$\text{Var}(b_0) = \frac{\sigma^2}{N} \qquad \text{Var}(b_1) = \frac{\sigma^2}{\sum_{u=1}^{N}x_u^2} = \frac{\sigma^2}{N[11]}$$

where $[11] = \sum_{u=1}^{N}x_u^2/N$ is the second-order moment of the design (see Eq. 2.44 in Chapter 2). Substituting these expressions for $\text{Var}(b_0)$ and $\text{Var}(b_1)$ into Eq. 6.10 and performing the integration on only the variance portion of J, the formula for the average variance, V, is

$$V = \frac{1}{2}\int_{-1}^{1}\left(1 + \frac{x^2}{[11]}\right)dx$$
$$= 1 + \frac{1}{3[11]} \tag{6.11}$$

To obtain an expression for the average squared bias in Eq. 6.10, it is necessary first to obtain an expression for $E[\hat{Y}(x)]$. Now, in general, with a fitted model of the form

$$\hat{Y}(x) = \mathbf{f}'(x)\mathbf{b}$$

the expectation of $\hat{Y}(x)$ is

$$E[\hat{Y}(x)] = \mathbf{f}'(x)E(\mathbf{b}) = \mathbf{f}'(x)(\mathbf{X}_1'\mathbf{X}_1)^{-1}\mathbf{X}_1'E(\mathbf{Y})$$

where the elements in $\mathbf{f}'(x)$ are the same as the elements in a row of the matrix $\mathbf{X}_1$. For the case considered in this section, the expectation of $\mathbf{Y}$ in Eq. 6.9, over N observations, is

$$E[\mathbf{Y}] = \mathbf{X}_1\boldsymbol{\beta}_1 + \mathbf{X}_2\boldsymbol{\beta}_2$$

where the matrices X_1 and X_2 are given by

$$X_1 = \begin{bmatrix} 1 & x_1 \\ 1 & x_2 \\ \vdots & \vdots \\ 1 & x_N \end{bmatrix} \qquad X_2 = \begin{bmatrix} x_1^2 \\ x_2^2 \\ \vdots \\ x_N^2 \end{bmatrix} \tag{6.12}$$

and the vectors β_1 and β_2 are, respectively

$$\beta_1 = \begin{bmatrix} \beta_0 \\ \beta_1 \end{bmatrix} \qquad \beta_2 = [\beta_2]$$

Then the expectation of $\hat{Y}(x)$ is

$$E[\hat{Y}(x)] = f'(x)(X_1'X_1)^{-1}X_1'(X_1\beta_1 + X_2\beta_2)$$
$$= f'(x)\beta_1 + f'(x)A\beta_2$$

where the matrix $A = (X_1'X_1)^{-1}X_1'X_2$ is the alias matrix.

The form of the matrix A for our example, given the forms of the matrices X_1 and X_2 in Eq. 6.12, is

$$A = \begin{bmatrix} 1/N & 0 \\ 0 & 1/(N[11]) \end{bmatrix} \begin{bmatrix} N[11] \\ N[111] \end{bmatrix} = \begin{bmatrix} [11] \\ [111]/[11] \end{bmatrix}$$

where $[111] = \sum_{u=1}^{N} x_u^3/N$. Hence,

$$E[\hat{Y}(x)] = f'(x) \begin{bmatrix} \beta_0 \\ \beta_1 \end{bmatrix} + f'(x)A\beta_2$$
$$= \beta_0 + \beta_1 x + \beta_2\{[11]^2 + [111]x\}/[11]$$

If we substitute this expression for $E[\hat{Y}(x)]$ in the bias portion of J in Eq. 6.10, the average squared bias can be written as

$$B = \frac{N\beta_2^2}{2\sigma^2} \int_{-1}^{1} \{([11]^2 + [111]x)/[11] - x^2\}^2 dx$$

$$= \frac{N\beta_2^2}{\sigma^2} \{[11]^2 + ([111]^2/3[11]^2) - 2[11]/3 + 1/5\} \tag{6.13}$$

The expressions (Eqs. 6.11 and 6.13) for V and B, respectively, are

$$V = 1 + \frac{1}{3[11]}$$

$$B = \frac{N\beta_2^2}{\sigma^2} \left\{ [11]^2 + \frac{[111]^2}{3[11]^2} - \frac{2[11]}{3} + \frac{1}{5} \right\} \tag{6.14}$$

While V does not contain β_2 at all and depends only on the design moment $[11]$, B does depend on both β_2 and the moments $[11]$ and $[111]$. Thus, if we wish to minimize $J = V + B$, we shall require knowledge of the value of β_2 or at least something about the magnitude of the ratio β_2/σ, the curvature of the true surface in R to the standard deviation of an observation.

From the expressions or formulas for V and B, respectively, there are clues as to how we might proceed in selecting a design to minimize V or B. For example, to minimize V in Eq. 6.14, we would seek a design for which the value of $[11] = (1/N) \sum_{u=1}^{N} x_u^2$ is as large as possible within the region R.

To minimize B in Eq. 6.14 on the other hand, one can begin by considering only designs for which the third-order design moment, $[111]$, is equal to zero. When this is done, B is expressed as

$$B = \frac{N\beta_2^2}{\sigma^2} \{([11] - 1/3)^2 + 4/45\}$$

(We shall explain shortly how to choose a design that has a zero third-order moment.) Minimization of B is achieved by selecting a design for which $[11] = 1/3$. This action corresponds to reducing the size or spread of the design points to guard against the possibility of picking up curvature of the true surface in R. But this reduction is in contradiction to what needs to be done to minimize V. Hence, if we turn our attention to $J = V + B$, we see that any attempt to minimize J by minimizing both V and B simultaneously will require compromising our selection of a value for the second moment, $[11]$, which is a measure of the spread of the design points. This compromise in choosing a value for $[11]$ is further evidenced by writing J, with $[111]$ set equal to zero, as the sum of V and B in Eq. 6.14,

$$J = 1 + \frac{1}{3[11]} + \frac{N\beta_2^2}{\sigma^2} \{([11] - 1/3)^2 + 4/45\} \tag{6.15}$$

and noticing that $[11]$ appears in both the denominator and in the numerator of separate terms of J.

Finally, in the expression in Eq. 6.15 for J, the quantity $N\beta_2^2/\sigma^2 = [\beta_2/(\sigma/\sqrt{N})]^2$ is the square of the ratio of β_2 to the sampling error, $\sigma/\sqrt{N}$. The two extreme cases we have considered thus far are:

1. When the curvature is very small relative to the sampling error, then J is minimized by making $[11]$ large (i.e., by minimizing V).
2. When $\beta_2/(\sigma/\sqrt{N})$ is large, then J is minimized by setting $[11] = 1/3$ (i.e., by minimizing B).

Since these extreme cases lead to widely different conclusions regarding the optimal choice of $[11]$, we are naturally interested in finding the value of $[11]$ that minimizes J when both V and B are nonzero.

How then do we proceed in seeking a design to minimize J? To address the spread-of-the-design-points problem when both V and B are nonzero, it is helpful to express V and B as functions of [11]. Denoting the separate contributions as functions of [11] by writing them as $V([11])$ and $B([11])$, respectively, then J in Eq. 6.15 becomes

$$J = V([11]) + \frac{N\beta_2^2}{\sigma^2} B([11]) \tag{6.16}$$

For every fixed value of $\beta_2/(\sigma/\sqrt{N})$, there will be a J-minimizing value of [11] from which the corresponding values of $V = V([11])$ and $B = (N\beta_2^2/\sigma^2)B([11])$ can be calculated. In utilizing this approach, Box and Draper (1959) suggested selecting the particular value of [11] from the set of minimizing values that forces $V = gB$ where g is any desired positive constant value (actually g is a measure of the ratio of the variance to the bias). For example, when the value of $\beta_2/(\sigma/\sqrt{N})$ is approximately 4.5 (actually 4.499) and $g = 1$ (i.e., $V = B$), the optimal value of [11] is 0.388, for which the value of J in Eq. 6.16 is $J = 3.718$. As another example, when $V = 4B$ and $\beta_2/(\sigma/\sqrt{N}) = 1.822$, then [11] = 0.519 and $J = 2.052$. In this latter case where the V contribution to J is four times the B contribution to J, the resulting value of J is very close to the value of J (2.296) when $V = 0$ and $B > 0$.

The underlying message from the results just stated is that the bias contribution to J, which arises from underfitting the true surface (of order 2) using a fitted model of only order 1 and $k = 1$, is the more important component of the two (V or B), when concentrating on the choice of the optimal spread of the design points. This is true for the case of k variables as well (see Section 6.2.3). Specific values of [11] and J for several values of $\sqrt{N}\beta_2/\sigma$ and the values of $g = 1/2, 1, 2, 4$ and 10, corresponding to the optimal design and values of J for the all-bias design ($V = 0, B > 0$) when fitting a first-order model with $k = 1$, are listed in Table 6.1

Thus far in this section, we have discussed the determination of the spread of the points of the optimal design over the interval $[-1, +1]$ for minimizing V, B and $J = V + B$. We have not said anything about the specific locations of the points except to say that when minimizing B, the third-order moment of the

Table 6.1 Values of J Corresponding to the Optimal and All-Bias Designs for Values of the Ratio $g = V/B$ and $\sqrt{N}\beta_2/\sigma$ ($k = 1$)

$g = V/B$	$\sqrt{N}\beta_2/\sigma$	$[11]_{\text{optimal}}$	J_{optimal}	$J_{\text{all bias } (V=0, B>0)}$
1/2	6.540	0.363	5.755	5.799
1	4.499	0.388	3.718	3.798
2	2.994	0.433	2.656	2.797
4	1.822	0.519	2.052	2.296
10	0.501	1.000	1.467	2.022

design should be zero, which means the points are to be positioned symmetrically about the midpoint ($x = 0$) of the interval $[-1, +1]$. The following is a partial listing of suggested designs where each suggestion is based on the size of the contributions of V and B in J. The designs are illustrated in Figure 6.1 and further satisfy $\sum_{u=1}^{N} x_u = 0$.

1. *All-variance design* ($V > 0$, $B = 0$). If N is even, collect $N/2$ observations at each of the end points $x_u = -1$ and $x_u = +1$. If N is odd, collect one observation at $x_u = 0$ and $(N - 1)/2$ observations at $x_u = -1$ and $x_u = +1$.
2. *All-bias design* ($V = 0$, $B > 0$). If N is even and only two levels of x_u are to be used, then $N/2$ observations are collected at $x_u = -0.58$ and at $x_u = +0.58$. If N is odd, collect one observation at $x_u = 0$ and collect $(N - 1)/2$ observations at $x_u = \pm\sqrt{N/[3(N - 1)]}$. If three levels of x_u are desired and N is a multiple of 3, then $N/3$ observations are collected at $x_u = -\sqrt{0.5}, 0, +\sqrt{0.5}$.
3. $V = B$. If N is even and only two levels of x_u are to be used, then $N/2$ observations are collected at $x_u = \pm 0.623$. If N is odd and one observation is to be collected at $x_u = 0$ and $(N - 1)/2$ observations are to be collected at each of the other settings of x_u, then these settings will depend on the size of N, for example, if $N = 3$, $x_u = \pm 0.763$; $N = 5$, $x_u = \pm 0.696$; $N = 7$, $x_u = \pm 0.672$. In all cases, we obtain the settings of x for which $[11] = 0.388$. If N is a multiple of 3 and $N/3$ observations are to be collected at each level of x_u, then the settings are $x_u = -0.763, 0, +0.763$.

a)

b)

c)

Figure 6.1 One-dimensional designs where R is the two-unit interval $[-1, +1]$; • denotes that N is even and ○ denotes that N is a multiple of 3. (a) All-variance design ($V > 0$, $B = 0$). (b) All-bias design ($V = 0$, $B > 0$). (c) $V = B$ design ($V > 0$, $B > 0$).

6.2.2   The *J* Criterion for *k* Input Variables

The use of the average MSE $[\hat{Y}(\mathbf{x})]$ as a design criterion is easily extended to the case of k input variables. Let us first alter our notation slightly from that of the general linear model, $Y_u = \mathbf{f}'(\mathbf{x}_u)\boldsymbol{\beta} + \varepsilon_u$, where for the fitted model we had $\hat{Y}(\mathbf{x}) = \mathbf{f}'(\mathbf{x})\mathbf{b}$, and instead write the fitted model of order d_1 in the k coded variables over the region of interest R as

$$\hat{Y}(\mathbf{x}) = \mathbf{x}_1'\mathbf{b} \tag{6.17}$$

If $d_1 = 1$, then $\mathbf{x}_1 = (1, x_1, x_2, \ldots, x_k)'$ and $\mathbf{b} = (b_0, b_1, \ldots, b_k)'$. The variance of $\hat{Y}$ at the point $\mathbf{x} = (x_1, x_2, \ldots, x_k)'$ is

$$\mathrm{Var}[\hat{Y}(\mathbf{x})] = \mathbf{x}_1'(\mathbf{X}_1'\mathbf{X}_1)^{-1}\mathbf{x}_1\sigma^2 \tag{6.18}$$

where the elements in a row of the $N \times (k + 1)$ matrix $\mathbf{X}_1$ are the same as in $\mathbf{x}_1'$. Now, let us assume the true surface is represented by a model of order d_2 where $d_2 > d_1$,

$$\eta(\mathbf{x}) = \mathbf{x}_1'\boldsymbol{\beta}_1 + \mathbf{x}_2'\boldsymbol{\beta}_2 \tag{6.19}$$

For example, if $d_1 = 1$ and $d_2 = 2$, then $\mathbf{x}_2 = (x_1^2, x_2^2, \ldots, x_k^2, x_1x_2, x_1x_3, \ldots, x_{k-1}x_k)'$ and $\boldsymbol{\beta}_2 = (\beta_{11}, \beta_{22}, \ldots, \beta_{kk}, \beta_{12}, \beta_{13}, \ldots, \beta_{k-1,k})'$.

Now, let us recall the expression in Eq. 6.7 for J

$$J = \frac{N\Omega}{\sigma^2}\int_R \mathrm{Var}[\hat{Y}(\mathbf{x})]d\mathbf{x} + \frac{N\Omega}{\sigma^2}\int_R \{E[\hat{Y}(\mathbf{x})] - \eta(\mathbf{x})\}^2 d\mathbf{x}$$

$$= V + B \tag{6.20}$$

From Eq. 6.18, the average variance, V, is

$$V = N\Omega\int_R \mathbf{x}_1'(\mathbf{X}_1'\mathbf{X}_1)^{-1}\mathbf{x}_1 d\mathbf{x} \tag{6.21}$$

To obtain an expression for the average squared bias, B, we recall from the previous section that the expectation of the estimate of the response at the point $\mathbf{x}$ is expressible as $E[\hat{Y}(\mathbf{x})] = \mathbf{x}_1'E(\mathbf{b})$, where

$$E(\mathbf{b}) = \boldsymbol{\beta}_1 + \mathbf{A}\boldsymbol{\beta}_2$$

and $\mathbf{A} = (\mathbf{X}_1'\mathbf{X}_1)^{-1}\mathbf{X}_1'\mathbf{X}_2$. The expectation of $\hat{Y}$ at the point $\mathbf{x} = (x_1, x_2, \ldots, x_k)'$ is, therefore,

$$E[\hat{Y}(\mathbf{x})] = \mathbf{x}_1'\boldsymbol{\beta}_1 + \mathbf{x}_1'\mathbf{A}\boldsymbol{\beta}_2$$

so that the average squared bias, B, in Eq. 6.20 is

$$B = \frac{N\Omega}{\sigma^2} \int_R \{\mathbf{x}_1'\boldsymbol{\beta}_1 + \mathbf{x}_1'\mathbf{A}\boldsymbol{\beta}_2 - \mathbf{x}_1'\boldsymbol{\beta}_1 - \mathbf{x}_2'\boldsymbol{\beta}_2\}^2 d\mathbf{x}$$

$$= \frac{N\Omega}{\sigma^2} \int_R \boldsymbol{\beta}_2'[\mathbf{A}'\mathbf{x}_1 - \mathbf{x}_2][\mathbf{x}_1'\mathbf{A} - \mathbf{x}_2']\boldsymbol{\beta}_2 \, d\mathbf{x} \qquad (6.22)$$

The expressions in Eqs. 6.21 and 6.22 for V and B, respectively, involve the averaging of quadratic forms over the region R. To obtain final formulas for V and B which involve certain properties of the design under consideration, let us turn our attention for the moment to describing the shape of R in the k-variable case.

Recall in the single-input variable case of the previous section that the region R was the two-unit interval $[-1, +1]$. This type of region is easily extended in the k-variable case to the cuboidal region $\mathbf{R} = \{\mathbf{x} = (x_1, x_2, \ldots, x_k)': -1 \leq x_i \leq +1, i = 1, 2, \ldots, k\}$. However, in most of the development work by Box and others involving response surface designs, the region R in the coded variables has been defined as spherical or ellipsoidal. In particular, it is mathematically convenient to denote R as a unit sphere with center defined at $\mathbf{x} = (0, 0, \ldots, 0)'$, that is,

$$\sum_{i=1}^{k} x_{ui}^2 \leq 1 \qquad u = 1, 2, \ldots, N \qquad (6.23)$$

By defining the region of interest as a sphere, this has resulted in the development of designs with certain appealing characteristics, as we shall see. Hence, we shall assume that R is a sphere of radius ρ, that is, $\sum_{i=1}^{k} x_{ui}^2 \leq \rho^2$. If one wishes to refer to R as the unit sphere, one need only set $\rho = 1$ in the formulas that follow. (The consideration of R as cuboidal is addressed briefly in Section 6.2.3 and later in Section 6.3.2.) As for the shape of the region in the original variables, X_i, if the coded variables x_i are defined as in Eq. 4.2 in Chapter 4, that is, $x_{ui} = (X_{ui} - \bar{X}_i)/s_{X_i}$, $i = 1, 2, \ldots, k$, and if all the scale constants, s_{X_i}, are equal in value, then the corresponding region in the uncoded variables X_i is a sphere centered at $(\bar{X}_1, \bar{X}_2, \ldots, \bar{X}_k)$. In most cases, on the other hand, the s_{X_i} are not equal so that the region in the X_i is an ellipsoidal region.

In specifying the region of interest R as a sphere of radius ρ, Dirichlet multiple integrals can be used for integrating certain terms in the expressions for V and B in Eqs. 6.21 and 6.22, respectively, over the region R. The integration produces the moments of the region R by recognizing the following property of the integral:

$$\int_R x_1^{\delta_1} x_2^{\delta_2} \cdots x_k^{\delta_k} d\mathbf{x}$$

$$= \frac{\Gamma\left(\dfrac{\delta_1 + 1}{2}\right) \Gamma\left(\dfrac{\delta_2 + 1}{2}\right) \cdots \Gamma\left(\dfrac{\delta_k + 1}{2}\right)}{\Gamma\left\{\displaystyle\sum_{i=1}^{k} \dfrac{(\delta_i + 1)}{2} + 1\right\}} \rho^{\sum_{i=1}^{k}(\delta_i+1)} \qquad (6.24)$$

where the values of δ_i are even integers, and $\Gamma(\cdot)$ refers to the gamma function, that is,

$$\Gamma(\delta) = \int_0^{\infty} x^{\delta-1} e^{-x} dx \qquad (\delta > 0)$$

If any of the values of δ_i are odd, the value of the integral in Eq. 6.24 is zero.

Continuing our development toward obtaining final expressions for V and B, we now introduce matrices whose elements are the moments of the spherical region R. Let us define a matrix μ_{ij}, the elements of which are the region moments of R up to and including order $d_i + d_j$ ($d_i < d_j$) and is of the form

$$\mu_{ij} = \Omega \int_R \mathbf{x}_i \mathbf{x}_j' \, d\mathbf{x} \qquad (6.25)$$

where $\int$(matrix)$d\mathbf{x}$ means the corresponding matrix of integrals. The vector $\mathbf{x}_i$ is made up of terms required for the polynomial model of order d_i and the vector $\mathbf{x}_j$ consists of the additional high-order terms required to represent the polynomial of order d_j. In Eq. 6.25, R is the spherical region of radius ρ and $\Omega^{-1} = \int_R d\mathbf{x}$. For example, recall from Eq. 6.17 that the fitted model was of order d_1 while the true surface model was of order d_2, so that $i = 1$ and $j = 2$. In this case, the moment matrices of interest are μ_{11}, μ_{12} ($= \mu_{21}'$) and μ_{22}.

To determine the forms of the moment matrices of interest for our example where $d_1 = 1$ and $d_2 = 2$, first we note that Ω in Eq. 6.25 is

$$\Omega^{-1} = \int_R d\mathbf{x}$$

$$= \frac{[\Gamma(1/2)]^k}{\Gamma(k/2 + 1)} \rho^k \qquad (6.26)$$

Now, since $d_1 = 1$ and $d_2 = 2$ so that $\mathbf{x}_1 = (1, x_1, x_2, \ldots, x_k)'$ and $\mathbf{x}_2 = (x_1^2, x_2^2, \ldots, x_k^2, x_1x_2, \ldots, x_{k-1}x_k)'$, then from Eq. 6.25 and by performing the integration element by element on the entries in the matrix $\mathbf{x}_1\mathbf{x}_1'$, we get

$$
\boldsymbol{\mu}_{11} = \Omega \int_R
\begin{bmatrix}
1 & x_1 & x_2 & \cdots & x_k \\
x_1 & x_1^2 & x_1x_2 & \cdots & x_1x_k \\
x_2 & x_2x_1 & x_2^2 & \cdots & x_2x_k \\
\vdots & \vdots & \vdots & \ddots & \vdots \\
x_k & x_kx_1 & x_kx_2 & \cdots & x_k^2
\end{bmatrix} d\mathbf{x}
$$

$$
=
\begin{bmatrix}
1 & \mathbf{0}' \\
\mathbf{0} & \left(\dfrac{\rho^2}{k+2} \right) \mathbf{I}_k
\end{bmatrix}
\tag{6.27}
$$

where $\mathbf{I}_k$ is the identity matrix of order $k \times k$. In addition, the matrices $\boldsymbol{\mu}_{12}$ and $\boldsymbol{\mu}_{22}$ are

$$
\boldsymbol{\mu}_{12} = \Omega \int_R \mathbf{x}_1\mathbf{x}_2' d\mathbf{x}
$$

$$
= \frac{\rho^2}{k+2}
\begin{bmatrix}
\mathbf{1}_k' & \mathbf{0} \\
\mathbf{0} & \mathbf{0}
\end{bmatrix}
\tag{6.28}
$$

where $\mathbf{1}_k'$ is a $1 \times k$ vector of ones, and

$$
\boldsymbol{\mu}_{22} = \Omega \int_R \mathbf{x}_2\mathbf{x}_2' d\mathbf{x}
$$

$$
= \frac{\rho^4}{(k+2)(k+4)}
\begin{bmatrix}
2\mathbf{I}_k + \mathbf{1}_k\mathbf{1}_k' & \mathbf{0} \\
\mathbf{0} & \mathbf{I}_{\binom{k}{2}}
\end{bmatrix}
\tag{6.29}
$$

Let us return to the expression in Eq. 6.21 for V:

$$
V = N\Omega \int_R \mathbf{x}_1'(\mathbf{X}_1'\mathbf{X}_1)^{-1}\mathbf{x}_1 \, d\mathbf{x}
$$

$$
= N\Omega \int_R \text{trace}[(\mathbf{X}_1'\mathbf{X}_1)^{-1}\mathbf{x}_1\mathbf{x}_1'] d\mathbf{x}
\tag{6.30}
$$

Since $(\mathbf{X}_1'\mathbf{X}_1)^{-1}$ is constant, the general formula for the average variance, no matter what the values of d_1 and d_2 are, is

$$V = \text{trace}\left[N(\mathbf{X}_1'\mathbf{X}_1)^{-1}\left\{\Omega\int_R \mathbf{x}_1\mathbf{x}_1'\,d\mathbf{x}\right\}\right]$$

$$= \text{trace}[N(\mathbf{X}_1'\mathbf{X}_1)^{-1}\boldsymbol{\mu}_{11}] \qquad (6.31)$$

For the particular case where $d_1 = 1$, $\boldsymbol{\mu}_{11}$ is defined in Eq. 6.27.

Turning our attention to B in Eq. 6.22, we have

$$B = \frac{N\Omega}{\sigma^2}\int_R \boldsymbol{\beta}_2'[\mathbf{A}'\mathbf{x}_1 - \mathbf{x}_2][\mathbf{x}_1'\mathbf{A} - \mathbf{x}_2']\boldsymbol{\beta}_2\,d\mathbf{x}$$

$$= \frac{N\Omega}{\sigma^2}\int_R \boldsymbol{\beta}_2'[\mathbf{A}'\mathbf{x}_1\mathbf{x}_1'\mathbf{A} - \mathbf{x}_2\mathbf{x}_1'\mathbf{A} - \mathbf{A}'\mathbf{x}_1\mathbf{x}_2' + \mathbf{x}_2\mathbf{x}_2']\boldsymbol{\beta}_2\,d\mathbf{x}$$

$$= \frac{N}{\sigma^2}\boldsymbol{\beta}_2'\boldsymbol{\Delta}\boldsymbol{\beta}_2 \qquad (6.32)$$

where $\boldsymbol{\Delta} = \mathbf{A}'\boldsymbol{\mu}_{11}\mathbf{A} - \boldsymbol{\mu}_{12}'\mathbf{A} - \mathbf{A}'\boldsymbol{\mu}_{12} + \boldsymbol{\mu}_{22}$. Further simplification of the expression for B is possible by adding and subtracting $\boldsymbol{\mu}_{12}'\boldsymbol{\mu}_{11}^{-1}\boldsymbol{\mu}_{12}$ from $\boldsymbol{\Delta}$ to produce

$$B = \frac{N}{\sigma^2}\boldsymbol{\beta}_2'[(\boldsymbol{\mu}_{22} - \boldsymbol{\mu}_{12}'\boldsymbol{\mu}_{11}^{-1}\boldsymbol{\mu}_{12}) + (\mathbf{A} - \boldsymbol{\mu}_{11}^{-1}\boldsymbol{\mu}_{12})'\boldsymbol{\mu}_{11}(\mathbf{A} - \boldsymbol{\mu}_{11}^{-1}\boldsymbol{\mu}_{12})]\boldsymbol{\beta}_2$$

$$(6.33)$$

If we now set $\mathbf{A} = \boldsymbol{\mu}_{11}^{-1}\boldsymbol{\mu}_{12}$, the second term in the brackets on the right-hand side of the equality sign in Eq. 6.33 is eliminated. Box and Draper (1959) state that, in particular, B is minimized when $(1/N)(\mathbf{X}_1'\mathbf{X}_1) = \boldsymbol{\mu}_{11}$ and $(1/N)(\mathbf{X}_1'\mathbf{X}_2) = \boldsymbol{\mu}_{12}$, which is just a statement that the moments of the design equal the moments of the region R up to and including order $d_1 + d_2$. In this case, B in Eq. 6.33 becomes

$$B = \frac{N}{\sigma^2}\boldsymbol{\beta}_2'(\boldsymbol{\mu}_{22} - \boldsymbol{\mu}_{12}'\boldsymbol{\mu}_{11}^{-1}\boldsymbol{\mu}_{12})\boldsymbol{\beta}_2 \qquad (6.34)$$

The formulas in Eqs. 6.31 and 6.33 for V and B, respectively, are quite general and can be written in more specific terms once the elements of the vectors $\mathbf{x}_1$ and $\mathbf{x}_2$ are known. Since we would like to talk about designs (i.e., the specific design coordinate settings of the values of the x_i) that minimize V and B and this is only possible by knowing the order of the fitted model and the assumed form of the true surface, we shall now discuss the following two cases: (1) the fitted model is of order $d_1 = 1$ but the true surface is of order $d_2 = 2$, and (2) the fitted model is of order $d_1 = 2$ but the true surface is of order

$d_2 = 3$. In both cases, we shall concentrate on designs that minimize V and B separately. The minimization of $J = V + B$ is discussed in Section 6.2.4.

6.2.3 Some Specific First-Order Design Arrangements for Minimizing *V* and *B* Separately

Recall from Chapter 3 that when we speak of a first-order design, we mean an arrangement of points which permit the separate estimation of the parameters in the first-order model

$$\mathbf{Y} = \mathbf{X}_1\boldsymbol{\beta}_1 + \boldsymbol{\varepsilon} \tag{6.35}$$

where over N observations,

$$\mathbf{X}_1 = \begin{bmatrix} 1 \\ 1 \\ \vdots \\ 1 \end{bmatrix} : \mathbf{D} \quad \text{and} \quad \mathbf{D} = \begin{bmatrix} x_{11} & x_{12} & \cdots & x_{1k} \\ x_{21} & x_{22} & \cdots & x_{2k} \\ \vdots & \vdots & & \vdots \\ x_{N1} & x_{N2} & \cdots & x_{Nk} \end{bmatrix} \tag{6.36}$$

The matrix $\mathbf{D}$ is the design matrix.

In attempting to obtain a best distribution of experimental points, we shall keep the following three considerations in mind. The arrangement must be such that

1. The variances of the estimated parameters are as small as possible.
2. The biases in the estimated parameters which might occur if the first-order model is representationally inadequate should be as small as possible.
3. Since these designs are the first step in any type of sequential experimentation, we ought to concentrate our efforts only on first-order designs which can easily be augmented by the addition of points to second-order designs.

Furthermore, we shall only seek design arrangements for which the third-order moments of the design are equal to zero. The reason behind seeking designs with this property will become clearer when we discuss the minimization of B.

In what follows, three types of designs will be considered and for all designs, it will be shown that the third-order moments vanish. The first type of design is the scaled two-level factorial, where a row of the $\mathbf{X}_1$ matrix is $\mathbf{x}_1' = (1, \pm c, \pm c, \ldots, \pm c)$ and where the number of points is $N \geq k + 1$. The second type of design is an octahedron or axial-type design, where the N rows of the matrix $\mathbf{X}_1 = [\mathbf{1}:\mathbf{D}]$ look like $\mathbf{x}_{11}' = (1, \pm ac, 0, 0, \ldots, 0)$, $\mathbf{x}_{21}' = (1, 0, \pm ac, 0, \ldots, 0)$, $\ldots$, $\mathbf{x}_{N1}' = (1, 0, 0, \ldots, \pm ac)$. The third type of design considered is the double-simplex, (as in Figure 4.9 in Chapter 4, minus the center point), where there are $N = 2(k + 1)$ design points. This design is made up of two regular k-dimensional simplex designs centered at $\mathbf{x} = \mathbf{0}$. For each design, the design matrix is of the form

$$D = c \begin{bmatrix} F \\ -F \end{bmatrix} \tag{6.37}$$

where $-F$ denotes the settings in the bottom half of D and these settings are the negative replicates of the settings in the top half of D. The quantity c is a scale factor such that $[ii] = c^2$, $i = 1, 2, \ldots, k$, is not too large for the design points to fit in the region R. We shall from time to time refer to c as the *radius multiplier* since the design points in D, defined by Eq. 6.37, lie on a sphere of radius $c\sqrt{k}$.

The design matrix D in Eq. 6.37 is $N \times k$. For the matrix F in the top half of D, all we require is that the number of rows (e.g., $N/2$ where N is even) be greater than or equal to $(k + 1)/2$. The k columns of F are orthogonal so that in F (and in $-F$),

$$\sum_{u=1}^{N/2} x_{ui} x_{uj} = 0 \qquad i, j = 1, 2, \ldots, k; \ i \neq j$$

$$\sum_{u=1}^{N/2} x_{ui}^2 = \left(\frac{N}{2}\right) c^2$$

Our choice of using first-order designs in which the bottom half of the design matrix is the negative replicate of the top half of the design matrix assures us that the third-order moments of the design are zero. This property of the design is shown by introducing a matrix Z of order $(N/2) \times [k(k + 1)/2]$ where the elements in its uth row $(u = 1, 2, \ldots, N/2)$ are $x_{u1}^2, x_{u2}^2, \ldots, x_{uk}^2, x_{u1}x_{u2}$, $x_{u1}x_{u3}, \ldots, x_{u_{k-1}}x_{u_k}$. These are the squares and cross-products of the elements of the uth row of F $(u = 1, 2, \ldots, N/2)$. Now, the matrix of third-order moments can be written as

$$c^2 D' \begin{bmatrix} Z \\ Z \end{bmatrix} = c^3 [F' : -F'] \begin{bmatrix} Z \\ Z \end{bmatrix} \tag{6.38}$$

which is equal to the zero matrix. We shall see the implication of Eq. 6.38 when we talk about the form of the alias matrix in our attempt to minimize B.

Minimize V when B is assumed to be zero.   When fitting a first-order model of the form in Eq. 6.35 where the design matrix D is shown in Eq. 6.36, we learned in Section 3.3 in Chapter 3 that the variances of the parameter estimates in the fitted model are minimized if the columns of D are orthogonal to one another. By forcing this orthogonality property on the columns of F, (thus on D as well), the matrix $(X_1'X_1)^{-1}$ in Eq. 6.31 for V will be diagonal, that is $(X_1'X_1)^{-1} = N^{-1} \operatorname{diag}(1, 1/[11], 1/[22], \ldots, 1/[kk])$. Furthermore, when $d_1 = 1$ and with the matrix μ_{11} defined as in Eq. 6.27, the expression in Eq. 6.31 for

V simplifies to

$$V = 1 + \frac{\rho^2}{(k + 2)} \sum_{i=1}^{k} \frac{1}{[ii]} \tag{6.39}$$

Since ρ^2 is fixed, we see that to minimize V, the largest value of $[ii]$ must be selected. This is achieved if all the design points lie on the boundary of the largest sphere centered at $x = 0$ that will fit in the experimental region R. Therefore, if the matrix D is as specified in Eq. 6.37, which of course fixes all of the $[ii]$ equal to $[ii] = c^2$, and if we set the value of the radius multiplier equal to

$$c = \frac{\rho}{\sqrt{k}} \tag{6.40}$$

then the average variance criterion in Eq. 6.39 is minimized and its value is $\min V = 1 + k^2/(k + 2)$. Such a design is called an *all-variance* design.

Minimize B when V is assumed to be zero. The expression for the average squared bias B was rewritten in Eq. 6.34 by setting $A = \mu_{11}^{-1}\mu_{12}$ in Eq. 6.33, that is, by forcing $(1/N)(X_1'X_1) = \mu_{11}$ and $(1/N)(X_1'X_2) = \mu_{12}$. What this implies when $d_1 = 1$ and $d_2 = 2$ is that the matrix of design moments and the matrix of second- and third-order moments are equal to the corresponding matrices of region moments.

To see what impact setting these matrices equal to one another has on the value of the radius multiplier, let us use a design matrix of the form in Eq. 6.37, that is,

$$D = c \begin{bmatrix} F \\ -F \end{bmatrix}$$

so that

$$\frac{1}{N}(X_1'X_1) = \begin{bmatrix} 1 & 0' \\ 0 & c^2 I_k \end{bmatrix} \tag{6.41}$$

If we force $(X_1'X_1)/N$ equal to μ_{11} in Eq. 6.27, that is, $c^2 = \rho^2/(k + 2)$, then the value of the radius multiplier is

$$c = \frac{\rho}{\sqrt{k + 2}} \tag{6.42}$$

and the resulting design is called an *all-bias* design.

Example 6.1. For a double-simplex example, let $k = 2$ and define the first-order model to be fitted as

$$Y = \beta_0 + \beta_1 x_1 + \beta_2 x_2 + \varepsilon \tag{6.43}$$

In matrix notation

$$Y = X_1\beta_1 + \varepsilon \tag{6.44}$$

where $\beta_1 = (\beta_0, \beta_1, \beta_2)'$ and a row of X_1 is $x'_{u1} = (1, x_{u1}, x_{u2})$. Let the true surface be represented by the second-order model

$$E[Y] = X_1\beta_1 + X_2\beta_2 \tag{6.45}$$

where $\beta_2 = (\beta_{11}, \beta_{22}, \beta_{12})'$ and a row of X_2 is $x'_{u2} = (x^2_{u1}, x^2_{u2}, x_{u1}x_{u2})$.

For estimating the parameters in the model in Eq. 6.43, suppose we choose to position the design points on a sphere (circle) of radius ρ, and set the radius of the region R equal to unity. Since the design is a double-simplex, that is, the top half of D is a three-point simplex and the bottom half of D is a negative replicate of the top three rows of D, then one possibility for the matrix X_1, along with the resulting matrix X_2 is

$$X_1 = \begin{bmatrix} 1 & \rho & 0 \\ 1 & \dfrac{-\rho}{2} & \dfrac{\sqrt{3}\rho}{2} \\ 1 & \dfrac{-\rho}{2} & \dfrac{-\sqrt{3}\rho}{2} \\ 1 & -\rho & 0 \\ 1 & \dfrac{\rho}{2} & \dfrac{-\sqrt{3}\rho}{2} \\ 1 & \dfrac{\rho}{2} & \dfrac{\sqrt{3}\rho}{2} \end{bmatrix} \qquad X_2 = \begin{bmatrix} \rho^2 & 0 & 0 \\ \dfrac{\rho^2}{4} & \dfrac{3\rho^2}{4} & \dfrac{-\sqrt{3}\rho^2}{4} \\ \dfrac{\rho^2}{4} & \dfrac{3\rho^2}{4} & \dfrac{\sqrt{3}\rho^2}{4} \\ \rho^2 & 0 & 0 \\ \dfrac{\rho^2}{4} & \dfrac{3\rho^2}{4} & \dfrac{-\sqrt{3}\rho^2}{4} \\ \dfrac{\rho^2}{4} & \dfrac{3\rho^2}{4} & \dfrac{\sqrt{3}\rho^2}{4} \end{bmatrix} \tag{6.46}$$

Then

$$(X_1'X_1)^{-1} = \begin{bmatrix} \dfrac{1}{6} & 0 & 0 \\ 0 & \dfrac{1}{3\rho^2} & 0 \\ 0 & 0 & \dfrac{1}{3\rho^2} \end{bmatrix} \qquad X_1'X_2 = \begin{bmatrix} 3\rho^2 & 3\rho^2 & 0 \\ 0 & 0 & 0 \\ 0 & 0 & 0 \end{bmatrix}$$

and the alias matrix is

$$A = (X_1'X_1)^{-1}X_1'X_2 = \begin{bmatrix} \dfrac{\rho^2}{2} & \dfrac{\rho^2}{2} & 0 \\ 0 & 0 & 0 \\ 0 & 0 & 0 \end{bmatrix} \tag{6.47}$$

With $\boldsymbol{\beta}_1$ and $\boldsymbol{\beta}_2$ defined in Eq. 6.45, the expected values of the estimates of the parameters in the first-order model 6.43 are

$$E(\mathbf{b}) = \boldsymbol{\beta}_1 + \mathbf{A}\boldsymbol{\beta}_2$$

We note from the form of **A** in Eq. 6.47 that b_0 is the only estimate biased by second-order effects and that the smaller the radius ρ, the smaller the amount of bias.

With the double-simplex design in Eq. 6.46, the all-variance design is defined by setting $\rho = 1$ in $\mathbf{X}_1$. The all-bias design has $\rho = 1/\sqrt{2}$ in $\mathbf{X}_1$. These designs are displayed in Figure 6.2.

Example 6.2. As an example of scaled two-level factorial, let us continue with $k = 2$, but now the region R is a circle of radius ρ and let the matrix $\mathbf{F}$ in $\mathbf{D}$ of Eq. 6.37 be of the form

$$\mathbf{F} = \begin{bmatrix} -1 & -1 \\ 1 & -1 \end{bmatrix} \tag{6.48}$$

Then the all-variance design is one for which we set $c = \rho/\sqrt{2}$ in $\mathbf{D}$ so that the points will be on the perimeter of the circle of radius ρ. The all-bias design, on

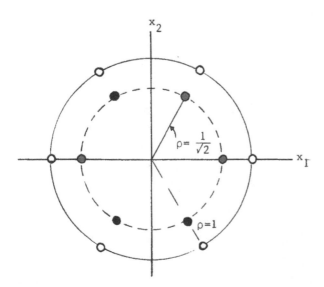

Figure 6.2 Double-simplex all-variance design (○), and all-bias design (●).

the other hand, is obtained by setting $c = \rho/2$ in $\mathbf{D}$, in which case the points are positioned on the perimeter of a circle of radius $\rho/\sqrt{2}$. In both cases, the design is a two-level factorial arrangement. With the inclusion of center point replicates $\mathbf{x}_1' = (1, 0, 0, \ldots, 0)$, a test for curvature can be made (see Section 5.3.1 in Chapter 5).

Before we proceed to the next section to discuss second-order designs, we remark that had the region of interest R, been defined as cuboidal, $-1 \le x_i \le 1$ for $i = 1, 2, \ldots, k$ then $\Omega^{-1} = 2^k$ in Eq. 6.26 and the moment matrices μ_{11}, μ_{12}, and μ_{22} would be of the same form as in Eqs. 6.27–6.29, except that now

$$\mu_{11} = \text{diag}\left(1, \frac{1}{3}, \frac{1}{3}, \ldots, \frac{1}{3}\right) \qquad \mu_{12} = \frac{1}{3}\begin{bmatrix} \mathbf{1}_k' & 0 \\ 0 & 0 \end{bmatrix}$$

$$\mu_{22} = \frac{1}{9}\begin{bmatrix} \left(\frac{4}{5}\right)\mathbf{I}_k + \mathbf{1}_k\mathbf{1}_k' & 0 \\ 0 & \mathbf{I}_{\binom{k}{2}} \end{bmatrix} \tag{6.49}$$

Now, if we consider either of the two point sets in k dimensions specified by

1. $(\pm c, \pm c, \ldots, \pm c)$, a "cube" containing $N = 2^k$ points
2. $(\pm a, 0, \ldots, 0), \ldots, (0, 0, \ldots, \pm a)$,
   an "octahedron" containing $N = 2k$ points $\tag{6.50}$

then the design moment conditions to be satisfied for the all-bias design are, for $i = 1, 2, \ldots, k$,

$$\sum_{u=1}^{N} x_{ui}^2 = \frac{N}{3}$$

$$\sum_{u=1}^{N} x_{ui} = \sum_{u=1}^{N} x_{ui}x_{uj} = \sum_{u=1}^{N} x_{ui}^3 = \sum_{u=1}^{N} x_{ui}^2 x_{uj} = 0 \qquad i \ne j \tag{6.51}$$

When for point set 1., $c = 1/\sqrt{3}$ or for set 2., $a = \sqrt{k/3}$, each satisfies the conditions in Eq. 6.51. When R is the sphere of radius ρ, the same point sets could be used for the all-bias design with conditions

1. $c = \{N\rho^2/(2^k(k+2))\}^{1/2} = \rho/\sqrt{k+2}$
2. $a = \{N\rho^2/(2k+4)\}^{1/2} = \rho\sqrt{k/(k+2)}$

If $\rho = 1$, that is, the region R is the unit sphere, the cuboidal designs have a wider spread than the spherical designs when $k > 1$. Thus, when choosing R to be cuboidal, we are expressing an interest in the "corners" of the cuboidal region which lie outside the sphere, and hence the all-bias design is slightly larger in this latter case.

6.2.4 Some Specific Second-Order Designs for Minimizing *V* and *B*

We shall now discuss the use of the *J* criterion of Section 6.2.2 for the case where the fitted model in *k* variables is of the second order ($d_1 = 2$), yet the true surface is better represented by a cubic polynomial ($d_2 = 3$). The expressions for the fitted model and the true surface are written as

$$\hat{Y}(\mathbf{x}) = \mathbf{x}_1'\mathbf{b} \qquad \text{and} \qquad \eta(\mathbf{x}) = \mathbf{x}_1'\boldsymbol{\beta}_1 + \mathbf{x}_2'\boldsymbol{\beta}_2 \qquad (6.52)$$

where

$$\mathbf{x}_1 = (1, x_1, x_2, \ldots, x_k, x_1^2, \ldots, x_k^2, x_1 x_2, \ldots, x_{k-1} x_k)'$$

$$\boldsymbol{\beta}_1 = (\beta_0, \beta_1, \ldots, \beta_k, \beta_{11}, \ldots, \beta_{kk}, \beta_{12}, \ldots, \beta_{k-1,k})'$$

$$\mathbf{x}_2 = (x_1^3, x_1 x_2^2, \ldots, x_1 x_k^2, x_2^3, x_2 x_1^2, \ldots, x_2 x_k^2, \ldots, x_{k-1} x_k^2, x_1 x_2 x_3, \ldots,$$
$$x_{k-2} x_{k-1} x_k)'$$

$$\boldsymbol{\beta}_2 = (\beta_{111}, \beta_{122}, \ldots, \beta_{1kk}, \beta_{222}, \beta_{211}, \ldots, \beta_{2kk}, \ldots, \beta_{k-1kk}, \beta_{123}, \ldots,$$
$$\beta_{k-2,k-1,k})'$$

and **b** is the least squares estimator of $\boldsymbol{\beta}_1$. For example, when $k = 2$ and by writing the coefficient estimates in **b** as b_i we have

$$\hat{Y}(\mathbf{x}) = b_0 + b_1 x_1 + b_2 x_2 + b_{11} x_1^2 + b_{22} x_2^2 + b_{12} x_1 x_2$$

and

$$\eta(\mathbf{x}) = \beta_0 + \beta_1 x_1 + \beta_2 x_2 + \beta_{11} x_1^2 + \beta_{22} x_2^2 + \beta_{12} x_1 x_2 + \beta_{111} x_1^3 + \beta_{122} x_1 x_2^2$$
$$+ \beta_{222} x_2^3 + \beta_{112} x_1^2 x_2$$

We shall once again assume the design is centered at the origin $(0, 0, \ldots, 0)$ of a spherical region, *R*, of radius ρ, in the scaled variables $x_1, x_2, \ldots, x_k$. Furthermore, for ease in illustrating the form of the moment matrix of the design, we shall consider only designs which belong to the class of second-order rotatable designs.

Minimization of *B*. In the development of the general formulas for *V* and *B*, the expression in Eq. 6.33 for *B* was simplified to Eq. 6.34 by insisting that the design moments satisfy

$$\mathbf{A} = \boldsymbol{\mu}_{11}^{-1} \boldsymbol{\mu}_{12} \qquad (6.53)$$

One way to ensure the equality in Eq. 6.53 is to choose a second-order design for which

$$\frac{1}{N}(X_1'X_1) = \mu_{11} \qquad \frac{1}{N}(X_1'X_2) = \mu_{12} \tag{6.54}$$

that is, the design moments up to the fifth order equal the corresponding region moments. These latter conditions are only sufficient for minimizing B alone, and although they specify the exact sizes of the design moments in terms of the region moments, they are not necessary to achieve minimum bias. However, we shall work with conditions in Eq. 6.54 in minimizing B alone.

For a second-order rotatable design with fifth-order moments equal to zero, the forms of the design moment matrices are

$$\frac{1}{N}(X_1'X_1) = \begin{bmatrix} 1 & 0 & \lambda_2 1_k' & 0 \\ 0 & \lambda_2 I_k & 0 & 0 \\ \lambda_2 1_k & 0 & \lambda_4(2I_k + J_k) & 0 \\ 0 & 0 & 0 & \lambda_4 I_{\binom{k}{2}} \end{bmatrix}$$

$$\frac{1}{N}(X_1'X_2) =$$

$$\lambda_4 \begin{bmatrix} 0 & 0 & 0 & 0 \\ 3\ 1\ 1\ \dots\ 1 & & & 0 \\ & 3\ 1\ 1\ \dots\ 1 & & 0 \\ & & \ddots & 3\ 1\ 1\ \dots\ 1\ 0 \\ 0 & 0 & 0 & 0 \end{bmatrix}$$

where $\lambda_2 = (1/N)\sum_{u=1}^{N} x_{ui}^2$, $\lambda_4 = (1/N)\sum_{u=1}^{N} x_{ui}^2 x_{uj}^2$, $i \neq j$, $3\lambda_4 = (1/N)\sum_{u=1}^{N} x_{ui}^4$. In $(1/N)(X_1'X_1)$, I_k is a $k \times k$ identity matrix, J_k is a $k \times k$ matrix of ones, and 1_k is a $k \times 1$ column vector of ones.

The region moment matrices μ_{11} and μ_{12}, respectively, are identical in form to the $(1/N)X_1'X_1$ and $(1/N)X_1'X_2$ matrices,

$$\mu_{11} = \Omega \int_R x_1 x_1' dx = \begin{bmatrix} 1 & 0 & u\rho^2 1_k' & 0 \\ 0 & u\rho^2 I_k & 0 & 0 \\ u\rho^2 1_k & 0 & v\rho^4(2I_k + J_k) & 0 \\ 0 & 0 & 0 & v\rho^4 I_{\binom{k}{2}} \end{bmatrix} \tag{6.55}$$

and

$$\mu_{12} = \Omega \int_R \mathbf{x}_1 \mathbf{x}_2' \, dx =$$

$$v\rho^4 \begin{bmatrix} 0 & & & & 0 & & & & 0 & 0 \\ 3 & 1 & 1 & \ldots & 1 & & & & & 0 \\ & & & & 3 & 1 & 1 & \ldots & 1 & & 0 \\ & & & & & & & \ddots & & & \\ & & & & & & & 3 & 1 & 1 & \ldots & 1 & 0 \\ 0 & & & & 0 & & & & 0 & 0 \end{bmatrix}$$

where $u(k + 2) = v(k + 2)(k + 4) = 1$. Note that the odd region moments up to and including order 5 are zero by Eq. 6.24.

It can be verified that with this choice of rotatable design, the equalities in Eq. 6.54 are satisfied provided that

$$\lambda_2 = \frac{\rho^2}{k + 2} \qquad \lambda_4 = \frac{\rho^4}{(k + 2)(k + 4)} \tag{6.56}$$

Minimization of V. In the general expression, $V = \text{trace}\{N(\mathbf{X}_1'\mathbf{X}_1)^{-1}\mu_{11}\}$ for the average variance, the matrix $N(\mathbf{X}_1'\mathbf{X}_1)^{-1}$ is the inverse of the moment matrix and is called the *precision matrix*. For $d_1 = 2$, its form is

$$N(\mathbf{X}_1'\mathbf{X}_1)^{-1} = \begin{bmatrix} e & 0 & f\mathbf{1}_k' & 0 \\ 0 & \dfrac{1}{\lambda_2}\mathbf{I}_k & 0 & 0 \\ f\mathbf{1}_k & 0 & (g - h)\mathbf{I}_k + h\mathbf{J}_k & 0 \\ 0 & 0 & 0 & \dfrac{1}{b}\mathbf{I}_{\binom{k}{2}} \end{bmatrix}$$

where $e = \lambda_4(k + 2)/U$, $f = -\lambda_2/U$, $g = \{\lambda_4(k + 1) - \lambda_2^2(k - 1)\}/(2\lambda_4 U)$, $h = (\lambda_2^2 - \lambda_4)/(2\lambda_4 U)$ and $U = \lambda_4(k + 2) - k\lambda_2^2$. Now, if the form of μ_{11} is as shown in Eq. 6.55, then the average variance, V, is

$$V = \text{trace}\{N(\mathbf{X}_1'\mathbf{X}_1)^{-1}\mu_{11}\}$$

$$= \frac{1}{\lambda_4(k + 2) - \lambda_2^2 k}\left[\lambda_4(k + 2) - k\rho^2\left(\lambda_2 - \frac{\lambda_4}{\lambda_2}\right)\right.$$

$$\left. + k\rho^4 \frac{\lambda_4(k + 1) - \lambda_2^2(k - 1)}{2\lambda_4(k + 4)}\right] \tag{6.57}$$

Minimization of V requires finding the values of both λ_2 and λ_4 that produce a low value for Eq. 6.57.

To illustrate the use of the ideas presented thus far in choosing a second-order design that minimizes B and V separately, let us select a central composite design with the design matrix partitioned in the following way:

$$
\mathbf{D} =
\begin{array}{c}
\begin{array}{cccc} x_1 & x_2 & \cdots & x_k \end{array} \\
\left[
\begin{array}{cccc}
\pm r & \pm r & \cdots & \pm r \\
-\alpha & 0 & \cdots & 0 \\
\alpha & 0 & & 0 \\
0 & -\alpha & & 0 \\
0 & \alpha & & 0 \\
0 & 0 & & 0 \\
\vdots & \vdots & \ddots & \vdots \\
0 & 0 & & 0 \\
0 & 0 & & -\alpha \\
0 & 0 & \cdots & \alpha \\
0 & 0 & \cdots & 0 \\
\vdots & \vdots & & \vdots \\
0 & 0 & \cdots & 0
\end{array}
\right]
\begin{array}{l}
\left.\vphantom{\begin{array}{c}a\end{array}}\right\} \quad 2^k \text{ factorial settings} \\[1.0em]
\left.\vphantom{\begin{array}{c}a\\a\\a\\a\\a\\a\\a\\a\\a\end{array}}\right\} \quad 2k \text{ axial settings} \\[3.0em]
\left.\vphantom{\begin{array}{c}a\\a\\a\end{array}}\right\} \quad n_0 \text{ center point replicates}
\end{array}
\end{array}
\tag{6.58}
$$

The objective here is to determine the particular values of r and α so that the design is rotatable along with satisfying Eq. 6.56 for λ_2 and λ_4 when minimizing B, or the values that minimize V in Eq. 6.57. For both cases, note that with $\mathbf{D}$ defined in Eq. 6.58, $N = 2^k + 2k + n_0$. Then

$$
\frac{1}{N}\sum_{u=1}^{N} x_{ui}^2 = \frac{1}{N}(2^k r^2 + 2\alpha^2) = \lambda_2
$$

$$
\frac{1}{N}\sum_{u=1}^{N} x_{ui}^2 x_{uj}^2 = \frac{1}{N}(2^k r^4) = \lambda_4 \qquad \alpha = 2^{k/4} r \text{ (for rotatability)}
$$

$$
\frac{1}{N}\sum_{u=1}^{N} x_{ui}^4 = \frac{1}{N}(2^k r^4 + 2\alpha^4)
$$

1. Minimizing B Only. To achieve the relationships concerning λ_2 and λ_4 in Eq. 6.56, we have for $k > 1$ and $\alpha = r2^{k/4}$,

$$
\left(\frac{1}{N}\right) 2^k r^4 = \frac{1}{N}\frac{(2^k + 2^{(k+2)/2})r^2 \rho^2}{k + 4}
$$

Solving for r, we find

$$
r = \rho \left[\frac{(1 + 2^{(2-k)/2})}{(k + 4)} \right]^{1/2}
\tag{6.59}
$$

For example, if ρ is set equal to unity, then for $k = 2$, $r = 0.577$; $k = 3$, $r = 0.494$; and for $k = 4$, $r = 0.433$.

2.   *Minimizing V Only.*   For any value of $k > 1$, suppose we position the points in the factorial portion of **D** on the boundary of a sphere of radius ρ^*. Then, $r = \rho^*/\sqrt{k}$ and $\alpha = 2^{k/4}\rho^*/\sqrt{k}$. Minimizing V is achieved by selecting $\rho^* = \rho$, the radius of the spherical region R.

Some comments on minimizing $J = V + B$.   In Section 6.2.1, it was shown for $k = 1$ that in setting up a first-order design, when in fact the true surface is quadratic, the minimization of J required knowing the value of the quadratic coefficient, or at least knowing the value of the ratio β_2/σ. In Box and Draper (1959) and Myers (1976: Ch. 9), tables are given that show for $k > 1$ that with any first-order rotatable design in which the third-order moments are zero, the best spread of the design points is a compromise between the all-variance design and the all-bias design where the compromise favors the all-bias design. A rough guideline stated by Box and Draper (1963) is to choose a design which has moments that are approximately 10 percent larger than the moments of the all-bias design.

When the fitted model is quadratic but the true surface is cubic, as shown in Eq. 6.52, minimization of $J = V + B$ where $V > 0$ and $B > 0$ is again impossible without knowledge of the values of the cubic parameters β_{111}, β_{122}, $\ldots$, β_{kkk}. Box and Draper (1963), however, present numerous curves depicting the values of $\lambda_2^{1/2} = [ii]^{1/2}$ and $\lambda = 3\lambda_4/\lambda_2^2 = [iiii]/[ii]^2$ for a second-order rotatable design with zero fifth-order moments that minimizes $J = V + B$ in situations where the ratio V/B ranges from zero to infinity and $k = 1, 2, 3, 4,$ and 5. Some of their findings are summarized as follows:

1.   For $k = 1$, as B approaches infinity (V/B goes to zero), the value of $\lambda_2^{1/2}$ approaches 0.606 and λ approaches 1.632. As the variance contribution to J becomes larger while the B contribution becomes smaller, the moment λ_2^2 and the ratio λ both increase. For example, when $V = 8B$, approximately, the best design is one for which $\lambda_2^{1/2} = 0.70$ and $\lambda = 2.0$. Such a design is much closer, however, to the all-bias design than to the all-variance design. When $V = B$, approximately, the best design has $\lambda_2^{1/2} = 0.621$ and $\lambda = 1.669$ and is very close to the all-bias design.

2.   For $k = 2, 3, 4,$ and 5, the values of $\lambda_2^{1/2}$ and λ for the best design change only slightly from those of the all-bias design for each value of k as B varies over the range of 0 to infinity. As B approaches infinity, the values of $\lambda_2^{1/2}$ for $k = 2$ to 5 range between 0.57 and 0.32, approximately, while the values of λ range between 1.87 and 2.30, approximately. As the value of V/B approaches infinity (i.e., approaches the all-variance situation), the best design is the largest possible one in which case $\lambda_2^{1/2}$ and λ are both made large.

6.3 OTHER PROCEDURES FOR ESTIMATING RESPONSE SURFACES THAT USE THE IMSE CRITERION

In the previous sections, the minimization of J, the IMSE of $\hat{Y}(\mathbf{x})$, with respect to the choice of experimental design depended on the relative magnitudes of the V and B contributions to J. The somewhat surprising conclusion, first reported by Box and Draper (1959), was that unless the V contribution was many times larger than the B contribution (e.g., $V \geq 4B$), the optimum designs were remarkably close to those obtained by ignoring V completely. Consequently, in using this approach for choosing a design, hereafter called the *Box-Draper approach*, the resulting all-bias design must have design moments that are equal to the corresponding region moments up to and including order $d_1 + d_2$ as shown in Eq. 6.54.

An alternative strategy to that of choosing a design that minimizes the bias due to specific higher-order terms of a true surface equation but which are omitted when fitting a model of order d_1 is to adopt a method of estimation aimed directly at minimizing the bias. Then, other criteria, such as the variance of prediction, could be minimized through the choice of design. In this section, we shall discuss two such methods, minimum bias estimation (MBE) and generalized least squares estimation (GE). MBE concentrates on minimizing the bias of the fitted model through the estimation of the parameters while GE minimizes the variance of $\hat{Y}(\mathbf{x})$ by weighting the parameter estimates. At the end of this section, we shall discuss very briefly the comparison of the three approaches, Box-Draper, MBE, and GE.

6.3.1 Minimum Bias Estimation

Minimum bias estimation (MBE) is a technique used for estimating the parameters in a fitted model of order d_1 so that the average squared bias contribution of J in Eq. 6.20 is minimized. Furthermore, provided only that the design satisfies a simple estimability condition, the minimum bias estimator can also achieve minimum average variance of $\hat{Y}(\mathbf{x})$ for any fixed design. MBE was introduced by Karson, Manson, and Hader (1969).

Let us assume the true surface in the k-dimensional space is represented by a polynomial of degree d_2, as in Eq. 6.19, in the coded variables of the form

$$\eta(\mathbf{x}) = \mathbf{x}_1'\boldsymbol{\beta}_1 + \mathbf{x}_2'\boldsymbol{\beta}_2 \tag{6.60}$$

The fitted model is of lower order d_1 $(d_1 < d_2)$ and is expressed as

$$\hat{Y}(\mathbf{x}) = \mathbf{x}_1'\mathbf{b}_1 \tag{6.61}$$

where $\mathbf{b}_1$ is an estimator of $\boldsymbol{\beta}_1$, not necessarily the least squares estimator. Since the fitted model in Eq. 6.61 is of order d_1, $E[\hat{Y}(\mathbf{x})]$ is a polynomial of degree

d_1, that is,

$$E[\hat{Y}(\mathbf{x})] = \mathbf{x}_1'\boldsymbol{\gamma}_1 \tag{6.62}$$

where the elements of $\boldsymbol{\gamma}_1$ are the expectations of the elements of $\mathbf{b}_1$ and these expectations will be shown to be functions of the elements of $\boldsymbol{\beta}_1$ and $\boldsymbol{\beta}_2$ in Eq. 6.60. Associated with $J = V + B$, the average squared bias is expressible as

$$B = \frac{N\Omega}{\sigma^2} \int_R \{E[\hat{Y}(\mathbf{x})] - \eta(\mathbf{x})\}^2 d\mathbf{x}$$

$$= \frac{N}{\sigma^2} \{(\boldsymbol{\gamma}_1 - \boldsymbol{\beta}_1)'\boldsymbol{\mu}_{11}(\boldsymbol{\gamma}_1 - \boldsymbol{\beta}_1) - 2(\boldsymbol{\gamma}_1 - \boldsymbol{\beta}_1)'\boldsymbol{\mu}_{12}\boldsymbol{\beta}_2 + \boldsymbol{\beta}_2'\boldsymbol{\mu}_{22}\boldsymbol{\beta}_2\} \tag{6.63}$$

where the region moment matrices $\boldsymbol{\mu}_{11}$, $\boldsymbol{\mu}_{12}$, and $\boldsymbol{\mu}_{22}$ are defined as before (see Eq. 6.25),

$$\boldsymbol{\mu}_{ij} = \Omega \int_R \mathbf{x}_i\mathbf{x}_j' d\mathbf{x} \qquad i \leq j = 1, 2 \tag{6.64}$$

To minimize B in Eq. 6.63, we first differentiate B with respect to $\boldsymbol{\gamma}_1$, and equate the results to zero,

$$\frac{\partial B}{\partial \boldsymbol{\gamma}_1} = 2\boldsymbol{\mu}_{11}(\boldsymbol{\gamma}_1 - \boldsymbol{\beta}_1) - 2\boldsymbol{\mu}_{12}\boldsymbol{\beta}_2 = 0$$

Solving for $\boldsymbol{\gamma}_1$, we obtain

$$\boldsymbol{\gamma}_1 = \boldsymbol{\beta}_1 + \boldsymbol{\mu}_{11}^{-1}\boldsymbol{\mu}_{12}\boldsymbol{\beta}_2 \tag{6.65}$$

Substituting the expression in Eq. 6.65 for $\boldsymbol{\gamma}_1$ into Eq. 6.63, the expression for the minimized value of B is

$$\text{Min } B = \frac{N}{\sigma^2} \boldsymbol{\beta}_2'\{\boldsymbol{\mu}_{22} - \boldsymbol{\mu}_{12}'\boldsymbol{\mu}_{11}^{-1}\boldsymbol{\mu}_{12}\}\boldsymbol{\beta}_2 \tag{6.66}$$

Thus, a necessary and sufficient condition for the minimization of B, using a fitted model of order d_1, is simply

$$E[\hat{Y}(\mathbf{x})] = E[\mathbf{x}_1'\mathbf{b}_1] = \mathbf{x}_1'\tilde{\mathbf{A}}\boldsymbol{\beta} \tag{6.67}$$

where

$$\tilde{\mathbf{A}} = [\mathbf{I}_p : \boldsymbol{\mu}_{11}^{-1}\boldsymbol{\mu}_{12}] \qquad \text{and} \qquad \boldsymbol{\beta} = \begin{bmatrix} \boldsymbol{\beta}_1 \\ \boldsymbol{\beta}_2 \end{bmatrix} \tag{6.68}$$

where p is the number of elements in $\boldsymbol{\beta}_1$. From Eq. 6.67 we may conclude that $\tilde{\mathbf{A}}\boldsymbol{\beta}$ must be estimable for the minimization of B.

To see that the minimized value of B is attainable by our fitting a model of order d_1 such that $\tilde{A}\beta$ is estimable, let us write the vector of estimates b_1 in Eq. 6.61 as a linear combination of the N observations

$$b_1 = LY \tag{6.69}$$

where L is a $p \times N$ matrix of constants, one form of which is shown below. Since $E(b_1) = LE(Y) = L[X_1\beta_1 + X_2\beta_2] = L[X_1 : X_2]\beta = LX\beta$, where $X = [X_1 : X_2]$ and X_1, X_2 are, respectively, the matrices corresponding to β_1 and β_2 in the model for the true mean response, the matrix L in Eq. 6.69 must satisfy $LX = \tilde{A}$ from Eq. 6.67. A class of designs satisfying this condition is the class of all-bias designs which, previously in Eq. 6.53, were shown to satisfy the condition

$$(X_1'X_1)^{-1}X_1'X_2 = \mu_{11}^{-1}\mu_{12}$$

It can be easily verified that the matrix L defined by

$$L = (X_1'X_1)^{-1}X_1' \tag{6.70}$$

does satisfy $LX = \tilde{A}$. In this special case, the standard least squares estimator of β_1 minimizes B.

Having obtained a fitted model of the form in Eq. 6.61, where b_1 is expressed in Eq. 6.69, the average variance of $\hat{Y}(x)$ is

$$V = \frac{N\Omega}{\sigma^2} \int_R \text{Var}[\hat{Y}(x)]dx \tag{6.71}$$

Since $E[\hat{Y}(x)] = x_1'\tilde{A}\beta$, then for a fixed design, V in Eq. 6.71 is minimized when b_1 is the unbiased least squares estimator of $\tilde{A}\beta$, that is,

$$b_1 = \tilde{A}(X'X)^{-1}X'Y$$

In this case, L has the form

$$L = \tilde{A}(X'X)^{-1}X' \tag{6.72}$$

which satisfies the condition $LX = \tilde{A}$. Note that in Eq. 6.72 the X matrix is assumed to be of full-column rank. In general, this may not be the case and, consequently, $X'X$ will be singular. In this case, a generalized inverse, $(X'X)^-$, can be used instead of $(X'X)^{-1}$. For a definition of a generalized inverse of a matrix and a description of its properties, the reader is referred to Searle (1971; Chapter 1).

Since the variance of $\hat{Y}(x)$ is

$$\text{Var}[\hat{Y}(x)] = x_1'\tilde{A}(X'X)^{-1}\tilde{A}'x_1\sigma^2$$

then the expression for V in Eq. 6.71 is

$$V = N \text{ trace}\{\tilde{\mathbf{A}}(\mathbf{X}'\mathbf{X})^{-1}\tilde{\mathbf{A}}'\boldsymbol{\mu}_{11}\} \tag{6.73}$$

Before we illustrate the method of MBE with an example, we point out that the vector of estimates (Eq. 6.69), where the matrix **L** is shown in Eq. 6.72, involves the full **X** matrix which includes columns for the $\mathbf{x}_2'$ variables in $\eta(\mathbf{x})$ of Eq. 6.60. In other words, although the fitted model of order d_1 contains only the variables specified in $\mathbf{x}_1$, to minimize both the variance and the bias of the fitted model requires knowing the form of the true model over the region R. Thus, if $\mathbf{X}'\mathbf{X}$ is nonsingular, all of the elements of $\boldsymbol{\beta}$ in $\boldsymbol{\beta}_1$ and $\boldsymbol{\beta}_2$ of Eq. 6.60 are estimated by standard least squares and put into

$$\hat{Y}(\mathbf{x}) = \mathbf{x}_1'\mathbf{b}_1$$

by taking $\mathbf{b}_1 = \tilde{\mathbf{A}}\boldsymbol{\beta}$ where $\tilde{\mathbf{A}}$ and $\boldsymbol{\beta}$ are shown in Eq. 6.68 and $\hat{\boldsymbol{\beta}} = (\mathbf{X}'\mathbf{X})^{-1}\mathbf{X}'\mathbf{Y}$.

Example 6.3. Let us illustrate the method of MBE for the case of fitting a simple linear regression model ($k = 1$) but where the true surface is better expressed by the parabola

$$\eta(x) = \beta_0 + \beta_1 x + \beta_2 x^2$$
$$= \mathbf{x}_1'\boldsymbol{\beta}_1 + x_2'\boldsymbol{\beta}_2 \tag{6.74}$$

In Eq. 6.74, $\mathbf{x}_1 = (1, x)'$, $x_2 = x^2$, $\boldsymbol{\beta}_1 = (\beta_0, \beta_1)'$ and $\boldsymbol{\beta}_2 = \beta_2$. The fitted model is

$$\hat{Y}(x) = b_0 + b_1 x$$
$$= \mathbf{x}_1'\mathbf{b}_1$$

where

$$\mathbf{b}_1 = (b_0, b_1)' \tag{6.75}$$

Suppose we choose to take observations at each of the three settings, $x = 0$, $x = \pm l$ and solve for the value of l that minimizes V in Eq. 6.73. The **X**, **Y** and $(\mathbf{X}'\mathbf{X})^{-1}$ matrices respectively, are

$$\mathbf{X} = [\mathbf{X}_1 : \mathbf{X}_2] = \begin{bmatrix} 1 & -l & l^2 \\ 1 & 0 & 0 \\ 1 & l & l^2 \end{bmatrix} \qquad \mathbf{Y} = \begin{bmatrix} Y_1 \\ Y_2 \\ Y_3 \end{bmatrix}$$

$$(\mathbf{X}'\mathbf{X})^{-1} = \begin{bmatrix} 1 & 0 & \dfrac{-1}{l^2} \\ 0 & \dfrac{1}{2l^2} & 0 \\ \dfrac{-1}{l^2} & 0 & \dfrac{3}{2l^4} \end{bmatrix} \tag{6.76}$$

If we choose as our experimental region $R = [-1, +1]$, then

$$\Omega^{-1} = 2 \qquad \mu_{11} = \begin{bmatrix} 1 & 0 \\ 0 & \dfrac{1}{3} \end{bmatrix}$$

$$\mu_{12} = \begin{bmatrix} \dfrac{1}{3} \\ 0 \end{bmatrix} \qquad \mu_{22} = \dfrac{1}{5} \tag{6.77}$$

The minimum value of B, from Eq. 6.66, is

$$\text{Min } B = \left(\dfrac{3}{\sigma^2}\right) \beta_2^2 \left(\dfrac{1}{5} - \dfrac{1}{9}\right)$$

$$= \left(\dfrac{4}{15}\right) \left(\dfrac{\beta_2^2}{\sigma^2}\right) \tag{6.78}$$

which is independent of l.

The formulas for the coefficient estimates in the fitted model, expressed as functions of the observations, are

$$\mathbf{b}_1 = \mathbf{LY} = \tilde{\mathbf{A}}(\mathbf{X'X})^{-1}\mathbf{X'Y}$$

$$= \dfrac{1}{6l^2} \begin{bmatrix} Y_1 + Y_3 + 2(3l^2 - 1)Y_2 \\ 3l(Y_3 - Y_1) \end{bmatrix}$$

since

$$\tilde{\mathbf{A}} = \begin{bmatrix} 1 & 0 & \dfrac{1}{3} \\ 0 & 1 & 0 \end{bmatrix} \tag{6.79}$$

The expression in Eq. 6.73 for the average variance, V, becomes

$$V = 3 - \dfrac{3}{2l^2} + \dfrac{1}{2l^4}$$

and its minimum value, obtained by solving $\partial V/\partial l = 0$ for l, is $\text{Min } V = 15/8$, and this minimum occurs at $l = \sqrt{2/3}$.

Two other examples are presented by Karson, Manson, and Hader (1969) and consist of the following, for $k = 1$:

1. $\hat{Y}(x)$ is of the first order, $\eta(x)$ is quadratic, and one observation is collected at each of the four points $x = \pm l_1$, $x = \pm l_2$.
2. $\hat{Y}(x)$ is of the first order, $\eta(x)$ is cubic, and one observation is collected at each of the four points $x = \pm l_1$, $x = \pm l_2$.

In each case, they show that the expression for $J = V + B$ obtained through MBE can lead to a value lower than the value of J obtained by the Box-Draper least square approach.

In a follow-up paper, Karson (1970) introduced a criterion whereby, when fitting a model of order d_1 using MBE, one is protected against the possibility that the true surface in R is really of order $d_1 + m$ ($m \geq 1$). The proposed criterion considers different values of m (e.g., $m = 1$ and $m = 2$) and chooses the design that makes the corresponding minimum bias estimators identical. The designs which produce the same minimum bias estimators for different values of m also make the corresponding values of V equal.

Let us illustrate Karson's dual design criterion for $k = 1$ where the fitted model is of order $d_1 = 1$ and the region of interest R is the interval $-1 \leq x \leq 1$. We shall require that the designs be symmetric about the center $x = 0$ and that the odd design moments be equal to zero.

The fitted model is of the form

$$\hat{Y}(x) = \mathbf{x}_1'\mathbf{b}_1 = b_0 + b_1 x$$

where b_0 and b_1 are determined from $\mathbf{b}_1 = \tilde{\mathbf{A}}(\mathbf{X}'\mathbf{X})^{-1}\mathbf{X}'\mathbf{Y}$. If the true surface is of order $d_1 + m$, then the form of the matrix $\mathbf{X}$ depends on the value of m, that is, if $m = 1$, a row of the matrix $\mathbf{X}$ contains the elements $1, x, x^2$, and if $m = 2$, a row of $\mathbf{X}$ contains the elements $1, x, x^2$ and x^3.

Suppose we decide by our choice of design to protect against the dual possibility that $m = 1$ or $m = 2$. This means in using MBE, the design we select must satisfy the property, $\tilde{\mathbf{A}}(\mathbf{X}'\mathbf{X})^{-1}\mathbf{X}'\mathbf{Y}$ for $m = 1$ be equal to $\tilde{\mathbf{A}}(\mathbf{X}'\mathbf{X})^{-1}\mathbf{X}'\mathbf{Y}$ for $m = 2$. To show that this is possible, for $m = 1$, let us write the even design moments as $[2] = (1/N) \sum_{u=1}^{N} x_u^2$ and $[4] = (1/N) \sum_{u=1}^{N} x_u^4$. (Since $k = 1$, it would be easier and more convenient here to adopt this notation. Note that in Chapter 2 the same two moments would have been denoted by $[1^2]$ and $[1^4]$, respectively). Then, the matrices $\mathbf{X}$ and $(\mathbf{X}'\mathbf{X})^{-1}$ are

$$\mathbf{X}_{m=1} = \begin{bmatrix} 1 & x_1 & x_1^2 \\ 1 & x_2 & x_2^2 \\ \vdots & \vdots & \vdots \\ 1 & x_N & x_N^2 \end{bmatrix}$$

$$(\mathbf{X}'\mathbf{X})_{m=1}^{-1} = \frac{1}{N[2]([4] - [2]^2)} \begin{bmatrix} [2][4] & 0 & -[2]^2 \\ 0 & [4] - [2]^2 & 0 \\ -[2]^2 & 0 & [2] \end{bmatrix} \quad (6.80)$$

Recalling from Eq. 6.79 the form of

$$\tilde{\mathbf{A}} = \begin{bmatrix} 1 & 0 & \dfrac{1}{3} \\ 0 & 1 & 0 \end{bmatrix}$$

and letting $\mathbf{b}_1 = \mathbf{L}_{m=1}\mathbf{Y}$, where

$$\mathbf{L}_{m=1} = \begin{bmatrix} \mathbf{L}_0' \\ \mathbf{L}_1' \end{bmatrix} = \tilde{\mathbf{A}}(\mathbf{X}'\mathbf{X})_{m=1}^{-1}\mathbf{X}_{m=1}'$$

we get

$$b_0 = \mathbf{L}_0'\mathbf{Y}$$

$$= \frac{1}{D}\left\{\left([4] - \frac{[2]}{3}\right)\sum_{u=1}^{N} Y_u + \left(\frac{1}{3} - [2]\right)\sum_{u=1}^{N} x_u^2 Y_u\right\} \tag{6.81}$$

$$b_1 = \mathbf{L}_1'\mathbf{Y} = \left(\frac{1}{N[2]}\right)\sum_{u=1}^{N} x_u Y_u \tag{6.82}$$

where $D = N([4] - [2]^2)$.

Now, for $m = 2$, we have

$$\mathbf{X}_{m=2} = \begin{bmatrix} 1 & x_1 & x_1^2 & x_1^3 \\ 1 & x_2 & x_2^2 & x_2^3 \\ \vdots & \vdots & \vdots & \vdots \\ 1 & x_N & x_N^2 & x_N^3 \end{bmatrix}$$

$$(\mathbf{X}'\mathbf{X})_{m=2} = \begin{bmatrix} (\mathbf{X}'\mathbf{X})_{m=1} & \mathbf{t}_1 \\ \mathbf{t}_1' & N[6] \end{bmatrix} \tag{6.83}$$

where $\mathbf{t}_1 = N(0, [4], 0)'$ and $[6] = (1/N)\sum_{u=1}^{N} x_u^6$. (In Chapter 2 this was denoted by $[1^6]$.) The matrix $\tilde{\mathbf{A}}$ is

$$\tilde{\mathbf{A}} = \begin{bmatrix} 1 & 0 & \dfrac{1}{3} & 0 \\ 0 & 1 & 0 & \dfrac{3}{5} \end{bmatrix}$$

To obtain a nonsingular $\mathbf{X}'\mathbf{X}$ for $m = 2$, suppose there are four distinct levels of x in the interval $-1 \leq x \leq 1$, say, for example, $x = \pm l_1$ and $\pm l_2$ where $l_i \leq 1$, $i = 1, 2$. Then, for $m = 2$,

$$(\mathbf{X}'\mathbf{X})_{m=2}^{-1} = \begin{bmatrix} \mathbf{Q} & \mathbf{R} \\ \mathbf{R}' & c_1 \end{bmatrix}$$

where c_1 is the positive scalar, $c_1 = [2]/\{N([2][6] - [4]^2)\}$, the matrix $\mathbf{Q}$ is $\mathbf{Q} = (\mathbf{X}'\mathbf{X})_{m=1}^{-1}(\mathbf{I}_3 + \mathbf{t}_1 c_1 \mathbf{t}_1'(\mathbf{X}'\mathbf{X})_{m=1}^{-1})$ and $\mathbf{R} = -(\mathbf{X}'\mathbf{X})_{m=1}^{-1}\mathbf{t}_1 c_1$. Carrying out the matrix multiplication and letting $\mathbf{b}_1 = \mathbf{L}_{m=2}\mathbf{Y}$ where

$$\mathbf{L}_{m=2} = \begin{bmatrix} \mathbf{L}_2' \\ \mathbf{L}_3' \end{bmatrix} = \tilde{\mathbf{A}}(\mathbf{X}'\mathbf{X})_{m=2}^{-1}\mathbf{X}_{m=2}' \tag{6.84}$$

we have

$$b_0 = L_2'Y = \left(\frac{[4] - \frac{[2]}{3}}{N([4] - [2]^2)}\right) \sum_{u=1}^{N} Y_u + \left(\frac{-[2] + \frac{1}{3}}{N([4] - [2]^2)}\right) \sum_{u=1}^{N} x_u^2 Y_u \qquad (6.85)$$

$$b_1 = L_3'Y = \left(\frac{1}{N[2]} - \frac{(\frac{3}{5} - \frac{[4]}{[2]})[4]c_1}{[2]}\right) \sum_{u=1}^{N} x_u Y_u + \left[\frac{(\frac{3}{5} - \frac{[4]}{[2]})c_1}{5[2]}\right] \sum_{u=1}^{N} x_u^3 Y_u \qquad (6.86)$$

Equating the formulas in Eqs. 6.81 and 6.85 for b_0 as well as the expressions in Eqs. 6.82 and 6.86 for b_1, we conclude that the design moments, [2] and [4], must satisfy the condition, $[4] = (3/5)[2]$, which produces the estimators

$$b_0 = \left(\frac{4}{15N(\frac{3}{5} - [2])}\right) \sum_{u=1}^{N} Y_u + \left(\frac{\frac{1}{3} - [2]}{N[2](\frac{3}{5} - [2])}\right) \sum_{u=1}^{N} x_u^2 Y_u$$

$$b_1 = \left(\frac{1}{N[2]}\right) \sum_{u=1}^{N} x_u Y_u$$

For example, if $x = \pm l_1$ and $x = \pm l_2$, as in Figure 6.3, then $[2] = (1/2)(l_1^2 + l_2^2) = (1/2)a$, where $a = l_1^2 + l_2^2$, and

$$b_0 = \frac{2 \sum_{u=1}^{4} Y_u}{18 - 15a} + \frac{5(2 - 3a)\{l_1^2(Y_2 + Y_3) + l_2^2(Y_1 + Y_4)\}}{6a(6 - 5a)}$$

$$b_1 = \frac{1}{2a} \{l_2(Y_4 - Y_1) + l_1(Y_3 - Y_2)\}$$

The average variance, for $m = 1$ is

$$V = \left(\frac{2}{9}\right) \left[\frac{9b - 6a + 2}{2b - a^2}\right] + \frac{2}{3a} \qquad (6.87)$$

where $b = l_1^4 + l_2^4$. For $m = 2$ we have

$$V = \left(\frac{2}{9}\right) \left[\frac{9b - 6a + 2}{2b - a^2}\right] + \left(\frac{2}{75}\right) \left[\frac{75ab - 25a^3 - 60b + 18a}{(2b - a^2)(a - b)}\right] \qquad (6.88)$$

Figure 6.3 Four-point design where $x = \pm l_1, \pm l_2$, so that $a = l_1^2 + l_2^2$ and $b = l_1^4 + l_2^4$ in V. The response values at the points are denoted by $Y_1, Y_2, Y_3,$ and Y_4.

Substituting the protection requirement $b = (3/5)a$ into the Eqs. 6.87 and 6.88 produces the single expression for the average variance

$$V = \frac{4(14 - 9a)}{9a(6 - 5a)}$$

and the minimum value of V is 1.88806 when $a = 0.81186$.

Karson (1970) lists protection conditions on the design moments for the following cases: $d_1 = 1$, $m = 1, 2, 3$, and 4; $d_1 = 2$, $m = 1, 2, 3$; and $d_1 = 3$, $m = 1, 2$. In addition, the two-variable problem ($k = 2$) where the fitted model is $\hat{Y}(\mathbf{x}) = b_0 + b_1 x_1 + b_2 x_2$, $m = 1$ and 2 and the region R is cuboidal is presented. The extension to the k-variable case is illustrated where the fitted model is of the first order and $m = 1$ and 2 where R is again cuboidal of the form $-1 \le x_i \le 1$, $i = 1, 2, \ldots, k$.

6.3.2 A Generalized Estimator for the Estimated Response Function

Another method of estimating the response surface using a polynomial model involves first obtaining the fitted model of order d_1 by standard least squares and then weighting one or more of the coefficient estimates in the fitted model to reduce the mean squared error of $\hat{Y}(\mathbf{x})$. To illustrate, suppose we choose to fit the simple linear regression equation, over N observations,

$$\mathbf{Y} = \mathbf{X}\boldsymbol{\beta} + \boldsymbol{\varepsilon}$$

where $\mathbf{Y}$ is $N \times 1$, $\mathbf{X} = [\mathbf{1}:\mathbf{x}]$ is an $N \times 2$ matrix with uth row $(1, x_u)$, $\boldsymbol{\beta} = (\beta_0, \beta_1)'$ is a 2×1 vector where the estimator of β_0 is $\bar{Y}$, and $\boldsymbol{\varepsilon}$ is the $N \times 1$ vector of errors satisfying the assumptions $E(\boldsymbol{\varepsilon}) = \mathbf{0}$, $E(\boldsymbol{\varepsilon}\boldsymbol{\varepsilon}') = \sigma^2 \mathbf{I}_N$. An estimator of β_1, which we denote by $b_1 = \mathbf{l}'\mathbf{Y}$, and which has minimum mean squared error, $\mathrm{MSE}(b_1) = \mathbf{l}'\mathbf{l}\sigma^2 + (1 - \mathbf{l}'\mathbf{x})^2 \beta_1^2$, is of the form $b_1 = \kappa \hat{\beta}_1$, where $\hat{\beta}_1 = (\mathbf{x}'\mathbf{x})^{-1}\mathbf{x}'\mathbf{Y}$ and $\kappa = \beta_1^2 / \{\beta_1^2 + (\mathbf{x}'\mathbf{x})^{-1}\sigma^2\}$.

In applying the above ideas to the multiple variable case, Kupper and Meydrech (1973, 1974) suggested that when fitting a model of order d_1 in the k variables $x_1, x_2, \ldots, x_k$,

$$\hat{Y}(\mathbf{x}) = \mathbf{x}_1'\mathbf{b}_1 \tag{6.89}$$

but the true surface $\eta(\mathbf{x}) = \mathbf{x}_1'\boldsymbol{\beta}_1 + \mathbf{x}_2'\boldsymbol{\beta}_2$ is of order d_2 ($>d_1$) (see Eq. 6.60), one should use the generalized estimator (GE)

$$\mathbf{b}_1 = \mathbf{K}\hat{\boldsymbol{\beta}}_1 = \mathbf{K}(\mathbf{X}_1'\mathbf{X}_1)^{-1}\mathbf{X}_1'\mathbf{Y} \tag{6.90}$$

In Eq. 6.90, $\mathbf{K}$ is a $p \times p$ diagonal matrix of appropriately chosen constants and $\mathbf{X}_1$ is the $N \times p$ matrix of values taken by the terms of $\mathbf{x}_1$ for the N

experimental combinations of the coded variables. The value of p is $(k + d_1)!/(k!d_1!)$.

With a fitted model of the form in Eq. 6.89, where $\mathbf{b}_1$ is defined by Eq. 6.90, the integrated variance, V, and integrated squared bias, B, take the respective forms

$$V = N\Omega \int_R \mathbf{x}_1' \mathbf{K}(\mathbf{X}_1'\mathbf{X}_1)^{-1}\mathbf{K}\mathbf{x}_1 d\mathbf{x} \tag{6.91}$$

$$B = \frac{N\Omega}{\sigma^2} \int_R \{\mathbf{x}_1'\mathbf{K}(\boldsymbol{\beta}_1 + \mathbf{A}\boldsymbol{\beta}_2) - (\mathbf{x}_1'\boldsymbol{\beta}_1 + \mathbf{x}_2'\boldsymbol{\beta}_2)\}^2 d\mathbf{x} \tag{6.92}$$

where $\mathbf{A} = (\mathbf{X}_1'\mathbf{X}_1)^{-1}\mathbf{X}_1'\mathbf{X}_2$ and $\Omega^{-1} = \int_R d\mathbf{x}$.

In general, the particular choice of $\mathbf{K}$ which minimizes the sum of Eqs. 6.91 and 6.92 will depend on the unknown elements of $\boldsymbol{\beta}_1$ and $\boldsymbol{\beta}_2$ and on σ^2. However, if the parameter space of $\boldsymbol{\beta}_1$ can be restricted by specifying bounds for at least one of the elements of $\boldsymbol{\beta}_1$ relative to σ^2/N, it is possible to find a set of values of $\mathbf{K}$ that make $J = V + B$ smaller in value than when $\mathbf{K} = \mathbf{I}$ for any choice of experimental design. In fact, the particular choice of $\mathbf{K}$ from the set of values of $\mathbf{K}$ is that one which minimizes the maximum value of J over the restricted parameter space.

Before we discuss a method for choosing the matrix $\mathbf{K}$, let us write out the formulas in Eqs. 6.91 and 6.92 for V and B, respectively, in the case of k variables where $d_1 = 1$ and $d_2 = 2$. The fitted model of order one is

$$\hat{Y}(\mathbf{x}) = \mathbf{x}_1'\mathbf{K}\hat{\boldsymbol{\beta}}_1 \tag{6.93}$$

where $\mathbf{x}_1 = (1, x_1, x_2, \ldots, x_k)'$, $\mathbf{K} = \text{diag}(1, \kappa_1, \kappa_2, \ldots, \kappa_k)$ and $\hat{\boldsymbol{\beta}}_1 = (\mathbf{X}_1'\mathbf{X}_1)^{-1}\mathbf{X}_1'\mathbf{Y}$. The true response is expressed as

$$\eta(\mathbf{x}) = \mathbf{x}_1'\boldsymbol{\beta}_1 + \mathbf{x}_2'\boldsymbol{\beta}_2 \tag{6.94}$$

where $\mathbf{x}_2 = (x_1^2, \ldots, x_k^2, x_1x_2, \ldots, x_{k-1}x_k)'$ and $\boldsymbol{\beta}_2 = (\beta_{11}, \ldots, \beta_{kk}, \beta_{12}, \ldots, \beta_{k-1,k})'$. Now, from Eq. 6.31, it follows that

$$V = N \, \text{trace}[\mathbf{K}(\mathbf{X}_1'\mathbf{X}_1)^{-1}\mathbf{K}\boldsymbol{\mu}_{11}] \tag{6.95}$$

where $\boldsymbol{\mu}_{11} = \Omega \int_R \mathbf{x}_1\mathbf{x}_1' d\mathbf{x}$ is the matrix of region moments up to order $2d_1$.

To determine what is gained through the properties of $\hat{Y}(\mathbf{x})$ by using Eq. 6.93, we compare the integrated properties of the generalized estimator to those obtained using standard least squares (i.e., $\mathbf{K} = \mathbf{I}$). We can define the experimental region as a hypersphere (Box and Draper, 1959) or as a cuboid (Draper and Lawrence, 1965). In Section 6.2.2 we developed formulas for V and B where the fitted model was studied over a spherical region. The consideration of R as being cuboidal was addressed in Eqs. 6.49–6.51. In order to be able to contrast the formulas for V and B for the two types of regions of interest, we shall now consider the region to be cuboidal in shape and defined by $-1 \le x_i \le 1$, $i = 1, 2, \ldots, k$. The cuboidal region will be denoted by $\check{R}$ to distinguish it from the hypersphere which we denote by R.

To express the integrated squared bias, B, in terms of the design moments, the region moments, and as a function of the parameters in Eq. 6.94, let us standardize the elements of $\boldsymbol{\beta}_1$ and $\boldsymbol{\beta}_2$ by expressing them relative to the sampling error $\sigma/\sqrt{N}$ by writing them as $\boldsymbol{\alpha}_1 = (\sqrt{N}/\sigma)\boldsymbol{\beta}_1$ and $\boldsymbol{\alpha}_2 = (\sqrt{N}/\sigma)\boldsymbol{\beta}_2$. Then B in Eq. 6.92 is written as

$$B = (K\boldsymbol{\alpha}_1 + KA\boldsymbol{\alpha}_2 - \boldsymbol{\alpha}_1)'\boldsymbol{\mu}_{11}(K\boldsymbol{\alpha}_1 + KA\boldsymbol{\alpha}_2 - \boldsymbol{\alpha}_1)$$
$$- 2\boldsymbol{\alpha}_2'\boldsymbol{\mu}_{21}(K\boldsymbol{\alpha}_1 + KA\boldsymbol{\alpha}_2 - \boldsymbol{\alpha}_1) + \boldsymbol{\alpha}_2'\boldsymbol{\mu}_{22}\boldsymbol{\alpha}_2 \tag{6.96}$$

where for the cuboidal region of interest, $\check{R}$, $\Omega^{-1} = 2^k$ and

$$\boldsymbol{\mu}_{11} = \Omega \int_{\check{R}} \mathbf{x}_1\mathbf{x}_1' \, dx = \text{diag}\left(1, \frac{1}{3}, \ldots, \frac{1}{3}\right)$$

$$\boldsymbol{\mu}_{21} = \Omega \int_{\check{R}} \mathbf{x}_2\mathbf{x}_1' \, dx = \frac{1}{3}\begin{bmatrix} \mathbf{1}_k & 0 \\ 0 & 0 \end{bmatrix}$$

$$\boldsymbol{\mu}_{22} = \Omega \int_{\check{R}} \mathbf{x}_2\mathbf{x}_2' \, dx = \frac{1}{9}\begin{bmatrix} \dfrac{4}{5}\mathbf{I}_k + \mathbf{1}_k\mathbf{1}_k' & 0 \\ 0 & \mathbf{I}_{\binom{k}{2}} \end{bmatrix} \tag{6.97}$$

In Eq. 6.97, $\mathbf{1}_k$ is a $k \times 1$ column vector of ones and $\mathbf{I}_k$ is the identity matrix of order $k \times k$. Denoting the moments of the design using the standard notation

$$[ij] = \frac{\sum_{u=1}^{N} x_{ui}x_{uj}}{N}, \quad [ijl] = \frac{\sum_{u=1}^{N} x_{ui}x_{uj}x_{ul}}{N} \quad 1 \le i, j, l \le k$$

it follows from Eqs. 6.95 and 6.96

$$V = 1 + \frac{\sum_{i=1}^{k} \kappa_i^2 d^{ii}}{3} \tag{6.98}$$

$$B = \frac{1}{3}\sum_{i=1}^{k} \alpha_i^2(\kappa_i - 1)^2 + \left(\sum_{i=1}^{k}\sum_{\substack{j=i}}^{k} [ij]\alpha_{ij} - \frac{\sum_{i=1}^{k}\alpha_{ii}}{3}\right)^2$$
$$+ \frac{2}{3}\sum_{l=1}^{k} \kappa_l(\kappa_l - 1)\alpha_l \sum_{i=1}^{k}\sum_{\substack{j=i}}^{k} \alpha_{ij} \sum_{h=1}^{k} d^{lh}[hij]$$
$$+ \frac{1}{3}\sum_{l=1}^{k}\left(\kappa_l\sum_{i=1}^{k}\sum_{\substack{j=i}}^{k} \alpha_{ij} \sum_{h=1}^{k} d^{lh}[hij]\right)^2 + \frac{4}{45}\sum_{i=1}^{k} \alpha_{ii}^2$$
$$+ \frac{1}{9}\sum_{i=1}^{k-1}\sum_{j=i+1}^{k} \alpha_{ij}^2 \tag{6.99}$$

where d^{ij} represents the (i, j)th element of $N(\mathbf{X}_1'\mathbf{X}_1)^{-1}$. The corresponding formulas for V and B using standard least squares would be obtained by setting $\kappa_i = 1$, $i = 1, 2, \ldots, k$ in Eqs. 6.98 and 6.99.

Comparisons can now be made between the expressions for V and B in Eqs. 6.98 and 6.99, respectively, obtained with the generalized estimator and those for V and B obtained by standard least squares. In fact, Draper and Lawrence (1965), using standard least squares in fitting a first-order model over a cuboidal region of interest (i.e., letting $d_1 = 1$ and $d_2 = 2$) and considering only designs subject to the following moment conditions:

$$[i] = [iii] = [ij] = [iij] = [ijl] = 0 \quad i, j, l = 1, 2, \ldots, k$$
$$[ii] = \text{constant} \tag{6.100}$$

obtained the following expressions for V and B:

$$V_{\text{LS}} = 1 + \frac{k}{3[ii]} \tag{6.101}$$

where $[ii]^{-1} = d^{ii}$

$$B_{\text{LS}} = \left([ii] - \frac{1}{3}\right)^2 \left(\sum_{i=1}^{k} \alpha_{ii}\right)^2 + \left(\frac{4}{45}\right) \sum_{i=1}^{k} \alpha_{ii}^2 + \left(\frac{1}{9}\right) \sum_{i=1}^{k-1} \sum_{j=i+1}^{k} \alpha_{ij}^2 \tag{6.102}$$

The subscript LS for V_{LS} and B_{LS} denotes the use of least squares. If the moment restrictions in Eq. 6.100 are likewise forced on the d^{ij} and $[ij]$, etc. in Eqs. 6.98 and 6.99, then the formulas for V and B, respectively, using the generalized estimator, simplify to

$$V_{\text{GE}} = 1 + \left(\frac{1}{3[ii]}\right) \sum_{i=1}^{k} \kappa_i^2 \tag{6.103}$$

$$B_{\text{GE}} = \frac{1}{3} \sum_{i=1}^{k} \alpha_i^2 (\kappa_i - 1)^2 + \left([ii] - \frac{1}{3}\right)^2 \left(\sum_{i=1}^{k} \alpha_{ii}\right)^2$$
$$+ \frac{4}{45} \sum_{i=1}^{k} \alpha_{ii}^2 + \left(\frac{1}{9}\right) \sum_{i=1}^{k-1} \sum_{j=i+1}^{k} \alpha_{ij}^2 \tag{6.104}$$

Note that the formulas in Eqs. 6.101 and 6.103 for V differ only by the presence of the κ_i in the latter and Eqs. 6.102 and 6.104 for B are identical with the exception of the extra summation $(1/3) \sum \alpha_i^2 (\kappa_i - 1)^2$ in the latter.

A valid comparison can be made between $J_{\text{GE}} = V_{\text{GE}} + B_{\text{GE}}$ of the generalized estimator and $J_{\text{LS}} = V_{\text{LS}} + B_{\text{LS}}$ for the least squares estimator by taking the difference $J_{\text{GE}} - J_{\text{LS}}$ and concentrating on the quantities in the difference that make it nonzero. The difference is

$$J_{GE} - J_{LS} = \left(\frac{1}{3}\right) \sum_{i=1}^{k} \left\{ \frac{(\kappa_i^2 - 1)}{[ii]} + \alpha_i^2(\kappa_i - 1)^2 \right\} \qquad (6.105)$$

and the gain in terms of a reduction in the value of J by using the generalized estimator is noted when the difference in Eq. 6.105 is negative.

The ith term of Eq. 6.105 is negative for κ_i satisfying

$$\frac{[ii]\alpha_i^2 - 1}{1 + [ii]\alpha_i^2} < \kappa_i < 1 \qquad (6.106)$$

Since α_i is unknown, if we are able to say that α_i^2 is bounded above by some positive constant M_i, say, then any κ_i satisfying

$$\frac{[ii]M_i - 1}{1 + [ii]M_i} \leq \kappa_i \leq \frac{[ii]M_i}{1 + [ii]M_i} \qquad (6.107)$$

will also satisfy Eq. 6.106. Note that the value of κ_i which minimizes J_{GE} is $[ii]\alpha_i^2/(1 + [ii]\alpha_i^2)$ and this value of κ_i is never greater than $[ii]M_i/(1 + [ii]M_i)$ when $\alpha_i^2 < M_i$.

With the generalized estimator in Eq. 6.93 then, one has a prediction equation with a lower IMSE than the standard least squares estimator if one can specify an upper bound for at least one of the α_i^2. The improvement in the value of J_{GE} to J_{LS} increases with an increasing number of upper bounds $M_1, M_2, \ldots$, that can be specified. Furthermore, a considerable amount of flexibility exists in the choice of the value of κ_i as shown by Kupper and Meydrech (1973) where they suggest setting

$$\kappa_i = \frac{[ii]M_i}{1 + [ii]M_i} \qquad \kappa_j = 1 \text{ for } j \neq i \qquad (6.108)$$

in which case Eq. 6.105 becomes

$$J_{GE} - J_{LS} = -\frac{1 + [ii](2M_i - \alpha_i^2)}{3[ii](1 + [ii]M_i)^2} \qquad (6.109)$$

The difference in Eq. 6.109 is negative even if the actual value of α_i^2 is as large as $2M_i$. Throughout this development, the improvement in J_{GE} over J_{LS} has not depended on the actual size of the design moment $[ii]$.

In a follow-up paper, Kupper and Meydrech (1974) present some experimental design considerations for obtaining the generalized estimator. For $k = 1$, they show that when fitting a first-order model and the true surface is quadratic, if an upper bound M_2 can be specified for $\alpha_2^2 = N\beta_2^2/\sigma^2$, the gain in using the generalized estimator over that of least squares is

$$J_{GE} - J_{LS} = \frac{1}{3} \left[\frac{\alpha_1^2 + [11]_{GE} M_1^2}{(1 + [11]_{GE} M_1)^2} - \frac{1}{[11]_{LS}} \right]$$
$$+ \alpha_2^2 \left[\left([11]_{GE} - \frac{1}{3}\right)^2 - \left([11]_{LS} - \frac{1}{3}\right)^2 \right] \qquad (6.110)$$

where $[11]_{GE}$ and $[11]_{LS}$ are, respectively, the pure second-order moments of the design for the generalized estimator and the least squares design, respectively, where the GE design is based on the information that $\alpha_2^2 \leq M_2$. Kupper and Meydrech go on to show that a particular choce of $[11]_{GE}$ exists which minimizes the maximum of $J_{GE} = V_{GE} + B_{GE}$ over the restricted parameter space. The value of the minimizing second moment satisfies

$$\frac{1}{3} < [11]_{GE} < [11]_{LS} < [11]_{GE} + \frac{1}{M_1} \qquad (6.111)$$

where $M_1 = \alpha_1^2$. With the inequalities on $[11]_{GE}$ and $[11]_{LS}$ in Eq. 6.111, the second term on the right-hand side of the equality sign in 6.110 is always negative. Furthermore, by replacing $[11]_{LS}$ in the first term on the right-hand side of Eq. 6.110 by its upper bound $[11]_{GE} + (1/M_1)$, then the first term is also negative when $\alpha_1^2 < M_1$. Hence, when upper bounds for α_1^2 and α_2^2 can be specified, however conservative, an optimal choice for a design can be made for obtaining the generalized estimator which will have a smaller integrated MSE than the least squares estimator.

6.3.3 Summary Comments on the Differences Between the Three Estimation Procedures

In this section we shall list and comment briefly on the differences between the expression for IMSE $[\hat{Y}(x)]$ corresponding to the least squares estimator (LS), the minimum bias estimator (MBE), and generalized estimator (GE). For brevity, we shall limit our discussion to the single variable case where $d_1 = 1$, $d_2 = 2$, and R is the interval $[-1, 1]$.

The true surface is expressed by the quadratic polynomial

$$\eta = \beta_0 + \beta_1 x + \beta_2 x^2$$

while the estimating polynomial is of the first degree

$$\hat{Y}(x) = b_0 + b_1 x$$

Let us consider two symmetrical designs centered about $x = 0$; a two-point design where $x = \pm l$, and, a three-point design where $x = 0$, $x = \pm l$. Then only the even design moments, denoted by

$$[2] = \left(\frac{1}{N}\right) \sum_{u=1}^{N} x_u^2 \qquad [4] = \left(\frac{1}{N}\right) \sum_{u=1}^{N} x_u^4$$

will be nonzero. Furthermore, by letting

$$\alpha_i^2 = \frac{N\beta_i^2}{\sigma^2} \qquad i = 0, 1, 2$$

the average or integrated variance and integrated squared bias contributions corresponding to each of the three procedures are

$$V_{LS} = 1 + (3[2])^{-1} \qquad B_{LS} = \alpha_2^2 \left[\left([2] - \frac{1}{3}\right)^2 + \frac{4}{45} \right] \qquad (6.112)$$

$$V_{MBE} = V_{LS} + \frac{([2] - \frac{1}{3})^2}{[4] - [2]^2} \qquad B_{MBE} = \frac{4\alpha_2^2}{45} \qquad (6.113)$$

$$V_{GE} = \kappa_0^2 + \kappa_1^2 (3[2])^{-1}$$

$$B_{GE} = \left[(\kappa_0 - 1)\alpha_0 + \left(\kappa_0[2] - \frac{1}{3}\right)\alpha_2 \right]^2 + \frac{1}{3}[(\kappa_1 - 1)\alpha_1]^2 + \frac{4\alpha_2^2}{45} \qquad (6.114)$$

where in B_{GE} of Eq. 6.114, we have not set the value of κ_0 in the matrix **K** of Eq. 6.93 equal to unity.

For the generalized estimation procedure, the values of κ_0 and κ_1 which result in a minimum value of $J_{GE} = V_{GE} + B_{GE}$ are

$$\kappa_0 = \frac{(\alpha_0 + [2]\alpha_2)(\alpha_0 + \alpha_2/3)}{1 + (\alpha_0 + [2]\alpha_2)^2}$$

$$\kappa_1 = \frac{\alpha_1^2[2]}{1 + \alpha_1^2[2]} \qquad (6.115)$$

Since the values of κ_0 and κ_1 depend on the unknown parameters α_0, α_1 and α_2, upper bounds, M_1 and M_2, can be specified for α_1^2 and α_2^2, respectively, and used in place of α_1^2 and α_2^2 in Eq. 6.115. For the special case where $\kappa_0 = 1$, the value of κ_1 which results in the largest improvement in J_{GE} compared to J_{LS}, is known to lie in the interval

$$\frac{[2]M_1 - 1}{1 + [2]M_1} \leq \kappa_1 \leq \frac{[2]M_1}{1 + [2]M_1} \qquad (6.116)$$

The comparison between J_{GE} and J_{LS} is made by taking the difference $J_{GE} - J_{LS}$ and subject only to specifying an upper bound M_1 for α_1^2, the difference is less than zero for any κ_1 satisfying Eq. 6.116. This result holds for any design for which $[1] = [111] = 0$ and $[2]_{GE} = [2]_{LS}$.

In a similar manner, J_{GE} can be compared to $J_{MBE} = V_{MBE} + B_{MBE}$. However, since to obtain a value for V_{MBE} requires $([4] - [2]^2) > 0$, the design must contain at least three distinct settings of x. When it does and is used for both procedures so that $[2]_{GE} = [2]_{MBE}$, $[4]_{GE} = [4]_{MBE}$, and further if $\kappa_0 = 1$, then

$$
\begin{aligned}
J_{GE} - J_{MBE} &= \frac{(\kappa_1^2 - 1)}{3[2]} + \frac{1}{3}\, \alpha_1^2(\kappa_1 - 1)^2 \\
&+ \left([2] - \frac{1}{3}\right)^2 \left[\alpha_2^2 - \frac{1}{[4] - [2]^2}\right]
\end{aligned}
\tag{6.117}
$$

The difference in Eq. 6.117 is less than zero when bounds M_1 and M_2 are chosen such that $\alpha_1^2 < M_1$ and $\alpha_2^2 < M_2$ and those bounds are substituted into Eq. 6.117 along with any κ_1 satisfying Eq. 6.116 provided the design chosen satisfies

$$
0 < [4] - [2]^2 \leq \frac{1}{M_2}
$$

The minimized value of J_{GE} when compared to both J_{LS} and J_{MBE} is achieved as a result of a reduction in the value of the V component at the expense of a slight increase in the value of the B component.

In closing this section, we remark that a comparison of the three estimation procedures (least squares, minimum bias, and generalized estimation) was made by Ott and Cornell (1974) using the unintegrated variance and bias properties of the fitted first-order model, $\hat{Y}(x)$. Measured at points equally spaced across the two-unit interval $[-1, 1]$, they found that the least squares estimator, $\hat{Y}(x)_{LS}$, fitted to the all-bias design ($[2]$ or $[11] = \frac{1}{3}$), possessed a lower viarance profile near the center ($x = 0$) of the region than did the minimum bias estimator, $\hat{Y}(x)_{MBE}$, when fitted to the minimum variance design ($x = \pm \sqrt{2/3}$, $x = 0$ making $[11] = 4/9$). The minimum bias estimator, $\hat{Y}(x)_{MBE}$, had lower variance near the outer limits of the region, however (a result of the larger second-order moment with the MBE design). The generalized estimator, $\hat{Y}(x)_{GE}$, fitted to the least squares all-bias design possessed a lower variance than both $\hat{Y}(x)_{LS}$ and $\hat{Y}(x)_{MBE}$. On a more practical note, however, there seems to be little difference in the three estimation procedures and the advantages of one procedure over the others can really be classified as academic.

EXERCISES

6.1   Show that the IMSE

$$
J = \frac{N\Omega}{\sigma^2} \int_R E[\hat{Y}(\mathbf{x}) - \eta(\mathbf{x})]^2 d\mathbf{x}
$$

can be expressed as the average variance of $\hat{Y}(\mathbf{x})$ and the average squared bias of $\hat{Y}(\mathbf{x})$, or $J = V + B$, as shown in Eq. 6.7.

6.2   When fitting a first-order model in three input variables over a region of interest that is the unit sphere and the true surface is actually of the second order, show that the all-bias design is an orthogonal design with points spread out from the center of the region a distance of $\sqrt{3/5}$.

6.3   Suppose when fitting a first-order model in $k = 2$ variables, one chooses a double-simplex design of the form

x_1	x_2
$\dfrac{\sqrt{3}\rho}{2}$	$\dfrac{\rho}{2}$
$\dfrac{-\sqrt{3}\rho}{2}$	$\dfrac{\rho}{2}$
0	$-\rho$
$\dfrac{-\sqrt{3}\rho}{2}$	$\dfrac{-\rho}{2}$
0	ρ

where ρ is the radius of the circle whose perimeter the points are placed on. Determine the biases in the estimates of the parameters in the first-order model if the true surface is really quadratic, that is, $\eta = \beta_0 + \beta_1 x_1 + \beta_2 x_2 + \beta_{11} x_1^2 + \beta_{22} x_2^2 + \beta_{12} x_1 x_2$. Are these biases different from the biases associated with the design in Eq. 6.46?

6.4   Suppose for $d_1 = 2$ and $d_2 = 3$, the k-dimensional region R is cuboidal and defined by $-1 \leq x_i \leq +1$, $i = 1, 2, \ldots, k$. Find the conditions concerning the design moments,

$$m_1 = \left(\frac{1}{N}\right) \sum_{u=1}^{N} x_{ui}^2 \qquad m_2 = \left(\frac{1}{N}\right) \sum_{u=1}^{N} x_{ui}^4$$

$$m_3 = \left(\frac{1}{N}\right) \sum_{u=1}^{N} x_{ui}^2 x_{uj}^2 \qquad i \neq j$$

that ensure $\mathbf{A} = \boldsymbol{\mu}_{11}^{-1} \boldsymbol{\mu}_{12}$. Note that since R is cuboidal rather than spherical, a rotatable design with moment conditions $m_2 = 3m_3$ is not necessarily desirable. Hint: You might try combining a cube containing 2^k points with an octahedron containing $2k$ points resulting in a symmetrical design with odd moments zero and $m_2 = 1.8m_3$.

6.5   Consider for $k = 1$ the four-point design, $x = \pm l_1, \pm l_2$, where $l_1 < l_2 < 1$. Show when fitting the model, $\hat{Y}(x) = b_0 + b_1 x$, using the method of minimum bias estimation, that if the true surface is quadratic, the average variance of $\hat{Y}(x)$, over the region of interest $[-1, +1]$, is

$$V = \left(\frac{2}{9}\right)\left(\frac{9b - 6a + 2}{2b - a^2}\right) + \frac{2}{3a}$$

where $a = l_1^2 + l_2^2$ and $b = l_1^4 + l_2^4$. For what values of l_1 and l_2 is V minimized? What is the value of Min V?

6.6 An experimenter desires to model fungal growth (Y) as a function of incubation temperature (T) using the first-order model, $Y = \beta_0 + \beta_1 x + \varepsilon$, where x is a coded variable for T. The temperature range is $125 \leq T \leq 175$ and up to six settings of temperature can be performed in the allotted time if only one growth reading is recorded at each setting of temperature. It is suspected, however, that over the interval $125 \leq T \leq 175$, fungal growth might be better modeled with a quadratic polynomial. Suggest appropriate designs for fitting the first-order model (i.e., give the actual temperature settings) using each of the three procedures

1. Least squares (Box and Draper)
2. Minimum bias estimation
3. Generalized estimation

Comment on the differences among the designs obtained in 1., 2., and 3.

REFERENCES AND BIBLIOGRAPHY

Box, G. E. P. and N. R. Draper (1959). A Basis for the Selection of a Response Surface Design, *J. Amer. Statist. Assoc.*, **54**, 622–654.

Box, G. E. P. and N. R. Draper (1963). The Choice of a Second Order Rotatable Design, *Biometrika*, **50**, 335–352.

Draper, N. R. and W. E. Lawrence (1965). Designs Which Minimize Model Inadequacies: Cuboidal Regions of Interest, *Biometrika*, **52**, 111–118.

Karson, M. J. (1970). Design Criterion for Minimum Bias Estimation of Response Surfaces, *J. Amer. Statist. Assoc.*, **65**, 1565–1572.

Karson, M. J., A. R. Manson, and R. J. Hader (1969). Minimum Bias Estimation and Experimental Design for Response Surfaces, *Technometrics*, **11**, 461–475.

Kupper, L. L. and E. F. Meydrech (1973). A New Approach to Mean Squared Error Estimation of Response Surfaces, *Biometrika*, **60**, 573–579.

Kupper, L. L. and E. F. Meydrech (1974). Experimental Design Considerations Based on a New Approach to Mean Square Error Estimation of Response Surfaces, *J. Amer. Statist. Assoc.*, **69**, 461–463.

Myers, R. H. (1976). *Response Surface Methodology*, Blacksburg, VA: Author.

Ott, R. L. and J. A. Cornell (1974). A Comparison of Methods Which Utilize the Integrated M.S.E. Criterion for Constructing Response Surface Designs, *Commun. Statist. Theor. Meth. A*, **3**, 1053–1068.

Searle, S. R. (1971). *Linear Models*, New York: John Wiley.

Thompson, W. O. (1973). Secondary Criterion in the Selection of Minimum Bias Designs in Two Variables, *Technometrics*, **15**, 319–328.

7

Analysis of Multiresponse Experiments

7.1 INTRODUCTION

In the previous chapters, we discussed experiments involving single response variables. We refer to these experiments as *single-response experiments*. Quite often, however, research workers obtain measurements associated with several response variables. An experiment in which a number of responses are measured simultaneously for each setting of a group of input variables is called a *multiresponse experiment*.

Examples of multiresponse experiments are numerous. An industrial engineer may want to study the influence of cutting speed and depth of cut on the life of a tool and the rate at which it loses metal. A food technologist may be interested in determining optimum combinations of the various ingredients of a product on the basis of acceptability, nutritional value, economics, and other considerations. In another situation, a medical researcher studying the effects of complexing agents on the yield of a certain antibiotic may also be interested in the product cost. Hill and Hunter (1966) cite several papers in which multiple responses are investigated.

The analysis of data from a multiresponse experiment requires careful consideration of the multivariate nature of the data. In other words, the response variables should not be investigated individually and independently of one another. Interrelationships that may exist among the responses can render such univariate investigations meaningless. For example, if we desire to optimize

251

several response functions simultaneously, it would be futile to obtain separate individual optima. Optimal conditions for one response may be far from optimal or even physically impractical for the others. Also, in the design and analysis of multiresponse experiments, a design suitable for one response may produce unsatisfactory results for the remaining responses. In this case, any design criterion should be based on perceiving the responses as a group rather than as individual entities.

In this chapter, we discuss methods suitable for the analysis of multiresponse experiments. These include the estimation of parameters from multiresponse models, the design and analysis of multiresponse experiments, the testing of lack of fit of a multiresponse model, and the simultaneous optimization of several response functions.

7.2 PARAMETER ESTIMATION

One of the objectives in a multiresponse experiment is the simultaneous modeling of the behavior of the response variables as a function of the input variables within some region of interest. The model associated with such a function is called a *multiresponse model*. This section is concerned with the estimation of the parameters of this model. We first consider linear multiresponse models.

7.2.1 The Linear Multiresponse Model

Let N be the number of experimental runs and r be the number of response variables which can be measured for each setting of a group of k coded variables, $x_1, x_2, \ldots, x_k$. We assume that the response variables can be represented by polynomial regression models in the values of x_j within a certain region R. Hence, the ith response model can be written in vector form as

$$\mathbf{Y}_i = \mathbf{Z}_i \boldsymbol{\beta}_i + \boldsymbol{\varepsilon}_i \qquad i = 1, 2, \ldots, r \tag{7.1}$$

where $\mathbf{Y}_i$ is an $N \times 1$ vector of observations on the ith response, $\mathbf{Z}_i$ is an $N \times p_i$ matrix of rank p_i of known functions of the settings of the coded variables, $\boldsymbol{\beta}_i$ is a $p_i \times 1$ vector of unknown constant parameters, and $\boldsymbol{\varepsilon}_i$ is a random error vector associated with the ith response ($i = 1, 2, \ldots, r$). We also assume that

$$E(\boldsymbol{\varepsilon}_i) = 0$$

$$\text{Var}(\boldsymbol{\varepsilon}_i) = \sigma_{ii} \mathbf{I}_N \qquad i = 1, 2, \ldots, r \tag{7.2}$$

$$\text{Cov}(\boldsymbol{\varepsilon}_i, \boldsymbol{\varepsilon}_j) = \sigma_{ij} \mathbf{I}_N \qquad i, j = 1, 2, \ldots, r; \, i \neq j$$

The $r \times r$ matrix whose (i, j)th element is $\sigma_{ij}(i, j = 1, 2, \ldots, r)$ will be denoted by $\boldsymbol{\Sigma}$.

The r equations given in Eq. 7.1 may be represented as

$$\tilde{\mathbf{Y}} = \mathbf{Z}\boldsymbol{\beta} + \boldsymbol{\varepsilon} \tag{7.3}$$

where $\check{\mathbf{Y}} = [\mathbf{Y}_1' : \mathbf{Y}_2' :\ldots: \mathbf{Y}_r']'$, $\boldsymbol{\beta} = [\boldsymbol{\beta}_1' : \boldsymbol{\beta}_2' :\ldots: \boldsymbol{\beta}_r']'$, $\boldsymbol{\varepsilon} = [\boldsymbol{\varepsilon}_1' : \boldsymbol{\varepsilon}_2' :\ldots: \boldsymbol{\varepsilon}_r']'$, and $\mathbf{Z}$ is the block-diagonal matrix, $\mathrm{diag}(\mathbf{Z}_1, \mathbf{Z}_2, \ldots, \mathbf{Z}_r)$. For example, with $r = 3$ responses, model 7.3 appears as

$$
\begin{bmatrix}
Y_{11} \\ Y_{21} \\ \cdot \\ \cdot \\ \cdot \\ Y_{N1} \\ \cdots \\ Y_{12} \\ Y_{22} \\ \cdot \\ \cdot \\ \cdot \\ Y_{N2} \\ \cdots \\ Y_{13} \\ Y_{23} \\ \cdot \\ \cdot \\ \cdot \\ Y_{N3}
\end{bmatrix}
=
\begin{bmatrix}
& \underset{N \times p_1}{\mathbf{Z}_1} & \mathbf{0} & \mathbf{0} \\
& \mathbf{0} & \underset{N \times p_2}{\mathbf{Z}_2} & \mathbf{0} \\
& \mathbf{0} & \mathbf{0} & \underset{N \times p_3}{\mathbf{Z}_3}
\end{bmatrix}
\begin{bmatrix}
\beta_{01} \\ \beta_{11} \\ \cdot \\ \cdot \\ \beta_{p_11} \\ \cdots \\ \beta_{02} \\ \beta_{12} \\ \cdot \\ \cdot \\ \beta_{p_22} \\ \cdots \\ \beta_{03} \\ \beta_{13} \\ \cdot \\ \cdot \\ \beta_{p_33}
\end{bmatrix}
+
\begin{bmatrix}
\varepsilon_{11} \\ \varepsilon_{21} \\ \cdot \\ \cdot \\ \varepsilon_{N1} \\ \cdots \\ \varepsilon_{12} \\ \varepsilon_{22} \\ \cdot \\ \cdot \\ \varepsilon_{N2} \\ \cdots \\ \varepsilon_{13} \\ \varepsilon_{23} \\ \cdot \\ \cdot \\ \varepsilon_{N3}
\end{bmatrix}
$$

$(3N) \times 1 \qquad (3N) \times (p_1 + p_2 + p_3) \qquad (p_1 + p_2 + p_3) \times 1 \qquad (3N) \times 1$

where Y_{ui} and ε_{ui} are, respectively, the uth elements of $\mathbf{Y}_i$ and $\boldsymbol{\varepsilon}_i$ ($u = 1, 2, \ldots, N$; $i = 1, 2, \ldots, r$), and β_{li} is the lth element of $\boldsymbol{\beta}_i$ ($l = 0, 1, \ldots, p_i$; $i = 1, 2, \ldots, r$).

From the formula in Eq. 7.2 it can be seen that $\boldsymbol{\varepsilon}$ has the variance-covariance matrix

$$\mathrm{Var}(\boldsymbol{\varepsilon}) = \boldsymbol{\Sigma} \otimes \mathbf{I}_N \tag{7.4}$$

where $\otimes$ is a symbol for the direct (or Kronecker) product or matrices. By definition, the direct product of two matrices, $\mathbf{A}$ and $\mathbf{B}$, of orders $n_1 \times n_2$ and $m_1 \times m_2$, respectively, is the $n_1 m_1 \times n_2 m_2$ matrix $\mathbf{A} \otimes \mathbf{B}$, which is partitioned as $[a_{ij}\mathbf{B}]$, where a_{ij} is the (i, j)th element of the matrix $\mathbf{A}$. The best linear unbiased estimate (BLUE) of $\boldsymbol{\beta}$ is given by

$$\hat{\boldsymbol{\beta}} = (\mathbf{Z}'\boldsymbol{\Delta}^{-1}\mathbf{Z})^{-1}\mathbf{Z}'\boldsymbol{\Delta}^{-1}\check{\mathbf{Y}} \tag{7.5}$$

where $\boldsymbol{\Delta} = \boldsymbol{\Sigma} \otimes \mathbf{I}_N$, hence $\boldsymbol{\Delta}^{-1} = \boldsymbol{\Sigma}^{-1} \otimes \mathbf{I}_N$. The variance-covariance matrix of $\hat{\boldsymbol{\beta}}$ is

$$\mathrm{Var}(\hat{\boldsymbol{\beta}}) = (\mathbf{Z}'\boldsymbol{\Delta}^{-1}\mathbf{Z})^{-1} \tag{7.6}$$

Equations 7.5 and 7.6 require knowledge of Σ. If Σ is unknown, as is usually the case, then an estimate of β can be obtained by replacing Σ in Eq. 7.5 by an estimate $\hat{\Sigma}$ provided that this estimate is nonsingular. One such estimate was proposed by Zellner (1962) and is given by $\hat{\Sigma} = (\hat{\sigma}_{ij})$, where

$$\hat{\sigma}_{ij} = \frac{\mathbf{Y}_i'[\mathbf{I}_N - \mathbf{Z}_i(\mathbf{Z}_i'\mathbf{Z}_i)^{-1}\mathbf{Z}_i'][\mathbf{I}_N - \mathbf{Z}_j(\mathbf{Z}_j'\mathbf{Z}_j)^{-1}\mathbf{Z}_j']\mathbf{Y}_j}{N}$$

$$i, j = 1, 2, \ldots, r \tag{7.7}$$

We note that $\hat{\sigma}_{ij}$ is computed from the residual vectors which result from an ordinary least-squares fit of the ith and jth single-response models to their respective data sets. Using this estimate of Σ in Eq. 7.5, we get the estimate

$$\beta^* = (\mathbf{Z}'\Delta^{*-1}\mathbf{Z})^{-1}\mathbf{Z}'\Delta^{*-1}\hat{\mathbf{Y}} \tag{7.8}$$

where $\Delta^* = \hat{\Sigma} \otimes \mathbf{I}_N$. Zellner referred to β^* as the two-stage Aitken estimator. He showed that β^* is biased but that the bias is at most of order $1/N$. The asymptotic variance-covariance matrix of β^* is the same as that of $\hat{\beta}$. Furthermore, under certain conditions, the asymptotic distribution of $\sqrt{N}(\beta^* - \beta)$ is normal. Kakwani (1967) has shown that β^* can be unbiased provided that its mean exists and that the random error vectors $\varepsilon_1, \varepsilon_2, \ldots, \varepsilon_r$ have a continuous symmetric distribution. In particular, if in Eq. 7.1, $\mathbf{Z}_i = \mathbf{X}_0$ for $i = 1, 2, \ldots, r$, then $\mathbf{Z} = \mathbf{I}_r \otimes \mathbf{X}_0$. In this case, it is easy to show that

$$\hat{\beta} = [\mathbf{I}_r \otimes (\mathbf{X}_0'\mathbf{X}_0)^{-1}\mathbf{X}_0']\hat{\mathbf{Y}}$$

Thus the BLUE of β does not depend on Σ. Hence, the BLUE of β_i is the same as the ordinary least-squares estimator obtained from fitting the ith model individually $(i = 1, 2, \ldots, r)$.

7.2.2 The General Multiresponse Model

Suppose that the ith response value at the uth experimental run is represented by the model

$$Y_{ui} = f_i(\mathbf{x}_u, \beta) + \varepsilon_{ui} \qquad u = 1, 2, \ldots, N$$

$$i = 1, 2, \ldots, r \tag{7.9}$$

where $\mathbf{x}_u$ is the vector $(x_{u1}, x_{u2}, \ldots, x_{uk})'$ with x_{uj} being the uth level of the jth coded variable $(u = 1, 2, \ldots, N; j = 1, 2, \ldots, k)$, β is a vector of unknown parameters, ε_{ui} is a random error, and f_i is a function of known form for the ith response and is assumed to be continuous. The same assumptions stated in Eq. 7.2 concerning the random errors are made here. If f_i is linear in the elements of β, then the model in Eq. 7.9 is reduced to the type discussed in Section 7.2.1.

In general, f_i is nonlinear in the elements of β. In this case, an estimate of β can be obtained by using a weighted nonlinear least-squares fitting procedure; the weighting factors being the elements of the inverse $\Sigma^{-1} = (\sigma^{ij})$ of the variance-covariance matrix of the random errors, provided that Σ is known. On the other hand, when Σ is unknown, Box and Draper (1965) developed a method to estimate β, which does not require knowledge of Σ. Their procedure involves Bayesian techniques combined with the assumption that the random errors are normally distributed. Box and Draper apply the Bayes theorem under noninformative prior distributions for β and Σ. The marginal posterior density of β is then found and maximized with respect to β. This amounts to minimizing the determinant $|V(\beta)|$, where

$$V(\beta) = (Y - F)'(Y - F) \tag{7.10}$$

where $Y = |Y_1 : Y_2 : \cdots : Y_r]$ is the data matrix of order $N \times r$ and F is the $N \times r$ matrix whose (u, i)th element is $f_i(x_u, \beta)$. Minimization of the determinant $|V(\beta)|$ can be carried out by using the modified Newton algorithm of Stewart and Sorensen (1976), or by using the controlled random search procedure by Price (1977). We shall hereafter refer to the process of estimating β by minimizing this determinant as the Box-Draper estimation criterion. An extension of this criterion to the case in which some of the multiresponse data are missing was given by Box et al. (1970). The missing data were treated as additional parameters in the Bayesian solution. Stewart and Sorensen (1981) handled this case more directly by using the joint posterior density function of β and Σ in which only the actual observed multiresponse values appear. Estimates of β and Σ were then obtained by maximizing the latter density function.

Linear Dependencies Among the Responses.　The Box-Draper estimation criterion can lead to meaningless results when exact linear relationships exist among the responses. Usually, these relationships occur as a consequence of some physical or chemical laws. For example, in a particular chemical mechanism, a certain linear relationship must exist among the amounts of the substituents in all of the experimental runs in order to maintain the carbon balance. Relationships of this kind are called *stoichiometric* (see Box et al., 1973).

Suppose, for example, that m linear relationships exist among the responses and are represented as

$$B(Y_{u1}, Y_{u2}, \ldots, Y_{ur})' = c \qquad u = 1, 2, \ldots, N \tag{7.11}$$

where B is an $m \times r$ matrix of rank $m < r$ of constant coefficients, $(Y_{u1}, Y_{u2}, \ldots, Y_{ur})$ is the uth row of the $N \times r$ data matrix Y, and c is an $m \times 1$ vector of constants. Equation 7.11 can be expressed as

$$BY' = 1'_N \otimes c \tag{7.12}$$

Since $E(\mathbf{Y}) = \mathbf{F}$ by Eq. 7.9, then from Eq. 7.12 we have

$$E(\mathbf{BY}') = \mathbf{BF}' = \mathbf{1}_N' \otimes \mathbf{c} \tag{7.13}$$

From Eq. 7.12 and 7.13, we conclude that

$$\mathbf{B}(\mathbf{Y}' - \mathbf{F}') = \mathbf{0} \tag{7.14}$$

where $\mathbf{0}$ is a zero matrix of order $m \times N$. Equation 7.14 shows that there exist m linear relationships among the columns of the matrix $\mathbf{Y} - \mathbf{F}$, hence the matrix $\mathbf{V}(\boldsymbol{\beta})$ in Eq. 7.10 must be singular, that is, $|\mathbf{V}(\boldsymbol{\beta})| = 0$ for all values of the parameter vector $\boldsymbol{\beta}$. Any attempt to minimize the determinant of $\mathbf{V}(\boldsymbol{\beta})$ in this case would be meaningless.

In numerical work, generally the observed response values are rounded off to a certain number of decimal places. Such rounding errors can cause $|\mathbf{V}(\boldsymbol{\beta})|$ to be different from zero as the parameter values are changed even when exact linear relationships exist among the responses. Under these circumstances, any attempt to estimate $\boldsymbol{\beta}$ by the Box-Draper estimation criterion will produce inaccurate results. Box et al. (1973) discussed this problem in detail and suggested that the multiresponse data should first be scrutinized and checked to see if linear relationships of the type described earlier exist prior to the application of the Box-Draper estimation criterion. This can be accomplished by an examination of the eigenvalues of the matrix $\mathbf{DD}'$, where $\mathbf{D}$ is an $r \times N$ matrix given by

$$\mathbf{D} = \mathbf{Y}' \left[\mathbf{I}_N - \frac{\mathbf{1}_N \mathbf{1}_N'}{N} \right] \tag{7.15}$$

where $\mathbf{1}_N$ is a vector of ones of order $N \times 1$. We shall later see that m linearly independent relationships exist among the responses if and only if $\mathbf{DD}'$ has a zero eigenvalue of multiplicity m. Here also, rounding errors in the response values may prevent an eigenvalue of $\mathbf{DD}'$ from being exactly zero even when the responses are linearly related. Small eigenvalues of $\mathbf{DD}'$ should, therefore, be subjected to further analysis to determine if they are truly zero.

Detecting Linear Dependencies by Eigenvalue Analysis. Let us suppose that rounding errors in the response values exist and are distributed independently and uniformly over the interval $(-\delta, \delta)$. The quantity δ is equal to one-half of the last digit reported when all the multiresponse values are rounded to the same number of significant figures. Let λ denote a small eigenvalue of $\mathbf{DD}'$ which would be zero if it were not for the rounding errors. Under the assumptions made earlier concerning the rounding error and if this error is the only error present, then when δ is small enough, the expected value of λ, denoted by μ_λ, is approximately given by

$$\mu_\lambda = (N - 1)\sigma_{\text{re}}^2 \tag{7.16}$$

where $\sigma_{re}^2 = \delta^2/3$ is the rounding error variance (see Box et al., 1973: 44). In addition to μ_λ, some knowledge of σ_λ, the standard deviation of λ, is needed in order to determine the deviation of a computed eigenvalue of $\mathbf{DD}'$ from μ_λ in units of σ_λ. It was shown by Khuri and Conlon (1981) that

$$\sigma_\lambda^2 \le \left[\frac{9Nr}{5} + Nr(Nr - 1) - (N - 1)^2\right] \sigma_{re}^4 \tag{7.17}$$

approximately and for a sufficiently small δ. We thus conclude that when δ is small, the expected value of λ is of the same order of magnitude as $(N - 1)\sigma_{re}^2$ and the variance of λ is of order of magnitude not exceeding the upper bound given in Eq. 7.17. Hence, if a computed value of an eigenvalue of $\mathbf{DD}'$ falls within four to five values of σ_λ from μ_λ, where σ_λ is given approximately by the square root of the upper bound in Eq. 7.17 and μ_λ is given in Eq. 7.16, then that eigenvalue can be regarded as corresponding to a zero eigenvalue, an indication that a linear relationship exists among the responses. When m small eigenvalues of $\mathbf{DD}'$ are labeled as zero by this procedure, then $\mathbf{DD}'$ is described as having a zero eigenvalue of multiplicity m.

Identifying the Linear Dependencies by Eigenvector Analysis. The eigenvalue analysis determines whether or not linear relationships exist among the responses. It does not, however, identify what these relationships are. Such identification requires an examination of the eigenvectors of $\mathbf{DD}'$ which correspond to a zero eigenvalue. Any such eigenvector, η, must satisfy the equation

$$\mathbf{DD}'\eta = 0 \tag{7.18}$$

which is equivalent to

$$\mathbf{D}'\eta = 0 \tag{7.19}$$

From Eq. 7.15 and 7.19, we then have

$$\left(\mathbf{I}_N - \frac{\mathbf{1}_N\mathbf{1}_N'}{N}\right)\mathbf{Y}\eta = 0$$

which can be written as

$$\mathbf{Y}\eta = \gamma\mathbf{1}_N \tag{7.20}$$

where $\gamma = \Sigma_{i=1}^r \eta_i \bar{Y}_i$; $\eta_1, \eta_2, \ldots, \eta_r$ are the elements of η, and $\bar{Y}_i$ is the average of the N values of the ith response $(i = 1, 2, \ldots, r)$.

From Eq. 7.20, we see that the elements of η define a linear relationship among the r responses of the form

$$\sum_{i=1}^r Y_{ui}\eta_i = \gamma \qquad u = 1, 2, \ldots, N \tag{7.21}$$

with the constant γ being the same in all N experimental runs. If the zero eigenvalue is of multiplicity m, then there will be m corresponding linearly independent eigenvectors, and hence, m linearly independent relationships among the responses. These eigenvectors can be found by applying Theorem 14 in Chapter 2 to the matrix DD', that is,

$$DD' = P\Lambda P' \tag{7.22}$$

where Λ is a diagonal matrix of eigenvalues of DD' with the first m diagonal elements equal to zero, and P is an orthogonal matrix of eigenvectors, the first m columns of which correspond to the zero eigenvalue.

When certain eigenvalues of DD' are regarded as zero by the eigenvalue analysis, their corresponding eigenvectors will only represent approximate relationships among the responses. An investigation of the mean and variance, in the presence of rounding error, of the elements of these eigenvectors can be carried out in a manner similar to the eigenvalue analysis. This investigation, however, is more complicated and falls beyond the scope of this book. Furthermore, even if we gain access to the true eigenvectors associated with a zero eigenvalue, they may not represent meaningful relationships from the technical point of view. A nonsingular transformation of these eigenvectors may be needed for that purpose. It is, therefore, imperative to have a good knowledge of the technical aspects of the experiment and the conditions under which the response values were obtained.

Example 7.1.   We consider an example based on a chemical experiment conducted by Fuguitt and Hawkins (1947) and reported in Box et al. (1973). In this example, the concentrations of five chemical products were observed at eight timed intervals at 189.5°C. The data are reproduced in Table 7.1. The eigenvalues and associated eigenvectors of the 5×5 DD' matrix, where $D = Y'[I_8 - 1_8 1_8'/8]$, are listed in Table 7.2.

Since the response values in Table 7.1 were rounded off to one decimal place, $\delta = 0.05$, hence, the rounding error variance, σ_{re}^2, is $\delta^2/3 = 0.00083$. By applying Eqs. 7.16 and 7.17, we find that $\mu_\lambda = 0.0058$ and $\sigma_\lambda \leq 0.033$. Using 0.033 as a conservative estimate of σ_λ we can see that the two small eigenvalues, λ_1 and λ_2, in Table 7.2 fall within one standard deviation from μ_λ, whereas the next smallest eigenvalue, λ_3, is at least 36 standard deviations away. We thus conclude that λ_1 and λ_2 do indeed correspond to a zero eigenvalue of DD' of multiplicity $m = 2$. The corresponding eigenvectors, η_1 and η_2, in Table 7.2 reveal the following linear relationships among the five resposnes (see Eq. 7.21):

$$-0.169Y_1 - 0.211Y_2 - 0.161Y_3 + 0.931Y_4 - 0.185Y_5 = -16.773 \tag{7.23}$$

$$0.476Y_1 + 0.490Y_2 + 0.435Y_3 + 0.364Y_4 + 0.459Y_5 = 47.581 \tag{7.24}$$

Table 7.1 Concentration Versus Time Data for the Isomerization of α-Pinene at 189.5°C

Time (min)	Y_1 % α-Pinene	Y_2 % Dipentene	Y_3 % Allo-Ocimene	Y_4 % Pyronene	Y_5 % Dimer
1230	88.4	7.3	2.3	0.4	1.8
3060	76.4	15.6	4.5	0.7	2.8
4920	65.1	23.1	5.3	1.1	5.8
7800	50.4	32.9	6.0	1.5	9.3
10680	37.5	42.7	6.0	1.9	12.0
15030	25.9	49.1	5.9	2.2	17.0
22620	14.0	57.4	5.1	2.6	21.0
36420	4.5	63.1	3.8	2.9	25.7

Source: G. E. P. Box, W. G. Hunter, J. F. MacGregor, and J. Erjavec (1973). Reproduced with permission of the American Statistical Association.

The true linear relationships among the five responses were reported by Box et al. (1973) as

$$0.03Y_1 + Y_4 = 3 \tag{7.25}$$

$$Y_1 + Y_2 + Y_3 + Y_4 + Y_5 = 100 \tag{7.26}$$

The latter two relationships can be approximately obtained by applying a non-singular linear transformation to Eqs. 7.23 and 7.24 given by the matrix **Q**,

Table 7.2 Eigenvalues and Eigenvectors of **DD'**

Eigenvalues of **DD'**

λ_1	λ_2	λ_3	λ_4	λ_5
0.0013	0.0168	1.21	25.0	9660

Eigenvectors of **DD'**

η_1	η_2	η_3	η_4	η_5
−0.169	0.476	−0.296	0.057	0.809
−0.211	0.490	−0.611	−0.224	−0.540
−0.161	0.435	0.640	−0.612	−0.013
0.931	0.364	−0.010	0.004	−0.024
−0.185	0.459	0.360	0.756	−0.231

Source: G. E. P. Box, W. G. Hunter, J. F. MacGregor, and J. Erjavec (1973). Reproduced with permission of the American Statistical Association.

where

$$Q = \begin{bmatrix} 0.922 & 0.390 \\ 0.222 & 2.180 \end{bmatrix}$$

7.2.3 A Procedure for Dropping Responses That Are Linear Functions of Others

Once linear relationships in the responses are identified by the eigenvalue-eigenvector analysis, the next step is to consider dropping those responses which depend linearly on the others. The Box-Draper estimation criterion can then be applied to the remaining responses. Unfortunately, because of rounding errors, the values and precision of the parameter estimates will depend on which responses are dropped. There does not seem to be a clear-cut solution to this problem. One possible procedure is the following: Let η_1, η_2, ..., η_m be m linearly independent eigenvectors associated with a zero eigenvalue of DD'. Then as in Eq. 7.19, we have

$$D'T = 0 \qquad (7.27)$$

where $T = [\eta_1 : \eta_2 : \cdots : \eta_m]$ and 0 is a zero matrix of order $N \times m$. In the presence of rounding errors, the term on the right-hand side of Eq. 7.27 may be a nonzero matrix, E, of small elements. We can thus rewrite Eq. 7.27 as

$$D'T = E \qquad (7.28)$$

Without loss of generality, we assume that T is partitioned as $T = [T_1' : T_2']'$, where T_1 is of order $m \times m$ and rank m. Correspondingly, D' can be partitioned as $D' = [D_1' : D_2']$. From Eq. 7.28 it follows that

$$D_1' = -D_2'T_2T_1^{-1} + ET_1^{-1} \qquad (7.29)$$

We note that since the ith column of D' is $Y_i - \bar{Y}_i 1_N$ (see Eq. 7.15), $i = 1, 2, \ldots, r$, then D_1' consists of m columns of response vectors corrected for their means. A reasonable criterion for dropping m responses is the reduction of the effect of rounding error in Eq. 7.29, which prevents truly linearly dependent responses from achieving exact linear relationships as in Eq. 7.27. To accomplish this, we choose T_1 such that $\|T_1^{-1}\|$ is minimum in the class of all nonsingular $m \times m$ submatrices of T, where $\|T_1^{-1}\|$ is the Euclidean norm of T_1^{-1} and is equal to the square root of the trace of $T_1^{-1}T_1^{-1'}$. It is easy to see that this choice of T_1 leads to a reduction in the size of $\|ET_1^{-1}\|$ since

$$\|ET_1^{-1}\| \leq \|E\|\|T_1^{-1}\|$$

The responses that may be dropped are, therefore, those corresponding to the columns of D_1' that match the rows of the optimal T_1 submatrix. The user, however, should exercise caution and keep technical as well as theoretical considerations in perspective before deciding which responses to drop.

Example 7.2. Consider again Example 7.1. From Table 7.2, the **T** matrix is

$$
\mathbf{T} = \begin{bmatrix} -0.169 & -0.211 & -0.161 & 0.931 & -0.185 \\ 0.476 & 0.490 & 0.435 & 0.364 & 0.459 \end{bmatrix}'
$$

There are ten 2×2 nonsingular submatrices of **T**. These submatrices will be identified by the responses which correspond to the rows of **T**. Values of $\|\mathbf{T}_1^{-1}\|$ for each of the ten submatrices are given in Table 7.3.

We note that the pairs of responses with corresponding small $\|\mathbf{T}_1^{-1}\|$ values are $\{Y_1, Y_4\}$, $\{Y_2, Y_4\}$, $\{Y_3, Y_4\}$, and $\{Y_4, Y_5\}$. Any of these four pairs are candidates for deletion. We also note that Y_4 appears in all four pairs. It is, therefore, the first candidate to be dropped. It turned out that this response was a fabricated response; because of experimental difficulties, Y_4 was not measured independently but was rather assumed to satisfy Eq. 7.25 (see Box et al., 1973: 41).

7.2.4 The Effect of Response Scaling on the Detection of Linear Dependencies

In Section 7.2.2 the rounding errors in all the response values were assumed to have the same uniform distribution over $(-\delta, \delta)$. In general, however, the rounding errors for the different responses may vary considerably in a manner that cannot be explained by a single uniform distribution. For example, the response variables may have distinct physical entities with different units of measurement,

Table 7.3 Values of $\|\mathbf{T}_1^{-1}\|$ for Pairs of Responses Listed in Table 7.1

Responses		$\|\mathbf{T}_1^{-1}\|$
Y_1	Y_2	41.682
Y_1	Y_3	221.216
Y_1	Y_4	2.219
Y_1	Y_5	70.714
Y_2	Y_3	54.377
Y_2	Y_4	2.126
Y_2	Y_5	121.283
Y_3	Y_4	2.377
Y_3	Y_5	102.771
Y_4	Y_5	2.253

and possibly widely different orders of magnitude. It would therefore be more realistic to assume that the rounding errors for the ith response are distributed uniformly over $(-\delta_i, \delta_i)$ with the δ_i's not necessarily equal ($i = 1, 2, \ldots, r$). In this case, the eigenvalue analysis in Section 7.2.2 should be modified accordingly. We now show how this can be accomplished.

In order to avoid problems resulting from having responses with different units of measurement and/or sizably different orders of magnitude, the responses should be scaled before any eigenvalue analysis is carried out. For this purpose, let ψ be an $r \times r$ diagonal matrix whose ith diagonal element is of the form

$$\psi_i = \left[\sum_{u=1}^{N} (Y_{ui} - \bar{Y}_i)^2 \right]^{1/2}, \qquad i = 1, 2, \ldots, r. \tag{7.30}$$

The responses are scaled using the linear transformation

$$\mathbf{H} = \mathbf{Y}\psi^{-1}$$

Let the $r \times N$ matrix $\mathbf{A}$ be defined as

$$\mathbf{A} = \mathbf{H}' \left(\mathbf{I}_N - \frac{1}{N} \mathbf{J}_N \right) \tag{7.31}$$

This matrix will be used in place of $\mathbf{D}$ in a modified eigenvalue analysis.

From (7.31) we have

$$\mathbf{AA}' = \mathbf{H}' \left(\mathbf{I}_N - \frac{1}{N} \mathbf{J}_N \right) \mathbf{H}$$
$$= \psi^{-1}\mathbf{DD}'\psi^{-1}$$

Note that each row of $\mathbf{A}$ has a unit length. Hence, the diagonal elements of $\mathbf{AA}'$ are equal to unity, and its off-diagonal elements do not exceed 1 in absolute value.

If the responses are linearly related, then, when rounding errors are present, both $\mathbf{DD}'$ and $\mathbf{AA}'$ will have eigenvalues close to zero. In this case, the rows of $\mathbf{D}$, as well as those of $\mathbf{A}$, will be nearly linearly dependent, that is, multicollinear. In other words, both $\mathbf{D}'$ and $\mathbf{A}'$ will be ill-conditioned. The degrees of their ill-conditioning, however, can be quite different.

To measure the degrees of ill-conditioning of $\mathbf{A}'$ and $\mathbf{D}'$, the condition number κ can be used, which for $\mathbf{A}'$ and $\mathbf{D}'$ has the values

$$\kappa(\mathbf{A}') = \left[\frac{e_{\max}(\mathbf{AA}')}{e_{\min}(\mathbf{AA}')} \right]^{1/2}$$

$$\kappa(\mathbf{D}') = \left[\frac{e_{\max}(\mathbf{DD}')}{e_{\min}(\mathbf{DD}')} \right]^{1/2}$$

respectively. Here, $e_{max}(\cdot)$ and $e_{min}(\cdot)$ denote the largest and smallest eigenvalues, respectively, of a symmetric matrix. The larger the condition number, the more ill-conditioned the matrix.

The unit row scaling of **A** has the distinct advantage of reducing its condition number. This was shown to be the case by Van der Sluis (1969) (see also Belsley et al.,1980, pp. 184–185, and Stewart, 1986). As a result, the matrix **A** is less sensitive to rounding errors than **D**. It is therefore more suitable for detecting linear dependencies among the responses.

Let us now proceed to determine the expected value and variance of a small eigenvalue of **AA′** in the presence of rounding errors. Let $\mathbf{\Delta}_i$ be the vector of rounding errors for the ith response $(i = 1, 2, \ldots, r)$. The elements of $\mathbf{\Delta}_i$ are assumed to be independently distributed as uniform random variates over $(-\delta_i, \delta_i)$, $i = 1, 2, \ldots, r$. Thus,

$$E(\mathbf{\Delta}_i) = \mathbf{0}, \qquad i = 1, 2, \ldots, r$$

$$\mathrm{Var}(\mathbf{\Delta}_i) = \frac{\delta_i^2}{3}\, \mathbf{I}_N, \qquad i = 1, 2, \ldots, r$$

We also assume that the $\mathbf{\Delta}_i$'s are statistically independent. Note that if the ith response values are rounded off to e_i decimal places, then $\delta_i = 5[10^{-(e_i+1)}]$, $i = 1, 2, \ldots, r$

Suppose now that **AA′** has m small eigenvalues suspected of being equal to zero if no rounding errors were present. Let these eigenvalues be denoted by $\lambda_1, \lambda_2, \ldots, \lambda_m$. Furthermore, let $\mathbf{v}_1, \mathbf{v}_2, \ldots, \mathbf{v}_m$ be a corresponding set of orthonormal eigenvectors. Then, under the assumption that the rounding errors are independent, and, for the ith response, uniformly distributed over $(-\delta_i, \delta_i)$, $i = 1, 2, \ldots, r$, the expected value of λ_j is approximately given by

$$E(\lambda_j) = \frac{N-2}{3} \sum_{i=1}^{r} \frac{\delta_i^2 v_{ij}^2}{\psi_i^2}, \qquad j = 1, 2, \ldots, m \tag{7.32}$$

where v_{ij} is the ith element of $\mathbf{v}_j$ and ψ_i is defined in (7.30). Furthermore,

$$\mathrm{Var}(\lambda_j) \leq \left[\frac{N}{5} + \frac{N(N-1)}{9}\right] \sum_{i=1}^{r} \tau_i^2 \delta_i^4 + \frac{2N^2}{9} \sum_{i<l}^{r}\sum \tau_i \tau_l \delta_i^2 \delta_l^2 \\ - \frac{(N-2)^2}{9} \left[\sum_{i=1}^{r} \frac{\delta_i^2 v_{ij}^2}{\psi_i^2}\right]^2, \qquad j = 1, 2, \ldots, m \tag{7.33}$$

approximately, where τ_i is given by

$$\tau_i = \frac{1}{\psi_i^4} [(N-1)\psi_i^2 + N\bar{Y}_i^2], \qquad i = 1, 2, \ldots, r$$

The proofs of (7.32) and (7.33) are given in Khuri (1990a, Theorem 2.1). The eigenvalue analysis can now be carried out using (7.32) and (7.33) in lieu of

(7.16) and (7.17), respectively. More specifically, let u_j denote the upper bound in (7.33). Consider the ratio,

$$t_j = \frac{\lambda_j - E(\lambda_j)}{\sqrt{u_j}}, \qquad j = 1, 2, \ldots, m \tag{7.34}$$

A small value of $|t_j|$ indicates that λ_j corresponds to a zero eigenvalue of $\mathbf{AA}'$, and hence a linear relationship exists among the responses. As in Section 7.2.2, the elements of the corresponding eigenvector $\mathbf{v}_j$ can be used to identify such a relationship.

Example 7.3.   Consider the hospital data set reported in Myers (1990, pp. 130–133). Initially, five input variables were used to predict the manpower needs for U.S. Naval hospitals. In the current example these five variables are treated as responses, namely,

Y_1 = average daily patient load
Y_2 = monthly X-ray exposures
Y_3 = monthly occupied bed days
Y_4 = eligible population in the area per thousand
Y_5 = average length of patients' stay in days

Table 7.4 gives values of these variables taken from 17 Naval hospitals at various sites around the world.

From Table 7.4 we note that the rounding errors are not equal, with $\delta_1 = 0.005$, $\delta_2 = 0.5$, $\delta_3 = 0.005$, $\delta_4 = 0.05$, and $\delta_5 = 0.005$. The eigenvalues of $\mathbf{AA}'$ and corresponding orthonormal eigenvectors are shown in Table 7.5. The smallest eigenvalue is $\lambda_1 = 0.000054$, and from formula (7.34), the corresponding value of t_1 is 0.4812. The next smallest eigenvalue is 0.040712, for which $t_2 = 363.02$. We conclude that λ_1 corresponds to a zero eigenvalue with multiplicity 1. The elements of the corresponding eigenvector, $\mathbf{v}_1$, can be used to define an approximate linear relationship among the five responses. We note that the responses Y_1, Y_3, and Y_4 are most involved in this relationship.

For the sake of comparison, let us now apply the standard eigenvalue analysis, as described in Section 7.2.2, to the same data set. Using formula (7.15), we find that the eigenvalues of $\mathbf{DD}'$ are 8.882, 58.002, 19809.5, 6.5×10^7, and 7.6×10^9. Clearly, none of these can be suspected of being equal to zero. It should be noted, however, that since the rounding errors are different, it would be more appropriate to make them equal first. This can be accomplished, for example, by dividing Y_2 by 100 and Y_4 by 10 (this causes all the δ_i's to be equal to 0.005). In this case, the eigenvalues of $\mathbf{DD}'$ become 8.422, 52.046, 232.781, 1.28×10^5, and 3.86×10^8. Again, none of these is considered equal to zero. On the other hand, the eigenvalues of $\mathbf{AA}'$ remain unchanged by such a trans-

Table 7.4 Data for Example 7.3

Site	Y_1	Y_2	Y_3	Y_4	Y_5
1	15.57	2463	472.92	18.0	4.45
2	44.02	2048	1339.75	9.5	6.92
3	20.42	3940	620.25	12.8	4.28
4	18.74	6505	568.33	36.7	3.90
5	49.20	5723	1497.60	35.7	5.50
6	44.92	11520	1365.83	24.0	4.60
7	55.48	5779	1687.00	43.3	5.62
8	59.28	5969	1639.92	46.7	5.15
9	94.39	8461	2872.33	78.7	6.18
10	128.02	20106	3655.08	180.5	6.15
11	96.00	13313	2912.00	60.9	5.88
12	131.42	10771	3921.00	103.7	4.88
13	127.21	15543	3865.67	126.8	5.50
14	252.90	36194	7684.10	157.7	7.00
15	409.20	34703	12446.33	169.4	10.78
16	463.70	39204	14098.40	331.4	7.05
17	510.22	86533	15524.00	371.6	6.35

Source: R. H. Myers (1990). Reproduced with permission of PWS-KENT Publishing Company.

formation of the responses. Thus, the standard eigenvalue analysis based on the **DD'** matrix fails here to detect any linear dependencies among the responses.

This example clearly demonstrates the advantage of scaling the responses. An eigenvalue analysis applied directly to the original multiresponse data may fail to detect linear relationships that may exist among the responses.

7.2.5 Further Remarks Concerning the Box-Draper Estimation Criterion

In general, the eigenvalue analysis is effective in avoiding nonsensical results when using the Box-Draper estimation criterion. However, there can be situations which require further scrutiny. Consider, for example, the following two examples.

Example 7.4. Suppose that there are m linearly independent relationships of the type given in Eq. 7.11, except that the right-hand side of this equation is a vector c_u ($u = 1, 2, \ldots, N$), which may vary from one run to another, that is,

$$\mathbf{B}(Y_{u1}, Y_{u2}, \ldots, Y_{ur})' = c_u \qquad u = 1, 2, \ldots, N \tag{7.35}$$

Table 7.5 Eigenvalues and Orthonormal Eigenvectors of $\mathbf{AA'}$

Eigenvalues	4.19712	0.667484	0.094633	0.040712	0.000054
Orthonormal	−0.485286	0.002029	−0.166232	−0.468191	−0.719484
eigenvectors	−0.453235	0.335605	0.804239	0.187470	−0.001158
	−0.484977	0.000855	−0.153960	−0.509262	0.694079
	−0.460969	0.310800	−0.537199	0.633860	0.023440
	−0.333737	−0.889251	0.115239	0.290733	0.006781

Equation 7.35 can be written as

$$\mathbf{BY'} = \mathbf{C} \tag{7.36}$$

where $\mathbf{Y}$ is the $N \times r$ data matrix and $\mathbf{C}$ is the $m \times N$ matrix $[\mathbf{c}_1 : \mathbf{c}_2 : \ldots : \mathbf{c}_N]$. By postmultiplying the terms in Eq. 7.36 with $\mathbf{I}_N - \mathbf{1}_N\mathbf{1}_N'/N$, and recalling from Eq. 7.15 that $\mathbf{D} = \mathbf{Y'}[\mathbf{I}_N - \mathbf{1}_N\mathbf{1}_N'/N]$, we obtain

$$\mathbf{BD} = \mathbf{C} - (\mathbf{C}\mathbf{1}_N/N)\mathbf{1}_N' \tag{7.37}$$

Since the elements of each column of the second matrix term on the right-hand side of Eq. 7.37 are the means of the m rows of $\mathbf{C}$, the N columns of this matrix term are identical. Hence, $\mathbf{BD}$ is equal to the zero matrix if and only if the columns of $\mathbf{C}$ are identical, that is, if and only if $\mathbf{c}_u = \mathbf{c}$ for $u = 1, 2, \ldots, N$. In this case $\mathbf{D'B'} = \mathbf{0}$, or equivalently, $\mathbf{DD'B'} = \mathbf{0}$ which means that the matrix $\mathbf{DD'}$ has a zero eigenvalue of multiplicity m.

Now, if the columns of $\mathbf{C}$ are not identical, then from the above argument we may conclude that $\mathbf{DD'}$ has no zero eigenvalues, that is, it is nonsingular. On the other hand, whether or not the columns of $\mathbf{C}$ are equal, if the multiresponse model in Eq. 7.9 is correct, then from Eq. 7.36 we must have

$$\mathbf{BF'} = \mathbf{C} \tag{7.38}$$

where $\mathbf{F}$ is the $N \times r$ matrix in Eq. 7.10. From Eqs. 7.36 and 7.38, we obtain the same equation as in Eq. 7.14. As before, we may conclude that the matrix $\mathbf{V}(\boldsymbol{\beta})$ in Eq. 7.10 is singular. In fact, because of Eq. 7.14, the rank of this matrix is $r - m$, hence, it must have a zero eigenvalue of multiplicity m. We thus have a situation where a singularity exists in $\mathbf{V}(\boldsymbol{\beta})$ even though the $\mathbf{DD'}$ matrix is nonsingular. In other words, the eigenvalue analysis in this case does not alert the experimenter to the existence of such a singularity.

Example 7.5 Suppose that there are m linearly independent relationships among the responses as in Eqs. 7.11 or 7.12. Then as in Example 7.4, the matrix $\mathbf{DD'}$ will have a zero eigenvalue of multiplicity m. If some of the models in Eq. 7.9 suffer from lack of fit, then Eq. 7.13 and, hence, Eq. 7.14 will not be valid. If none of the rows of the matrix $\mathbf{B}(\mathbf{Y}' - \mathbf{F}')$ is zero, then $\mathbf{Y} - \mathbf{F}$ will be of rank r. In this case, the matrix $\mathbf{V}(\boldsymbol{\beta})$ will be nonsingular even though $\mathbf{DD}'$ is singular. Thus, in the presence of lack of fit, it might be possible to apply the Box-Draper estimation criterion in spite of the singularity in $\mathbf{DD}'$. It is, however, questionable that the use of this criterion will be meaningful in this case, especially when lack of fit is significant.

From Examples 7.4 and 7.5 we conclude that a careful determination of the type of linear relationships among the responses must be made in order to know whether the values of $\mathbf{c}_u$ in Eq. 7.35 can be equal or vary from one run to another. Furthermore, the models should be checked for lack of fit before the Box-Draper criterion is used. A formal test for lack of fit of a linear multiresponse model will be given in Section 7.4. For more details about Examples 7.4 and 7.5, the interested reader is referred to McLean et al. (1979).

7.3 THE DESIGN OF MULTIRESPONSE EXPERIMENTS

The design problem in the multiresponse case is more complex than in the case of a single response. A design efficient for one response may not be efficient for the other responses. In a multiresponse situation, therefore, the choice of a design should be based on a criterion which incorporates measures of efficiency pertaining to all the responses.

So far, the development of multiresponse designs has been lagging in comparison with the development of single-response designs. This is partly attributed to the complexity associated with the former designs and to the availability of fewer design criteria than in the single-response case.

Draper and Hunter (1966) obtained a design criterion for the estimation of parameters in multiresponse models of the form given in Eq. 7.9. This criterion can be used to select n additional experimental runs after N runs have already been performed. Draper and Hunter adopted a Bayesian approach using a noninformative prior distribution for $\boldsymbol{\beta}$. The posterior distribution for $\boldsymbol{\beta}$, after the initial N experimental runs, serves as a prior distribution in order to determine the posterior distribution after the subsequent n runs. The posterior density of $\boldsymbol{\beta}$ obtained after $N + n$ runs is then maximized with respect to $\boldsymbol{\beta}$ and with respect to the n runs to be chosen. Draper and Hunter showed that the latter runs can be determined as a result of maximizing the determinant $|\mathbf{U}|$, where $\mathbf{U}$ is the following $p \times p$ matrix with p being the number of elements in $\boldsymbol{\beta}$:

$$\mathbf{U} = \sum_{i=1}^{r} \sum_{j=1}^{r} \sigma^{ij} \mathbf{L}_i' \mathbf{L}_j$$

where σ^{ij} is the (i, j)th element of the inverse of the variance-covariance matrix, Σ, of the r responses, and $\mathbf{L}_i$ is the $n \times p$ matrix

$$
\mathbf{L}_i = \begin{bmatrix} \kappa_{1i}^{(1)} & \kappa_{1i}^{(2)} & \cdots & \kappa_{1i}^{(p)} \\ \kappa_{2i}^{(1)} & \kappa_{2i}^{(2)} & \cdots & \kappa_{2i}^{(p)} \\ \vdots & \vdots & & \vdots \\ \kappa_{ni}^{(1)} & \kappa_{ni}^{(2)} & \cdots & \kappa_{ni}^{(p)} \end{bmatrix} \qquad i = 1, 2, \ldots, r
$$

Here $\kappa_{ui}^{(t)}$ is the partial derivative of f_i in Eq. 7.9 with respect to β_t, evaluated at $\mathbf{x}_{N+u}$ and $\boldsymbol{\beta} = \hat{\boldsymbol{\beta}}$, that is,

$$
\kappa_{ui}^{(t)} = \left. \frac{\partial f_i(\mathbf{x}_{N+u}, \boldsymbol{\beta})}{\partial \beta_t} \right|_{\boldsymbol{\beta} = \hat{\boldsymbol{\beta}}} \qquad \begin{array}{l} i = 1, 2, \ldots, r \\ t = 1, 2, \ldots, p \\ u = 1, 2, \ldots, n \end{array}
$$

where β_t is the tth element of $\boldsymbol{\beta}$ and $\hat{\boldsymbol{\beta}}$ is the maximum likelihood estimate of $\boldsymbol{\beta}$ based on the observations collected at the initial N experimental runs.

We note that the above criterion requires knowledge of the value of Σ. As an example of the use of this criterion, suppose that in Eq. 7.9, $k = 1$, $p = 2$, $r = 2$, and that

$$
f_1(\mathbf{x}_{N+u}, \boldsymbol{\beta}) = e^{-\beta_1 x_{N+u}} - e^{-\beta_2 x_{N+u}} \qquad u = 1, 2
$$
$$
f_2(\mathbf{x}_{N+u}, \boldsymbol{\beta}) = 1 - e^{-\beta_2 x_{N+u}}
$$

Then,

$$
\mathbf{L}_1 = \begin{bmatrix} -x_{N+1} e^{-\hat{\beta}_1 x_{N+1}} & x_{N+1} e^{-\hat{\beta}_2 x_{N+1}} \\ -x_{N+2} e^{-\hat{\beta}_1 x_{N+2}} & x_{N+2} e^{-\hat{\beta}_2 x_{N+2}} \end{bmatrix}
$$

$$
\mathbf{L}_2 = \begin{bmatrix} 0 & x_{N+1} e^{-\hat{\beta}_2 x_{N+1}} \\ 0 & x_{N+2} e^{-\hat{\beta}_2 x_{N+2}} \end{bmatrix}
$$

The design criterion described earlier selects the design points x_{N+1} and x_{N+2} so that the determinant

$$
|\mathbf{U}| = |\sigma^{11} \mathbf{L}_1' \mathbf{L}_1 + \sigma^{22} \mathbf{L}_2' \mathbf{L}_2 + \sigma^{12} (\mathbf{L}_1' \mathbf{L}_2 + \mathbf{L}_2' \mathbf{L}_1)|
$$

is maximized.

The rest of this section will be devoted to discussing design criteria for linear multiresponse models of the form given in Eq. 7.1.

7.3.1 Designs for Parameter Estimation in Linear Multiresponse Models

Let us again consider the linear response models given in Eq. 7.1 where the random error vectors are assumed to satisfy the conditions described in Eq. 7.2.

The least squares estimator of $\boldsymbol{\beta}$ in the corresponding multiresponse model 7.3 is given by $\hat{\boldsymbol{\beta}}$ as in Eq. 7.5. The predicted response value for the ith response ($i = 1, 2, \ldots, r$) at a point $\mathbf{x} = (x_1, x_2, \ldots, x_k)'$ in an experimental region R can be represented as

$$\hat{Y}_i(\mathbf{x}) = \mathbf{z}_i'(\mathbf{x})\hat{\boldsymbol{\beta}}_i \qquad i = 1, 2, \ldots, r$$

where $\mathbf{z}_i'(\mathbf{x})$ is of the same form as a row of the matrix $\mathbf{Z}_i$, but is evaluated at the point $\mathbf{x}$. If $\hat{\mathbf{Y}}(\mathbf{x}) = (\hat{Y}_1(\mathbf{x}), \hat{Y}_2(\mathbf{x}), \ldots, \hat{Y}_r(\mathbf{x}))'$ is the predicted response vector at the point $\mathbf{x}$, then

$$\hat{\mathbf{Y}}(\mathbf{x}) = \boldsymbol{\phi}'(\mathbf{x})\hat{\boldsymbol{\beta}} \tag{7.39}$$

where $\boldsymbol{\phi}'(\mathbf{x})$ is the $r \times p(p = \Sigma_{i=1}^r p_i$, where p_i is the number of elements in $\boldsymbol{\beta}_i$) block-diagonal matrix $\boldsymbol{\phi}'(\mathbf{x}) = \text{diag}(\mathbf{z}_1'(\mathbf{x}), \mathbf{z}_2'(\mathbf{x}), \ldots, \mathbf{z}_r'(\mathbf{x}))$.

The variance-covariance matrix of $\hat{\mathbf{Y}}(\mathbf{x})$ is thus of the form

$$\text{Var}[\hat{\mathbf{Y}}(\mathbf{x})] = \boldsymbol{\phi}'(\mathbf{x})(\mathbf{Z}'\boldsymbol{\Delta}^{-1}\mathbf{Z})^{-1}\boldsymbol{\phi}(\mathbf{x}) \tag{7.40}$$

where $\boldsymbol{\Delta}$ and $\boldsymbol{\Sigma} \otimes \mathbf{I}_N$ and $\boldsymbol{\Sigma} = (\sigma_{ij})$ is the variance-covariance matrix for the r responses. If we denote by $\mathbf{D}_N$ the $N \times k$ design matrix for the r responses, then the moment matrix, $\mathbf{M}(\mathbf{D}_N, \boldsymbol{\Sigma})$ for the multiresponse model 7.3 is given by

$$\mathbf{M}(\mathbf{D}_N, \boldsymbol{\Sigma}) = \left(\frac{1}{N}\right) \mathbf{Z}'\boldsymbol{\Delta}^{-1}\mathbf{Z} \tag{7.41}$$

We define the matrix $\mathbf{V}(\mathbf{x}, \mathbf{D}_N, \boldsymbol{\Sigma})$ as

$$\mathbf{V}(\mathbf{x}, \mathbf{D}_N, \boldsymbol{\Sigma}) = \boldsymbol{\phi}'(\mathbf{x})\mathbf{M}^{-1}(\mathbf{D}_N, \boldsymbol{\Sigma})\boldsymbol{\phi}(\mathbf{x}) \tag{7.42}$$

This matrix is called the *prediction variance matrix*. We note from Eqs. 7.40, 7.41, and 7.42 that

$$\mathbf{V}(\mathbf{x}, \mathbf{D}_N, \boldsymbol{\Sigma}) = N \text{ Var}[\hat{\mathbf{Y}}(\mathbf{x})]$$

If the random error vectors in Eq. 7.1 are normally distributed, then when $\boldsymbol{\Sigma}$ is a known matrix, a $(1 - \alpha)100\%$ confidence region for $\boldsymbol{\beta}$ is the ellipsoid

$$\{\boldsymbol{\beta}: N(\hat{\boldsymbol{\beta}} - \boldsymbol{\beta})'\mathbf{M}(\mathbf{D}_N, \boldsymbol{\Sigma})(\hat{\boldsymbol{\beta}} - \boldsymbol{\beta}) \le \chi_{\alpha,p}^2\}$$

where $\chi_{\alpha,p}^2$ is the upper $100\alpha\%$ point of the chi-square distribution with p degrees of freedom. To obtain an efficient estimator for $\boldsymbol{\beta}$, the volume of this ellipsoid should be as small as possible. Since this volume is proportional to $|\mathbf{M}(\mathbf{D}_N, \boldsymbol{\Sigma})|^{-1/2}$, the design $\mathbf{D}_N$ is chosen so that $|\mathbf{M}(\mathbf{D}_N, \boldsymbol{\Sigma})|$ attains its maximum value within the region R. Such a design is D-optimal and reduces to the usual D-optimal design in the single-response case (see Section 12.2 in Chapter 12).

The *D*-optimal design chosen in the manner described above is a discrete design (see Section 12.2 in Chapter 12 for a definition of discrete designs) of a predetermined number of runs, namely N. The construction of this design requires the determination of Nk values, namely, the elements of the matrix $\mathbf{D}_N$, which maximize $|\mathbf{M}(\mathbf{D}_N, \Sigma)|$. This, of course, can be a very difficult task, especially when Nk is large. A sequentially generated *D*-optimal design, as we shall soon see, is computationally more feasible.

The Sequential Generation of a *D*-Optimal Design When Σ Is Known.   Fedorov (1972) introduced an algorithm for the construction of a *D*-optimal multiresponse design using a sequential procedure in which design points are chosen one at a time. A *D*-optimal design obtained through this procedure is of the continuous type (see Section 12.2 in Chapter 12 for a definition of continuous designs) and maximizes $|\mathbf{M}(\zeta, \Sigma)|$ in the class of all design measures ζ which can be defined over the region R (for a definition of a design measure, see Section 12.2 in Chapter 12; also Wynn 1970, Kiefer 1959, Fedorov 1972). For a given design measure ζ, defined on R, the moment matrix $\mathbf{M}(\zeta, \Sigma)$ is the $p \times p$ matrix

$$\mathbf{M}(\zeta, \Sigma) = \int_R \phi(\mathbf{x})\Sigma^{-1}\phi'(\mathbf{x}) \, d\zeta$$

where $\phi(\mathbf{x})$ is the block-diagonal matrix in Eq. 7.39. In particular, if ζ is the discrete design $\mathbf{D}_N$ mentioned earlier, then

$$\mathbf{M}(\mathbf{D}_N, \Sigma) = \left(\frac{1}{N}\right) \sum_{u=1}^{N} \phi(\mathbf{x}_u)\Sigma^{-1}\phi'(\mathbf{x}_u)$$

where $\mathbf{x}_1, \mathbf{x}_2, \ldots, \mathbf{x}_N$ are the design points. It is easy to show that $\mathbf{M}(\mathbf{D}_N, \Sigma)$ can be equivalently written as in Eq. 7.41.

Fedorov's (1972) algorithm is based on Theorem 7.1.

Theorem 7.1 (The Equivalence Theorem).   Let $\bar{H}$ denote the class of all design measures on R and let $\mathbf{V}(\mathbf{x}, \zeta, \Sigma)$ be the corresponding prediction variance matrix defined as

$$\mathbf{V}(\mathbf{x}, \zeta \, \Sigma) = \phi'(\mathbf{x})\mathbf{M}^{-1}(\zeta, \Sigma)\phi(\mathbf{x})$$

Then the following assertions are equivalent:

1.   A design measure $\zeta^* \in \bar{H}$ is *D*-optimal, that is,

$$|\mathbf{M}(\zeta^*, \Sigma)| = \sup_{\zeta \in \bar{H}} |\mathbf{M}(\zeta, \Sigma)|$$

2. $\sup_{\mathbf{x} \in R} \mathrm{tr}[\Sigma^{-1}\mathbf{V}(\mathbf{x}, \zeta^*, \Sigma)] = \inf_{\zeta \in \bar{H}}\{\sup_{\mathbf{x} \in R} \mathrm{tr}[\Sigma^{-1}\mathbf{V}(\mathbf{x}, \zeta, \Sigma)]\}$
3. $\sup_{\mathbf{x} \in R} \mathrm{tr}[\Sigma^{-1}\mathbf{V}(\mathbf{x}, \zeta^*, \Sigma)] = p$

where p is the total number of parameters in the multiresponse model in Eq. 7.3. The proof of this theorem is given in Chapter 5 of Fedorov (1972).

According to Theorem 7.1, a design ζ^* is D-optimal if and only if it minimizes the supremum of the trace of the matrix $\Sigma^{-1}V(x, \zeta, \Sigma)$ over the region R, or if and only if the minimum value of this supremum is equal to the number of parameters in the multiresponse model. The last condition in this theorem will be utilized to develop a D-optimal design according to Fedorov's algorithm. This is done as follows:

1. Select an initial design $\mathbf{D}_{N_0}$ consisting of N_0 points such that the matrix $\mathbf{M}(\mathbf{D}_{N_0}, \Sigma)$ is nonsingular.
2. Augment $\mathbf{D}_{N_0}$ with a point $\mathbf{x}_{N_0+1}$ which maximizes $\text{tr}[\Sigma^{-1}V(\mathbf{x}, \mathbf{D}_{N_0}, \Sigma)]$ over the region R, to obtain the design $\mathbf{D}_{N_1}$, $N_1 = N_0 + 1$.
3. Continue this process and obtain the nested sequence of designs $\mathbf{D}_{N_0}$, $\mathbf{D}_{N_1}, \ldots, \mathbf{D}_{N_i}, \ldots$, where $N_i = N_0 + i$ $(i = 1, 2, \ldots)$ and $\mathbf{D}_{N_i}$ is obtained from $\mathbf{D}_{N_{i-1}}$ by adding the point $\mathbf{x}_{N_i}$ which maximizes $\text{tr}[\Sigma^{-1}V(\mathbf{x}, \mathbf{D}_{N_{i-1}}, \Sigma)]$ over the region R.
4. Stop this process when a point $\mathbf{x}_{N^*}$ in R is reached for which

$$\text{tr}[\Sigma^{-1}V(\mathbf{x}_{N^*}, \mathbf{D}_{N^*-1}, \Sigma)] - p < \nu \qquad (7.43)$$

where ν is some small positive number chosen a priori. Inequality 7.43 is based on the fact that the sequence $\{\sup_{\mathbf{x}\in R} \text{tr}[\Sigma^{-1}V(\mathbf{x}, \mathbf{D}_{N_i}, \Sigma)]\}_{i=0}^{\infty}$ converges to p from above as can be deduced from assertion 3 of Theorem 7.1.

The Sequential Generation of a D-Optimal Design When Σ Is Not Known. When Σ is not known, an estimate of it can be used in the sequential procedure. Recall earlier in Eq. 7.7 that an estimator, $\hat{\Sigma}$, of Σ was proposed by Zellner (1962) and is given by $\hat{\Sigma} = (\hat{\sigma}_{ij})$, where

$$\hat{\sigma}_{ij} = \frac{1}{N} (\mathbf{Y}_i - \mathbf{Z}_i\tilde{\boldsymbol{\beta}}_i)'(\mathbf{Y}_j - \mathbf{Z}_j\tilde{\boldsymbol{\beta}}_j) \qquad i, j = 1, 2, \ldots, r \qquad (7.44)$$

where $\tilde{\boldsymbol{\beta}}_i = (\mathbf{Z}_i'\mathbf{Z}_i)^{-1}\mathbf{Z}_i'\mathbf{Y}_i$ is the usual least squares estimator of $\boldsymbol{\beta}_i$ in the single-response model 7.1 $(i = 1, 2, \ldots, r)$. Thus, $\hat{\sigma}_{ij}$ is a multiple of the dot product of the residual vectors obtained from fitting two single-response models to their respective data. This estimator converges in probability to Σ (see Zellner, 1962, for more details). The sequential procedure described earlier can be modified as follows:

1. Select an initial design $\mathbf{D}_{N_0}$ such that $\mathbf{M}(\mathbf{D}_{N_0}, \mathbf{I}_r)$ is nonsingular, where $\mathbf{I}_r$

is the identity matrix of order $r \times r$. Obtain observations on all r responses at each experimental run as specified by a row of $\mathbf{D}_{N_0}$.

2. The data obtained in step 1 are used to obtain the estimate $\hat{\Sigma}_{N_0}$ according to Eq. 7.44.

3. Augment $\mathbf{D}_{N_0}$ with a point $\mathbf{x}_{N_0+1}$ which maximizes $\text{tr}[\hat{\Sigma}_{N_0}^{-1}\mathbf{V}(\mathbf{x}, \mathbf{D}_{N_0}, \hat{\Sigma}_{N_0})]$ over the region R; then obtain the design $\mathbf{D}_{N_1}$, $N_1 = N_0 + 1$.

4. Obtain observations on all r responses at the point $\mathbf{x}_{N_0+1}$ and undate the earlier estimate of Σ by using $\hat{\Sigma}_{N_1}$, which is based on data collected at all of the $N_1 = N_0 + 1$ experimental runs.

5. Repeat step 3. with $\mathbf{D}_{N_1}$ and $\hat{\Sigma}_{N_1}$ used instead of $\mathbf{D}_{N_0}$ and $\hat{\Sigma}_{N_0}$, respectively. By continuing this process, we obtain the nested sequence of designs, $\mathbf{D}_{N_0}, \mathbf{D}_{N_1}, \ldots, \mathbf{D}_{N_i}, \ldots$, where $N_i = N_0 + i$ and $\mathbf{D}_{N_i}$ is obtained from $\mathbf{D}_{N_{i-1}}$ by augmenting it with the point $\mathbf{x}_{N_i}$ ($i = 1, 2, \ldots$).

6. Stop this process when a point $\mathbf{x}_{N^*}$ is found that satisfies

$$\text{tr}[\hat{\Sigma}_{N^*-1}^{-1}\mathbf{V}(\mathbf{x}_{N^*}, \mathbf{D}_{N^*-1}, \hat{\Sigma}_{N^*-1})] - p < \nu \tag{7.45}$$

where ν is some specified small positive number. The convergence of the sequence $\{\text{tr}[\hat{\Sigma}_{N_i}^{-1}\mathbf{V}(\mathbf{x}_{N_{i+1}}, \mathbf{D}_{N_i}, \hat{\Sigma}_{N_i})]\}_{i=0}^{\infty}$ to p is guaranteed by a theorem in Wijesinha (1984). See also Wijesinha and Khuri (1987a).

We note that in both sequential procedures, when Σ is known and also when it is unknown, the dimension of the optimization problem at each sequential step is equal to k, the number of input variables in the multiresponse model. Hence, the optimization process is computationally manageable unless k is very large.

7.3.2 Multiresponse Rotatability

In the previous chapters, rotatability was defined in terms of a single-response model. This design property can easily be extended to a linear multiresponse model.

By definition, a design for fitting the model in Eq. 7.3 is multiresponse rotatable if the prediction variance matrix, $\text{Var}[\hat{\mathbf{Y}}(\mathbf{x})]$, in formula 7.40, is constant at all points $\mathbf{x}$ that are equidistant from the design center. In this case the variances of all the predicted responses, as well as their covariances, are constant on spheres (hyperspheres, in general) centered at the origin (design center). Thus the correlation coefficients among all the predicted responses are constant on such spheres. The same applies to the variance of any linear combination of the predicted responses. This is a desirable feature since certain linear combinations of the responses may be of special interest. For example, it may be experimentally difficult to obtain independent measurements on some response. If, however, this response is linearly related to the remaining $r - 1$ responses (because of physical or chemical laws), then its value can be determined in terms of those responses (see Box et al., 1973, Sec. 6).

A Necessary and Sufficient Condition for Multiresponse Rotatability. Let us suppose that the ith response model is represented by a complete polynomial of degree d_i in $x_1, x_2, \ldots, x_k$ ($i = 1, 2, \ldots, r$). Let $d = \max(d_1, d_2, \ldots, d_r)$. Then, a necessary and sufficient condition for a design to be multiresponse rotatable is that its design moments through order $2d$ be of the form

$$[1^{\theta_1}2^{\theta_2} \cdots k^{\theta_k}] = \begin{cases} 0, & \text{if any } \theta_i \text{ is odd} \\ \dfrac{\lambda_\theta \Pi_{i=1}^k (\theta_i)!}{2^{\theta/2} \Pi_{i=1}^k \left(\dfrac{\theta_i}{2}\right)!}, & \text{if all } \theta_i \text{ are even} \end{cases}$$

where $\theta = \Sigma_{i=1}^k \theta_i \leq 2d$, and λ_θ is a constant depending on θ. The derivation of this result is given in Khuri (1990b).

We note that this condition is identical to Eq. 2.B.1 in Appendix 2B concerning rotatability for a single-response model of order d. We conclude that multiresponse rotatability can be achieved if and only if the design is rotatable for a single-response model having the highest degree among all the response models under consideration.

For example, a design for fitting the three models

$$Y_1 = \beta_{10} + \beta_{11}x_1 + \beta_{12}x_2 + \varepsilon_1$$

$$Y_2 = \beta_{20} + \beta_{21}x_1 + \beta_{22}x_2 + \beta_{211}x_1^2 + \beta_{222}x_2^2 + \beta_{212}x_1x_2 + \varepsilon_2$$

$$Y_3 = \beta_{30} + \beta_{31}x_1 + \beta_{32}x_2 + \beta_{311}x_1^2 + \beta_{322}x_2^2 + \beta_{312}x_1x_2 + \varepsilon_3$$

is multiresponse rotatable if and only if it is rotatable for the Y_2 (or Y_3) model.

7.3.3 Other Design Criteria for Linear Multiresponse Models

Detection of model inadequacy is an important consideration in multiresponse modeling. In Section 7.4 we present a multivariate test of lack of fit of a linear multiresponse model. The power of this test depends, among other things, on the design used to fit the multiresponse model. Hence, by a proper choice of design it is possible to increase the power of the lack of fit test for a specified departure from the fitted model. Multiresponse designs based on the power criterion are discussed in Wijesinha and Khuri (1987b).

Another criterion for the choice of a multiresponse design is the robustness criterion. Some multivariate tests that are used to test the significance of the parameters of a multiresponse model are sensitive to nonnormality of the random errors. These tests can be made less sensitive, or robust, to nonnormality by a proper choice of design, as was shown in Mardia (1971). A procedure for the construction of designs satisfying the robustness criterion is given in Wijesinha and Khuri (1991).

7.4 A TEST FOR LACK OF FIT OF A LINEAR MULTIRESPONSE MODEL

Let us consider the linear models described in Eq. 7.1, which we write again for convenience

$$\mathbf{Y}_i = \mathbf{Z}_i \boldsymbol{\beta}_i + \boldsymbol{\varepsilon}_i \qquad i = 1, 2, \ldots, r \tag{7.46}$$

where $\mathbf{Z}_i$ is $N \times p_i$ of rank p_i. A multivariate formulation of these models can be written as

$$\mathbf{Y} = \mathbf{W}\boldsymbol{\Gamma} + \mathbf{e} \tag{7.47}$$

where $\mathbf{Y} = [\mathbf{Y}_1 : \mathbf{Y}_2 : \ldots : \mathbf{Y}_r]$, $\mathbf{W} = [\mathbf{Z}_1 : \mathbf{Z}_2 : \ldots : \mathbf{Z}_r]$, $\boldsymbol{\Gamma} = \text{diag}(\boldsymbol{\beta}_1, \boldsymbol{\beta}_2, \ldots, \boldsymbol{\beta}_r)$, and $\mathbf{e} = [\boldsymbol{\varepsilon}_1 : \boldsymbol{\varepsilon}_2 : \ldots : \boldsymbol{\varepsilon}_r]$. We assume that the rows of $\mathbf{e}$ are independently distributed as $N(\mathbf{0}, \boldsymbol{\Sigma})$, where $\boldsymbol{\Sigma}$ is the variance-covariance matrix for the r responses (see Eq. 7.2). We also assume that the design used to fit these models allows for repeated runs to be taken on all r responses at some points in the experimental region. Without loss of generality, we consider that the repeated runs are taken at each of the first n points of the design $(1 \leq n < N)$.

The model in Eq. 7.47 is said to suffer from lack of fit if it fails to provide adequate representation of the true means of the r responses over the experimental region. Checking inadequacy of such a model requires proper recognition of the multivariate nature of the data. In particular, it would be inappropriate to apply the lack of fit test described in Section 2.6 in Chapter 2 to the individual response models in Eq. 7.46. This procedure does not take into account correlations that may exist among the responses. Obviously, when the responses are correlated, lack of fit in one response can influence the quality of fit of the other responses. We now show how to develop a multivariate lack of fit test for a multiresponse model of the form given in Eq. 7.47.

7.4.1 The Development of the Multivariate Lack of Fit Test

Let $\mathbf{c} = (c_1, c_2, \ldots, c_r)'$ be an arbitrary nonzero $r \times 1$ vector. From Eq. 7.47 we obtain the model

$$\mathbf{Y}_c = \mathbf{W}\boldsymbol{\gamma}_c + \boldsymbol{\varepsilon}_c \tag{7.48}$$

where $\mathbf{Y}_c = \mathbf{Yc}$ is a vector of N observations on the univariate response $Y_c = \sum_{i=1}^{r} c_i Y_i$, $\boldsymbol{\gamma}_c = \boldsymbol{\Gamma}\mathbf{c}$, $\boldsymbol{\varepsilon}_c = \mathbf{ec}$. Note that $\boldsymbol{\varepsilon}_c$ has the multivariate normal distribution $N(\mathbf{0}, \sigma_c^2 \mathbf{I}_N)$, where $\sigma_c^2 = \mathbf{c}'\boldsymbol{\Sigma}\mathbf{c}$.

The multiresponse model in Eq. 7.47 is adequate if and only if the univariate models in Eq. 7.48 are adequate for all $\mathbf{c} \neq \mathbf{0}$. Equivalently, if for some $\mathbf{c} \neq \mathbf{0}$

model 7.48 is inadequate, then the multiresponse model 7.47 is also inadequate. Since $\mathbf{Y}_c$ has the multivariate normal distribution $N(\mathbf{0}, \sigma_c^2 \mathbf{I}_N)$ and contains replicated observations on the univariate response Y_c at each of the first n design points, then the lack of fit test described in Section 2.6 in Chapter 2 can be used to check the adequacy of Eq. 7.48. For $\mathbf{c} \neq \mathbf{0}$ the test statistic can be written as

$$F(\mathbf{c}) = \frac{\nu_{PE} SS_{LOF}(\mathbf{c})}{\nu_{LOF} SS_{PE}(\mathbf{c})}$$

where $SS_{PE}(\mathbf{c})$ and $SS_{LOF}(\mathbf{c})$ are the pure error and lack of fit sums of squares, respectively, and ν_{PE} and ν_{LOF} are their corresponding degrees of freedom. These sums of squares can be expressed as

$$SS_{PE}(\mathbf{c}) = \mathbf{c}'\mathbf{Y}'\mathbf{K}\mathbf{Y}\mathbf{c} \tag{7.49}$$

$$SS_{LOF}(\mathbf{c}) = \mathbf{c}'\mathbf{Y}'[\mathbf{I}_N - \mathbf{W}(\mathbf{W}'\mathbf{W})^-\mathbf{W}' - \mathbf{K}]\mathbf{Y}\mathbf{c} \tag{7.50}$$

where $\mathbf{K} = \text{diag}(\mathbf{K}_1, \mathbf{K}_2, \ldots, \mathbf{K}_n, 0)$ and $\mathbf{K}_i$ is the matrix

$$\mathbf{K}_i = \mathbf{I}_{\nu_i} - \left(\frac{1}{\nu_i}\right) \mathbf{J}_{\nu_i} \qquad i = 1, 2, \ldots, n \tag{7.51}$$

where $\mathbf{J}_{\nu_i}$ is the matrix of ones of order $\nu_i \times \nu_i$ and ν_i is the number of observations at the ith repeat-runs site ($i = 1, 2, \ldots, n$). Note that a generalized inverse of $\mathbf{W}'\mathbf{W}$ is used in Eq. 7.50 since $\mathbf{W}$ is not necessarily of full column rank. We shall denote the matrices of the quadratic forms in Eqs. 7.49 and 7.50 by $\mathbf{G}_2$ and $\mathbf{G}_1$, respectively, and label them as the pure error and lack of fit matrices. We thus have

$$\mathbf{G}_1 = \mathbf{Y}'[\mathbf{I}_N - \mathbf{W}(\mathbf{W}'\mathbf{W})^-\mathbf{W}' - \mathbf{K}]\mathbf{Y}$$

$$\mathbf{G}_2 = \mathbf{Y}'\mathbf{K}\mathbf{Y}$$

The statistic $F(\mathbf{c})$ can be rewritten as

$$F(\mathbf{c}) = \frac{\nu_{PE} \mathbf{c}'\mathbf{G}_1\mathbf{c}}{\nu_{LOF} \mathbf{c}'\mathbf{G}_2\mathbf{c}}$$

If model 7.48 is correct, then $F(\mathbf{c})$ has the F-distribution with ν_{LOF} and ν_{PE} degrees of freedom. A large value of $F(\mathbf{c})$, or equivalently, a large value of $\mathbf{c}'\mathbf{G}_1\mathbf{c}/\mathbf{c}'\mathbf{G}_2\mathbf{c}$ leads us to believe that Eq. 7.48 is inadequate. Since Eq. 7.47 is considered inadequate if and only if at least one of the models in Eq. 7.48 is inadequate for some $\mathbf{c} \neq \mathbf{0}$, then model 7.47 has a significant lack of fit if $\max(\mathbf{c}'\mathbf{G}_1\mathbf{c}/\mathbf{c}'\mathbf{G}_2\mathbf{c})$ exceeds a certain critical value. But $\max(\mathbf{c}'\mathbf{G}_1\mathbf{c}/\mathbf{c}'\mathbf{G}_2\mathbf{c}) = e_{max}(\mathbf{G}_2^{-1}\mathbf{G}_1)$, where $e_{max}(\mathbf{G}_2^{-1}\mathbf{G}_1)$ denotes the largest eigenvalue of the $r \times r$ matrix $\mathbf{G}_2^{-1}\mathbf{G}_1$ (see Roy et al., 1971: Chapter 4). If λ_α denotes the upper $100\alpha\%$ point of the distribution of $e_{max}(\mathbf{G}_2^{-1}\mathbf{G}_1)$ when model 7.47 is correct, then a significant lack of fit can be detected at the α-level if

$$e_{\max}(G_2^{-1}G_1) \geq \lambda_\alpha \tag{7.52}$$

The lack of fit test given by inequality 7.52 is a multivariate analog of the usual lack of fit test for a single response and is known as Roy's largest root test. Tables for the critical value λ_α are available in Morrison (1976), Roy et al. (1971), and Seber (1984).

Other multivariate test statistics can also be used to test lack of fit of model 7.47. These include Wilks's likelihood ratio, $|G_2|/|G_1 + G_2|$, Hotelling-Lawley's trace, $\text{tr}(G_2^{-1}G_1)$, and Pillai's trace, $\text{tr}[G_1(G_1 + G_2)^{-1}]$. Small values of Wilks's likelihood ratio are significant, whereas large values of the latter two statistics are significant. Tables for the critical values of these three tests are available in Seber (1984).

More recently, Levy and Neill (1990) considered additional multivariate lack-of-fit tests based on the determinant and trace of $G_2^{-1}G_1$. They compared the power functions of these tests by simulation and extended the lack-of-fit test to situations where no independent replicates are available on all the responses.

7.4.2 The Selection of Subsets of the Responses Which Contribute to Lack of Fit

A significant multivariate lack of fit test is an indication that there exists at least one linear combination of the responses, given by some nonzero value of c, for which model 7.48 is inadequate. In particular, the vector c^* which maximizes $c'G_1c/c'G_2c$ is such a vector. By differentiating this ratio with respect to the vector c and equating the derivative to zero, it can be easily seen that c^* satisfies the equation

$$\left[G_1 - \max_c \left(\frac{c'G_1c}{c'G_2c} \right) G_2 \right] c^* = 0$$

or equivalently, since G_2 is nonsingular,

$$\left[G_2^{-1}G_1 - \max_c \left(\frac{c'G_1c}{c'G_2c} \right) I_r \right] c^* = 0$$

Since

$$\max_c \left(\frac{c'G_1c}{c'G_2c} \right) = e_{\max}(G_2^{-1}G_1)$$

where $e_{\max}(G_2^{-1}G_1)$ represents the largest eigenvalue of the $r \times r$ matrix $G_2^{-1}G_1$, then

$$[G_1 - e_{\max}(G_2^{-1}G_1)G_2]c^* = 0 \tag{7.53}$$

Thus c^* is an eigenvector of $G_2^{-1}G_1$ corresponding to its largest eigenvalue, $e_{max}(G_2^{-1}G_1)$. The elements of c^* produce the linear combination

$$Y_{c^*} = \sum_{i=1}^{r} c_i^* Y_i \qquad (7.54)$$

The responses involved in this linear combination are considered to have some influence on lack of fit. Since the responses may be measured in different units, the linear combination in Eq. 7.54 should be expressed in terms of standardized response variables as

$$Y_{c^*} = \sum_{i=1}^{r} d_i^* z_i \qquad (7.55)$$

where $z_i = Y_i / \|Y_i\|$ with $\|Y_i\|$ being the Euclidean norm $(Y_i'Y_i)^{1/2}$ of the vector Y_i of observations on the ith response, and $d_i^* = c_i^* \|Y_i\|$ ($i = 1, 2, \ldots, r$).

The size of each of the d_i^* coefficients determines the contribution of the responses to lack of fit. Large absolute values of d_i^* are associated with responses that are influential with respect to lack of fit. A subset of the responses can be checked to see if it suffices by itself to produce a significant lack of fit. This is determined by whether or not $e_{max}(G_2^{-1}G_1)$, computed on the basis of that subset only, exceeds the critical value λ_α in Eq. 7.52. It is easy to see that when all nonempty subsets of the r responses are examined on that basis, the probability of falsely detecting a significant subset does not exceed the value α (see Khuri, 1985: 216). These are simultaneous tests of significance and the use of the same critical value, λ_α, in these tests helps control the Type I family error rate at a level not exceeding α.

In summary, the determination of subsets of the responses which are influential contributors to lack of fit can be arrived at by an examination of the elements of the eigenvector corresponding to $e_{max}(G_2^{-1}G_1)$ followed by an inspection of the values of the latter quantity for all nonempty subsets of the responses. For more details concerning this procedure and the multivariate lack of fit test, see Khuri (1985).

7.4.3 A Numerical Example

Richert et al. (1974) investigated the effects of heating temperature (X_1), pH level (X_2), redox potential (X_3), sodium oxalate (X_4), and sodium lauryl sulfate (X_5) on foaming properties of whey protein concentrates (WPC). These products are of considerable interest to the food industry because of their potential value as functional food ingredients. Measurements were made on three responses, namely, $Y_1 =$ whipping time (the total elapsed time required to produce peaks of foam formed during whipping of a liquid sample), $Y_2 =$ maximum overrun

[overrun is determined by weighing 5 oz. paper cups of foam and unwhipped liquid sample and calculating by the expression: % overrun = 100 (weight of liquid − weight of foam)/weight of foam], and Y_3 = percent soluble protein. The design used was a central composite design which consisted of a one-half fraction of a 2^5 factorial design, 10 axial points, and five center point replications. The original and coded levels of the five input variables are given in Table 7.6. The design, in coded form, and the multiresponse data are given in Table 7.7.

The fitted model, in the coded variables, for each of the three responses is a quadratic model of the form

$$Y = \beta_0 + \sum_{i=1}^{5} \beta_i x_i + \sum_{i=1}^{5} \beta_{ii} x_i^2 + \sum_{i<j=2}^{5} \beta_{ij} x_i x_j + \varepsilon$$

In this case, the matrix $\mathbf{W}$ which appears in Eq. 7.47 is of order 31×63 and is partitioned as $\mathbf{W} = [\mathbf{Z}_1 : \mathbf{Z}_2 : \mathbf{Z}_3]$ with $\mathbf{Z}_1 = \mathbf{Z}_2 = \mathbf{Z}_3$. The matrix $\mathbf{K}$ in Eq. 7.49 is equal to $\mathbf{K} = \text{diag}(\mathbf{K}_1, \mathbf{0})$, where $\mathbf{K}_1 = \mathbf{I}_5 - (1/5)\mathbf{J}_5$ and $\mathbf{0}$ is a zero matrix of order 26×26. The pure error and lack of fit degrees of freedom are, respectively, $\nu_{PE} = 4$ and $\nu_{LOF} = 6$.

The value of Roy's largest root test statistic given in Eq. 7.52 is $e_{\max}(\mathbf{G}_2^{-1}\mathbf{G}_1) = 245.518$. The critical value λ_α for this test can be obtained from the generalized beta distribution table given in Foster (1957). This table gives values of x_α, the upper $100\alpha\%$ point of the distribution of the largest eigenvalue of the matrix $(\mathbf{G}_1 + \mathbf{G}_2)^{-1}\mathbf{G}_1$ when model 7.47 is correct. The relationship between λ_α and x_α is given by

$$\lambda_\alpha = \frac{x_\alpha}{1 - x_\alpha}$$

Table 7.6 The Original and Coded Levels of the Input Variables

Variable		Coded Levels				
		−2	−1	0	1	2
Heating temperature (°C)	X_1	65.0	70.0	75.0	80.0	85.0
pH	X_2	4.0	5.0	6.0	7.0	8.0
Redox potential (volt)	X_3	−0.025	0.075	0.175	0.275	0.375
Sodium oxalate (molar)	X_4	0.0	0.0125	0.025	0.0375	0.05
Sodium lauryl sulfate (%)	X_5	0.0	0.05	0.10	0.15	0.20

Source: A. I. Khuri (1985). Reproduced with permission of the American Statistical Association and the Institute of Food Technologies.

Table 7.7 Experimental Design (coded) and the Multiresponse Data

x_1	x_2	x_3	x_4	x_5	Y_1 (min)	Y_2 (%)	Y_3 (%)
0	0	0	0	0	3.5	1179	104
0	0	0	0	0	3.5	1183	107
0	0	0	0	0	4.0	1120	104
0	0	0	0	0	3.5	1180	101
0	0	0	0	0	3.0	1195	103
−1	−1	−1	−1	1	4.75	1082	81.4
1	−1	−1	−1	−1	4.0	824	69.6
−1	1	−1	−1	−1	5.0	953	105
1	1	−1	−1	1	9.5	759	81.2
−1	−1	1	−1	−1	4.0	1163	80.8
1	−1	1	−1	1	5.0	839	76.3
−1	1	1	−1	1	3.0	1343	103
1	1	1	−1	−1	7.0	736	76.9
−1	−1	−1	1	−1	5.25	1027	87.2
1	−1	−1	1	1	5.0	836	74.0
−1	1	−1	1	1	3.0	1272	98.5
1	1	−1	1	−1	6.5	825	94.1
−1	−1	1	1	1	3.25	1363	95.9
1	−1	1	1	−1	5.0	855	76.8
−1	1	1	1	−1	2.75	1284	100
1	1	1	1	1	5.0	851	104
−2	0	0	0	0	3.75	1283	100
2	0	0	0	0	11.0	651	50.5
0	−2	0	0	0	4.5	1217	71.2
0	2	0	0	0	4.0	982	101
0	0	−2	0	0	5.0	884	85.8
0	0	2	0	0	3.75	1147	103
0	0	0	−2	0	3.75	1081	104
0	0	0	2	0	4.75	1036	89.4
0	0	0	0	−2	4.0	1213	105
0	0	0	0	2	3.5	1103	113

Source: A. I. Khuri (1985). Reproduced with permission of the American Statistical Association and the Institute of Food Technologies.

At the $\alpha = 0.10$ level of significance, $x_\alpha = 0.9884$, hence $\lambda_\alpha = 85.21$. Consequently, a significant lack of fit can be detected at the 10% level.

In order to assess the contributions of the various responses to lack of fit, we follow the procedure outlined in Section 7.4.2. Here the eigenvector $\mathbf{c}^*$ corresponding to the eigenvalue $e_{max}(\mathbf{G}_2^{-1}\mathbf{G}_1) = 245.518$ is

$$\mathbf{c}^* = (3.2659, 0.0385, -0.0904)'$$

The Euclidean norms of the response vectors $\mathbf{Y}_1$, $\mathbf{Y}_2$, and $\mathbf{Y}_3$ are $\|\mathbf{Y}_1\| = 27.60$, $\|\mathbf{Y}_2\| = 5929.27$, and $\|\mathbf{Y}_3\| = 517.49$. Thus, in terms of the standardized response variables, z_1, z_2, and z_3, the linear combination in Eq. 7.55 is written as

$$
\begin{aligned}
Y_{c^*} &= 90.139z_1 + 228.277z_2 - 46.781z_3 \\
&\sim 0.395z_1 + z_2 - 0.205z_3
\end{aligned}
\tag{7.56}
$$

where the symbol $\sim$ indicates that the two linear combinations are proportional.

From the size of the absolute values of the coefficients in Eq. 7.56 we can see that the responses Y_1 and Y_2 are the main contributors to lack of fit, with the latter being more influential than the former. Values of $e_{max}(\mathbf{G}_2^{-1}\mathbf{G}_1)$ were subsequently computed for all nonempty subsets of the three responses. The results are given in Table 7.8. In addition to the subset of all three responses, the only other significant subset at the $\alpha = 0.10$ level is the subset $\{Y_1, Y_2\}$, which supports our earlier finding. We conclude that the responses Y_1 and Y_2 together produce a significant lack of fit, whereas the other two pairs of responses, $\{Y_1, Y_3\}$ and $\{Y_2, Y_3\}$, do not appear to contribute as much to lack of fit. When considered individually, none of the three responses is sufficient to produce a significant lack of fit.

We note that the value of $e_{max}(\mathbf{G}_2^{-1}\mathbf{G}_1)$ for each of the individual-response

Table 7.8 Values of $e_{max}(\mathbf{G}_2^{-1}\mathbf{G}_1)$ for All Nonempty Subsets of the Three Responses

Subset	$e_{max}(\mathbf{G}_2^{-1}\mathbf{G}_1)$	Critical Value ($\lambda_{0.10}$)
Y_1, Y_2, Y_3	245.518*	85.21
Y_1, Y_2	214.307*	85.21
Y_1, Y_3	45.532	85.21
Y_2, Y_3	32.107	85.21
Y_1	14.936	85.21
Y_2	19.573	85.21
Y_3	28.495	85.21

*Significant at the 10% level.
Source: A. I. Khuri (1985). Reproduced with permission of the American Statistical Association.

subsets is the ratio of the lack of fit sum of squares to the pure error sum of squares that result from the analysis of each of the three fitted response models. Hence, if each such ratio is multiplied by $\nu_{PE}/\nu_{LOF} = 2/3$, we would obtain the value of the F statistic that would result from applying the univariate lack of fit test to each of the three responses.

7.5 GENERAL LINEAR HYPOTHESIS

Consider testing the general linear hypothesis

$$H_0: \Omega\Gamma = \Omega_0$$

where Γ is the block-diagonal matrix of parameter vectors associated with the linear multiresponse model in Eq. 7.47; Ω and Ω_0 are known matrices. It is assumed that Ω is of full row rank q. If the rows of Ω belong to the row space of W in Eq. 7.47, then H_0 is said to be a testable hypothesis. In this case, Roy's largest root, given by $e_{\max}(S_h S_e^{-1})$, can be used as a test statistic, where

$$S_h = [\Omega(W'W)^- W'Y - \Omega_0]'[\Omega(W'W)^-\Omega']^{-1}[\Omega(W'W)^- W'Y - \Omega_0]$$

$$S_e = Y'[I_N - W(W'W)^- W']Y$$

This test requires that $N \geq r + \text{rank} (W)$. As in Section 7.4.1, other multivariate test statistics can also be used to test H_0, namely, Wilks's likelihood ratio, $|S_e|/|S_h + S_e|$, Hotelling-Lawley's trace, $\text{tr}(S_h S_e^{-1})$, and Pillai's trace, $\text{tr}[S_h(S_e + S_h)^{-1}]$.

In particular, if the β_i vectors in Eq. 7.46 are of the same order, then it is possible to test that they are equal. For this purpose, let us first multiply the two sides in Eq. 7.47 on the right by the $r \times (r - 1)$ matrix C^* given by

$$
C^* =
\begin{bmatrix}
1 & 1 & & 1 \\
-1 & 0 & & 0 \\
0 & -1 & & 0 \\
0 & 0 & \cdots & \cdot \\
\cdot & \cdot & & \cdot \\
\cdot & \cdot & & \cdot \\
\cdot & \cdot & & \cdot \\
0 & 0 & & -1
\end{bmatrix}
$$

We get

$$YC^* = W\Gamma C^* + eC^*$$

Now, the hypothesis that the β_i's are equal can be expressed as

$$\Omega\Gamma C^* = 0$$

where, in this case, $\Omega = [\mathbf{I}_{p_0}:\mathbf{I}_{p_0}:\ldots:\mathbf{I}_{p_0}]$ with p_0 being the common value of p_i, the number of elements of $\boldsymbol{\beta}_i$ $(i = 1, 2, \ldots, r)$.

Alternatively, we can test the equality of corresponding subvectors of $\boldsymbol{\beta}_i (i = 1, 2, \ldots, r)$. This is accomplished by a proper choice of the matrix Ω. More details concerning these tests can be found in Khuri (1986).

7.6 OPTIMIZATION OF A MULTIRESPONSE FUNCTION

We recall from Chapter 5 that one of the primary objectives of response surface methodology is the determination of operating conditions on a set of input variables that result in an optimum response. In a multiresponse situation several response variables are considered, and the optimization problem is more complex than in the single-response case. The main difficulty stems from the fact that when two or more response variables are under investigation simultaneously, the meaning of optimum becomes unclear since there is no unique way to order multivariate values of a multiresponse function. Furthermore, conditions which are optimal for one response may be far from optimal or even physically impractical for the other responses. Only rarely do all response variables achieve their respective optima at the same set of conditions. Heuristically, one might consider superimposing contours of all response variables and then pinpoint a region where conditions can be "near" optimal for all the responses (see, for example, Lind et al. 1960). This procedure, although simple and straightforward, has its limitations in large systems involving several input variables and several responses. Furthermore, it is difficult to identify one set of conditions (or one point in the experimental region) as being optimal with such a procedure.

7.6.1 Optimization Associated with a Dual Response System

Myers and Carter (1973) considered an optimization problem associated with a dual response system consisting of two responses; one is called a primary response, and the other is called a secondary response. These authors developed an algorithm for the determination of conditions on a set of input variables which maximize (or minimize) the primary response subject to the condition that the secondary response takes on some specified or desirable values. In other words, the secondary response function imposes certain constraints on the optimization of the primary response function.

Suppose that the fitted models over the region of k coded input variables, $x_1, x_2, \ldots, x_k$, for the primary and secondary (or constraint) response functions are of the second order and are expressible as

$$\hat{Y}_p(\mathbf{x}) = b_{01} + \mathbf{x}'\mathbf{b}_1 + \mathbf{x}'\mathbf{B}_1\mathbf{x}$$

$$\hat{Y}_s(\mathbf{x}) = b_{02} + \mathbf{x}'\mathbf{b}_2 + \mathbf{x}'\mathbf{B}_2\mathbf{x}$$

respectively, where $\mathbf{x} = (x_1, x_2, \ldots, x_k)'$. The solution to the above-mentioned problem is similar to that of ridge analysis, which was discussed in Section 5.7 in Chapter 5. Here conditions on $\mathbf{x}$ need be found which optimize $\hat{Y}_p$ subject to $\hat{Y}_s = \tau$, where τ belongs to a set of desirable values of the secondary response. Using the method of Lagrange multipliers, we consider the function

$$L = b_{01} + \mathbf{x}'\mathbf{b}_1 + \mathbf{x}'\mathbf{B}_1\mathbf{x} - \mu(b_{02} + \mathbf{x}'\mathbf{b}_2 + \mathbf{x}'\mathbf{B}_2\mathbf{x} - \tau)$$

where μ is a Lagrange multiplier. Differentiating L with respect to each x_i and equating the derivatives to zero, we get

$$2(\mathbf{B}_1 - \mu\mathbf{B}_2)\mathbf{x} = \mu\mathbf{b}_2 - \mathbf{b}_1$$

The nature of the "stationary point" generated by the preceding equation depends on the properties of the matrix of second-order partial derivatives of L with respect to the values of x_i, namely,

$$\mathbf{M}(\mathbf{x}) = 2(\mathbf{B}_1 - \mu\mathbf{B}_2)$$

Let us consider the following examples.

Example 7.6: $\mathbf{B}_2$ Is Positive Definite. Since $\mathbf{B}_1$ and $\mathbf{B}_2$ are symmetric and $\mathbf{B}_2$ is positive definite, then there exists a nonsingular matrix, $\mathbf{S}$, such that (see Searle, 1982: 312)

$$\mathbf{S}'\mathbf{B}_1\mathbf{S} = \text{diag}(\tilde{\lambda}_1, \tilde{\lambda}_2, \ldots, \tilde{\lambda}_k)$$

$$\mathbf{S}'\mathbf{B}_2\mathbf{S} = \mathbf{I}_k$$

where $\mathbf{I}_k$ is the identity matrix of order $k \times k$ and the values of $\tilde{\lambda}_i$ are the eigenvalues of the symmetric matrix $\tilde{\mathbf{T}}_1$ arranged in ascending order, where $\tilde{\mathbf{T}}_1$ is given by

$$\tilde{\mathbf{T}}_1 = \Lambda_2^{-1/2}\mathbf{Q}_2'\mathbf{B}_1\mathbf{Q}_2\Lambda_2^{-1/2}$$

Here $\mathbf{Q}_2$ is an orthogonal matrix and Λ_2 is a diagonal matrix such that

$$\mathbf{Q}_2'\mathbf{B}_2\mathbf{Q}_2 = \Lambda_2$$

The matrix $\Lambda_2^{-1/2}$ is a diagonal matrix with diagonal elements equal to the reciprocals of the square roots of the corresponding diagonal elements of Λ_2.

The matrix $\mathbf{M}(\mathbf{x})$ can then be written as

$$\mathbf{M}(\mathbf{x}) = 2\mathbf{S}'^{-1}[\text{diag}(\tilde{\lambda}_1, \tilde{\lambda}_2, \ldots, \tilde{\lambda}_k) - \mu\mathbf{I}_k]\mathbf{S}^{-1}$$

Consequently, if $\mu > \tilde{\lambda}_k$, where $\tilde{\lambda}_k$ is the largest eigenvalue of the matrix $\tilde{\mathbf{T}}_1$, then $\mathbf{M}(\mathbf{x})$ becomes negative definite (positive definite if $\mu < \tilde{\lambda}_1$, where $\tilde{\lambda}_1$ is the smallest eigenvalue of $\tilde{\mathbf{T}}_1$), and the corresponding stationary point, $\mathbf{x}_0$, gives rise to an absolute maximum (absolute minimum for $\mu < \tilde{\lambda}_1$) of $\hat{Y}_p$ subject to being on a surface of the secondary response given by

$$\hat{Y}_s(\mathbf{x}_0) = b_{02} + \mathbf{x}_0'\mathbf{b}_2 + \mathbf{x}_0'\mathbf{B}_2\mathbf{x}_0$$

In a typical situation, by choosing several values of $\mu > \tilde{\lambda}_k$ ($\mu < \tilde{\lambda}_1$), we can generate values of $x_1, x_2, \ldots, x_k$ which represent points of constrained maximum (minimum) for the primary response. By plotting the values of each of $x_1, x_2, \ldots, x_k$ against the corresponding values of $\hat{Y}_s$, we obtain plots that can be used to determine operating conditions which maximize $\hat{Y}_p$ conditional on $\hat{Y}_s$ having certain desirable values.

Example 7.7: $\mathbf{B}_2$ Is Negative Definite. In this example $-\mathbf{B}_2$ is positive definite and the matrix $\mathbf{M}(\mathbf{x})$ is now of the form

$$\mathbf{M}(\mathbf{x}) = 2\mathbf{S}'^{-1}[-\text{diag}(\tilde{\lambda}_1, \tilde{\lambda}_2, \ldots, \tilde{\lambda}_k) + \mu\mathbf{I}_k]\mathbf{S}^{-1}$$

By following a similar argument as in Example 7.6, it can be seen that a value of μ larger than the largest eigenvalue of $\tilde{\mathbf{T}}_1$ leads to an absolute minimum of $\hat{Y}_p$ subject to a constraint on $\hat{Y}_s$. On the other hand, values of μ which are smaller than the smallest eigenvalue of $\tilde{\mathbf{T}}_1$ lead to absolute maxima of $\hat{Y}_p$.

Example 7.8: $\mathbf{B}_2$ Is Indefinite. In this example it is not always possible to have a solution to the dual optimization problem. This is because the constrained optimization region, as determined by the values of $\hat{Y}_s$, is no longer bounded. For more details concerning this case, see Section 4 in Myers and Carter (1973).

A more general version of Myers and Carter's (1973) dual response problem was given by Biles (1975), who described a procedure for optimizing a primary response function subject to maintaining a set of secondary response functions within specified ranges. Biles's procedure employs a modification of the method of steepest ascent which was described in Chapter 5.

7.6.2 Optimization Using the Desirability Function Approach

Harrington (1965) introduced an analytic technique for the optimization of a multiresponse function based on the concept of utility or desirability of a property associated with a given response function. Each predicted response function, $\hat{Y}_i$ ($i = 1, 2, \ldots, r$), is transformed to a desirability function, h_i, $0 \leq h_i \leq 1$, such that h_i increases as the desirability of the corresponding property increases. Harrington proposed exponential-type transformations such as $h_i = \exp(-|\hat{Y}_i|^s)$, where s is a positive number specified by the user. Derringer and Suich (1980) introduced more general transformations that offer the user greater flexibility in the setting of desirabilities. These transformations require the user to specify minimum and maximum acceptable values for each response in addition to specifying certain parameters to be chosen so that the transformations express the user's own evaluation of desirability. A great deal of subjectivity can, therefore, be introduced in the setting of the desirability functions.

The individual desirability functions for the various responses under consideration are then incorporated into a single function which gives the overall assessment of the desirability of the combined responses. Both Harrington (1965) and Derringer and Suich (1980) used the geometric mean of the values of h_i, namely, $[h_1 \times h_2 \times \cdots \times h_r]^{1/r}$, where r is the number of responses, as the overall desirability function. This function has values inside the interval [0, 1] and approaches unity as the desirability of the responses increases. The geometric mean was chosen because it vanishes whenever any h_i is equal to zero, that is, if at least one of the response variables is unacceptable.

Once the overall desirability has been set up as a function of the input variables, the next step is to maximize this function over the experimental region using univariate optimization techniques. The optimization process is relatively simple since the overall desirability is a well-behaved continuous function of the input variables.

7.6.3 Optimization Using the Generalized Distance Approach

One of the basic characteristics of the desirability function approach is the subjectivity on the part of the user in evaluating desirability. Improperly assessed desirabilities can lead to inaccurate results. Furthermore, this approach does not take into account heterogeneity of the variances of the responses or any correlations that may exist among them. These are important considerations in any simultaneous optimization procedure involving the responses. Therefore, we shall present another procedure for multiresponse optimization which avoids these difficulties. We shall assume that all response functions in a multiresponse system depend on the same set of input variables, namely, $x_1, x_2, \ldots, x_k$, and can be represented by polynomial regression models of the same degree within the experimental region.

Let r be the number of responses of interest. Then under the above assumptions, the model for the ith univariate response can be written as

$$\mathbf{Y}_i = \mathbf{X}_0 \boldsymbol{\beta}_i + \boldsymbol{\varepsilon}_i \qquad i = 1, 2, \ldots, r$$

where $\mathbf{X}_0$ is $N \times p_0$ of rank p_0 and the values of the $\boldsymbol{\varepsilon}_i$s have means zero and a variance-covariance structure as in Eq. 7.2. An estimate of the variance-covariance matrix $\boldsymbol{\Sigma}$ is given by

$$\hat{\boldsymbol{\Sigma}} = \frac{\mathbf{Y}'[\mathbf{I}_N - \mathbf{X}_0(\mathbf{X}_0'\mathbf{X}_0)^{-1}\mathbf{X}_0']\mathbf{Y}}{N - p_0} \tag{7.57}$$

where $\mathbf{Y} = [\mathbf{Y}_1 : \mathbf{Y}_2 : \ldots : \mathbf{Y}_r]$. Without loss of generality we can assume that all r responses are not linearly related; otheriwse, those responses that are linear functions of the others can be dropped using the eigenvalue-eigenvector analysis shown in Section 7.2.3.

The prediction equation for the ith response is

$$\hat{Y}_i(\mathbf{x}) = \mathbf{z}'(\mathbf{x})\hat{\boldsymbol{\beta}}_i \qquad i = 1, 2, \ldots, r \tag{7.58}$$

where $\mathbf{x} = (x_1, x_2, \ldots, x_k)'$ is the vector of coded input variables, $\mathbf{z}'(\mathbf{x})$ is vector of the same form as a row of the matrix $\mathbf{X}_0$ evaluated at the point $\mathbf{x}$, $\hat{\boldsymbol{\beta}}_i$ is the least squares estimator of $\boldsymbol{\beta}_i$ and is equal to $(\mathbf{X}_0'\mathbf{X}_0)^{-1}\mathbf{X}_0'\mathbf{Y}_i$ ($i = 1, 2, \ldots, r$). It follows that

$$\text{Var}[\hat{Y}_i(\mathbf{x})] = \mathbf{z}'(\mathbf{x})(\mathbf{X}_0'\mathbf{X}_0)^{-1}\mathbf{z}(\mathbf{x})\sigma_{ii} \qquad i = 1, 2, \ldots, r$$

$$\text{Cov}[\hat{Y}_i(\mathbf{x}), \hat{Y}_j(\mathbf{x})] = \mathbf{z}'(\mathbf{x})(\mathbf{X}_0'\mathbf{X}_0)^{-1}\mathbf{z}(\mathbf{x})\sigma_{ij} \qquad i, j = 1, 2, \ldots, r, i \neq j$$

where σ_{ij} is the (i, j)th element of the variance-covariance matrix Σ. Hence,

$$\text{Var}[\hat{\mathbf{Y}}(\mathbf{x})] = \mathbf{z}'(\mathbf{x})(\mathbf{X}_0'\mathbf{X}_0)^{-1}\mathbf{z}(\mathbf{x})\Sigma \tag{7.59}$$

where $\hat{\mathbf{Y}}(\mathbf{x}) = (\hat{Y}_1(\mathbf{x}), \hat{Y}_2(\mathbf{x}), \ldots, \hat{Y}_r(\mathbf{x}))'$ is the vector of predicted responses at the point $\mathbf{x}$. An unbiased estimator of $\text{Var}[\hat{\mathbf{Y}}(\mathbf{x})]$ is given by

$$\widehat{\text{Var}}[\hat{\mathbf{Y}}(\mathbf{x})] = \mathbf{z}'(\mathbf{x})(\mathbf{X}_0'\mathbf{X}_0)^{-1}\mathbf{z}(\mathbf{x})\hat{\Sigma} \tag{7.60}$$

where $\hat{\Sigma}$ is given in Eq. 7.57.

Let $\hat{\phi}_i$ be the optimum value of $\hat{Y}_i(\mathbf{x})$ optimized individually over the experimental region ($i = 1, 2, \ldots, r$), and let $\hat{\boldsymbol{\phi}} = (\hat{\phi}_1, \hat{\phi}_2, \ldots, \hat{\phi}_r)'$. If these individual optima are attained at the same set, $\mathbf{x}$, of operating conditions, then an "ideal" optimum is said to be achieved. In this case the problem of multiresponse optimization is obviously solved and no further work is needed. As was stated earlier in this section, such an ideal optimum rarely exists. In more general situations we might consider finding compromising conditions on the input variables that are somewhat favorable to all responses. By that we mean conditions under which the multiresponse function deviates as little as possible from the ideal optimum, that is, at such conditions we have a *near optimum* for each of the r predicted response functions in Eq. 7.58. Such a deviation of the compromising conditions from the ideal optimum can be formulated by means of a distance function which measures the distance of $\hat{\mathbf{Y}}(\mathbf{x})$, considered as a point in the r-dimensional Euclidean space, from $\hat{\boldsymbol{\phi}}$, the vector of individual optima. We denote this distance function by $\rho[\hat{\mathbf{Y}}(\mathbf{x}), \hat{\boldsymbol{\phi}}]$. The multiresponse optimization then involves finding conditions on $\mathbf{x}$ that minimize this distance function over the experimental region.

The distance function ρ can be chosen in a variety of ways. One possible choice is the weighted distance

$$\rho[\hat{Y}(x), \hat{\phi}] = [(\hat{Y}(x) - \hat{\phi})' \{Var[\hat{Y}(x)]\}^{-1} (\hat{Y}(x) - \hat{\phi})]^{1/2} \tag{7.61}$$

Using the estimate given in Eq. 7.60 for the variance-covariance matrix of $\hat{Y}(x)$, we get

$$\rho_1[\hat{Y}(x), \hat{\phi}] = \left[\frac{(\hat{Y}(x) - \hat{\phi})'\hat{\Sigma}^{-1}(\hat{Y}(x) - \hat{\phi})}{z'(x)(X_0'X_0)^{-1}z(x)} \right]^{1/2} \tag{7.62}$$

If the variance-covariance matrix is diagonal (this is either known or based on a statistical test), then the distance function ρ_1 can be written as

$$\rho_2[\hat{Y}(x), \hat{\phi}] = \left[\sum_{i=1}^{r} \frac{(\hat{Y}_i(x) - \hat{\phi}_i)^2}{\{\hat{\sigma}_{ii}z'(x)(X_0'X_0)^{-1}z(x)\}} \right]^{1/2} \tag{7.63}$$

where $\hat{\sigma}_{ii}$ is the ith diagonal element of $\hat{\Sigma}$ ($i = 1, 2, \ldots, r$).

Another distance function that has some intuitive appeal, particularly to those who like to consider relative changes from the individual optima, is

$$\rho_3[\hat{Y}(x), \hat{\phi}] = \left[\sum_{i=1}^{r} \frac{(\hat{Y}_i(x) - \hat{\phi}_i)^2}{\hat{\phi}_i^2} \right]^{1/2} \tag{7.64}$$

We refer to any of the aforementioned distance functions as a generalized distance. It is simply a measure of the distance of $\hat{Y}(x)$ from $\hat{\phi}$ after it has been properly scaled. If x_0 is the point in the experimental region at which $\rho[\hat{Y}(x), \hat{\phi}]$ attains its absolute minimum, and if m_0 is the value of this minimum, then at x_0 the experimental conditions can be described as being near optimal for each of the r response functions. The smaller the value of m_0, the closer these conditions are to representing an "ideal" optimum in the sense described earlier.

In the development of the generalized distance, $\rho[\hat{Y}(x), \hat{\phi}]$, no account has been taken of the fact that the elements of $\hat{\phi}$, being the individual optimum values of $\hat{Y}_1(x), \hat{Y}_2(x), \ldots, \hat{Y}_r(x)$, are random variables. Any variation associated with $\hat{\phi}$ can influence the determination of the optimum of the multiresponse function. The variability of $\hat{\phi}$ must, therefore, be taken into consideration. This can be conveniently accomplished as follows: let ζ_i be the optimum value of the true mean of the ith response, optimized individually over the experimental region ($i = 1, 2, \ldots, r$) and let $\zeta = (\zeta_1, \zeta_2, \ldots, \zeta_r)'$. If D_ζ is a confidence region for ζ, then whenever $\zeta \in D_\zeta$,

$$\rho[\hat{Y}(x), \zeta] \leq \max_{\xi \in D_\zeta} \rho[\hat{Y}(x), \xi] \tag{7.65}$$

The right-hand side of Eq. 7.65, serves as an overestimate of $\rho[\hat{\mathbf{Y}}(\mathbf{x}), \boldsymbol{\zeta}]$, the function that should be minimized rather than $\rho[\hat{\mathbf{Y}}(\mathbf{x}), \hat{\boldsymbol{\phi}}]$. Since $\boldsymbol{\zeta}$ is unknown, we adopt a conservative approach and minimize the function on the right-hand side of Eq. 7.65 over the experimental region. This function is regarded, as a generalized distance measure which accounts for the randomness of $\hat{\boldsymbol{\phi}}$.

In Khuri and Conlon (1981) it is shown that the inequalities

$$\gamma_{1i} \le \zeta_i \le \gamma_{2i} \qquad i = 1, 2, \ldots, r \tag{7.66}$$

hold simultaneously with an approximate confidence coefficient of at least $1 - \alpha^*$, where

$$\gamma_{1i} = \hat{\phi}_i - g_i(\mathbf{X}_0, \hat{\boldsymbol{\xi}}_i) \sqrt{\mathrm{MS}_i} t_{\alpha/2, N-p}$$

$$\gamma_{2i} = \hat{\phi}_i + g_i(\mathbf{X}_0, \hat{\boldsymbol{\xi}}_i) \sqrt{\mathrm{MS}_i} t_{\alpha/2, N-p} \qquad i = 1, 2, \ldots, r \tag{7.67}$$

and $\alpha^* = 1 - (1 - \alpha)^r$. In Eq. 7.67, MS_i is the error mean square for the ith response given by

$$\mathrm{MS}_i = \frac{\mathbf{Y}_i'[\mathbf{I}_N - \mathbf{X}_0(\mathbf{X}_0'\mathbf{X}_0)^{-1}\mathbf{X}_0']\mathbf{Y}_i}{N - p_0}$$

$\hat{\boldsymbol{\xi}}_i$ is the point at which $\hat{Y}_i(\mathbf{x})$ attains its individual optimum $\hat{\phi}_i$, and

$$g_i(\mathbf{X}_0, \hat{\boldsymbol{\xi}}_i) = [\mathbf{z}'(\hat{\boldsymbol{\xi}}_i)(\mathbf{X}_0'\mathbf{X}_0)^{-1}\mathbf{z}(\hat{\boldsymbol{\xi}}_i)]^{1/2} \qquad i = 1, 2, \ldots, r$$

The double inequalities in Eq. 7.66 determine a rectangular confidence region for $\boldsymbol{\zeta}$ with an approximate confidence coefficient of at least $1 - \alpha^*$. We shall consider this rectangular confidence region to be the region D_ζ mentioned earlier.

In summary, the steps to be taken to optimize a multiresponse function consisting of r responses are as follows:

1. The predicted equations in Eq. 7.58 are obtained using the multiresponse data and the method of least squares.
2. A distance measure ρ is chosen using Eqs. 7.62, 7.63, or 7.64.
3. The predicted responses are optimized individually over the experimental region R to obtain the vector $\hat{\boldsymbol{\phi}}$ of extimated individual optima. For example, if the prediction equations 7.58 are second-order models, then the method of ridge analysis described in Chapter 5 can be used for the determination of the elements of $\hat{\boldsymbol{\phi}}$. For more general polynomial models, the controlled random search optimization procedure (Price 1977 and Conlon 1985) can be used instead.
4. A rectangular confidence region D_ζ is constructed using Inequalities 7.66 and Eq. 7.67.
5. The distance function $\rho[\hat{\mathbf{Y}}(\mathbf{x}), \boldsymbol{\xi}]$ chosen in step 2 is maximized with respect to $\boldsymbol{\xi} \in D_\zeta$ for a fixed $\mathbf{x}$ in R. Here Carroll's (1961) created response surface

technique, a constrained gradient technique based on using the method of steepest ascent within specified boundaries, can be used for that purpose.

6. The maximum obtained in step 5, being a function of x alone, is minimized over R. Here again the controlled random search procedure (Price 1977 and Conlon 1985) can be employed effectively.

Note that if the variation associated with the elements of $\hat{\phi}$, the vector of individual optima, is ignored, then an approximate multiresponse optimum can be obtained by replacing steps 4−6 with a direct minimization over R of the distance measure chosen in step 2.

The computer code needed to implement the aforementioned optimization procedure is described in Conlon and Khuri (1992). A complete electronic copy of this paper, along with examples and the code, can be downloaded from the Internet at ftp://ftp.stat.ufl.edu/pub/mr.tar.Z

7.6.4 A Numerical Example

Schmidt et al. (1979) investigated the effects of cysteine (X_1) and calcium chloride (X_2) on the textural and water-holding characteristics of dialyzed whey protein concentrates (WPC) gel systems. These characteristics are measured by hardness (Y_1), cohesiveness (Y_2), springiness (Y_3), and compressible water (Y_4). We shall consider a simultaneous maximization of these four responses.

A central composite design with five center point replications was used. The design settings in the original and coded variables, and the multiresponse data, are given in Table 7.9. A second-order model was fitted to each of the four sets of response values. The regression coefficient estimates, their standard errors, and the coefficients of determination, R^2, are given in Table 7.10.

The experimental region R in the coded variables space consists of the interior and boundary of a circle of radius $\sqrt{2}$ centered at the origin. The rectangular confidence region D_ζ described in 7.66 and 7.67 with $1 - \alpha^* = 0.95$ is given in Table 7.11, which also includes the individual maxima $\hat{\phi}_1$, $\hat{\phi}_2$, $\hat{\phi}_3$, and $\hat{\phi}_4$ for the four responses along with their respective locations in the experimental region. These maxima can be easily obtained by the method of ridge analysis described in Chapter 5.

Using the distance measure ρ_1 given in Eq. 7.62, we can directly obtain simultaneous maxima for the four response functions by minimizing $\rho_1[\hat{Y}(x), \hat{\phi}]$, where $\hat{\phi} = (2.69, 0.68, 1.90, 0.72)'$, as can be seen from Table 7.11. The location at which this function attains its minimum and the corresponding simultaneous maxima for the responses are given in Table 7.12. Let us now use the more conservative distance measure given by the right-hand side of Eq. 7.65 with D_ζ as described in Table 7.11. The minimum value of this latter distance function is attained at the point $(-0.57, -1.29)$ and the corresponding simultaneous maxima are given in Table 7.12.

Table 7.9 Experimental Design and Response Values

Design (original)		Design (coded)		Responses			
X_1 (mM)*	X_2 (mM)	x_1	x_2	Y_1 (kg)	Y_2	Y_3 (mm)	Y_4 (g)
8.0	6.5	−1	−1	2.48	0.55	1.95	0.22
34.0	6.5	1	−1	0.91	0.52	1.37	0.67
8.0	25.9	−1	1	0.71	0.67	1.74	0.57
34.0	25.9	1	1	0.41	0.36	1.20	0.69
2.6	16.2	−1.414	0	2.28	0.59	1.75	0.33
39.4	16.2	1.414	0	0.35	0.31	1.13	0.67
21.0	2.5	0	−1.414	2.14	0.54	1.68	0.42
21.0	29.9	0	1.414	0.78	0.51	1.51	0.57
21.0	16.2	0	0	1.50	0.66	1.80	0.44
21.0	16.2	0	0	1.66	0.66	1.79	0.50
21.0	16.2	0	0	1.48	0.66	1.79	0.50
21.0	16.2	0	0	1.41	0.66	1.77	0.43
21.0	16.2	0	0	1.58	0.66	1.73	0.47

* milliMolar
Source: A. I. Khuri and M. Conlon (1981). Reproduced with permission of the American Statistical Association.

Table 7.10 The Least Squares Fit and Regression Estimates

	Regression Coefficients			
Model Term	Y_1	Y_2	Y_3	Y_4
constant	1.526(.069)*	0.660(.007)	1.776(.017)	0.468(.014)
x_1	−0.575(.071)	−0.092(.008)	−0.250(.018)	0.131(.014)
x_2	−0.524(.071)	−0.010(.008)	−0.078(.018)	0.073(.014)
x_1^2	−0.171(.076)	−0.096(.008)	−0.156(.019)	0.026(.015)
x_2^2	−0.098(.076)	−0.058(.008)	−0.079(.019)	0.024(.015)
$x_1 x_2$	0.318(.100)	−0.070(.011)	0.010(.025)	−0.083(.020)
Coefficient of Determination	0.95	0.98	0.98	0.95

* The number in parentheses is the standard error.
Source: A. I. Khuri and M. Conlon (1981). Reproduced with permission of the American Statistical Association.

Table 7.11 The Individual Maxima and the Region D_ζ

Response	Individual Maximum	Location of Maximum		D_ζ	
		x_1	x_2	γ_1	γ_2
$\hat{Y}_1$	2.69	−0.97	−1.05	2.16	3.22
$\hat{Y}_2$	0.68	−0.57	0.26	0.65	0.71
$\hat{Y}_3$	1.90	−0.82	−0.54	1.82	1.98
$\hat{Y}_4$	0.72	1.39	−0.35	0.61	0.83

Source: A. I. Khuri and M. Conlon (1981). Reproduced with permission of the American Statistical Association.

From Table 7.12 we note that the simultaneous maximum values obtained by minimizing ρ_1 and those obtained by minimizing the conservative distance measure are slightly different. This can be attributed to low variation associated with the individual maxima. We conclude that a combination of cysteine and calcium chloride at the levels specified by the optimal values of x_1 and x_2 (obtained by minimizing ρ_1 or the conservative distance measure) leads to a simultaneous maximization of the four response functions over the experimental region.

A visual comparison between the location of the multiresponse maximum, using the conservative measure, and the locations of the individual maxima for the four responses can be made by inspecting Figures 7.1–7.4.

Table 7.12 Simultaneous Maxima

Distance Measure		ρ_1	Conservative Distance
Minimum Distance		235.72	312.89
Simultaneous Maxima	$\hat{Y}_1$	2.47	2.54
	$\hat{Y}_2$	0.55	0.55
	$\hat{Y}_3$	1.83	1.84
	$\hat{Y}_4$	0.31	0.29
Location of Maxima	x_1	−0.46	−0.57
	x_2	−1.38	−1.29

Source: A. I. Khuri and M. Conlon (1981). Reproduced with permission of the American Statistical Association.

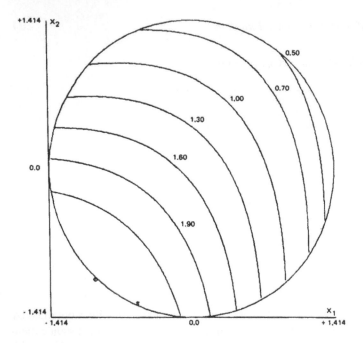

Figure 7.1 Individual (◇) versus simultaneous (∗) maxima of $\hat{Y}_1$. Source: A. I. Khuri and M. Conlon (1981). Reproduced with permission of the American Statistical Association.

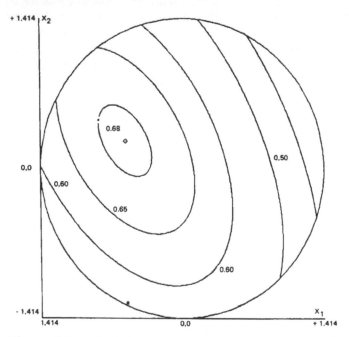

Figure 7.2 Individual (◇) versus simultaneous (∗) maxima of $\hat{Y}_2$. Source: A. I. Khuri and M. Conlon (1981). Reproduced with permission of the American Statistical Association.

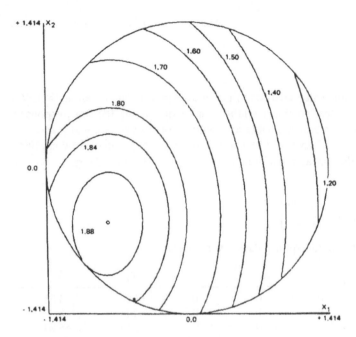

Figure 7.3 Individual ($\Diamond$) versus simultaneous (*) maxima of $\hat{Y}_3$. Source: A. I. Khuri and M. Conlon (1981). Reproduced with permission of the American Statistical Association.

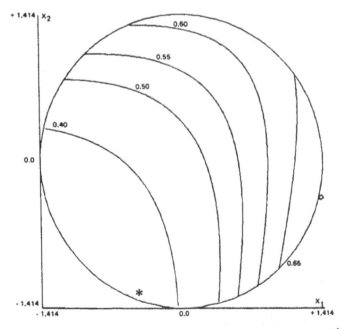

Figure 7.4 Individual ($\Diamond$) versus simultaneous (*) maxima of $\hat{Y}_4$. Source: A. I. Khuri and M. Conlon (1981). Reproduced with permission of the American Statistical Association.

EXERCISES

7.1   Apply the eigenvalue analysis to the multiresponse data given in Table 7.9 and conclude that no linear relationships exist among the four responses. Note that the data in this table are rounded off to two decimal places

7.2   Verify that Eqs. 7.25 and 7.26 can be approximately obtained by applying a nonsingular linear transformation to Eqs. 7.23 and 7.24 given by the matrix

$$Q = \begin{bmatrix} 0.922 & 0.390 \\ 0.222 & 2.180 \end{bmatrix}$$

7.3   Consider the multiresponse data given below:

x_1	x_2	x_3	x_4	Y_1	Y_2	Y_3
0	0	0	0	61.01	44.88	24.03
0	0	0	0	62.25	46.77	26.33
0	0	0	0	63.01	45.04	27.50
0	0	0	0	61.59	47.18	30.25
0	0	0	0	63.66	48.09	31.27
−1	−1	−1	−1	58.92	58.95	34.44
1	−1	−1	−1	79.32	54.36	41.70
−1	1	−1	−1	46.72	52.71	25.56
1	1	−1	−1	72.65	35.01	25.30
−1	−1	1	−1	46.32	58.92	27.41
1	−1	1	−1	65.62	53.91	32.72
−1	1	1	−1	34.19	57.02	20.01
1	1	1	−1	58.24	46.70	28.67
−1	−1	−1	1	70.66	41.82	29.95
1	−1	−1	1	85.05	30.41	24.93
−1	1	−1	1	62.01	36.77	24.70
1	1	−1	1	76.77	34.95	29.12
−1	−1	1	1	61.12	64.16	38.66
1	−1	1	1	75.11	44.30	34.20
−1	1	1	1	51.19	65.86	36.09
1	1	1	1	65.66	38.05	27.65

The fitted models in the coded variables, x_1, x_2, x_3 and x_4 are

$$Y_1 = \beta_{10} + \sum_{i=1}^{4} \beta_{1i}x_i + \beta_{114}x_1x_4 + \varepsilon_1$$

$$Y_2 = \beta_{20} + \beta_{21}x_1 + \beta_{23}x_3 + \beta_{24}x_4 + \beta_{234}x_3x_4 + \varepsilon_2$$

$$Y_3 = \beta_{30} + \beta_{32}x_2 + \beta_{314}x_1x_4 + \beta_{324}x_2x_4 + \beta_{334}x_3x_4 + \varepsilon_3$$

(a) Apply Roy's largest root test to check for lack of fit of the fitted multiresponse model. Use $\alpha = 0.10$.

(b) Check all nonempty subsets of the three responses to determine their contribution to lack of fit, if any.

7.4 Consider a dual response system consisting of a primary response $\hat{Y}_p$ and a secondary response $\hat{Y}_s$. The experimental region is given by

$$-2.0 \le x_i \le 2.0 \qquad i = 1, 2, 3$$

The predicted response models are

$$\hat{Y}_p(\mathbf{x}) = 62.01 + 8.95x_1 + 5.36x_2 + 6.29x_3 - 7.20x_1^2 - 7.75x_2^2$$
$$- 12.15x_3^2 - 12.86x_1x_2 - 18.91x_1x_3 - 13.92x_2x_3$$

$$\hat{Y}_s(\mathbf{x}) = 55.29 + 3.95x_1 + 7.23x_2 + 2.55x_3 + 5.22x_1^2 + 5.61x_2^2$$
$$+ 4.19x_3^2 + 9.12x_1x_2 + 2.31x_1x_3 + 4.02x_2x_3$$

Conduct a dual response analysis as outlined in Section 7.6.1 and determine one set of operating conditions on x_1, x_2, and x_3 which gives rise to an absolute maximum on $\hat{Y}_p$ subject to the condition that $55 \le \hat{Y}_s \le 70$.

7.5 Washing treatments were evaluated for quality improvement of minced mullet flesh. The effects of three input variables; washing temperature (X_1), washing time (X_2), and washing ratio (X_3), on four response variables were investigated. The response variables include springiness (Y_1), thiobarbituric acid (TBA) number (Y_2), percent cooking loss (Y_3), and whiteness index (Y_4). The design settings in the original and coded variables and the multiresponse data are given below

Original Input Variables						Responses			
Temp. (°C)	Washing Time (min)	Washing Ratio	Coded Input Variables			Springiness (mm)	TBA No.	Cooking Loss (%)	Whiteness Index
X_1	X_2	X_3	x_1	x_2	x_3	Y_1	Y_2	Y_3	Y_4
26.0	2.8	18.0	−1	−1	−1	1.83	29.31	29.50	50.36
40.0	2.8	18.0	1	−1	−1	1.73	39.32	19.40	48.16
26.0	8.2	18.0	−1	1	−1	1.85	25.16	25.70	50.72
40.0	8.2	18.0	1	1	−1	1.67	40.81	27.10	49.69
26.0	2.8	27.0	−1	−1	1	1.86	29.82	21.40	50.09
40.0	2.8	27.0	1	−1	1	1.77	32.20	24.00	50.61
26.0	8.2	27.0	−1	1	1	1.88	22.01	19.60	50.36
40.0	8.2	27.0	1	1	1	1.66	40.02	25.10	50.42
21.2	5.5	22.5	−1.682	0	0	1.81	33.00	24.20	29.31
44.8	5.5	22.5	1.682	0	0	1.37	51.59	30.60	50.67
33.0	1.0	22.5	0	−1.682	0	1.85	20.35	20.90	48.75

Original Input Variables						Responses			
Temp. (°C)	Washing Time (min)	Washing Ratio	Coded Input Variables			Springiness (mm)	TBA No.	Cooking Loss (%)	Whiteness Index
33.0	10.0	22.5	0	1.682	0	1.92	20.53	18.90	52.70
33.0	5.5	14.9	0	0	−1.682	1.88	23.85	23.00	50.19
33.0	5.5	30.1	0	0	1.682	1.90	20.16	21.20	50.86
33.0	5.5	22.5	0	0	0	1.89	21.72	18.50	50.84
33.0	5.5	22.5	0	0	0	1.88	21.21	18.60	50.93
33.0	5.5	22.5	0	0	0	1.87	21.55	16.80	50.98

Source: C. L. Tseo, J. C. Deng, J. A. Cornell, A. I. Khuri, and R. H. Schmidt (1983). Reprinted from *Journal of Food Science*. Copyright © by Institute of Food Technologies.

(a) Fit each response data to a complete second-order model in x_1, x_2, and x_3.

(b) Suppose it is of interest to simultaneously maximize Y_1, Y_4 and minimize Y_2 and Y_3.

    (i) Obtain the corresponding individual optima for the four responses using the method of ridge analysis

    (ii) Set up an expression for the distance function ρ_1 given in Eq. 7.62 in terms of the coded input variables.

    (iii) Use an optimization algorithm of your choice to minimize the distance function ρ_1 in (ii) over the experimental region. Obtain the location of the simultaneous multiresponse optimum and the corresponding optimum response values and compare these with those obtained under ridge analysis applied to the individual responses.

7.6 A study was conducted to estimate the relationship between weight and girth in young cattle for the purpose of predicting weight from girth, which can be accurately and easily measured. Researchers were also interested in knowing whether or not the parameters of the girth–weight relationship remained unchanged during the early growth period of cattle. The weight and girth of each of 60 calves were measured at 4, 6, and 7 weeks of age. The resulting data are given below.

Since girth is a linear measure and weight is nearly proportional to volume, weight can be modeled as a linear function of a power of girth. Thus, $Y = \log(\text{weight})$ is regressed against $x = \log(\text{girth})$ using a simple linear regression model. Let x_i and Y_i, $i = 1, 2, 3$, denote the values of x and Y corresponding to 4, 6, and 7 weeks, respectively. The model for the ith set of measurements ($i = 1, 2, 3$) is therefore of the form $Y_i = \beta_{i0} + \beta_{i1}x_i + \varepsilon_i$.

(a) Test the hypothesis

$$H_0: \beta_1 = \beta_2 = \beta_3,$$

where $\beta_i = (\beta_{i0}, \beta_{i1})'$, $i = 1, 2, 3$. Use $\alpha = 0.01$.

(b) Do we have sufficient evidence to conclude that the rate of change of log(weight) with respect to log(girth) is not the same at weeks 4, 6, and 7? Use $\alpha = 0.10$.

	Week 4		Week 6		Week 7	
Calf	Girth[a]	Weight[b]	Girth	Weight	Girth	Weight
1	81.5	100	88.5	128	91.5	136
2	82.5	102	88.5	120	92.5	142
3	86.0	120	91.5	140	93.0	150
4	89.0	111	89.5	132	93.0	148
5	83.5	98	89.0	118	92.0	124
6	84.5	110	89.5	138	92.5	146
7	79.5	94	85.5	118	90.5	124
8	83.0	110	89.0	138	93.5	146
9	80.0	94	86.5	118	89.0	128
10	83.0	96	89.0	118	89.0	126
11	84.0	104	88.5	126	92.0	138
12	82.0	114	90.0	140	92.5	144
13	80.0	91	85.5	112	87.5	122
14	86.5	111	94.0	138	95.5	150
15	84.0	104	91.0	134	92.5	144
16	80.0	100	90.0	128	91.0	140
17	79.0	89	89.0	116	91.5	124
18	82.0	100	89.0	122	90.5	136
19	84.0	109	93.0	134	92.0	144
20	85.5	118	95.5	144	93.5	152
21	80.5	105	92.0	126	89.5	134
22	85.5	112	89.0	128	95.0	140
23	79.0	94	86.0	118	89.0	132
24	74.0	80	81.5	106	84.5	120
25	80.0	100	89.0	126	90.0	138
26	87.0	115	90.5	136	95.5	150
27	82.5	116	89.0	126	95.0	142
28	86.0	121	90.5	142	94.0	158
29	85.0	99	89.5	116	91.5	130
30	79.5	86	84.5	110	85.0	116
31	83.5	102	85.5	122	89.5	140
32	87.0	108	91.0	132	95.0	148
33	81.0	94	85.0	112	92.0	131

	Week 4		Week 6		Week 7	
Calf	Girth[a]	Weight[b]	Girth	Weight	Girth	Weight
34	78.0	80	82.0	92	84.5	112
35	85.0	109	90.5	128	94.5	144
36	82.5	93	89.5	132	92.0	142
37	76.0	70	79.5	90	86.0	104
38	76.0	73	80.0	96	84.5	112
39	83.0	104	92.0	118	89.0	126
40	87.0	104	87.5	120	89.0	132
41	91.5	124	92.0	155	96.0	162
42	84.0	106	85.5	124	88.5	130
43	82.5	92	84.0	104	88.5	112
44	86.5	96	89.0	126	89.5	135
45	83.0	96	84.5	118	87.0	120
46	79.5	90	85.0	116	87.0	116
47	81.5	102	85.0	122	86.0	126
48	74.0	70	78.5	81	82.0	88
49	80.5	90	85.0	112	86.0	115
50	81.0	88	86.0	108	85.0	114
51	87.0	120	91.0	138	90.0	138
52	77.0	80	79.5	99	81.0	104
53	77.0	86	80.5	112	80.0	103
54	84.5	108	88.0	124	87.5	137
55	84.0	98	89.0	120	90.5	125
56	81.5	90	85.0	108	85.5	120
57	77.0	76	81.0	101	81.5	110
58	79.5	92	82.0	113	87.5	122
59	75.0	70	78.5	85	79.0	92
60	79.5	88	84.0	107	83.0	116

[a] Girth in centimeters.
[b] Weight in pounds.

REFERENCES AND BIBLIOGRAPHY

Belsley, D. A., E. Kuh, and R. E. Welsch (1980). *Regression Diagnostics*, New York: John Wiley.

Biles, W. E. (1975). A Response Surface Method for Experimental Optimization of Multi-Response Processes, *Ind. Eng. Chem., Process Des. Dev.*, **14**, 152–158.

Box, G. E. P. and N. R. Draper (1965). The Bayesian Estimation of Common Parameters from Several Responses, *Biometrika*, **52**, 355–365.

Box, G. E. P., W. G. Hunter, J. F. MacGregor, and J. Erjavec (1973). Some Problems Associated with the Analysis of Multiresponse Data, *Technometrics*, **15**, 33–51.

Box, M. J., N. R. Draper, and W. G. Hunter (1970). Missing Values in Multiresponse Nonlinear Model Fitting, *Technometrics*, **12**, 613–620.

Carroll, C. W. (1961). The Created Response Surface Technique for Optimizing Nonlinear, Restrained Systems, *Operations Research*, **9**, 169–185.

Conlon, M. (1985). Controlled Random Search Procedure for Function Minimization, *Technical Report No. 236*, Dept. of Statistics, Univ. of Florida, Gainesville, FL.

Conlon, M. and A. I. Khuri (1992). Multiple Response Optimization, *Technical Report*, Dept. of Statistics, University of Florida, Gainesville, FL.

Derringer, G. and R. Suich (1980). Simultaneous Optimization of Several Response Variables, *J. Qual. Technol.*, **12**, 214–219.

Draper, N. R. and W. G. Hunter (1966). Design of Experiments for Parameter Estimation in Multiresponse Situations, *Biometrika*, **53**, 525–533.

Federov, V. V. (1972). *Theory of Optimal Experiments*, New York: Academic Press.

Foster, F. G. (1957). Upper Percentage Points of the Generalized Beta Distribution II, *Biometrika*, **44**, 441–453.

Fuguitt, R. E. and J. E. Hawkins (1947). Rate of Thermal Isomerization of α-Pinene in the Liquid Phase, *J. Amer. Chem. Soc.*, **69**, 319–322.

Harrington, E. C. (1965). The Desirability Function, *Ind. Qual. Control*, **21**, 494–498.

Hill, W. J. and W. G. Hunter (1966). A Review of Response Surface Methodology: A Literature Survey, *Technometrics*, **8**, 571–590.

Kakwani, N. C. (1967). The Unbiasedness of Zellner's Seemingly Unrelated Regression Equations Estimators, *J. Amer. Statist. Assoc.*, **62**, 141–142.

Khuri, A. I. (1985). A Test for Lack of Fit of a Linear Multiresponse Model, *Technometrics*, **27**, 213–218.

Khuri, A. I. (1986). Exact Tests for the Comparison of Correlated Response Models with an Unknown Dispersion Matrix, *Technometrics*, **28**, 347–357.

Khuri, A. I. (1990a). The Effect of Response Scaling on the Detection of Linear Dependencies Among Multiresponse Data, *Metrika*, **37**, 217–231.

Khuri, A. I. (1990b). Multiresponse Rotatability, *J. Statist. Plan. Inf.*, **25**, 1–6.

Khuri, A. I. and M. Conlon (1981). Simultaneous Optimization of Multiple Responses Represented by Polynomial Regression Functions, *Technometrics*, **23**, 363–375.

Kiefer, J. (1959). Optimum Experimental Designs, *J. Roy. Statist. Soc.*, **B21**, 272–319.

Levy, M. S. and J. W. Neill (1990). Testing for Lack of Fit in Linear Multiresponse Models Based on Exact or Near Replicates, *Commun. Statist. Theor. Meth. A*, **19**, 1987–2002.

Lind, E. E., J. Goldin, and J. B. Hickman (1960). Fitting Yield and Cost Response Surfaces, *Chem. Eng. Progr.*, **56**, 62–68.

Mardia, K. V. (1971). The Effect of Nonnormality on Some Multivariate Tests and Robustness to Nonnormality in the Linear Model, *Biometrika*, **58**, 105–121.

McLean, D. D., D. J. Pritchard, D. W. Bacon, and J. Downie (1979). Singularities in Multiresponse Modelling, *Technometrics*, **21**, 291–298.

Morrison, D. F. (1976). *Multivariate Statistical Methods*, 2nd ed., New York: McGraw-Hill.

Myers, R. H. (1990). *Classical and Modern Regression with Applications*, 2nd ed., Boston: PWS-Kent.

Myers, R. H. and W. H. Carter, Jr. (1973). Response Surface Techniques for Dual Response Systems, *Technometrics*, **15**, 301–317.

Price, W. L. (1977). A Controlled Random Search Procedure for Global Optimization, *The Computer J.*, **20**, 367–370.

Richert, S. H., C. V. Morr, and C. M. Cooney (1974). Effect of Heat and Other Factors Upon Foaming Properties of Whey Protein Concentrates, *J. Food Sci.*, **39**, 42–48.

Roy, S. N., R. Gnanadesikan, and J. N. Srivastava (1971). *Analysis and Design of Certain Quantitative Multiresponse Experiments*, Oxford: Pergamon Press.

Schmidt, R. H., B. L. Illingsworth, J. C. Deng, and J. A. Cornell (1979). Multiple Regression and Response Surface Analysis of the Effects of Calcium Chloride and Cysteine on Heat-Induced Whey Protein Gelation, *J. Agric. Food Chem.*, **27**, 529–532.

Searle, S. R. (1982). *Matrix Algebra Useful for Statistics*, New York: John Wiley.

Seber, G. A. F. (1984). *Multivariate Observations*, New York: John Wiley.

Stewart, G. W. (1986). Collinearity, Scaling, and Rounding Error. In: *Proceedings of the Seventeenth Symposium on the Interface, Computer Science and Statistics* (D. M. Allen, ed.), Amsterdam: Elsevier Science Publishers, pp. 193–198.

Stewart, W. E. and J. P. Sorensen (1976). Sensitivity and Regression of Multicomponent Reactor Models. In *Fourth International Symposium on Chemical Reaction Engineering*, Frankfurt: DECHEMA, I-12–I-20.

Stewart, W. E and J. P. Sorensen (1981). Bayesian Estimation of Common Parameters from Multiresponse Data with Missing Observations, *Technometrics*, **23**, 131–141.

Tseo, C. L., J. C. Deng, J. A. Cornell, A. I. Khuri, and R. H. Schmidt (1983). Effect of Washing Treatment on Quality of Minced Mullet Flesh, *J. Food Sci.*, **48**, 163–167.

Van der Sluis, A. (1969). Condition Numbers and Equilibrium of Matrices, *Numerische Mathematik*, **14**, 14–23.

Wijesinha, M. C. (1984). Design of Experiments for Multiresponse Models, Unpublished Ph.D. Thesis, Dept. of Statistics, Univ. of Florida, Gainesville, FL.

Wijesinha, M. C. and A. I. Khuri (1987a). The Sequential Generation of Multiresponse D-Optimal Designs When the Variance-Covariance Matrix Is Not Known, *Commun. Statist. Simul. Comput. B*, **16**, 239–259.

Wijesinha, M. C. and A. I. Khuri (1987b). Construction of Optimal Designs to Increase the Power of the Multiresponse Lack of Fit Test, *J. Statist. Plan. Inf.*, **16**, 179–192.

Wijesinha, M. C. and A. I. Khuri (1991). Robust Designs for First-Order Multiple Design Multivariate Models, *Commun. Statist. Theor. Meth. A*, **20**, 2987–2999.

Wynn, H. P. (1970). The Sequential Generation of D-Optimum Experimental Designs, *Ann. Math. Statist.*, **41**, 1655–1664.

Zellner, A. (1962). An Efficient Method of Estimating Seemingly Unrelated Regressions and Tests for Aggregation Bias, *J. Amer. Statist. Assoc.*, **57**, 348–368.

8

Response Surface Models with Block Effects

8.1 INTRODUCTION

In the previous chapters, it was implicitly assumed that the experimental trials performed in a response surface investigation were carried out under homogeneous conditions. This assumption, however, may not be valid in some experimental situations. For example, batches of raw material used in a production process may be different with regard to source, composition, or other characteristics. In another situation, changes in the experimental conditions may be difficult to avoid or control, particularly if a long period of time is needed to perform the experiment. In general, heterogeneity of conditions can be caused by one or several extraneous sources of variation. In such circumstances, the experimental trials should be carried out in groups, or blocks, within each of which a greater degree of homogeneity of conditions can be maintained. This amounts to a division of the corresponding response surface design into several blocks. As a result, a block effect must be added to the fitted model. This effect is discrete, that is, does not have values on a continuous scale, as is the case with the model's input variables.

As will be seen later in this chapter, the presence of a block effect in the model can affect the estimation of the mean response, as well as the determination of optimum conditions. The block effect may be fixed or random, depending on how the blocks are set up. Both cases will be addressed in this chapter.

In order to understand the general treatment of response surface models with block effects, let us first consider the special case of a second-order model.

8.2 SECOND-ORDER MODELS WITH BLOCK EFFECTS

Suppose that the mean response depends on k input variables, $x_1, x_2, \ldots, x_k$, and is of the form

$$\eta = \beta_0 + \sum_{i=1}^{k} \beta_i x_i + \sum_{i=1}^{k} \beta_{ii} x_i^2 + \sum_{i<j}\sum^{k} \beta_{ij} x_i x_j \tag{8.1}$$

Furthermore, suppose that the experimental runs are divided into b blocks. Consider the model

$$Y_u = \beta_0 + \sum_{i=1}^{k} \beta_i x_{ui} + \sum_{i=1}^{k} \beta_{ii} x_{ui}^2 + \sum_{i<j}\sum^{k} \beta_{ij} x_{ui} x_{uj}$$
$$+ \sum_{l=1}^{b} \delta_l z_{ul} + \varepsilon_u, \qquad u = 1, 2, \ldots, N$$

where Y_u denotes the observed response value at the uth experimental run, x_{ui} is the corresponding setting of the ith (coded) input variable, δ_l denotes the effect of the lth block ($l = 1, 2, \ldots, b$), z_{ul} is a "dummy" variable taking the value 1, if the uth trial is carried out in the lth block; otherwise, it is equal to zero, and ε_u is a random error. Alternatively, this model can be written as

$$Y_u = \beta_0' + \sum_{i=1}^{k} \beta_i x_{ui} + \sum_{i=1}^{k} \beta_{ii} x_{ui}^2 + \sum_{i<j}\sum^{k} \beta_{ij} x_{ui} x_{uj}$$
$$+ \sum_{l=1}^{b} \delta_l(z_{ul} - \bar{z}_l) + \varepsilon_u \tag{8.2}$$

where $\bar{z}_l = (1/N) \sum_{u=1}^{N} z_{ul}$, $l = 1, 2, \ldots, b$, and $\beta_0' = \beta_0 + \sum_{l=1}^{b} \delta_l \bar{z}_l$.

In general, the least-squares estimates of the linear and quadratic effects (polynomial effects) in Eq. 8.2 are not independent of those of the block effects. If, however, the polynomial effects can be estimated independently of the block effects, then the design used is said to block orthogonally. In this case, the least-squares estimates of β_i, β_{ii}, and β_{ij} in Eq. 8.2 will be free of the estimates of $\delta_l(l = 1, 2, \ldots, b)$. Using Eq. 8.2, it is easy to see that the conditions for orthogonal blocking are

$$\sum_{u=1}^{N} x_{ui}(z_{ul} - \bar{z}_l) = 0, \qquad i = 1, 2, \ldots, k; \ l = 1, 2, \ldots, b \tag{8.3}$$

$$\sum_{u=1}^{N} x_{ui}x_{uj}(z_{ul} - \bar{z}_l) = 0, \qquad i, j = 1, 2, \ldots, k; \ i \neq j; \ l = 1, 2, \ldots, b \qquad (8.4)$$

$$\sum_{u=1}^{N} x_{ui}^2(z_{ul} - \bar{z}_l) = 0, \qquad i = 1, 2, \ldots, k; \ l = 1, 2, \ldots, b \qquad (8.5)$$

Since, as we shall see later, it may be desirable for a design that blocks orthogonally also to be rotatable, we shall assume that the first-order moments as well as the mixed second-order moments of the design are zero; that is,

$$\sum_{u=1}^{N} x_{ui} = 0, \qquad i = 1, 2, \ldots, k$$

$$\sum_{u=1}^{N} x_{ui}x_{uj} = 0, \qquad i, j = 1, 2, \ldots, k; \ i \neq j \qquad (8.6)$$

From Eqs. 8.3–8.6, and by the nature of the z variables, we obtain the following conditions on the x_{ui} settings for orthogonal blocking:

$$\sum_{u(l)} x_{ui} = 0, \qquad i = 1, 2, \ldots, k; \ l = 1, 2, \ldots, b \qquad (8.7)$$

$$\sum_{u(l)} x_{ui}x_{uj} = 0, \qquad i, j = 1, 2, \ldots, k; \ i \neq j; \ l = 1, 2, \ldots, b \qquad (8.8)$$

$$\frac{\sum_{u(l)} x_{ui}^2}{\sum_{u=1}^{N} x_{ui}^2} = \frac{n_l}{N}, \qquad i = 1, 2, \ldots, k; \ l = 1, 2, \ldots, b \qquad (8.9)$$

where $\sum_{u(l)}$ denotes summation extended only over those values of u in the lth block, and n_l is the number of runs in the lth block ($l = 1, 2, \ldots, b$).

Thus, the conditions for orthogonal blocking in second-order models, as based on Eqs. 8.7–8.9, can be summarized as follows:

1. Conditions 8.7 and 8.8 imply that the column arrays associated with x_1, x_1, $\ldots$, x_k are orthogonal and sum to zero within each block. Hence, *each block must consist of a first-order orthogonal design*.
2. Condition 8.9 implies that the fraction of the total sum of squares for variable x_i ($i = 1, 2, \ldots, k$) in each block must be equal to the fraction of the total number of runs allotted to that block.

The central composite design (CCD) can be made to block orthogonally. Each of the factorial and axial portions of the design forms a first-order orthogonal design. These portions provide a basis for a first division of the CCD into two blocks. The number of center point replications will have to be determined to satisfy condition 8.9. The composition of these two blocks can be described as follows:

Block 1. A factorial portion consisting of $F = 2^k$ or 2^{k-m_0} points, where m_0 is a positive integer, in addition to n_{OF} center point replications.

Block 2. An axial portion consisting of $2k$ points plus n_{OA} center point replications. From Eq. 8.9 the axial setting α must have the value

$$\alpha = \sqrt{\frac{F(2k + n_{OA})}{2(F + n_{OF})}}$$ (8.10)

Furthermore, if such a design is also required to be rotatable, the condition of which is $\alpha = F^{1/4}$ (see Eq. 4.18), then from Eq. 8.10 we must have the relationship

$$\sqrt{F} = \frac{2(F + n_{OF})}{2k + n_{OA}}$$

or, equivalently, the relationship

$$2F - \sqrt{F}(2k + n_{OA}) + 2n_{OF} = 0$$ (8.11)

We note from Eq. 8.11 that for some values of k, it is not always possible to find a rotatable CCD that blocks orthogonally. A necessary condition for the satisfaction of Eq. 8.11 is

$$(2k + n_{OA})^2 - 16n_{OF} \geq 0$$

For those values of k where Eq. 8.11 is not satisfied, we can achieve orthogonal blocking and near rotatability for a CCD. For example, when $k = 3$, $F = 8$, $n_{OF} = 4$, and $n_{OA} = 2$, we find from Eq. 8.10 that $\alpha = 1.633 \simeq 8^{1/4} = 1.68$, which is close to the rotatable setting.

Further subdivision of the axial portion into more than one block is not possible without violating the requirement that a block must consist of a first-order orthogonal design. The factorial portion, on the other hand, can be subdivided into more than one block by using fractional replications as long as the resulting fractional factorial designs are of resolution 3 or higher, and the number of center point replications is the same for all such fractional factorial blocks. The latter condition is needed for the satisfaction of condition 8.9. A list of orthogonal blocking arrangements for rotatable or near-rotatable central composite designs is given in Table 8.1.

Other second-order designs that block orthogonally are given in Box and Hunter (1957) and also in Myers (1976).

8.3 MODELS WITH FIXED BLOCK EFFECTS

As an extension of Section 8.2, let us now consider a general response surface model with an added block effect. Suppose that the mean response η is repre-

Table 8.1 Rotatable and Near-Rotatable Central Composite Designs with Orthogonal Blocking

k	2	3	4	5	5($\frac{1}{2}$rep)	6	6($\frac{1}{2}$rep)	7	7($\frac{1}{2}$rep)
Factorial blocks									
F^a	4	8	16	32	16	64	32	128	64
b_F^b	1	2	2	4	1	8	2	16	8
n_{0f}^c	3	2	2	2	6	1	4	1	1
n_f^d	7	6	10	10	22	9	20	9	9
Axial block									
n_{0i}	3	2	2	4	1	6	2	11	4
n_i	7	8	10	14	11	18	14	25	18
Value of α for orthogonal blocking	1.4142	1.6330	2.0000	2.3664	2.0000	2.8284	2.3664	3.3636	2.8284
Value of α for rotability	1.4142	1.6818	2.0000	2.3784	2.0000	2.8284	2.3784	3.3333	2.8284
N^e	14	20	30	54	33	90	54	169	80

[a]Number of points in factorial portion.
[b]Number of blocks in factorial portion.
[c]Number of center point replications in the lth block ($l = 1, 2, \ldots, b$).
[d]Total number of points in each block.
[e]Total number of points in the design.
Source: G. E. P. Box and J. S. Hunter (1957). Reproduced with permission of the Institute of Mathematical Statistics.

sented by a polynomial model of order d (≥ 1) of the form

$$\eta = \beta_0 + \mathbf{f}'(\mathbf{x})\boldsymbol{\beta} \tag{8.12}$$

where $\mathbf{x} = (x_1, x_2, \ldots, x_k)'$, $\mathbf{f}'(\mathbf{x})$ is a vector of order $1 \times p$ whose elements consist of powers and cross-products of powers of the x_i up to degree d, β_0 and $\boldsymbol{\beta} = (\beta_1, \beta_2, \ldots, \beta_p)'$ are unknown parameters.

In the presence of block effects, the observed response vector, $\mathbf{Y} = (Y_1, Y_2, \ldots, Y_N)'$, is represented by the model

$$\mathbf{Y} = \beta_0 \mathbf{1}_N + \mathbf{X}\boldsymbol{\beta} + \mathbf{Z}\boldsymbol{\delta} + \boldsymbol{\varepsilon} \tag{8.13}$$

where $\mathbf{1}_N$ is a vector of ones of order $N \times 1$, $\mathbf{X}$ is an $N \times p$ matrix whose uth row is $\mathbf{f}'(\mathbf{x}_u)$, with $\mathbf{x}_u$ being the vector of design settings at the uth experimental run ($u = 1, 2, \ldots, N$), $\boldsymbol{\delta} = (\delta_1, \delta_2, \ldots, \delta_b)'$, where δ_j denotes the effect of the jth block ($j = 1, 2, \ldots, b$), and $\mathbf{Z}$ is a block-diagonal matrix of the form

$$\mathbf{Z} = \text{diag}(\mathbf{1}_{n_1}, \mathbf{1}_{n_2}, \ldots, \mathbf{1}_{n_b})$$

where n_j is the size of the jth block ($j = 1, 2, \ldots, b$) such that $N = \sum_{j=1}^{b} n_j$. This model is a generalization of the model in Eq. 8.2. Here, the block effects are considered fixed; that is, $\delta_1, \delta_2, \ldots, \delta_b$ are unknown parameters. Such a consideration is justified whenever the blocks used in the experiment represent particular levels of an extraneous source of variation. The random error vector $\boldsymbol{\varepsilon}$ in Eq. 8.13 is assumed to have a zero mean and a variance-covariance matrix $\sigma_\varepsilon^2 \mathbf{I}_N$.

Note that the model in Eq. 8.13 is not of full rank since the columns of $\mathbf{Z}$ sum to $\mathbf{1}_N$, that is, $\mathbf{1}_N = \mathbf{Z}\mathbf{1}_b$. The model in Eq. 8.13 can therefore be written as

$$\mathbf{Y} = \mathbf{X}\boldsymbol{\beta} + \mathbf{Z}\boldsymbol{\tau} + \boldsymbol{\varepsilon} \tag{8.14}$$

where $\boldsymbol{\tau} = \beta_0 \mathbf{1}_b + \boldsymbol{\delta}$. If the columns of $\mathbf{Z}$ are linearly independent of those of $\mathbf{X}$, then model 8.14 is of full rank. Thus $\boldsymbol{\beta}$ and $\boldsymbol{\tau}$ can be uniquely estimated by using the method of ordinary least squares. It is not possible, however, to estimate β_0 independently of $\boldsymbol{\delta}$ unless a certain constraint is imposed on the elements of $\boldsymbol{\delta}$. For this purpose we assume that the block effects are constrained to sum to zero, that is,

$$\sum_{j=1}^{b} \delta_j = 0$$

In this case, β_0 is given by

$$\beta_0 = \frac{1}{b} \sum_{j=1}^{b} \tau_j$$

$$= \frac{1}{b} \mathbf{1}_b' \boldsymbol{\tau} \tag{8.15}$$

where τ_j is the jth element of τ ($j = 1, 2, \ldots, b$). The mean response in Eq. 8.12 can then be expressed as

$$\eta = \frac{1}{b} \mathbf{1}_b'\tau + \mathbf{f}'(\mathbf{x})\beta \tag{8.16}$$

Let us now write Eq. 8.14 in the form

$$\mathbf{Y} = \mathbf{W}\theta + \varepsilon \tag{8.17}$$

where $\mathbf{W} = [\mathbf{X}:\mathbf{Z}]$ and $\theta = (\beta', \tau')'$. The least-squares estimator of θ is given by $\hat{\theta} = (\hat{\beta}', \hat{\tau}')'$, where

$$\hat{\theta} = (\mathbf{W}'\mathbf{W})^{-1}\mathbf{W}'\mathbf{Y} \tag{8.18}$$

The variance-covariance matrix of $\hat{\theta}$ is

$$\mathrm{Var}(\hat{\theta}) = (\mathbf{W}'\mathbf{W})^{-1}\sigma_\varepsilon^2 \tag{8.19}$$

From Eqs. 8.16 and 8.18, the least-squares estimator of η is given by

$$\hat{\eta}(\mathbf{x}) = \mathbf{g}'(\mathbf{x})\hat{\theta} \tag{8.20}$$

where $\hat{\eta}(\mathbf{x})$ denotes the estimate of η at the point $\mathbf{x}$, and

$$\mathbf{g}'(\mathbf{x}) = \left[\mathbf{f}'(\mathbf{x}) : \frac{1}{b} \mathbf{1}_b' \right] \tag{8.21}$$

Hence, the prediction variance is equal to

$$\mathrm{Var}[\hat{\eta}(\mathbf{x})] = \mathbf{g}'(\mathbf{x})(\mathbf{W}'\mathbf{W})^{-1}\mathbf{g}(\mathbf{x})\sigma_\varepsilon^2 \tag{8.22}$$

8.3.1 Effect of Blocking on the Prediction Variance

Let $\hat{\eta}_0(\mathbf{x})$ denote the value of $\hat{\eta}(\mathbf{x})$ obtained when the block effects are zero, that is, $\delta = 0$ in Eq. 8.13. Consider the following theorem:

Theorem 8.1. For any point $\mathbf{x}$ in a region of interest R,

$$\mathrm{Var}[\hat{\eta}(\mathbf{x})] = \mathrm{Var}[\hat{\eta}_0(\mathbf{x})] + \mathbf{g}'(\mathbf{x})\mathbf{Q}\mathbf{g}(\mathbf{x})\sigma_\varepsilon^2 \tag{8.23}$$

where

$$\mathbf{Q} = (\mathbf{W}'\mathbf{W})^{-1} \mathbf{K}[\mathbf{K}'(\mathbf{W}'\mathbf{W})^{-1} \mathbf{K}]^{-1} \mathbf{K}'(\mathbf{W}'\mathbf{W})^{-1} \tag{8.24}$$

and $\mathbf{K}'$ is a matrix or order $(b - 1) \times (p + b)$ of the form $\mathbf{K}' = [\mathbf{0}:\mathbf{L}]$, where $\mathbf{0}$ is a zero matrix of order $(b - 1) \times p$ and $\mathbf{L} = [\mathbf{1}_{b-1}:-\mathbf{I}_{b-1}]$ (recall that p is the number of columns of $\mathbf{X}$ in Eq. 8.13).

The proof of Theorem 8.1 is given in Appendix 8A. Note that the matrix $\mathbf{Q}$ in Eq. 8.24 is positive semidefinite. Hence, $\mathbf{g}'(\mathbf{x})\mathbf{Q}\mathbf{g}(\mathbf{x}) \geq 0$ for all $\mathbf{x}$ in R. From

Eq. 8.23 we conclude that

$$\text{Var}[\hat{\eta}(\mathbf{x})] \geq \text{Var}[\hat{\eta}_0(\mathbf{x})] \tag{8.25}$$

for all $\mathbf{x}$ in R. This shows that the effect of blocking results in a possible increase in the prediction variance.

8.3.2 Tests of Significance Concerning the Elements of β

Let β_i denote the ith element of $\boldsymbol{\beta}$ in Eq. 8.14 ($i = 1, 2, \ldots, p$). Consider testing the hypothesis

$$H_0 : \beta_i = 0 \text{ versus } H_a : \beta_i \neq 0 \tag{8.26}$$

The Type III sum of squares for β_i is $R(\beta_i|\boldsymbol{\beta}^{(i)}, \boldsymbol{\tau})$, where $\boldsymbol{\beta}^{(i)}$ is the vector of the remaining $p - 1$ elements of $\boldsymbol{\beta}$. This particular sum of squares is the difference between the regression sum of squares for model 8.14 and the regression sum of squares for the reduced model obtained by deleting β_i. Thus,

$$R(\beta_i|\boldsymbol{\beta}^{(i)}, \boldsymbol{\tau}) = \mathbf{Y}'[\mathbf{W}(\mathbf{W}'\mathbf{W})^{-1}\mathbf{W}' - \mathbf{W}_i(\mathbf{W}_i'\mathbf{W}_i)^{-1}\mathbf{W}_i']\mathbf{Y}$$
$$i = 1, 2, \ldots, p \tag{8.27}$$

where $\mathbf{W}_i$ is the same as $\mathbf{W} = [\mathbf{X}:\mathbf{Z}]$, except that the column of $\mathbf{X}$ corresponding to β_i has been removed ($i = 1, 2, \ldots, p$). Using formula (79) in Searle (1971, p. 445), the expected value of $R(\beta_i|\boldsymbol{\beta}^{(i)}, \boldsymbol{\tau})$ can be written as

$$E[R(\beta_i|\boldsymbol{\beta}^{(i)},\boldsymbol{\tau})] = \beta_i^2\boldsymbol{\kappa}_i'[\mathbf{I}_N - \mathbf{W}_i(\mathbf{W}_i'\mathbf{W}_i)^{-1}\mathbf{W}_i']\boldsymbol{\kappa}_i + \sigma_\varepsilon^2,$$
$$i = 1, 2, \ldots, p \tag{8.28}$$

where $\boldsymbol{\kappa}_i$ denotes the column of $\mathbf{X}$ corresponding to β_i. Furthermore, under H_0, $R(\beta_i|\boldsymbol{\beta}^{(i)}, \boldsymbol{\tau})/\sigma_\varepsilon^2$ has the central chi-squared distribution with one degree of freedom, since the matrix $\mathbf{W}(\mathbf{W}'\mathbf{W})^{-1}\mathbf{W}' - \mathbf{W}_i(\mathbf{W}_i'\mathbf{W}_i)^{-1}\mathbf{W}_i'$ is idempotent of rank one (see Theorem 2 on page 57 in Searle, 1971). From Eq. 8.28 it follows that a test statistic for testing H_0 is given by

$$F = \frac{R(\beta_i|\boldsymbol{\beta}^{(i)}, \boldsymbol{\tau})}{\text{MS}_E}$$

which, under H_0, has the F-distribution with 1 and $N - p - b$ degrees of freedom. Here, MS_E is the residual mean square resulting from fitting the model in Eq. 8.14, and is equal to

$$\text{MS}_E = \frac{\mathbf{Y}'[\mathbf{I}_N - \mathbf{W}(\mathbf{W}'\mathbf{W})^{-1}\mathbf{W}']\mathbf{Y}}{N - p - b}$$

8.3.3 Determination of Optimum Conditions

To find the optimum response in the presence of block effects, we can use existing optimum-seeking techniques, which apply when $\delta = 0$ in Eq. 8.13, with some minor modifications. For example, if the model in Eq. 8.12 is of the second order ($d = 2$), then the method of ridge analysis, which was described in Section 5.7, can be used in conjunction with PROC RSREG in SAS (1989). This method is modified as follows:

1. Find the least-squares estimate of $\theta = (\beta', \tau')'$ using Eq. 8.18; then compute $\hat{\beta}_0 = (1/b)\mathbf{1}_b'\hat{\tau}$ using Eq. 8.15.
2. Compute $\mathbf{Y}^* = \mathbf{U}\hat{\zeta}$, where $\mathbf{U} = [\mathbf{1}_N : \mathbf{X}]$ and $\hat{\zeta} = (\hat{\beta}_0, \hat{\beta})'$
3. Apply PROC RSREG to the model in Eq. 8.12 using $\mathbf{Y}^*$, instead of $\mathbf{Y}$, as the vector of observations.

The optimum found in (3.) is the same as that of $\hat{\eta}(\mathbf{x})$ in Eq. 8.20 over the region of interest.

8.3.4 The Special Case of Orthogonal Blocking

We recall that in Section 8.2 block orthogonality was defined in conjunction with a second-order model. We shall now extend this definition to a model of order d (≥ 1). For this purpose, let us write the model in Eq. 8.13 as

$$\mathbf{Y} = \lambda_0 \mathbf{1}_N + \mathbf{X}\beta + \tilde{\mathbf{Z}}\delta + \varepsilon \tag{8.29}$$

where

$$\tilde{\mathbf{Z}} = \left(\mathbf{I}_N - \frac{1}{N}\mathbf{J}_N\right)\mathbf{Z}$$

$$\lambda_0 = \beta_0 + \frac{1}{N}\mathbf{1}_N'\mathbf{Z}\delta$$

Note that the elements of each column of $\tilde{\mathbf{Z}}$ sum to zero since $\tilde{\mathbf{Z}}'\mathbf{1}_N = 0$. In order for the least-squares estimates of the elements of β in Eq. 8.29 to be free of the estimate of δ, we must have $\mathbf{X}'\tilde{\mathbf{Z}} = 0$, that is,

$$\mathbf{X}'\left(\mathbf{I}_N - \frac{1}{N}\mathbf{J}_N\right)\mathbf{Z} = 0 \tag{8.30}$$

A design satisfying condition 8.30 is said to block orthogonally. This condition represents a generalization of the conditions in Eqs. 8.3–8.5 for a second-order model.

Designs that block orthogonally satisfy two interesting properties described in Theorems 8.2 and 8.3. The proofs of these theorems can be found in Khuri (1994, Appendix A).

Theorem 8.2. If the design blocks orthogonally, then the least-squares estimator, $\hat{\beta}$, of β in Eq. 8.14 is the same as that when the block effects (that is, the elements of δ in Eq. 8.13) are zero.

Theorem 8.3. If the design blocks orthogonally, then the prediction variance in Eq. 8.22 takes the form

$$\text{Var}[\hat{\eta}(\mathbf{x})] = \text{Var}[\hat{\eta}_0(\mathbf{x})] + \left(\frac{1}{b^2}\sum_{j=1}^{b}\frac{1}{n_j} - \frac{1}{N}\right)\sigma_e^2 \tag{8.31}$$

where $\hat{\eta}_0(\mathbf{x})$ is the value of $\hat{\eta}(\mathbf{x})$ when the block effects are zero and n_j is the size of the jth block ($j = 1, 2, \ldots, b$).

Note that Eq. 8.31 is a special case of Eq. 8.23 in the sense that $\mathbf{g}'(\mathbf{x})\mathbf{Q}\mathbf{g}(\mathbf{x})\sigma_e^2$ in the latter equation reduces to the second term in Eq. 8.31 when the design blocks orthogonally. In this case, the prediction variance at any point inside the region of interest exceeds $\text{Var}[\hat{\eta}_0(\mathbf{x})]$ by a constant amount that depends only on the sizes of the blocks. Furthermore, if the original design (that is, the design before it is divided into blocks) is rotatable, then $\text{Var}[\hat{\eta}_0(\mathbf{x})]$ will be constant at all points that are equidistant from the origin (design center). Consequently, from Eq. 8.31, the same is true of $\text{Var}[\hat{\eta}(\mathbf{x})]$. Thus, if a rotatable design blocks orthogonally, it remains rotatable after blocking. In other words, rotatability is preserved under orthogonal blocking.

It should be noted that even though the estimate of β in Eq. 8.13 does not depend on the estimates of the block effects when the design blocks orthogonally, the same is not true with regard to the intercept β_0. Its estimate depends on the value of $\hat{\tau}$, as can be seen from Eq. 8.15. It also depends on the type of restrictions imposed on the elements of δ.

8.3.5   A Numerical Example

A pilot plant study was conducted on a small continuous reactor. The input variables considered were

X_1 = flow rate in liters per hour
X_2 = concentration of catalyst
X_3 = temperature

and the response variable was Y = concentration of product. The basic design is a central composite with four center point replications and a replicated axial portion. The series of experimental runs were performed sequentially in four blocks. The data were originally given in Box and Draper (1987, p. 362) and are reproduced in Table 8.2. In this table, x_1, x_2, and x_3 denote the coded values of X_1, X_2, and X_3, respectively. A second-order model in x_1, x_2, and x_3 was fitted

Table 8.2 The Original Design and Response Values

Block	Experimental run	x_1	x_2	x_3	Concentration of Product Y
1	1	-1	-1	1	40.0
	2	1	-1	-1	18.6
	3	-1	1	-1	53.8
	4	1	1	1	64.2
	5	0	0	0	53.5
	6	0	0	0	52.7
2	7	-1	-1	-1	39.5
	8	1	-1	1	59.7
	9	-1	1	1	42.2
	10	1	1	-1	33.6
	11	0	0	0	54.1
	12	0	0	0	51.0
3	13	$-\sqrt{2}$	0	0	43.0
	14	$\sqrt{2}$	0	0	43.9
	15	0	$-\sqrt{2}$	0	47.0
	16	0	$\sqrt{2}$	0	62.8
	17	0	0	$-\sqrt{2}$	25.6
	18	0	0	$\sqrt{2}$	49.7
4	19	$-\sqrt{2}$	0	0	39.2
	20	$\sqrt{2}$	0	0	46.3
	21	0	$-\sqrt{2}$	0	44.9
	22	0	$\sqrt{2}$	0	58.1
	23	0	0	$-\sqrt{2}$	27.0
	24	0	0	$\sqrt{2}$	50.7

Source: G. E. P. Box and N. R. Draper (1987, p. 362). Reproduced with permission of John Wiley & Sons, Inc.

to the data. Note that the original design is rotatable and blocks orthogonally, as can be verified by applying Eqs. 8.7–8.9.

Using Eq. 8.20, the least-squares estimate of the mean response, $\eta(\mathbf{x})$, at a point $\mathbf{x}$ inside the region R (a sphere centered at the origin with a radius $\sqrt{3}$) is given by

$$\hat{\eta}(\mathbf{x}) = 51.796 + 0.745\, x_1 + 4.813\, x_2 + 8.012\, x_3 + 0.375\, x_1 x_2$$
$$+ 10.35\, x_1 x_3 - 2.825\, x_2 x_3 - 3.833\, x_1^2 + 1.217\, x_2^2 - 6.258\, x_3^2$$

$$(8.32)$$

One of the objectives of the experiment is the determination of operating conditions that give high concentrations of the product. For this purpose, the method of ridge analysis was applied to Eq. 8.32 using PROC RSREG in SAS (1989), as was described in Section 8.3.3. The maximum value of $\hat{\eta}(x)$ over the region R is equal to $\hat{\eta}_{max} = 62.49$, which is attained at the point whose coordinates are $x_1 = 0.561$, $x_2 = 1.155$, and $x_3 = 0.592$. From Eq. 8.22, the standard error of $\hat{\eta}_{max}$ in units of σ_ε is equal to 0.577.

Since the design blocks orthogonally, the prediction variance at a point x in R is given by Eq. 8.31, which, in this example takes the form

$$\text{Var}[\hat{\eta}(x)] = \text{Var}[\hat{\eta}_0(x)] + \left(\frac{1}{16} \sum_{j=1}^{4} \frac{1}{6} - \frac{1}{24} \right) \sigma_\varepsilon^2$$

$$= \text{Var}[\hat{\eta}_0(x)]$$

Thus, in this case, blocking causes no increase in the prediction variance.

In order to illustrate the effect of blocking on the estimation of the mean response, let us consider other blocking arrangements of the same 24 runs in Table 8.2. These arrangements are described in Table 8.3. Note that blocking arrangement number 3 is orthogonal, but the remaining five arrangements are

Table 8.3 Division of the Experimental Runs in Table 8.2 for the Blocking Arrangements

Blocking arrangement	Block 1	Block 2	Block 3	Block 4	Block 5
1	1, 2, 3, 4	7, 8, 9, 10	5, 6, 13, 14 15, 16, 17, 18	11, 12, 19, 20 21, 22, 23, 24	
2	1, 2, 3, 4	5, 6, 7, 8 9, 10, 11, 12	13, 14, 15, 16 17, 18	19, 20, 21 22, 23, 24	
3	1, 2, 3, 4 5, 6, 7, 8 9, 10, 11, 12	13, 14, 15 16, 17, 18	19, 20, 21 22, 23, 24		
4	1, 2, 5 6, 11, 12	3, 4, 7 8, 9, 10	13, 14, 15 16, 17, 18	19, 20, 21 22, 23, 24	
5	1, 2, 3	4, 5, 6	7, 8, 9 10, 11, 12 13, 14, 15	16, 17, 18 19, 20, 21 22, 23, 24	
6	1, 2, 3, 4 13	5, 10, 16, 18 23	6, 9, 11, 19 20	7, 8, 12, 15 24	14, 17, 21, 22

Source: A. I. Khuri (1994). Reproduced with permission of Journals Oxford Ltd.

Table 8.4 Least-Squares Estimates of the Parameters, $\hat{\eta}_{max}$, and Locations of Maxima for the Blocking Arrangements

	Blocking Arrangement					
	1	2	3	4	5	6
$\hat{\beta}_0$	54.369	51.963	51.453	51.111	52.883	51.979
$\hat{\beta}_1$	0.745	0.745	0.745	0.745	0.912	1.122
$\hat{\beta}_2$	4.813	4.813	4.813	4.984	5.141	4.799
$\hat{\beta}_3$	8.012	8.012	8.012	8.012	8.180	7.735
$\hat{\beta}_{12}$	0.375	0.375	0.375	0.375	0.710	0.428
$\hat{\beta}_{13}$	10.350	10.350	10.350	10.692	10.685	10.389
$\hat{\beta}_{23}$	−2.825	−2.825	−2.825	−2.825	−2.490	−2.878
$\hat{\beta}_{11}$	−4.863	−3.900	−3.833	−3.491	−4.395	−4.259
$\hat{\beta}_{22}$	0.188	1.150	1.217	1.559	0.882	1.386
$\hat{\beta}_{33}$	−7.288	−6.325	−6.258	−5.916	−6.366	−6.436
$\hat{\eta}_{max}$	63.005	62.524	62.147	62.805	64.076	62.790
x_1	0.561	0.561	0.561	0.591	0.623	0.477
x_2	1.155	1.155	1.155	1.126	1.078	1.237
x_3	0.592	0.592	0.592	0.619	0.671	0.492

Source: A. I. Khuri (1994). Reproduced with permission of Journals Oxford Ltd.

not. For each blocking arrangement, the least squares estimates of the parameters in the model of Eq. 8.1, where $k = 3$, are obtained using Eq. 8.18 and the same corresponding response values in Table 8.2. The results are shown in Table 8.4, which also contains the maximum values of $\hat{\eta}(\mathbf{x})$ over R, and the coordinates of the points at which these maxima are attained.

From Table 8.4 we note that except for β_0, the parameter estimates for arrangement number 3, which blocks orthogonally, are the same as those in Eq. 8.32. However, the same is not true for the remaining five arrangements. We also note that the values of $\hat{\eta}_{max}$ range from 62.147 to 64.076, an interesting difference considering that the same response values (from Table 8.2) were used to compute these maxima.

We conclude that the choice of the blocking arrangement for a response surface design can have a considerable effect on the estimation of the mean response. Since blocking results, in general, in an increase in the prediction variance, a decrease in the precision associated with estimating the mean response is expected.

8.4 MODELS WITH RANDOM BLOCK EFFECTS

There are experimental situations in which the block effects should be treated as random rather than fixed. For example, the raw material used in a production process may be obtained in batches selected at random over a period of time from the warehouse supply. Also, in the semiconductor industry, measurements may be taken on the resistance in computer chips using silicon wafers drawn from a large lot. In such situations, it would be appropriate to consider the effects of batches and wafers as random.

Let us again consider estimating the mean response given in Eq. 8.12. Suppose that the experimental runs were performed using b randomly selected blocks. In this case, δ, in the model of Eq. 8.13, is a random vector. We assume that it is distributed as $N(0, \sigma_\delta^2 I_b)$ independently of ε, which has the normal distribution $N(0, \sigma_\varepsilon^2 I_N)$. Since β is a vector of fixed unknown parameters, the model in Eq. 8.13 is called a *mixed-effects* model, or just a mixed model. The mean η and variance-covariance matrix Σ of Y in this model are given by

$$\eta = \beta_0 1_N + X\beta$$
$$= U\zeta, \tag{8.33}$$
$$\Sigma = \sigma_\delta^2 ZZ' + \sigma_\varepsilon^2 I_N$$
$$= \sigma_\varepsilon^2 A \tag{8.34}$$

where, if we recall, $U = [1_N : X]$, $\zeta = (\beta_0, \beta')'$, and

$$A = \text{diag}(A_1, A_2, \ldots, A_b) \tag{8.35}$$

where A_i is the matrix

$$A_i = \gamma J_{n_i} + I_{n_i}, \qquad i = 1, 2, \ldots, b \tag{8.36}$$

and γ is the ratio

$$\gamma = \frac{\sigma_\delta^2}{\sigma_\varepsilon^2} \tag{8.37}$$

8.4.1 Estimation of the Fixed Effects

By fixed effects we mean the elements of ζ in Eq. 8.33. If γ in Eq. 8.37 is known, then from Eqs. 8.33 and 8.34, the best linear unbiased estimator (BLUE) of ζ is obtained by the method of *generalized least squares*, and is of the form

$$\hat{\zeta}_g = (U' A^{-1} U)^{-1} U' A^{-1} Y \tag{8.38}$$

where the subscript g for $\hat{\zeta}_g$ represents the generalized least squares estimate. The variance-variance matrix of $\hat{\zeta}_g$ is

$$\text{Var}(\hat{\zeta}_g) = (U' A^{-1} U)^{-1} \sigma_\varepsilon^2 \tag{8.39}$$

In general, γ is unknown and should therefore be estimated by finding suitable estimates of the variance components, σ_δ^2 and σ_ε^2. Several methods for estimating variance components are available in the statistical literature (see, for example, Khuri and Sahai, 1985; Searle et al., 1992). The main ones include *analysis of variance* (ANOVA), also known as *Henderson's Method 3*, *minimum norm quadratic unbiased estimation* (MINQUE), *maximum likelihood* (ML), *and restricted maximum likelihood* (REML). The first two methods may produce negative estimates, but do not require the normality assumption. The latter two methods are more frequently used since they produce nonnegative estimates that have interesting asymptotic properties, but require the assumption of normality on the random effects. Furthermore, the values of these latter estimates can be easily obtained by using *PROC MIXED* in SAS (1992). A comprehensive coverage of ML and REML estimation techniques can be found in Searle et al. (1992, Chap. 6).

Utilizing either ML or REML estimation, let $\hat{\sigma}_\delta^2$ and $\hat{\sigma}_\varepsilon^2$ denote estimates of σ_δ^2 and σ_ε^2, respectively. An estimate of γ is given by $\hat{\gamma} = \hat{\sigma}_\delta^2/\hat{\sigma}_\varepsilon^2$. Using $\hat{\gamma}$ instead of γ in Eq. 8.38 yields the following estimate of ζ:

$$\hat{\zeta}_{ge} = (\mathbf{U}'\hat{\mathbf{A}}^{-1}\mathbf{U})^{-1}\mathbf{U}'\hat{\mathbf{A}}^{-1}\mathbf{Y} \tag{8.40}$$

where $\hat{\mathbf{A}}$ is obtained from Eqs. 8.35 and 8.36 by substituting $\hat{\gamma}$ for γ. We refer to $\hat{\zeta}_{ge}$ as the *estimated generalized least-squares* (EGLS) estimator of ζ. This estimator is no longer the BLUE of ζ. Kackar and Harville (1981) showed that it is still unbiased for ζ provided that its expectation is finite. From Eq. 8.39, an approximate expression for the estimated variance-covariance matrix of $\hat{\zeta}_{ge}$ is

$$\widehat{Var}(\hat{\zeta}_{ge}) = (\mathbf{U}'\hat{\mathbf{A}}^{-1}\mathbf{U})^{-1}\hat{\sigma}_\varepsilon^2 \tag{8.41}$$

Using Eq. 8.12, an estimate of the mean response (or predicted response) at a point $\mathbf{x}$ in the region of interest is given by

$$\hat{\eta}_{ge}(\mathbf{x}) = \mathbf{h}'(\mathbf{x})\hat{\zeta}_{ge} \tag{8.42}$$

where $\mathbf{h}'(\mathbf{x}) = [1, \mathbf{f}'(\mathbf{x})]$. Hence, the estimated prediction variance is approximately of the form

$$\widehat{Var}[\hat{\eta}_{ge}(\mathbf{x})] = \mathbf{h}'(\mathbf{x})(\mathbf{U}'\hat{\mathbf{A}}^{-1}\mathbf{U})^{-1}\mathbf{h}(\mathbf{x})\hat{\sigma}_\varepsilon^2 \tag{8.43}$$

8.4.2 Tests Concerning the Fixed Effects

Tests concerning the elements of $\zeta = (\beta_0, \boldsymbol{\beta}')'$ in the model of Eq. 8.33 are carried out using approximate t statistics. For example, a test statistic for testing the null hypothesis $H_0 : \beta_i = 0$ versus $H_a : \beta_i \neq 0$ of Eq. 8.26 is given by

$$t_i = \frac{\hat{\beta}_{ige}}{s_{ige}}, \qquad i = 1, 2, \ldots, p \tag{8.44}$$

where $\hat{\beta}_{ige}$ is the ith element of $\hat{\beta}_{ge}$, the EGLS estimator of β, which can be obtained from Eq. 8.40, and s_{ige} is the square root of the corresponding diagonal element of $\widehat{Var}(\hat{\zeta}_{ge})$ in Eq. 8.41 ($i = 1, 2, \ldots, p$). Under H_0, t_i has approximately the t-distribution with ν_i degrees of freedom, which must be estimated. Several methods are available for estimating ν_i. Some of these methods can be accessed by using PROC MIXED in SAS (1992, pp. 327–328).

8.4.3   Test Concerning the Block Effect

Let us now consider testing the following hypothesis concerning the random block effect:

$$H_0 : \sigma_\delta^2 = 0 \text{ versus } H_a : \sigma_\delta^2 \neq 0$$

The test is based on using $R(\delta|\zeta)$, the Type III sum of squares for δ in the model of Eq. 8.13, which is equal to

$$R(\delta|\zeta) = \mathbf{Y}'[\mathbf{W}(\mathbf{W}'\mathbf{W})^{-1}\mathbf{W}' - \mathbf{U}(\mathbf{U}'\mathbf{U})^{-1}\mathbf{U}']\mathbf{Y}$$

where, if we recall, $\mathbf{W} = [\mathbf{X} : \mathbf{Z}]$. The expected value of $R(\delta|\zeta)$ is given by (see formula 79 in Searle, 1971, p. 445)

$$E[R(\delta|\zeta)] = \sigma_\delta^2 \, tr\{\mathbf{Z}'[\mathbf{I}_N - \mathbf{U}(\mathbf{U}'\mathbf{U})^{-1}\mathbf{U}']\mathbf{Z}\} + (b - 1)\sigma_e^2$$

It follows that a test statistic for testing H_0 is given by the ratio

$$F = \frac{R(\delta|\zeta)}{(b - 1)\text{MS}_E} \tag{8.45}$$

where MS_E is the residual mean square used in Section 8.3.2. Under H_0, this statistic has the F-distribution with $b - 1$ and $N - p - b$ degrees of freedom. Note that this same test can be used when the block effect is fixed.

8.4.4   The Effect of Orthogonal Blocking

As in Section 8.3, the case of orthogonal blocking, described in Section 8.3.4, has certain desirable features. More specifically, let $\hat{\beta}_g$ denote the portion of $\hat{\zeta}_g$ corresponding to β in Eq. 8.38. If the orthogonal blocking condition in Eq. 8.30 is satisfied, then $\hat{\beta}_g$ is the same as $\hat{\beta}$, which is obtained from Eq. 8.18. In other words, the generalized least-squares estimator of β is the same as the estimator obtained when the block effect is fixed. The proof of this result can be found in Khuri (1992, Appendix A). Furthermore, by Theorem 8.2, $\hat{\beta}_g$ is also equal to the estimator obtained when the block effect is zero. Thus, when the design blocks orthogonally, the three estimators of β, namely, the generalized least-squares, the one that treats the block effect as fixed, and the one that ignores the block effect completely, are all equal. Therefore, there is no need to estimate the variance components or have knowledge of them in this case. As a result, all t tests concerning the fixed effects will be exact (see Eq. 8.44).

8.4.5 Example

In the manufacture of an 11-component adhesive for bonding galvanized steel bars together, temperature (X_1) and curing time (X_2) are believed to affect the shear strength of the bonding. An experiment was conducted to study the effects of these two factors using a 3^2 factorial design. The chosen levels of X_1 were 375°F, 400°F, and 450°F, and those of X_2 were 30, 35, and 40 seconds.

Aliquots of galvanized steel were selected on each of 12 dates (July 11, 16, 20; August 7, 8, 14, 20, 22; September 11, 24; October 3, 10) from the warehouse supply. Each batch of steel aliquots was selected at random on a given day from the supply on hand. The data are given in Table 8.5. In this table, x_1 and x_2 denote the coded values of X_1 and X_2, respectively, such that

$$x_1 = \frac{X_1 - 400}{25}$$

$$x_2 = \frac{X_2 - 35}{5}$$

Replications were taken at the origin (in the coded space) to obtain an estimate of pure error and to test for lack of fit (a complete second-order model in x_1 and x_2 was fitted to the data). These replications were taken at the very start (July 11) and then on August 7, September 11, and October 3. The latter three dates were chosen because each one of them represented the first time of sampling in the corresponding month. Since the numbers of replications are not the same, the design does not block orthogonally. This can be easily verified by checking that condition 8.9 is not satisfied.

The EGLS estimates of the coefficients of x_1, x_2, x_1^2, x_2^2, $x_1 x_2$ in the model, their corresponding standard errors, and test statistics values (see the formula in Eq. 8.44) are given in Table 8.6. We note that all the tests are significant except for the one concerning $x_1 x_2$. On the other hand, the test concerning the random block effect (see formula 8.45) has the value $F = 3.86$ with 11 and 101 degrees of freedom. The corresponding level of significance is 0.0001. This indicates a significant amount of variation among batches of steel aliquots.

8.5 MODELS WITH MIXED EFFECTS

Let us once more consider the mean response model given in Eq. 8.12. Suppose that in addition to the fixed polynomial effects, we have a random effect, as in Section 8.4, whose levels are a random sample from a much larger population. In the present section, we go one step further by assuming that there are also

Table 8.5 Design Settings and Shear Strength Values (psi)

| Coded settings[a] | | Block (batch) | | | | | | | | | | | |
Temperature x_1	Time x_2	1 (July 11)	2 (July 16)	3 (July 20)	4 (Aug. 7)	5 (Aug. 8)	6 (Aug. 14)	7 (Aug. 20)	8 (Aug. 22)	9 (Sep. 11)	10 (Sep. 24)	11 (Oct. 3)	12 (Oct. 10)
-1	-1	1,226	1,075	1,172	1,213	1,282	1,142	1,281	1,305	1,091	1,281	1,305	1,207
0	-1	1,898	1,790	1,804	1,961	1,940	1,699	1,833	1,774	1,588	1,992	2,011	1,742
2	-1	2,142	1,843	2,061	2,184	2,095	1,935	2,116	2,133	1,913	2,213	2,192	1,995
-1	0	1,472	1,121	1,506	1,606	1,572	1,608	1,502	1,580	1,343	1,691	1,584	1,486
0	0	2,010, 1,882	2,175	2,279	2,450, 2,355	2,291	2,374	2,471	2,393	2,205, 2,268	2,142	2,052, 2,032	2,339
		1,915, 2,106			2,420, 2,240					2,103		2,190	
2	0	2,352	2,274	2,168	2,298	2,147	2,413	2,430	2,440	2,093	2,208	2,201	2,216
-1	1	1,491	1,691	1,707	1,882	1,741	1,846	1,645	1,688	1,582	1,692	1,744	1,751
0	1	2,078	2,513	2,392	2,531	2,366	2,392	2,392	2,413	2,392	2,488	2,392	2,390
2	1	2,531	2,588	2,617	2,609	2,431	2,408	2,517	2,604	2,477	2,601	2,588	2,572

[a]The original design settings of x_1, corresponding to -1, 0, 2 are 375°F, 400°F, and 450°F, respectively; those for x_2 corresponding to -1, 0, 1 are 30, 35, and 40 sec, respectively.

Source: A. I. Khuri (1992). Reproduced with permission of the American Statistical Association.

Table 8.6 EGLS Estimates of the Fixed-Effects Parameters (Example 8.4.5)

Variable	Parameter Estimate	Standard Error	t-Ratio	P-Value
x_1	477.90	15.56	30.71	0.0001
x_2	255.90	13.82	18.52	0.0001
x_1^2	-204.80	11.10	-18.45	0.0001
x_2^2	-45.48	21.75	-2.09	0.039
$x_1 x_2$	-4.83	10.71	-0.45	0.653

Source: A. I. Khuri (1992). Reproduced with permission of the American Statistical Association.

interaction effects involving the random effect and the fixed polynomial effects. We thus have the model

$$\mathbf{Y} = \beta_0 \mathbf{1}_N + \mathbf{X}\boldsymbol{\beta} + \mathbf{Z}\boldsymbol{\delta} + \sum_{i=1}^{p} \mathbf{U}_i \boldsymbol{\Delta}_i + \boldsymbol{\varepsilon} \tag{8.46}$$

which is an extension of the model considered in Section 8.4. Here, β_0, $\mathbf{X}$, $\mathbf{Z}$, and $\boldsymbol{\beta}$ are the same as in Eq. 8.13; $\boldsymbol{\delta}$ is random and has the normal distribution $N(\mathbf{0}, \sigma_\delta^2 \mathbf{I}_b)$, where b is the number of levels (blocks, for example) of the random effect. The matrix $\mathbf{U}_i$ is of order $N \times b$ whose jth column is obtained by multiplying the elements of the ith column of $\mathbf{X}$ with the corresponding elements of the jth column of $\mathbf{Z}$ ($i = 1, 2, \ldots, p; j = 1, 2, \ldots, b$) (recall that p is the number of columns of $\mathbf{X}$). Furthermore, $\boldsymbol{\Delta}_i$ is a vector of interaction coefficients associated with the ith column of $\mathbf{X}$. This vector is random and is assumed to have the normal distribution $N(\mathbf{0}, \sigma_i^2 \mathbf{I}_b)$, $i = 1, 2, \ldots, p$. All random effects in the model are independent of one another and of the error vector $\boldsymbol{\varepsilon}$, which has the normal distribution $N(\mathbf{0}, \sigma_\varepsilon^2 \mathbf{I}_N)$. Thus the variance components associated with the mixed model in Eq. 8.46 are $\sigma_\delta^2, \sigma_1^2, \sigma_2^2, \ldots, \sigma_p^2, \sigma_\varepsilon^2$.

On the basis of the preceding assumptions, the mean response vector, $\boldsymbol{\eta}$, is given by Eq. 8.33, and the variance-covariance matrix, $\boldsymbol{\Sigma}$, of $\mathbf{Y}$ now has the form

$$\boldsymbol{\Sigma} = \sigma_\delta^2 \mathbf{Z}\mathbf{Z}' + \sum_{i=1}^{p} \sigma_i^2 \mathbf{U}_i \mathbf{U}_i' + \sigma_\varepsilon^2 \mathbf{I}_N \tag{8.47}$$

We note that, in this case, the elements of $\mathbf{Y}$ are statistically dependent since $\boldsymbol{\Sigma}$ is not diagonal, and their variances and covariances are functions of the design settings of the input variables through $\mathbf{U}_1, \mathbf{U}_2, \ldots, \mathbf{U}_p$. By contrast, the elements of $\boldsymbol{\Sigma}$ in Section 8.4 do not depend on the input variables (see Eq. 8.34). Hence, the presence of interaction effects in the model causes the process variance to not have the same value at all the experimental runs. This feature of the mixed model in Eq. 8.46 represents a marked departure from the traditional assumption of the constant response variance, which is usually made in a typical response surface exploration.

Let $\phi_i = \sigma_i^2/\sigma_\varepsilon^2$, $i = 1, 2, \ldots, p$. If $\phi_1, \phi_2, \ldots, \phi_p$, and γ are known, where γ is defined in Eq. 8.37, then from Eq. 8.47, the BLUE of ζ in Eq. 8.33 is given by

$$\hat{\zeta}_g^* = (\mathbf{U}'\boldsymbol{\Gamma}^{-1}\mathbf{U})^{-1}\mathbf{U}'\boldsymbol{\Gamma}^{-1}\mathbf{Y} \tag{8.48}$$

where

$$\boldsymbol{\Gamma} = \frac{1}{\sigma_\varepsilon^2}\boldsymbol{\Sigma}$$

$$= \gamma\mathbf{Z}\mathbf{Z}' + \sum_{i=1}^{p}\phi_i\mathbf{U}_i\mathbf{U}_i' + \mathbf{I}_n$$

The variance-covariance matrix of $\hat{\zeta}_g^*$ is

$$\mathrm{Var}(\hat{\zeta}_g^*) = (\mathbf{U}'\boldsymbol{\Gamma}^{-1}\mathbf{U})^{-1}\sigma_\varepsilon^2 \tag{8.49}$$

In general, the variance components, $\sigma_\delta^2, \sigma_1^2, \sigma_2^2, \ldots, \sigma_p^2, \sigma_\varepsilon^2$, are unknown. In this case, their ML or REML estimates can be used (see Section 8.4.1) in place of the true variance components in Eq. 8.48. Let such estimates be denoted by $\hat{\sigma}_\delta^2, \hat{\sigma}_1^2, \hat{\sigma}_2^2, \ldots, \hat{\sigma}_p^2, \hat{\sigma}_\varepsilon^2$, respectively. From Eq. 8.48 we obtain the following EGLS estimator of ζ:

$$\hat{\zeta}_{ge}^* = (\mathbf{U}'\hat{\boldsymbol{\Gamma}}^{-1}\mathbf{U})^{-1}\mathbf{U}'\hat{\boldsymbol{\Gamma}}^{-1}\mathbf{Y} \tag{8.50}$$

where

$$\hat{\boldsymbol{\Gamma}} = \hat{\gamma}\mathbf{Z}\mathbf{Z}' + \sum_{i=1}^{p}\hat{\phi}_i\mathbf{U}_i\mathbf{U}_i' + \mathbf{I}_N \tag{8.51}$$

and $\hat{\gamma} = \hat{\sigma}_\delta^2/\hat{\sigma}_\varepsilon^2$, $\hat{\phi}_i = \hat{\sigma}_i^2/\hat{\sigma}_\varepsilon^2$, $i = 1, 2, \ldots, p$. An approximate expression for the estimated variance-covariance matrix of $\hat{\zeta}_{ge}^*$ is given by (see Eq. 8.49)

$$\widehat{\mathrm{Var}}(\hat{\zeta}_{ge}^*) = (\mathbf{U}'\hat{\boldsymbol{\Gamma}}^{-1}\mathbf{U})^{-1}\hat{\sigma}_\varepsilon^2 \tag{8.52}$$

Using Eq. 8.12, an estimate of the mean response, $\eta(\mathbf{x})$, at a point $\mathbf{x}$ in a region of interest is

$$\hat{\eta}_{ge}^*(\mathbf{x}) = \mathbf{h}'(\mathbf{x})\hat{\zeta}_{ge}^* \tag{8.53}$$

where, if we recall, $\mathbf{h}'(\mathbf{x}) = [1, \mathbf{f}'(\mathbf{x})]$. This is similar to the estimate of $\eta(\mathbf{x})$ obtained in Section 8.4 (see Eq. 8.42). Accordingly, the estimated prediction variance is approximately of the form

$$\widehat{\mathrm{Var}}[\hat{\eta}_{ge}^*(\mathbf{x})] = \mathbf{h}'(\mathbf{x})(\mathbf{U}'\hat{\boldsymbol{\Gamma}}^{-1}\mathbf{U})^{-1}\mathbf{h}(\mathbf{x})\hat{\sigma}_\varepsilon^2 \tag{8.54}$$

Tests concerning the fixed effects, that is, the elements of ζ can be derived as in Section 8.4.2 by using approximate t statistics (see Eq. 8.44). Such tests are easily executable in PROC MIXED in SAS (1992). As for testing the significance of the random effects, F statistics based on Type III sums of squares can be obtained as in Section 8.4.3. For example, to test the hypothesis $H_0: \sigma_\delta^2 = 0$ versus $H_a: \sigma_\delta^2 \neq 0$, the test statistic is

$$F_\delta = \frac{R(\delta | \zeta, \Delta_1, \Delta_2, \ldots, \Delta_p)}{(b - 1)\text{MS}_E}$$

where MS_E is the residual mean square for model 8.46. The numerator of F_δ is the Type III sum of squares of δ with respect to model 8.46 (it is the difference between the regression sum of squares for this model and the regression sum of squares resulting from fitting a reduced model containing only the fixed effects and the interaction effects). Under H_0, F_δ has the F-distribution with $b - 1$ and $\nu = N - b - pb$ degrees of freedom. Similarly, the test statistic for the hypothesis, $H_{0i}: \sigma_i^2 = 0$ versus $H_{ai}: \sigma_i^2 \neq 0$, is

$$F_i = \frac{R(\Delta_i | \zeta, \delta, \Delta_1, \ldots, \Delta_{i-1}, \Delta_{i+1}, \ldots, \Delta_p)}{(b - 1)\text{MS}_E}, \qquad i = 1, 2, \ldots, p$$

which, under H_{0i}, has also the F-distribution with $b - 1$ and ν degrees of freedom.

8.5.1 Response Optimization

In Section 8.3.3, optimum conditions on the input variables were determined under the assumption of a fixed block effect. In that case, the process variance was constant. However, as was pointed out earlier in Section 8.5, in a mixed model such as the one in Eq. 8.46, the variance-covariance matrix depends on the settings of the input variables, and hence on the chosen design (see Eq. 8.47). This lack of homogeneity in the process variance should be taken into consideration when searching for the optimum since it affects the size of the prediction variance (see Eq. 8.54). Obviously, it would be desirable for the latter variance to be small throughout the region of experimentation. Optimization should therefore be carried out under certain constraints on the prediction variance.

Let us consider the expressions for the predicted response and the prediction variance given in Eqs. 8.53 and 8.54, respectively. The largest value taken by the prediction variance at the N points of the chosen design is denoted by ω_N. The minimum value of this variance over a region of interest, R, is denoted by Ψ_R. Thus, $\Psi_R \leq \omega_N$. Let us divide the interval (Ψ_R, ω_N) into m subintervals denoted by $C_1, C_2, \ldots, C_m$.

Now, in order to arrive at optimum operating conditions with "acceptable" values for the prediction variance, $\hat{\eta}_{ge}^*(\mathbf{x})$ in Eq. 8.53 can be optimized over R subject to the constraint that the prediction variance in Eq. 8.54 falls within C_i $(i = 1, 2, \ldots, m)$. The m optimal values of $\hat{\eta}_{ge}^*(\mathbf{x})$ obtained in this manner will present the experimenter with several options. A reasonable strategy in making such a choice is to decide on a proper trade-off between achieving a high value of the predicted response and keeping its variance at an acceptably low level. Thus the choice of m depends on how small the C_i intervals are desired to be; that is, it depends on the degree of precision of estimating the optimum response.

8.5.2 A Numerical Example

Let us consider again the example described in Section 8.4.5. The fitted model is the same as before, that is, a complete second-order model in x_1 and x_2. In addition, five interaction effects between the random batch effect and x_1, x_2, x_1^2, x_2^2, x_1x_2 are introduced into the model. Thus there are seven variance components associated with model 8.46, namely, σ_δ^2, σ_1^2, σ_2^2, σ_{11}^2, σ_{22}^2, σ_{12}^2, and σ_e^2. The middle five correspond to the interaction effects involving x_1, x_2, x_1^2, x_2^2, and x_1x_2, respectively.

Testing the significance of the six random effects, batch effect and five interaction effects, produced the following results: The variance components, σ_δ^2 and σ_2^2, are significantly different from zero with P-values 0.0014 and 0.0778, respectively. There is therefore a significant batch effect and a significant interaction with x_2. The tests for the variance components σ_{11}^2 and σ_{22}^2 have P-values equal to 0.4045 and 0.1312, respectively, which indicate a mild interaction with x_2^2. The variance components, σ_1^2 and σ_{12}^2, are nonsignificant with P-values 0.6822 and 0.9730, respectively. The REML estimates of the seven variance components are $\hat{\sigma}_\delta^2 = 4119.82$, $\hat{\sigma}_1^2 = 0.0$, $\hat{\sigma}_2^2 = 1712.09$, $\hat{\sigma}_{11}^2 = 61.46$, $\hat{\sigma}_{22}^2 = 1012.04$, $\hat{\sigma}_{12}^2 = 0.0$, and $\hat{\sigma}_e^2 = 11020.41$. Using these estimates in Eq. 8.51, we obtain

$$\hat{\Gamma} = 0.374\mathbf{ZZ}' + 0.155\mathbf{U}_2\mathbf{U}_2' + 0.006\mathbf{U}_{11}\mathbf{U}_{11}' + 0.092\mathbf{U}_{22}\mathbf{U}_{22}' + \mathbf{I}_N$$

where $N = 118$ and $\mathbf{U}_2$, $\mathbf{U}_{11}$, and $\mathbf{U}_{22}$ are the $\mathbf{U}$ matrices in Eq. 8.51 corresponding to σ_2^2, σ_{11}^2, and σ_{22}^2, respectively.

Since the interaction effects due to x_2, and to a lesser extent x_2^2, are significant, we conclude that the setting of x_2 (curing time) has a more pronounced effect on the amount of variation in the shear strength values than x_1 (temperature).

The EGLS estimates of the fixed effects and corresponding estimated standard errors are computed using Eqs. 8.50 and 8.52, respectively. The corresponding tests of significance are carried out by using approximate t tests (see Eq. 8.44). The results, which are shown in Table 8.7, were obtained by applying

Table 8.7 EGLS Estimates of the Fixed-Effects Parameters (Example 8.5.2)

Parameter	EGLS Estimate	Standard Error	t-Ratio	P-Value
β_0	2189.743	26.665	82.12	0.0000
β_1	478.305	14.431	33.14	0.0000
β_2	255.887	17.512	14.61	0.0000
β_{11}	-205.213	10.552	-19.45	0.0000
β_{22}	-46.379	22.186	-2.09	0.0403
β_{12}	-4.827	9.920	-0.49	0.6281

Source: A. I. Khuri (1996). Reproduced with permission of the American Society for Quality Control.

PROC MIXED in SAS (1992) to the data set in Table 8.5. For the reader's benefit, the SAS code used in this example is given in Appendix 8B.

From Table 8.7 we note that the parameters β_1, β_2, β_{11}, and β_{22} are significantly different from zero, but β_{12} is nonsignificant. The prediction equation is

$$\hat{\eta}_{ge}^{*}(\mathbf{x}) = 2189.743 + 478.305x_1 + 255.887x_2$$
$$- 205.213x_1^2 - 46.379x_2^2 - 4.827x_1x_2 \qquad (8.55)$$

Let us now proceed to determine optimum conditions on x_1 and x_2 that maximize $\hat{\eta}_{ge}^{*}(\mathbf{x})$ in Eq. 8.55 subject to the constraints described in Section 8.5.1. Using formula 8.54, the maximum of the estimated prediction variance at the 118 design points in Table 8.5 is $\omega_{118} = 1430.10$. The minimum of the same function over R is $\Psi_R = 628.56$. Here, R is the region $-1 \leq x_1 \leq 2$, $-1 \leq x_2 \leq 1$. The interval (628.56, 1430.10) is divided into $m = 17$ subintervals as shown in Table 8.8. For each subinterval, the maximum value of $\hat{\eta}_{ge}^{*}(\mathbf{x})$ is obtained over R subject to the constraint that the prediction variance falls within the bounds of the subinterval. The results are shown in Table 8.8.

With the help of Table 8.8 we can now make a subjective judgment regarding the proper trade-off between a high prediction value and a pleasingly small prediction variance. In light of this strategy, we find that for $x_1 = 1.3$ and $x_2 = 0.625$, the maximum response value is 2602.62 with a prediction variance 799.86, which falls inside the subinterval (750, 800). This represents a reasonable compromise. Another possible set of operating conditions is given by $x_1 = 1.4$ and $x_2 = 0.4$ with a maximum response 2549.38 and a prediction variance 749.77. The latter value falls inside the subinterval (700, 750). These compromise operating conditions can also be seen by examining the contour plots of $\hat{\eta}_{ge}^{*}(\mathbf{x})$ in Figure 8.1 along with the contour plots of the prediction variance in Figure 8.2.

Table 8.8 Maxima of the Predicted Response with Bounds on the Prediction Variance

Bounds on Prediction Variance	Maximum Prediction	Location of Maximum	
		x_1	x_2
(628.56, 650)	2149.04	−0.275	0.450
(650, 700)	2283.47	−0.025	0.450
(700, 750)	2549.38	1.400	0.400
(750, 800)	2602.62	1.300	0.625
(800, 850)	2620.91	1.150	0.700
(850, 900)	2634.55	1.150	0.775
(900, 950)	2643.36	1.150	0.825
(950, 1000)	2651.93	1.150	0.875
(1000, 1050)	2656.14	1.150	0.900
(1050, 1100)	2664.36	1.150	0.950
(1100, 1150)	2668.39	1.150	0.975
(1150, 1200)	2672.36	1.150	1.000
(1200, 1250)	2635.92	1.575	1.000
(1250, 1300)	2599.37	1.750	1.000
(1300, 1350)	2572.84	1.850	1.000
(1350, 1400)	2550.25	1.925	1.000
(1400, 1430.10)	2533.91	1.975	1.000

Source: A. I. Khuri (1996). Reproduced with permission of the American Society for Quality Control.

Appendix 8A   PROOF OF THEOREM 8.1

If $\delta = 0$, then $\tau = \beta_0 \mathbf{1}_b$ in Eq. 8.14. Thus, $\mathbf{L}\tau = 0$, where $\mathbf{L} = [\mathbf{1}_{b-1} : -\mathbf{I}_{b-1}]$, since the elements of τ are equal. Equivalently, if $\mathbf{K}$ is a matrix such that $\mathbf{K}' = [0:\mathbf{L}]$, where 0 is a zero matrix of order $(b - 1) \times p$, then

$$\mathbf{K}'\theta = 0 \tag{8A.1}$$

Let $\hat{\theta}_0$ denote the least-squares estimator of θ in Eq. 8.17 under the restriction given by Eq. 8A.1. Using a result in Searle (1971, formula 103, p. 113), this estimator is of the form

$$\hat{\theta}_0 = [\mathbf{I}_{p+b} - \mathbf{S}]\hat{\theta} \tag{8A.2}$$

where

$$\mathbf{S} = (\mathbf{W}'\mathbf{W})^{-1}\mathbf{K}[\mathbf{K}'(\mathbf{W}'\mathbf{W})^{-1}\mathbf{K}]^{-1}\mathbf{K}'$$

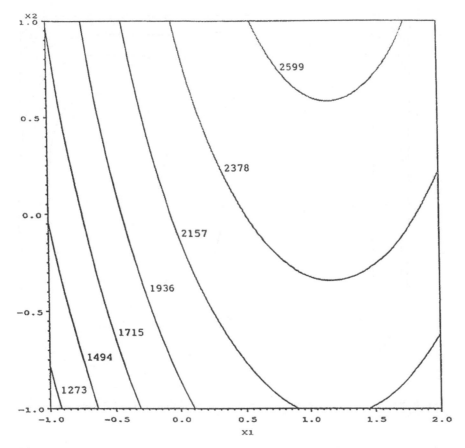

Figure 8.1 Contour plots of the predicted response. *Source*: A. I. Khuri (1996). Reproduced with permission of the American Society for Quality Control.

Hence, if $\hat{\eta}_0(\mathbf{x})$ denotes the value of $\hat{\eta}(\mathbf{x})$ obtained when $\boldsymbol{\delta} = \mathbf{0}$, then from Eq. 8.20,

$$\hat{\eta}_0(\mathbf{x}) = \mathbf{g}'(\mathbf{x})\hat{\boldsymbol{\theta}}_0$$

Using Eqs. 8.19 and 8A.2, the variance-covariance matrix of $\hat{\eta}_0(\mathbf{x})$ can be written as

$$\text{Var}[\hat{\eta}_0(\mathbf{x})] = \mathbf{g}'(\mathbf{x})(\mathbf{I}_{p+b} - \mathbf{S})(\mathbf{W}'\mathbf{W})^{-1}(\mathbf{I}_{p+b} - \mathbf{S})'\mathbf{g}(\mathbf{x})\sigma_\varepsilon^2$$

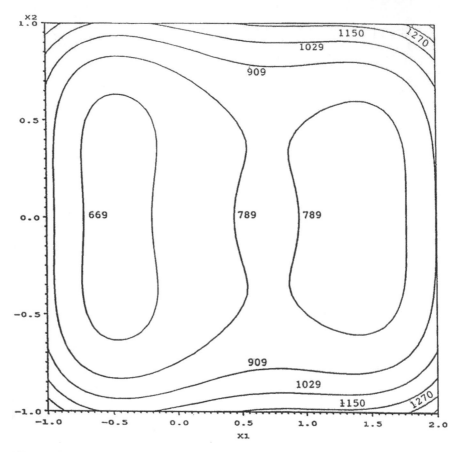

Figure 8.2 Contour plots of the prediction variance. *Source*: A. I. Khuri (1996). Reproduced with permission of the American Society for Quality Control.

It is easy to show that

$$\mathrm{Var}[\hat{\eta}_0(\mathbf{x})] = \mathbf{g}'(\mathbf{x})(\mathbf{W}'\mathbf{W})^{-1}\mathbf{g}(\mathbf{x})\sigma_\varepsilon^2 - \mathbf{g}'(\mathbf{x})\mathbf{Q}\mathbf{g}(\mathbf{x})\sigma_\varepsilon^2 \qquad (8\mathrm{A}.3)$$

where

$$\mathbf{Q} = (\mathbf{W}'\mathbf{W})^{-1}\mathbf{K}[\mathbf{K}'(\mathbf{W}'\mathbf{W})^{-1}\mathbf{K}]^{-1}\mathbf{K}'(\mathbf{W}'\mathbf{W})^{-1}$$

From Eqs. 8.22 and 8A.3, we conclude that

$$\mathrm{Var}[\hat{\eta}(\mathbf{x})] = \mathrm{Var}[\hat{\eta}_0(\mathbf{x})] + \mathbf{g}'(\mathbf{x})\mathbf{Q}\mathbf{g}(\mathbf{x})\sigma_\varepsilon^2$$

Appendix 8B   THE SAS CODE FOR EXAMPLE 8.5.2

```
DATA;
INPUT BATCH X1 X2 Y;
CARDS;
(include here the data given in Table 8.5)
PROC GLM;
CLASS BATCH;
MODEL Y = X1   X2   X1*X2   X1*X1   X2*X2   BATCH   X1*BATCH
X2*BATCH   X1*X2*BATCH   X1*X1*BATCH   X2*X2*BATCH;
PROC MIXED METHOD=REML;
CLASS BATCH;
MODEL Y = X1   X2   X1*X2   X1*X1   X2*X2/S;
RANDOM  BATCH  X1*BATCH  X2*BATCH   X1*X2*BATCH
X1*X1*BATCH   X2*X2*BATCH;
RUN;
```

Notes

1. The three PROC GLM statements are needed to obtain the Type III F-ratios for testing the significance of all the random effects in the MODEL statement.
2. The PROC MIXED code is given by the last four statements before the RUN statement. Note that in PROC MIXED the model contains only the fixed effects, which, in our case, make up the polynomial portion of the response surface model. The RANDOM statement specifies all effects that are to be treated as random in the model. These include the random batch effect and its interactions with x_1, x_2, $x_1 x_2$, x_1^2, and x_2^2. Note that on the basis of the Type III F-ratios (obtained from the PROC GLM output) for the random effects, the interactions with x_1 and $x_1 x_2$ are very insignificant and their REML estimates are equal to zero. These two interaction terms can therefore be dropped from the RANDOM statement.
3. PROC MIXED produced the results shown in Table 8.7 as well as the values of the REML estimates of the random effects' variance components.
4. The S option in the MODEL statement (in PROC MIXED) is needed to obtain the EGLS estimates of the parameters in Table 8.7.

EXERCISES

8.1 Verify that the Box-Behnken design described in Section 4.5.2 (Example 4.1) blocks orthogonally.
8.2 Consider the central composite design used in the example described in Section 4.6, along with the data given in Table 4.7. Does this design block

orthogonally? If not, can you increase the number of center point replications so that it does? If so, specify the number of such replications in each block.

8.3   Construct a design for fitting a second-order model in $k = 3$ input variables so that it contains three blocks, it blocks orthogonally, and each block contains seven design points.

8.4   Verify that the design shown in Table 8.2 blocks orthogonally. Is it rotatable as a blocked design?

8.5   Consider blocking arrangement number 6 in Table 8.3 along with the corresponding response values from Table 8.2

   (a)   Find the least-squares estimates of the parameters of the fitted second-order model and verify that they are equal to those shown in Table 8.4.

   (b)   Verify that the corresponding maximum predicted response is equal to 62.79. What is the value of the estimated prediction variance at the location of this maximum?

8.6   Consider the test concerning the block effect in Section 8.4.3. Show that under the null hypothesis $H_0: \sigma_\delta^2 = 0$, $R(\delta|\zeta)/\sigma_e^2$ has the chi-squared distribution with $b - 1$ degrees of freedom.

8.7   An experiment was conducted to determine the effects of storage conditions on the quality of a certain variety of apples. Apples were harvested from four different orchards, which were selected at random from a large agricultural district. The input variables considered were X_1 = number of weeks in storage after harvest and X_2 = storage temperature (°C). The quality of apples was measured by the response variable Y = amount of extractable juice (ml/100g). The experimental runs were carried out according to a 4×4 factorial design using apples from orchards 1, 2, and 3. However, only a 3×4 factorial design was used in orchard 4. The original design settings and the corresponding response values are given in the following table.

		Y			
X_1	X_2	Orchard 1	Orchard 2	Orchard 3	Orchard 4
1	0	74.1	73.9	73.5	75.0
2	0	73.0	75.2	75.5	70.1
3	0	74.5	74.5	72.9	75.9
4	0	72.5	73.5	72.5	
1	4	75.0	73.5	74.5	76.1
2	4	74.0	73.1	75.1	71.5
3	4	72.1	74.5	74.2	75.5

4	4	73.2	74.0	73.3	
1	15	72.0	74.2	71.1	71.9
2	15	70.5	71.2	71.9	68.9
3	15	69.5	71.9	69.1	69.1
4	15	70.7	72.5	68.3	
1	25	74.2	72.3	72.5	68.5
2	25	69.9	72.9	71.5	65.1
3	25	70.1	67.8	66.1	66.3
4	25	65.3	66.9	65.9	

(a) Suppose that the mean response is represented by the second-order model

$$\eta = \beta_0 + \beta_1 x_1 + \beta_2 x_2 + \beta_{11} x_1^2 + \beta_{22} x_2^2 + \beta_{12} x_1 x_2,$$

where x_1 and x_2 are the coded values of X_1 and X_2 given by

$$x_1 = \frac{X_1 - 2.4}{1.083}$$

$$x_2 = \frac{X_2 - 11.0}{9.772}$$

Obtain estimates of the parameters in this model using the given data. (Hint: Consider a mixed model with interaction effects).

(b) Test the significance of the fixed effects in the model.

(c) Test the significance of the random effects in the model.

(d) Use the fitted model in (a) to determine optimal values of X_1 and X_2 that yield maximum estimated amount of extractable juice over the experimental region.

REFERENCES AND BIBLIOGRAPHY

Box, G. E. P. and N. R. Draper (1987). *Empirical Model-Building and Response Surfaces*, New York: John Wiley.

Box, G. E. P. and J. S. Hunter (1957). Multi-Factor Experimental Designs for Exploring Response Surfaces, *Ann. Math. Statist.*, 28, 195–241.

Kackar, R. N. and D. A. Harville (1981). Unbiasedness of Two-Stage Estimation and Prediction Procedures for Mixed Linear Models, *Commun. Statist. Theor. Meth. A*, 10, 1249–1261.

Khuri, A. I. (1992). Response Surface Models with Random Block Effects, *Technometrics*, 34, 26–37.

Khuri, A. I. (1994). Effect of Blocking on the Estimation of a Response Surface, *J. Appl. Statist.*, 21, 305–316.

Khuri, A. I. (1996). Response Surface Models with Mixed Effects, *J. Quality Tech.*, 28 (in press).

Khuri, A. I. and H. Sahai (1985). Variance Components Analysis: A Selective Literature Survey, *Inter. Statist. Rev.*, 53, 279–300.

Myers, R. H. (1976). *Response Surface Methodology*, Blacksburg, VA: Author.

SAS Institute, Inc. (1989). *SAS/STAT User's Guide*, Vol. 2, Version 6, 4th ed., Cary, NC: Author.

SAS Institute, Inc. (1992). *SAS Technical Report P-229, SAS/STAT Software* (Release 6.07), Cary, NC: Author.

Searle, S. R. (1971). *Linear Models*, New York: John Wiley.

Searle, S. R., G. Casella, and C. E. McCulloch (1992). *Variance Components*, New York: John Wiley.

9

Mixture Designs and Analyses

9.1 INTRODUCTION

In the previous chapters experimental designs and the analyses of experimental data were presented for situations where the levels of each factor (input variable) were independent of the levels of all other variables. For example, in the peanut yield experiment discussed in Section 1.3.6, the input variables were two separate fertilizers and the levels used were the amounts (lbs per acre) $X_i = 100$, 200, 300, $i = 1$ and 2 of each. The nine experimental treatment combinations consisted of the pairs of values $(X_1, X_2) = (100, 100)$, $(100, 200)$, $(100, 300)$, ..., $(300, 300)$, ranging in the amounts from 200 to 900 pounds per acre.

In mixture experiments, the factors are ingredients of a mixture, and their levels are not independent. For example, consider two chemical pesticides (P_1 and P_2), each of which could be used for killing mites on strawberry plants by spraying the pesticides on the plants. Suppose one wished to determine if combinations of P_1 and P_2 exist that are more effective than either pesticide by itself. To remove the effect of the amount of pesticide applied, the experiment consists of tank-mixing five different combinations of P_1 and P_2 where all mixes or blends have the same volume. Expressed in percentages of this fixed volume, the five mixes are $(P_1, P_2) = (90\%, 10\%)$, $(75\%, 25\%)$, $(50\%, 50\%)$, $(25\%, 75\%)$, $(10\%, 90\%)$. Then, if the same amount of each of the five different blends is sprayed on five separate groups of plants, where each group of plants possesses the same number of mites, and after a specific period of time the percentage of mites killed is noted to differ from one group (or blend) to the next, we would infer that the mixtures differ in their effectiveness for killing the mites. The difference in the effectiveness of the five blends of P_1 and P_2 is a function only of the relative percentages of P_1 and P_2 in the blends and is not a function of the amount of pesticide applied. This is because for each blend the same amount of pesticide is sprayed on each group of plants. The dependency among the levels of the two pesticides is seen by recognizing that when increasing the percentage of P_1 while holding the volume of the tank mix fixed, the percentage of P_2 (or amount of P_2, in this case) decreases. Mixture experiments in which the total amount is varied in addition to varying the ingredient percentages are called *mixture–amount* experiments. Mixture–amount experiments are discussed in Piepel and Cornell (1987).

In mixture experiments, the measured response, such as the percentage of mites killed, is assumed to depend only on the relative proportions of the ingredients or components present in the mixture. Clearly, the main distinction between mixture experiments and independent-variable (temperature, time, etc.) experiments is that with the former, the input variables or components are nonnegative proportionate amounts of the mixture. For mixture experiments, if we let x_i represent the proportion of the ith component in the mixture where the number of components is q, then

$$x_i \geq 0 \qquad i = 1, 2, \ldots, q \tag{9.1}$$

and

$$\sum_{i=1}^{q} x_i = x_1 + x_2 + \cdots + x_q = 1.0 \tag{9.2}$$

The constraints in Eqs. 9.1 and 9.2 on the value of x_i make mixture experiments different from factorial experiments, for example, where in the latter the coded variables are often defined to take on integer values such as -1 and $+1$.

The experimental region or factor space of interest, which is defined by the values of x_i, is a regular $(q - 1)$-dimensional simplex. Since the proportions sum to unity as shown in Eq. 9.2, the values of the x_i are constrained proportions, and altering the proportion of one component in a mixture will cause a change in the proportion of at least one other component in the experimental region. For $q = 2$ components, the factor space is a straight line, for three components, $q = 3$, an equilateral triangle, and for four components, the factor space is a tetrahedron. Figure 9.1 displays the factor spaces for $q = 2$ and $q = 3$.

The coordinate system for mixture proportions is called a *simplex coordinate system*. With three components, for example, the vertices of the triangle represent single-component mixtures and are denoted by $x_i = 1$, $x_j = x_k = 0$ for $i, j, k = 1, 2,$ and 3, $i \neq j \neq k$, for which we would write $(x_1, x_2, x_3) = (1, 0, 0)$,

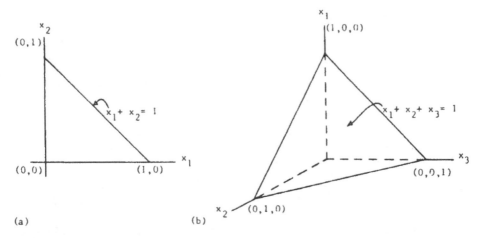

Figure 9.1 Simplex factor spaces for (a) $q = 2$ components, and (b) $q = 3$ components.

(0, 1, 0), (0, 0, 1). The interior points of the triangle represent mixtures where all of the component proportions are nonzero, that is, $x_1 > 0$, $x_2 > 0$, and $x_3 > 0$. The centroid of the triangle corresponds to the mixture with equal proportions $(\frac{1}{3}, \frac{1}{3}, \frac{1}{3})$ from each of the components.

In mixture problems, the purpose of the experimental program is to model the blending surface with some form of mathematical equation so that:

1. Predictions of the response for *any* mixture or combination of the ingredients can be made empirically, or
2. Some measure of the influence on the response of each component singly and in combination with the other components can be obtained.

We shall now discuss experimental designs for investigating the response surface over the entire simplex region. A natural choice for such a design is one with points that are positioned uniformly over the simplex factor space. A class of designs having this uniform coverage property is the $\{q, m\}$ simplex lattices.

9.2 THE SIMPLEX LATTICE ARRANGEMENTS AND THEIR ASSOCIATED MODELS

A $\{q, m\}$ simplex-lattice design for q components consists of points defined by the following coordinate settings: the proportions assumed by each component take the $m + 1$ equally spaced values from 0 to 1,

$$x_i = 0, \frac{1}{m}, \frac{2}{m}, \ldots, 1 \qquad i = 1, 2, \ldots, q \tag{9.3}$$

and all possible combinations (mixtures) of the components are considered, using the proportions in Eq. 9.3 for each component.

For a $q = 3$ component system, suppose each component can take the proportions $x_i = 0, \frac{1}{2}$, and 1, for $i = 1, 2$, and 3, which is the same as setting $m = 2$ in Eq. 9.3. Then the $\{3, 2\}$ simplex lattice consists of the six points on the boundary of the triangular factor space,

$$(x_1, x_2, x_3) = (1, 0, 0), (0, 1, 0), (0, 0, 1), \left(\frac{1}{2}, \frac{1}{2}, 0\right), \left(\frac{1}{2}, 0, \frac{1}{2}\right), \left(0, \frac{1}{2}, \frac{1}{2}\right)$$

The three vertices $(1, 0, 0)$, $(0, 1, 0)$, and $(0, 0, 1)$ represent the individual components, while the points $(\frac{1}{2}, \frac{1}{2}, 0)$, $(\frac{1}{2}, 0, \frac{1}{2})$, and $(0, \frac{1}{2}, \frac{1}{2})$ represent binary blends or two-component mixtures and are located at the midpoints of the three sides of the triangle. The $\{3, 2\}$ simplex lattice is shown in Figure 9.2

As another example, let the number of equally spaced levels (or proportions) for each component be four, that is $x_i = 0, \frac{1}{3}, \frac{2}{3}, 1$. Then all possible blends of the three components with these proportions are defined by the $\{3, m = 3\}$ simplex lattice and are

$$(x_1, x_2, x_3) = (1, 0, 0), (0, 1, 0), (0, 0, 1), \left(\frac{2}{3}, \frac{1}{3}, 0\right), \left(\frac{2}{3}, 0, \frac{1}{3}\right), \left(\frac{1}{3}, \frac{2}{3}, 0\right),$$

$$\left(\frac{1}{3}, 0, \frac{2}{3}\right), \left(\frac{1}{3}, \frac{1}{3}, \frac{1}{3}\right), \left(0, \frac{2}{3}, \frac{1}{3}\right), \left(0, \frac{1}{3}, \frac{2}{3}\right)$$

Each of the component proportions in every blend or mixture is a fractional number, and the sum of the fractions equals unity. When plotted as a lattice, these points form an array that is symmetrical with the respect to the vertices and sides of the simplex. The 10 points of a $\{3, 3\}$ simplex lattice are displayed in Figure 9.2, while Table 9.1 lists the number of points in a $\{q, m\}$ simplex lattice for values of q and m from $3 \leq q \leq 10$, $1 \leq m \leq 4$. The number of component combinations that belong to the $\{q, m\}$ simplex lattice is

$$\binom{q + m - 1}{m} = \frac{(q + m - 1)!}{m!(q - 1)!}$$

where $m!$ is "m factorial" and $m! = m(m - 1)(m - 2) \ldots (2)(1)$. For example, the $\{3, 2\}$ simplex lattice contains six points since

$$\binom{3 + 2 - 1}{2} = \frac{4!}{2!2!} = 6$$

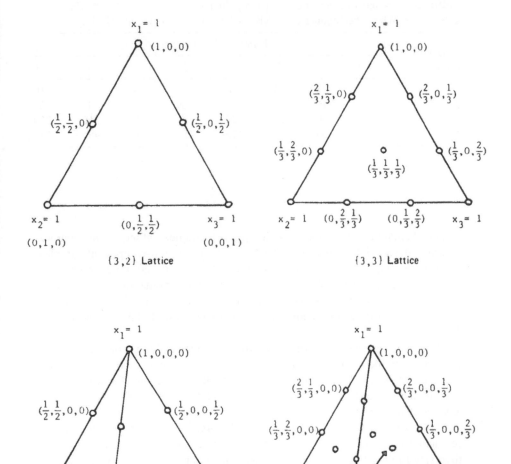

Figure 9.2 Some $\{3, m\}$ and $\{4, m\}$ simplex-lattice designs, $m = 2$ and 3, and the simplex coordinate values.

Table 9.1 Number of Points in the $\{q, m\}$ Simplex Lattice for $3 \leq q \leq 10$, $1 \leq m \leq 4$, Where the Number of Levels for Each Component is $m + 1$.

Degree of Model m	Number of Components							
	$q = 3$	4	5	6	7	8	9	10
1	3	4	5	6	7	8	9	10
2	6	10	15	21	28	36	45	55
3	10	20	35	56	84	120	165	220
4	15	35	70	126	210	330	495	715

An alternative arrangement to the $\{q, m\}$ simplex lattice is the simplex-centroid design introduced by Scheffé (1963). In a q-component simplex-centroid design, the number of points is $2^q - 1$. The design points correspond to the q permutations of $(1, 0, 0, \ldots, 0)$, the $\binom{q}{2}$ permutations of $(\frac{1}{2}, \frac{1}{2}, 0, 0, \ldots, 0)$, the $\binom{q}{3}$ permutations of $(\frac{1}{3}, \frac{1}{3}, \frac{1}{3}, 0, \ldots 0)$, $\ldots$, and the centroid point $(1/q, 1/q, \ldots, 1/q)$. A four-component simplex-centroid design consists of $2^4 - 1 = 15$ points.

Besides experimental regions, mixture experiments differ from the ordinary regression problems also in the form of the polynomial model to be fitted. Scheffé (1958; 1963) introduced the canonical polynomials for use with the simplex lattices and simplex-centroid designs, but these models can be fitted to any set of mixture data.

The canonical form of the mixture polynomial is derived by applying the restriction $x_1 + x_2 + \cdots + x_q = 1$ to the terms in the standard polynomial and then simplifying. For example, with two components, x_1 and x_2, where $x_1 + x_2 = 1$, the first-order polynomial is

$$\eta = \beta_0 + \beta_1 x_1 + \beta_2 x_2$$

Replacing β_0 by $\beta_0(x_1 + x_2 = 1)$ in η,

$$\eta = (\beta_0 + \beta_1)x_1 + (\beta_0 + \beta_2)x_2$$
$$= \beta_1' x_1 + \beta_2' x_2$$

so that the constant term β_0 is removed from the model. For the second-order polynomial in x_1 and x_2, we also replace x_1^2 by $x_1(1 - x_2)$ and x_2^2 by $x_2(1 - x_1)$

to get

$$\eta = \beta_0(x_1 + x_2) + \beta_1 x_1 + \beta_2 x_2 + \beta_{12} x_1 x_2$$
$$+ \beta_{11} x_1(1 - x_2) + \beta_{22} x_2(1 - x_1)$$
$$= (\beta_0 + \beta_1 + \beta_{11})x_1 + (\beta_0 + \beta_2 + \beta_{22})x_2$$
$$+ (\beta_{12} - \beta_{11} - \beta_{22})x_1 x_2$$
$$= \beta_1' x_1 + \beta_2' x_2 + \beta_{12}' x_1 x_2$$

so that the quadratic terms $\beta_{11} x_1^2$ and $\beta_{22} x_2^2$ are also removed from the model along with the constant term β_0. Thus the mixture models have fewer numbers of terms than the standard polynomials.

In general, the canonical forms of the mixture models (with the primes removed from the β_i) are:

Linear.

$$\eta = \sum_{i=1}^{q} \beta_i x_i \qquad (9.4)$$

Quadratic.

$$\eta = \sum_{i=1}^{q} \beta_i x_i + \sum_{i<j}^{q}\sum \beta_{ij} x_i x_j \qquad (9.5)$$

Full Cubic.

$$\eta = \sum_{i=1}^{q} \beta_i x_i + \sum_{i<j}^{q}\sum \beta_{ij} x_i x_j + \sum_{i<j}^{q}\sum \delta_{ij} x_i x_j (x_i - x_j)$$
$$+ \sum_{i<j<k}^{q}\sum\sum \beta_{ijk} x_i x_j x_k \qquad (9.6)$$

Special cubic.

$$\eta = \sum_{i=1}^{q} \beta_i x_i + \sum_{i<j}^{q}\sum \beta_{ij} x_i x_j + \sum_{i<j<k}^{q}\sum\sum \beta_{ijk} x_i x_j x_k \qquad (9.7)$$

The terms in the canonical polynomial models have simple interpretations. Geometrically, in Eqs. 9.4 through 9.7, the parameter β_i represents the expected response to the pure mixture $x_i = 1$, $x_j = 0$, $j \neq i$, and defines the height of the mixture surface at the simplex vertex denoted by $x_i = 1$. The portion $\sum_{i=1}^{q} \beta_i x_i$ of each of the models is called the *linear blending portion*, and if the blending of the components is strictly additive, then Eq. 9.4 is an appropriate representation of the surface. Figure 9.3 depicts a planar surface above the three component triangle where $\beta_2 > \beta_1 > \beta_3$.

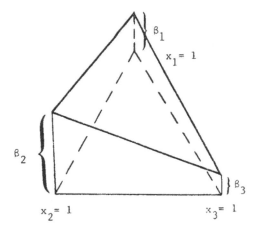

Figure 9.3   First-degree planar surface above the three-component triangle.

When there is curvature in the shape of the mixture surface owing to non-linear (often called synergistic or antagonistic) blending between pairs of components, the parameters β_{ij} in Eqs. 9.5–9.7 represent deviations of the surface from the plane defined by Eq. 9.4. Illustrated in Figure 9.4 is the deviation $\beta_{12}/4$ representing nonlinear blending of components 1 and 2 for the three component system. Higher-order terms (such as $\beta_{123}x_1x_2x_3$) in Eqs. 9.6 and 9.7 describe additional perturbations of the response surface beyond those described by first- and second-order terms.

When data are collected only at the points of the $\{q, m\}$ simplex lattice (similarly at the points of the simplex-centroid design), the estimates of the coefficients in the canonical polynomials are simple functions of the observed values of the response. This is because the number of terms in the respective models 9.4–9.7 is equal to the number of points in the corresponding lattice design. To show this, suppose we refer to the $\{3, 2\}$ simplex lattice of Figure 9.2 and define $\overline{Y}_i$ to be the average of r_i replicate observations collected at $x_i = 1$, $x_j = 0$, $i \neq j$, $i, j = 1, 2, 3$. Further, let $\overline{Y}_{ij}$ be the average of r_{ij} observations collected at the 50%:50% binary mixture ($x_i = \frac{1}{2}$, $x_j = \frac{1}{2}$, $x_k = 0$, for all $i < j < k = 1, 2, 3$) of components i and j. Then the estimates of the coefficients in model 9.5 for $q = 3$, which is the appropriate second-order ($m = 2$) model to be fitted to the $\{3, 2\}$ simplex lattice, are calculated using

$$b_i = \overline{Y}_i \qquad \text{and} \qquad b_{ij} = 4\overline{Y}_{ij} - 2(\overline{Y}_i + \overline{Y}_j) \qquad i, j \underset{i<j}{=} 1, 2, 3 \qquad (9.8)$$

where b_i estimates β_i and b_{ij} estimates β_{ij}, $i < j$. Note that in estimating β_{ij}, which is a measure of nonlinear blending between components i and j, only data col-

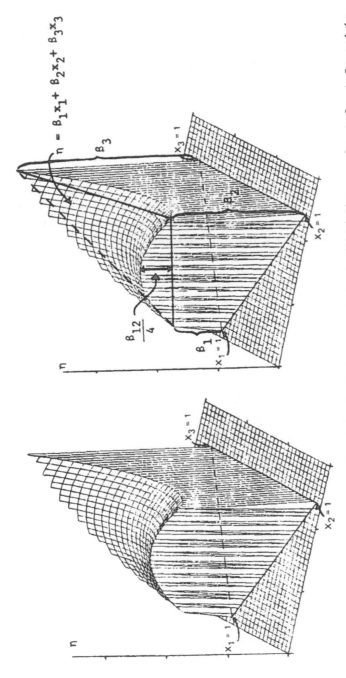

Figure 9.4 (Left) A quadratic surface above the three-component triangle. (Right) Plane $\eta = \beta_1 x_1 + \beta_2 x_2 + \beta_3 x_3$ and the deviation, $\beta_{12}/4$, of the surface above the plane.

lected along the edge of the triangle connecting the vertices $x_i = 1$, $x_l = 0$, $l \neq i$ and $x_j = 1$, $x_l = 0$, $l \neq j$ are used. Also, the scalar quantities 4 and 2 in the formula for b_{ij} do not depend on the values of r_i and r_{ij} but rather come from the values of x_i and x_j being equal to $\frac{1}{2}$ at the mid-edge point.

Upon obtaining the coefficient estimates b_i and b_{ij} in Eq. 9.8, the estimates are substituted for their respective parameters to produce the fitted model

$$\hat{Y}(\mathbf{x}) = \sum_{i=1}^{3} b_i x_i + \sum_{i<j}^{3} b_{ij} x_i x_j \qquad (9.9)$$

The fitted model in Eq. 9.9, as with any of the models 9.4–9.7 when fitted to data collected only at the points of the associated lattice designs, exactly describes the shape of the mixture surface at the points of the design. At locations in the triangle other than the design points, the fitted model may not adequately describe the shape of the surface for those blends of components. For example, in Figure 9.5, a mound shaped surface is displayed over the three-component triangle along with the strictly quadratic surface represented by the fitted model

$$\hat{Y}(\mathbf{x}) = 10x_1 + 8x_2 + 6x_3 + 20x_1x_2 - 12x_1x_3 - 12x_2x_3 \qquad (9.10)$$

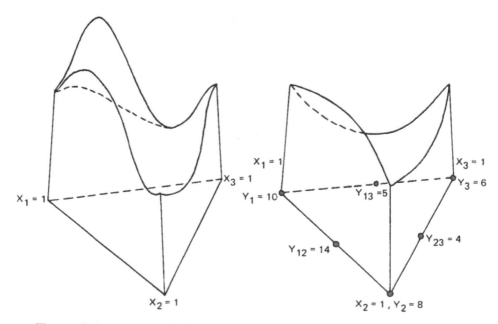

Figure 9.5 A mound-shaped surface modeled by the quadratic equation 9.10 which depicts the surface as being shaped like the one on the right.

The model in Eq. 9.10 was obtained using only the data (Y_i, Y_{ij}) at the six design points denoted by circles on the boundaries of the triangle. Although the two surfaces have identical heights at the design points, their shapes are different elsewhere inside and along the edges of the triangle. To be able to discover this difference in the shapes of the real and fitted surfaces, data must be collected at locations in the factor space in addition to the six points defined by the {3, 2} simplex lattice. When additional points are included in the design, and data are collected at all of the points, the formulas in Eqs. 9.8 for the estimates of the coefficients in Eq. 9.5 no longer are valid. What is required in this case is a multiple regression program to calculate the estimates of the coefficients using all of the data values.

Claiming that in most industrial settings mixture surfaces are more complicated in shape than that described by a quadratic polynomial, Cornell (1986) suggested two 10-point designs for fitting higher-order models with three-component systems. One design is the {3, 3} simplex lattice, and the other design is the three-component simplex-centroid design augmented with three interior points. The two designs, denoted by (a) and (b), respectively, are shown in Figure 9.6.

To data collected at the points of design (a), a sequential model fitting exercise might proceed as outlined in steps 1–4. If an estimate of the experimental error variance is available from replicated observations at one or more design points, then

1. Fit the linear model 9.4 and test for lack of fit where the number of degrees of freedom for lack of fit is $10 - 3 = 7$. If lack of fit is significant, go to 2.
2. Fit the quadratic model 9.5 and test for lack of fit where d.f. for lack of fit equals 4. If lack of fit is not significant, stop. If lack of fit is significant, go to 3.
3. Fit the special cubic model 9.7 and test for lack of fit where d.f. for lack of fit equals 3. If lack of fit is not significant, stop. If lack of fit is significant, go to 4.
4. Fit the full cubic model 9.6. This model is the most complete model for design (a).

To data collected at the points of design (b), a sequential model fitting exercise proceeds exactly as in steps 1, 2, and 3 with design (a), but the model in step 4 is the special quartic model

$$\eta = \sum_{i=1}^{3} \beta_i x_i + \sum_{i<j}^{3} \beta_{ij} x_i x_j$$
$$+ \beta_{1123} x_1^2 x_2 x_3 + \beta_{1223} x_1 x_2^2 x_3 + \beta_{1233} x_1 x_2 x_3^2 \qquad (9.11)$$

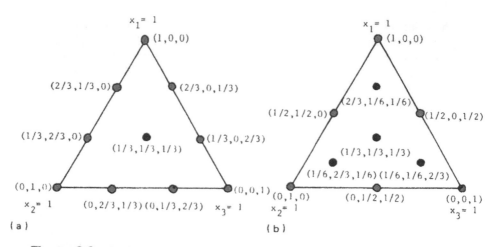

Figure 9.6 Two 10-point designs: design (a) is the {3, 3} simplex lattice, and design (b) is the simplex-centroid augmented with three interior points.

The special quartic model of Eq. 9.11 is especially useful for detecting asymmetry of the surface over the interior of the triangle that cannot be picked up by the terms in the full cubic model of Eq. 9.6.

For the purpose of developing an equation that can be used to predict response values for all blends of two and three components on or inside the triangle, designs (a) and (b) are distinctly different. Design (a) emphasizes the collection of data values along the edges of the triangle and therefore is particularly useful for studying the type of blending (linear, quadratic, or possibly cubic) that occurs between pairs of components. The cubic model 9.6 permits the detection of the presence of both synergistic (beneficial) and antagonistic (not beneficial) blending between pairs of components. An example of this type of blending is illustrated by the shape of the true surface along the x_1–x_2 edge of the triangle in Figure 9.5. Design (b) on the other hand, is the preferred design if the response to blends containing all three components with nonzero proportions is of interest. Furthermore, when following steps 1, 2, and 3 in a sequential model fitting exercise, design (b) is the more powerful arrangement for detecting the inadequacy of a lower-order model, particularly if the shape of the surface inside the triangle is different than the shape of the surface along the edges of the triangle. This is because design (b) has four points inside the triangle, whereas design (a) has only the single point at the centroid where all three components are present in nonzero proportions.

9.2.1 The Analysis of a Three-Component Sweetener Experiment

An experiment was set up to see if an artificial sweetener could be used in a popular athletic sport drink. The sweeteners were glycine (x_1), saccharin (x_2), and an enhancer (x_3). The amount of sweetener was held fixed in all blends at 4% of the total volume, which also was fixed for all blends.

Ten combinations of the three sweeteners were selected as defined by the points of design (b) in Figure 9.6. Replicate samples were made of the four interior blends. The data are listed in Table 9.2, where the response is a measure of "the intensity of sweetness aftertaste" using a scale of 1 (positively no aftertaste) to 30 (very extreme aftertaste). A high score (≥ 20) is undesirable, while a lower score (≤ 10) is desirable.

A sequential model fitting strategy as outlined in steps 1 through 4 was performed and the results are illustrated in Table 9.3. At each step the variation between the replicate scores at the four interior points was used as an estimate of the pure error variance for testing the presence of lack of fit of the model. The lack of fit sum of squares was calculated by subtracting the pure error sum of squares, 11.50, from the residual sum of squares. The estimate of the pure error variance is $s^2 = [(10 - 8)^2 + (23 - 20)^2 + (9 - 12)^2 + (5 - 6)^2]/(2 \times 4) = 2.875$ with 4 degrees of freedom.

The linear, quadratic, and special cubic fitted models were found to have significant lack of fit at the 0.05 level. The test for lack of fit of the fitted special quartic model was not significant at the 0.10 level. This model is

$$\hat{Y}(\mathbf{x}) = 12.27x_1 + 10.27x_2 + 6.27x_3 + 29.10x_1x_2 + 25.10x_1x_3 + 9.10x_2x_3$$
$$\quad\quad (1.98) \quad\quad (1.98) \quad\quad (1.98) \quad\quad (9.70) \quad\quad (9.70) \quad\quad (9.70)$$
$$+ 547.03x_1^2x_2x_3 - 388.97x_1x_2^2x_3 - 676.97x_1x_2x_3^2$$
$$\quad (158.98) \quad\quad\quad (158.98) \quad\quad\quad (158.98)$$

where the quantities in parentheses are the estimated standard errors of the coefficient estimates based on the error variance estimate MSE = residual mean square = 3.94 with 5 d.f.

In the fitted special quartic model, the estimates of the coefficients of the linear and quadratic terms are the approximate heights of the surface at the vertices and deviations of the surface from the plane of the three heights measured along the edges of the triangle, respectively. The estimates of the coefficients of the quartic terms reflect mounds and depressions or plateaus in the shape of the surface inside the triangle. For example, the estimate $b_{1123} = 547.03$, depicts a hill at the location $x_1 = 2/3$, $x_2 = x_3 = 1/6$ caused by the high aftertaste scores of 23 and 20 as seen in Table 9.2. The estimates $b_{1223} = -388.97$ and $b_{1233} = -676.97$ represent depressions or dips in the surface at the locations $x_1 =$

Table 9.2 Data from the Artificial Sweeteners Experiment

Glycine (x_1)	Components Saccharin (x_2)	Enhancer (x_3)	Intensity of Aftertaste Score (Y)
1	0	0	12
0	1	0	10
0	0	1	6
$\frac{1}{2}$	$\frac{1}{2}$	0	18
$\frac{1}{2}$	0	$\frac{1}{2}$	15
0	$\frac{1}{2}$	$\frac{1}{2}$	10
$\frac{1}{3}$	$\frac{1}{3}$	$\frac{1}{3}$	8,10
$\frac{2}{3}$	$\frac{1}{6}$	$\frac{1}{6}$	23,20
$\frac{1}{6}$	$\frac{2}{3}$	$\frac{1}{6}$	9,12
$\frac{1}{6}$	$\frac{1}{6}$	$\frac{2}{3}$	5,6

Table 9.3 Results of the Sequential Model Fitting Exercise

Model Form	ANOVA Summary Statistics Source	d.f.	SS	MS	F
Linear Eq. 9.4	Regression	2	195.52	97.76	
$R^2 = 0.5054$	Residual	11	191.34		
$R_A^2 = 0.4155^*$	Lack of fit	7	179.84	25.69	$8.94 > F_{0.05,7,4} = 6.09$
	Pure error	4	11.50	2.875	
Quadratic Eq. 9.5	Regression	5	232.14		
$R^2 = 0.6001$	Residual	8	154.72		
$R_A^2 = 0.3501$	Lack of fit	4	143.22	35.81	$12.45 > F_{0.05,4,4} = 6.39$
	Pure error	4	11.50	2.875	
Special Cubic Eq. 9.7	Regression	6	269.91		
$R^2 = 0.6977$	Residual	7	116.95		
$R_A^2 = 0.4386$	Lack of fit	3	105.45	35.15	$12.23 > F_{0.05,3,4} = 6.59$
	Pure error	4	11.50	2.875	
Special Quartic Eq. 9.11	Regression	8	367.14		
$R^2 = 0.9490$	Residual	5	19.72	3.94	
$R_A^2 = 0.8675$	Lack of fit	1	8.22	8.22	$2.86 < F_{0.10,1,4} = 4.54$
	Pure error	4	11.50	2.875	not significant

*With most computer programs, the adjusted R^2 value is calculated using $R_A^2 = 1 -$ Residual MS/ [Total SS/($N - 1$)], and is therefore model dependent.

$x_3 = 1/6$, $x_2 = 2/3$ and $x_1 = x_2 = 1/6$, $x_3 = 2/3$, respectively, where the dip at the latter location is the more extreme of the two. A contour plot of the estimated intensity of aftertaste surface, generated by the special quartic model, would suggest that acceptable blends (estimated score ≤ 10) of the three sweeteners are those where the proportion of enhancer lies in the range $0.33 \leq x_3 < 1$ and with equal proportions $x_1 = x_2 = (1 - x_3)/2$ of glycine and saccharin comprising the remaining 4% of the total sweetener volume.

9.3 CONSTRAINTS ON THE COMPONENT PROPORTIONS IN THE FORM OF UPPER AND LOWER BOUNDS ON THE VALUE OF X_i

In many mixture experiments, it is not possible to explore the total range, $0 \leq x_i \leq 1$, with all of the components. This is because one may require that at least a certain proportion of component i be present in all blends, thus eliminating the case where $x_i = 0$. Or one may insist that component j contribute at most a proportion $x_j \leq c_j < 1$. As an example, in the formulation of a rocket propellant consisting of the three components, fuel (x_1), oxidizer (x_2), and binder (x_3), each blend must contain at least 20% fuel, at least 40% oxidizer, and at least 20% binder but not more than 40% binder. This forces the following constraints on the component proportions

$$x_1 \geq 0.20 \qquad x_2 \geq 0.40 \qquad 0.40 \geq x_3 \geq 0.20 \qquad (9.12)$$

The upper bound 0.40 for x_3 is of course forced by the presence of lower bounds 0.20 and 0.40 for x_1 and x_2, respectively. In other words, since $x_3 = 1 - x_1 - x_2$, and upon substituting the lower bounds for x_1 and x_2, then component three can contribute at most $x_3 = 1 - 0.20 - 0.40 = 0.40$ and this happens when both x_1 and x_2 are at their minimum values. Similarly, x_1 and x_2 have implied upper bounds of 0.40 and 0.60, respectively.

Let us consider first those cases where lower bounds only are placed on the proportions of one or more of the components. As we mentioned previously, however, the placing of lower bounds creates implied upper bounds on the remaining x_j. These lower bounds are of the form

$$x_i \geq L_i \geq 0 \qquad i = 1, 2, \ldots, q \qquad (9.13)$$

and $\sum_{i=1}^{q} L_i < 1$. The resulting factor space of feasible mixtures is reduced to a subspace of the original simplex region and the subspace is also a simplex. For example, in Figure 9.7a, the upper triangular portion of the $q = 3$ component triangle is defined by the single constraint.

$$x_1 \geq L_1, x_2 + x_3 = 1 - L_1$$

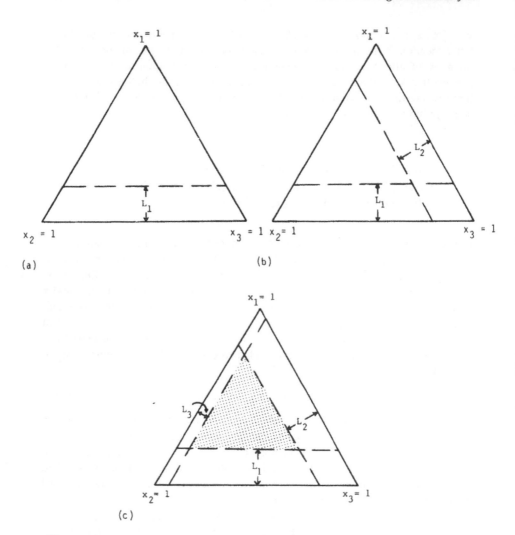

Figure 9.7 Reducing the size of the simplex factor space by forcing lower bounds L_1, L_2, and L_3 on x_1, x_2, and x_3, (a), (b), (c), respectively.

Now, forcing the lower bound L_2 on x_2 results in the region shown in Figure 9.7b; and by further requiring $x_3 \geq L_3$, this forces the region to be the shaded triangle in Figure 9.7c. In other words, the placing of lower bounds on x_1, x_2, and x_3 of the form

$$x_1 \geq L_1, \; x_2 \geq L_2, \; x_3 \geq L_3 \tag{9.14}$$

forces the feasible region of mixtures to the shaded triangular region in Figure 9.7c.

9.3.1 Introducing Pseudocomponents

In setting up designs for fitting a model over a subregion, an alternative system of coordinates is often used. Pseudocomponents x_i' are defined as

$$x_i' = \frac{x_i - L_i}{1 - L}, \quad L = \sum_{i=1}^{q} L_i < 1 \tag{9.15}$$

and $x_1' + x_2' + \cdots + x_q' = 1$. Since the orientation of the subregion is the same as the orientation of the original component triangle, the vertex of the subregion closest to the vertex $x_i = 1$, represents the $x_i' = 1$ point of the pseudocomponent system. In the original component system, the pseudocomponent coordinates $x_i' = 1, x_j' = 0, j \neq i$ correspond to the coordinates $x_i = L_i + 1 - L, x_j = L_j, j \neq i$; see Figure 9.8.

Once a design, whether it be a simplex lattice arrangement or a member from any other class of designs, has been set up in the x_i', the design coordinates of the x_i' are then mapped back to provide settings in the original components using

$$x_i = L_i + (1 - L)x_i' \tag{9.16}$$

Data are then collected on the blends in the original components, and these data are used to fit a model in the original components or a similar model form in the pseudocomponents.

To illustrate the use of pseudocomponents in setting up a design and fitting a model over a subregion defined by the placing of lower bounds on the x_i, let us refer once again to the rocket propellant example (Kurotori 1966) for which we set up the constraints in Eq. 9.12. Ignoring for the moment the upper bound 0.40 for x_3, the lower bounds are

$$x_1 \geq 0.20 \quad x_2 \geq 0.40 \quad x_3 \geq 0.20 \tag{9.17}$$

The subregion is like the one in Figure 9.8, but where now $L_1 = 0.20, L_2 = 0.40$ and $L_3 = 0.20$. Summing $L_1 + L_2 + L_3 = 0.20 + 0.40 + 0.20 = 0.80 = L$, the pseudocomponents x_1', x_2', and x_3' are

$$x_1' = \frac{x_1 - 0.20}{0.20} \quad x_2' = \frac{x_2 - 0.40}{0.20} \quad x_3' = \frac{x_3 - 0.20}{0.20} \tag{9.18}$$

where in the denominators, $0.20 = 1 - L = 1 - 0.80$. The vertices of the pseudocomponent triangle expressed in the original components using Eq. 9.16,

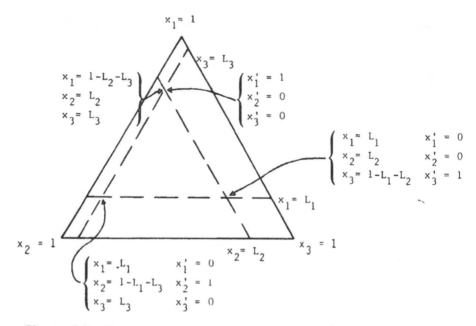

Figure 9.8 The coordinates of the vertices of the subregion expressed in the pseudocomponents (x_i') and the original components (x_i).

are

$$(x_1', x_2', x_3') = (1, 0, 0) \rightarrow (x_1, x_2, x_3) = (0.40, 0.40, 0.20)$$
$$= (0, 1, 0) \rightarrow \qquad\qquad = (0.20, 0.60, 0.20)$$
$$= (0, 0, 1) \rightarrow \qquad\qquad = (0.20, 0.40, 0.40)$$

Note that each so-called "pure" pseudocomponent corresponds to a "complete" blend in the original components.

Suppose it is desired to fit a seven-term special cubic model over the subregion. For the special cubic model, a simplex-centroid design could be used, but we shall augment the design with three additional interior points so that the design contains 10 points (design (b) in Figure 9.6). The design coordinates are defined in the pseudocomponents after which Eq. 9.16 is used to determine the settings in the original components.

Table 9.4 contains the 10 design settings in the pseudocomponents, the corresponding settings in the original components, and the observed response value at each of the 10 design points. The response is a modulus of elasticity rating for each rocket propellant blend. The blend of fuel (x_1), oxidizer (x_2) and binder (x_3) corresponding to the binary blend (0.50, 0.50, 0) of pseudocomponents 1

Table 9.4 Kurotori's Rocket Propellant Data

Design Point	x_1'	x_2'	x_3'	x_1	x_2	x_3	Elasticity
1	1	0	0	0.40	0.40	0.20	2650
2	0	1	0	0.20	0.60	0.20	2450
3	0	0	1	0.20	0.40	0.40	2350
4	0.50	0.50	0	0.30	0.50	0.20	2950
5	0.50	0	0.50	0.30	0.40	0.30	2750
6	0	0.50	0.50	0.20	0.50	0.30	2400
7	0.333	0.333	0.333	0.27	0.46	0.27	3000
8	0.667	0.166	0.166	0.33	0.43	0.23	2980
9	0.166	0.667	0.166	0.23	0.53	0.23	2770
10	0.166	0.166	0.667	0.23	0.43	0.33	2690

and 2, which is point 4 in Table 9.4, consists of the proportions

fuel: $\qquad x_1 = 0.20 + (0.20)0.50 = 0.30$

oxidizer: $\quad x_2 = 0.40 + (0.20)0.50 = 0.50$

binder: $\qquad x_3 = 0.20 + (0.20)0 \quad = 0.20$

The remaining nine blends of x_1, x_2, and x_3 in Table 9.4 were likewise found using Eq. 9.16.

Fitted to the 10 elasticity values in Table 9.4, the special cubic model in the pseudocomponents is

$$\hat{Y}(\mathbf{x}') = \underset{(10)}{2653x_1'} + \underset{(10)}{2446x_2'} + \underset{(10)}{2351x_3'} + \underset{(50)}{1597x_1'x_2'}$$
$$+ \underset{(50)}{1008x_1'x_3'} - \underset{(50)}{6x_2'x_3'} + \underset{(329)}{6141x_1'x_2'x_3'} \tag{9.19}$$

The quantities in parentheses below the coefficient estimates are the estimated standard errors of the coefficients, where the residual mean square, MSE = 105, is based on $10 - 7 = 3$ degrees of freedom. Because the number of data values ($N = 10$) exceeds the number of terms in the model ($p = 7$), the fitted Eq. 9.19 was obtained using a regression software package.

An equivalent model in the original components x_1, x_2, and x_3 is

$$\hat{Y}(\mathbf{x}) = \underset{(2870)}{35674x_1} + \underset{(1111)}{16991x_2} + \underset{(2870)}{50201x_3} - \underset{(8809)}{113595x_1x_2}$$
$$- \underset{(16999)}{281852x_1x_3} - \underset{(8809)}{153686x_2x_3} + \underset{(41134)}{767647x_1x_2x_3} \tag{9.20}$$

where again the quantities in parentheses below the coefficient estimates are the estimated standard errors of the coefficients based on MSE = 105. For predicting modulus of elasticity values (or generating a surface contour plot) over the constrained region defined in Eq. 9.17, Eqs. 9.19 and 9.20 would provide identical contour plots. The reason for this is that they are equivalent model forms; that is, they contain the same linear and nonlinear blending terms. And yet, the absolute values of the coefficient estimates in Eq. 9.20 are considerably greater in magnitude, with some even having opposite signs, than the estimates of the corresponding terms in Eq. 9.19.

The seemingly different surface representations of Eqs. 9.19 and 9.20 is explained in the following way. Equation 9.19 depicts the estimated elasticity surface directly above the constrained region in the metric of the pseudocomponents x_i', whereas Eq. 9.20 expresses the same surface but in the metric of the original components x_i, $i = 1, 2,$ and 3. As a result, the coefficient estimates 2653, 2446, and 2351 of the linear blending terms in Eq. 9.19 are estimates of the height of the surface directly above the pseudocomponent vertices $x_1' = 1$, $x_2' = x_3' = 0$; $x_1' = x_3' = 0$, $x_2' = 1$; and $x_1' = x_2' = 0$, $x_3' = 1$, respectively. The coefficient estimates 35674, 16991, and 50201 of the linear blending terms in Eq. 9.20, on the other hand, are extrapolated estimates of the heights of the surface at the vertices of the simplex region in the original components, that is, at $x_1 = 1$, $x_2 = x_3 = 0$; $x_1 = x_3 = 0$, $x_2 = 1$; and $x_1 = x_2 = 0$, $x_3 = 1$; respectively. Also, the coefficient estimates of the cross-product or nonlinear blending terms in Eq. 9.19 represent departures in the height of the estimated surface, above the pseudocomponent simplex, from the plane defined by the linear blending terms in Eq. 9.19, while the coefficient estimates of the nonlinear blending terms in Eq. 9.20 represent departures in the height of the extrapolated surface from the extrapolated plane defined by the linear blending terms in Eq. 9.20. It is important therefore to keep in mind the different metrics that Eqs. 9.19 and 9.20 are associated with, particularly when trying to describe the shape of the estimated surface directly above the constrained region. For this reason Eq. 9.19 is the preferred choice for describing the surface shape or for interpreting the component blending properties by looking at the signs and magnitudes of the coefficient estimates while both equations are equally effective for generating a surface plot (in two or three dimensions) or for plotting the response trace. Plotting the response trace is discussed in Section 9.3.3.

9.3.2 Extreme Vertices Designs

Now let us turn our attention to the placing of both lower (L_i) and upper (U_i) bounds on the x_i, that is,

$$0 < L_i \leq x_i \leq U_i < 1 \tag{9.21}$$

The feasible region of mixture blends defined in Eq. 9.21 is a convex polyhedron, which is more complicated in shape than the simplex-shaped subregion defined by placing only lower bounds (L_i) on the x_i. For example, suppose for $q = 3$ we have the following constraints on x_1, x_2, and x_3

$$0.20 \leq x_1 \leq 0.60 \qquad 0.10 \leq x_2 \leq 0.60 \qquad 0.10 \leq x_3 \leq 0.50 \qquad (9.22)$$

The convex subregion defined in Eq. 9.22 is shown in Figure 9.9.

The first step taken for the purpose of describing the shape of the constrained subregion is to enumerate the number of vertices of the subregion and to determine the coordinates of the vertices. This is because the vertices and convex combinations of some of the vertices are primary candidates for design points to be used in collecting data to fit various model forms.

Crosier (1984) presents a formula for calculating the number of vertices of any constrained region. The formula consists of summing subsets of the q components where a subset of components is defined by the sum of the ranges of the components in it. For example, suppose we have q components whose limits are defined by Eq. 9.21. Then, $R_i = U_i - L_i$ is the range of component i. Next, calculate the quantities (to be defined shortly),

$$R_L = 1 - \sum_{i=1}^{q} L_i \qquad \text{and} \qquad R_U = \sum_{i=1}^{q} U_i - 1 \qquad (9.23)$$

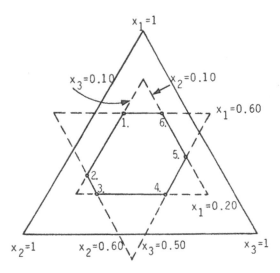

Figure 9.9 The subregion defined by the constraints $0.2 \leq x_1 \leq 0.6$, $0.1 \leq x_2 \leq 0.6$, $0.1 \leq x_3 \leq 0.5$.

and let

$$R_p = \min(R_L, R_U)$$

Define $C(q, r)$ to be the number of combinations of q components taken r at a time $(r \le q)$, where, of course, $C(q, r) = q!/r!(q - r)!$. Divide $C(q, r)$ into three disjoint groups,

Group 1: combinations of r components whose ranges sum to less than R_p
Group 2: combinations of r components whose ranges sum to R_p
Group 3: combinations of r components whose ranges sum to more than R_p

Denote the number of combinations in groups 1, 2, and 3 by $L(q, r)$, $E(q, r)$, and $G(q, r)$, respectively, where $C(q, r) = L(q, r) + E(q, r) + G(q, r)$. Then the number of vertices of the constrained region is

$$N_V = q + \sum_{r=1}^{q} L(q, r) \cdot (q - 2r) + \sum_{r=1}^{q} E(q, r) \cdot (1 - r) \qquad (9.24)$$

To illustrate the use of Eq. 9.24 for N_V, let us calculate the number of vertices of the constrained region defined in Eq. 9.22. Associated with the limits in Eq. 9.22, we have for $q = 3$,

$$R_1 = 0.60 - 0.20 = 0.40$$

$$R_2 = 0.60 - 0.10 = 0.50 \qquad R_3 = 0.50 - 0.10 = 0.40$$

and

$$R_L = 1 - (0.20 + 0.10 + 0.10) = 0.60$$

$$R_U = (0.60 + 0.60 + 0.50) - 1 = 0.70$$

so that $R_p = \min(0.60, 0.70) = 0.60$. The numbers of combinations $C(3, r)$ for $r = 1, 2, 3$ are $C(3, 1) = 3$, $C(3, 2) = 3$, and $C(3, 3) = 1$ and these combinations along with the sum of the component ranges are,

$$C(3, 1): x_1, R_1 = 0.40 \qquad x_2, R_2 = 0.50 \qquad x_3, R_3 = 0.40$$

$$C(3, 2): x_1, x_2, R_1 + R_2 = 0.90 \qquad x_1, x_3, R_1 + R_3 = 0.80$$

$$x_2, x_3, R_2 + R_3 = 0.90$$

$$C(3, 3): x_1, x_2, x_3, R_1 + R_2 + R_3 = 1.30$$

In group one are the combinations of components whose sum of the ranges is less than $R_p = 0.60$. Since each of the x_i, $i = 1, 2, 3$, has a range $R_i < R_p$, then $L(3, 1) = 3$ but $L(3, 2) = 0$ and $L(3, 3) = 0$. For group two, there are no combinations of components whose sum of ranges equals 0.60 so that $E(3, 1) = E(3, 2) = E(3, 3) = 0$. For group three, all pairs of x_i and x_j have sums of ranges that exceed $R_p = 0.60$ so that $G(3, 1) = 0$, $G(3, 2) = 3$ and $G(3, 3) = 1$. Substituting the values of $L(3, r)$ and $E(3, r)$ into Eq. 9.24, we find for the number of vertices of the constrained region in Eq. 9.22

$$N_V = q + \{L(3, 1) \cdot (3 - 2) + L(3, 2) \cdot (3 - 4) + L(3, 3) \cdot (3 - 6)\}$$
$$+ \{E(3, 1) \cdot (1 - 1) + E(3, 2) \cdot (1 - 2) + E(3, 3) \cdot (1 - 3)\}$$
$$= 3 + \{3 \cdot (1) + 0 \cdot (-1) + 0 \cdot (-3)\} + \{0 \cdot (0) + 0 \cdot (-1) + 0 \cdot (-2)\}$$
$$= 6$$

The coordinates of the six vertices drawn in Figure 9.9 will be calculated next.

Having determined the number of vertices of the constrained region using Eq. 9.24, the next order of business is to define the coordinates of the vertices in the system of the original components. For cases involving only three or four components, determining the coordinates of the vertices is straightforward; with three components, one need only draw the constrained region as in Figure 9.9. With four components, on the other hand, one can either draw the region and determine the coordinates from the figure or one can compute the vertices by listing the $2^4 = 16$ possible combinations of lower and upper bounds

$$
\begin{array}{llll}
(L_1, L_2, L_3, L_4) & (L_1, U_2, L_3, L_4) & (U_1, L_2, L_3, L_4) & (U_1, U_2, L_3, L_4) \\
(L_1, L_2, L_3, U_4) & (L_1, U_2, L_3, U_4) & (U_1, L_2, L_3, U_4) & (U_1, U_2, L_3, U_4) \\
(L_1, L_2, U_3, L_4) & (L_1, U_2, U_3, L_4) & (U_1, L_2, U_3, L_4) & (U_1, U_2, U_3, L_4) \\
(L_1, L_2, U_3, U_4) & (L_1, U_2, U_3, U_4) & (U_1, L_2, U_3, U_4) & (U_1, U_2, U_3, U_4)
\end{array}
$$

$$(9.25)$$

and select the combinations that are admissible, that is, that satisfy the constraint $x_1 + x_2 + x_3 + x_4 = 1$ where L_i or U_i is substituted for x_i, $i = 1, 2, 3$, and 4. The vertices of the region are not only those combinations in Eq. 9.25 that are admissible but also include all combinations in Eq. 9.25 where only one of the L_i or U_i has to be adjusted to make the combination admissible.

To illustrate the computing of the coordinates of the vertices by listing all 2^q combinations of lower and upper bounds and adjusting certain combinations to make them admissible, let us refer to the three component constrained region of Eq. 9.22 where the $2^3 = 8$ combinations are

$$
\begin{array}{llll}
1. & (L_1, L_2, L_3) = (0.2, 0.1, 0.1) & 5. & (U_1, L_2, L_3) = (0.6, 0.1, 0.1) \\
2. & (L_1, L_2, U_3) = (0.2, 0.1, 0.5) & 6. & (U_1, L_2, U_3) = (0.6, 0.1, 0.5) \\
3. & (L_1, U_2, L_3) = (0.2, 0.6, 0.1) & 7. & (U_1, U_2, L_3) = (0.6, 0.6, 0.1) \\
4. & (L_1, U_2, U_3) = (0.2, 0.6, 0.5) & 8. & (U_1, U_2, U_3) = (0.6, 0.6, 0.5)
\end{array}
$$

$$(9.26)$$

We determined previously that the number of vertices is six and yet it is interesting to observe that none of the combinations in Eq. 9.26 are admissible and therefore adjustments are necessary. The adjusted combinations are

9. 2. is adjusted by raising L_2 to 0.3 to get (0.2, 0.3, 0.5); point 4 in Figure 9.9.
10. 2. is adjusted by raising L_1 to 0.4 to get (0.4, 0.1, 0.5); point 5.
11. 3. is adjusted by raising L_3 to 0.2 to get (0.2, 0.6, 0.2); point 3.
12. 3. is adjusted by raising L_1 to 0.3 to get (0.3, 0.6, 0.1); point 2.
13. 4. is adjusted by lowering U_3 to 0.2 to get (0.2, 0.6, 0.2); point 3.
14. 4. is adjusted by lowering U_2 to 0.3 to get (0.2, 0.3, 0.5); point 4.
15. 5. is adjusted by raising L_3 to 0.3 to get (0.6, 0.1, 0.3); point 6.
16. 5. is adjusted by raising L_2 to 0.3 to get (0.6, 0.3, 0.1); point 1.

Since six points are already defined, any further adjustments would only produce repeats of points 1 to 6.

For mixture systems with five or more components, the number of vertices of the constrained region can be excessive. For example, the five-component case

$$0 \leq x_1 \leq 0.10 \qquad 0 \leq x_2 \leq 0.10 \qquad 0.05 \leq x_3 \leq 0.15$$

$$0.20 \leq x_4 \leq 0.40 \qquad 0.40 \leq x_5 \leq 0.60$$

results in a constrained region that contains 28 extreme vertices. Ten component problems are typically known to have regions containing up to 1000 extreme vertices.

In situations such as these, a computer software package having an algorithm that not only can compute the coordinates of the extreme vertices but is also capable of selecting a subset of the vertices to be used as design points is required. Some of the better known algorithms used for generating the extreme vertices of a constrained region are XVERT (Snee and Marquardt, 1974), UNIEXP (Goel, 1980), and XVERT1 (Nigam, Gupta, and Gupta, 1983). The XVERT algorithm also has the capability of selecting a subset of the total number of vertices to be used as a design for fitting a first-order mixture model of the form in Eq. 9.4. The subset of points chosen is based on a variance minimizing criterion such as minimum $|(\mathbf{X'X})^{-1}|$, minimum trace of $(\mathbf{X'X})^{-1}$, or the minimum of the maximum variance of $\hat{Y}(\mathbf{x})$. Two software packages that come to mind for generating the extreme vertices and then selecting a subset of them are ACED (Welch, 1985) and MIXSOFT (Piepel, 1990), but there are many others as well. The selection of several different 15-point designs by ACED, from a candidate list of 27 points, for supporting a second-order model in four components over a highly constrained region is presented in Cornell (1990). The designs differ because they are chosen using the *A*-, *D*-, *G*-, and *V*-optimality criteria. See Chapter 12 for a discussion of design optimality criteria.

The number of extreme vertices and convex combinations of the vertices to be used as design points will depend on the form of the mixture model to be fitted. If the first-order mixture model

$$Y = \sum_{i=1}^{q} \beta_i x_i + \varepsilon$$

is to be fitted, this can normally be accomplished using data taken from only q of the vertices. Usually, however, data are collected not only from q or more of the extreme vertices but from other locations as well. The other locations are convex combinations of the vertices such as the midpoints of the edges, centroids of the two-dimensional faces, etc., of the convex polyhedron. Data from these other locations can be used to estimate the experimental error variance, or if replicates are collected, then data from these other locations can be used for testing the lack of fit of the first-order model.

When fitting the second-order mixture model

$$Y = \sum_{i=1}^{q} \beta_i x_i + \sum_{i<j}^{q} \beta_{ij} x_i x_j + \varepsilon$$

generally support points consist of a subset of the extreme vertices of the region, edge centroids, or face centroids (if $q \geq 7$), and centroids of some of the r-dimensional flats (if $3 \leq q \leq 6$, then $2 \leq r \leq q - 1$; and if $q \geq 7$, then $3 \leq r \leq q - 1$). A centroid of an r-dimensional flat ($r \leq q - 1$) is the average of all the vertices which lie on the same constraint plane. To illustrate, in Figure 9.9, $q = 3$ so that $r = 2$. The candidate points for a design in which to collect data for fitting the second-order model

$$Y = \sum_{i=1}^{3} \beta_i x_i + \sum_{i<j}^{3} \beta_{ij} x_i x_j + \varepsilon$$

are the six vertices, the midpoints of the six edges (one-dimensional flats), and the centroid of the region itself (a two-dimensional flat). The coordinates of this latter centroid are approximated by averaging the six vertices for which we get $(x_1, x_2, x_3) = (0.383, 0.333, 0.283)$.

Another example is the four-component problem

$$0.89 \leq x_1 \leq 0.905 \qquad 0.02 \leq x_2 \leq 0.035$$
$$0.04 \leq x_3 \leq 0.065 \qquad 0.01 \leq x_4 \leq 0.02$$

The constrained region, drawn in Figure 9.10, possesses eight vertices, 13 edges, and seven faces (two-dimensional flats). Candidate points for fitting the 10-term second-order model

$$Y = \sum_{i=1}^{4} \beta_i x_i + \sum_{i<j}^{4} \beta_{ij} x_i x_j + \varepsilon$$

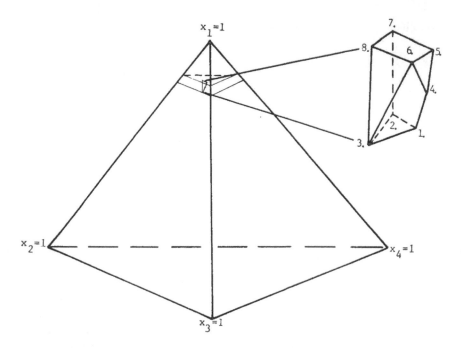

Figure 9.10 The four-component constrained region defined by the constraints $0.89 \leq x_1 \leq 0.905$, $0.02 \leq x_2 \leq 0.035$, $0.04 \leq x_3 \leq 0.065$, and $0.01 \leq x_4 \leq 0.02$.

are the extreme vertices, midpoints of the edges, centroids of the faces, and the overall centroid. The coordinates of the 29 candidate points are listed in Table 9.5.

The selection of the design points, for example, N points where $10 < N \leq 29$, could be made using an algorithm such as DETMAX (Mitchell, 1974). A variance minimizing algorithm like DETMAX would choose a subset of the vertices along with a subset of the midpoints of the edges depending on the value of N and most likely would select a subset of the centroids of the faces only after most of the vertices and edge midpoints had been chosen. This is because the model contains only linear terms, $\beta_i x_i$, and cross-product terms, $\beta_{ij} x_i x_j$, and the variances of the estimates of these coefficients are minimized using vertices and midpoints of the edges. If any of the four special cubic terms, $\beta_{ijk} x_i x_j x_k$, were added to the model, the centroids of the faces would then receive greater consideration.

Summarizing this section, when some or all of the mixture component proportions are constrained in value by placing lower bounds (L_i) on the x_i,

Table 9.5 Candidate Points for Fitting the 10-Term Second-Order Mixture Model

Pt.	Extreme Vertices[a] x_1	x_2	x_3	x_4
1.	0.89	0.025	0.065	0.02
2.	0.89	0.035	0.055	0.02
3.	0.89	0.035	0.065	0.01
4.	0.895	0.02	0.065	0.02
5.	0.905	0.02	0.055	0.02
6.	0.905	0.02	0.065	0.01
7.	0.905	0.035	0.04	0.02
8.	0.905	0.035	0.05	0.01

Pt.	Midpoints of Edges x_1	x_2	x_3	x_4	Pair of Vertices
9.	0.89	0.03	0.06	0.02	(1, 2)
10.	0.89	0.03	0.065	0.015	(1, 3)
11.	0.89	0.035	0.06	0.015	(2, 3)
12.	0.8925	0.0225	0.065	0.02	(1, 4)
13.	0.90	0.02	0.06	0.02	(4, 5)
14.	0.90	0.02	0.065	0.015	(4, 6)
15.	0.905	0.02	0.06	0.015	(5, 6)
16.	0.8975	0.0275	0.065	0.01	(3, 6)
17.	0.905	0.0275	0.0475	0.02	(5, 7)
18.	0.905	0.035	0.045	0.015	(7, 8)
19.	0.8975	0.035	0.0475	0.02	(2, 7)
20.	0.8975	0.035	0.0575	0.01	(3, 8)
21.	0.905	0.0275	0.0575	0.01	(6, 8)

Pt.	Centroids of Faces x_1	x_2	x_3	x_4	Combination of Vertices
22.	0.89	0.0317	0.0617	0.0166	(1, 2, 3)
23.	0.895	0.025	0.065	0.015	(1, 3, 4, 6)
24.	0.9017	0.02	0.0617	0.0166	(4, 5, 6)
25.	0.897	0.027	0.056	0.02	(1, 2, 4, 5, 7)
26.	0.905	0.0275	0.0525	0.015	(5, 6, 7, 8)
27.	0.8975	0.035	0.0525	0.015	(2, 3, 7, 8)
28.	0.90	0.03	0.06	0.01	(3, 6, 8)
29.	0.8981	0.0281	0.0575	0.0163	(all)

[a] Underlined proportions are L_i or U_i.

1. The experimental region or constrained region inside the original simplex is a simplex whose vertices are aligned with the vertices of the original simplex, and
2. The conventional pseudocomponent transformation in Eq. 9.15, sometimes called the L-pseudocomponent transformation, is appropriate.

Typically, designs are set up in the pseudocomponents, and the original component proportions are determined from Eq. 9.16. A model in the pseudocomponents is fitted to data collected at the design points. All of the standard summary calculations and statistics (ANOVA, R^2, R_A^2) are computed for the pseudocomponent model in the usual manner; see Section 9.2.1

When some or all of the component proportions are constrained by placing upper bounds (U_i) on the x_i,

3. The orientation of the experimental region is opposite that of the original simplex. For example, if instead of the three lower bounds in Eq. 9.14, x_1, x_2, and x_3 were constrained by the upper bounds

    $$U_1 \geq x_1 \qquad U_2 \geq x_2 \qquad U_3 \geq x_3$$

    where $\Sigma_{i=1}^3 U_i > 1$, then the resulting region is an inverted triangle. If the vertices of the inverted triangle extend beyond the boundaries of the original triangle, the experimental region is not an inverted simplex. If, however,

    $$\sum_{i=1}^3 U_i - U_{min} \leq 1$$

    where U_{min} is the minimum of the three upper bounds, then the inverted triangle lies entirely inside the original triangle and is the experimental region. The quantity in Eq. 9.23

    $$R_U = \sum_{i=1}^3 U_i - 1$$

    is the length of the axis of the U-pseudocomponent

    $$Z_i = \frac{U_i - x_i}{R_U}$$

    measured in the units of the original x_i components.

When some or all of the component proportions are constrained by placing lower and upper bounds on the x_i,

4. First the consistency of the upper and lower bounds is checked by comparing each $R_i = U_i - L_i$ against $R_L = 1 - \Sigma_{i=1}^q L_i$ and $R_U = \Sigma_{i=1}^q U_i - 1$; any $R_i > R_L$ denotes an inconsistent U_i, and any $R_i > R_U$ denotes an incon-

sistent lower bound. Inconsistent bounds are replaced by the respective implied bounds, i.e., $U_i^* = L_i + R_L$ and $L_i^* = U_i - R_U$ to force the set of constraints to be consistent.

5. The experimental region is the smaller of the simplex created by the lower bounds and the inverted simplex created by the upper bounds.

6. If any component range R_i is less than $R_p = \min(R_L, R_U)$, the experimental region is not a simplex, and calculating the number of extreme vertices, N_V, using Eq. 9.24 is the next order of business. Then the coordinates of the extreme vertices are determined so that all or some of the vertices along with some number of convex combinations of the vertices can be used as design points at which to collect data for fitting the proposed mixture model.

9.3.3 The Response Trace

We have seen previously from Eq. 9.20 that by looking at the coefficient estimates in a model expressed in the original components, it is very difficult, and in some cases impossible, to describe the shape of the estimated surface over a constrained subregion of the simplex. The reason for this is that the model in the original component proportions expresses the shape of the estimated surface above the constrained region as if it were extrapolated to the boundaries of the original simplex. Consequently, the shape of the extrapolated surface may appear very different, based on the signs and magnitudes of the coefficient estimates in the original components' model, from the shape of the surface above the subregion expressed in the pseudocomponents. Thus, the fitted model itself is what one has to work with when trying to understand the roles played by the separate mixture components.

A technique that has been used to measure changes in the estimated response that are brought about by changing the proportion of a single component while keeping the relative proportions of the other components fixed is the *response trace*. The response trace is a plot of estimated response values taken along certain directions. These directions were first defined by Cox (1971). These directions, one for each component, are referred to by Cornell (1990a) as *component rays*. With a three-component system, for example, the component rays or x_i-rays, $i = 1$, 2, and 3, are depicted by the dashed lines shown in Figure 9.11. The three x_i-rays intersect one another at a mixture blend **s**, called the *reference mixture*. It is from the reference mixture **s**, toward as well as away from each of the q vertices of the simplex, that the proportion of each component is incremented to produce the response trace.

The steps taken in constructing the response trace are:

1. Select the fitted model (linear, quadratic, special cubic, etc.) that you think is best for predicting the response over the experimental region. The experimental region may be a constrained region or the simplex itself.

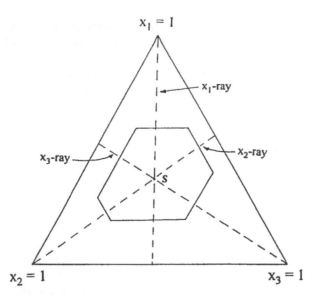

Figure 9.11 The x_i-rays, $i = 1, 2, 3$ drawn from the three vertices and intersecting at the reference mixture s. The constrained region is the region of Figure 9.9.

2. Select a reference mixture s inside the experimental region. Generally s will be the centroid of the experimental region.

3. On the x_i-ray and beginning at the reference mixture s, increment component i by an arbitrarily small amount Δ_i ($= 0.001$ or 0.01 or 0.02) by moving away from the reference mixture toward ($\Delta_i > 0$) the vertex $x_i = 1$, as well as away from ($\Delta_i < 0$) the vertex. For example, ten values of x_i defined on the x_i-ray where five are toward ($\Delta_i > 0$) the $x_i = 1$ vertex and five are in the opposite direction ($\Delta_i < 0$) from the vertex are

$$x_i = s_i + \Delta_i, \qquad x_i = s_i + 2\Delta_i, \ldots, x_i = s_i + 5\Delta_i \qquad (9.27)$$

where s_i is the ith element of s ($i = 1, 2, \ldots, q$). At the point on the x_i-ray where $x_i = s_i + l\Delta_i$, $l = 1, 2, \ldots$, the proportions for the other components x_j are defined as

$$x_j = s_j \frac{(1 - s_i - l\Delta_i)}{1 - s_i}, \qquad j = 1, 2, \ldots, q, j \neq i \qquad (9.28)$$

Equation 9.28 ensures that the relative proportions of components j and k, where $j, k \neq i$, remain the same along the x_i-ray as they were at the reference mixture s, that is, $x_j/x_k = s_j/s_k$, $j, k \neq i$.

4. Each value of x_i in Eq. 9.27 coupled with the values of the x_j's in Eq. 9.28, defines a point or blend on the x_i-ray. Choose some number of blends on the x_i-ray at which to obtain predicted values of the response remembering that *only blends inside and on the boundary of the experimental region are valid.* Generally the value of Δ_i used in step 3 is one that, with a sufficiently large l value in $x_i = s_i + l\Delta_i$, positions the blend closest to the $x_i = 1$ vertex on the boundary of the experimental region ($x_i = U_i$, $x_j = s_j (1 - U_i)/(1 - s_i)$). The blend ($\Delta_i < 0$) farthest away from the $x_i = 1$ vertex should be positioned on the boundary ($x_i = L_i$, $x_j = s_j (1 - L_i)/(1 - s_i)$) of the experimental region also. When this happens, the response trace is then plotted over the entire range of values, $L_i \leq x_i \leq U_i$, for the ith component.
5. Substitute the coordinates of each blend generated in step 4 into the fitted model and obtain a predicted response value for that blend. Plot the predicted response values against the value of x_i in Eq. 9.27.
6. Repeat steps 3 to 5 for the other components on the x_j-rays, $j = 1, 2, \ldots, q, j \neq i$. When this is done, there will be q plots.

A response trace for a three-component fish patty experiment is shown in Figure 9.12. The experiment is described in Section 9.4.1. The response plotted is the estimated texture of a fish patty where texture is defined in terms of the force in grams ($\times 10^{-3}$) required to puncture the surface of the patty, and ranged in value from 1.37 to 2.08.

The response trace in Figure 9.12 was produced from a fitted second-degree equation, Eq. 9.29. The design was a seven-point simplex-centroid arrangement, and two replicate texture values for each of seven blends of the three fish were fitted. The fitted model is

$$\hat{Y}(\mathbf{x}) = \underset{(0.05)}{2.05x_1} + \underset{(0.05)}{1.42x_2} + \underset{(0.05)}{1.88x_3} + \underset{(0.25)}{0.88x_1x_2} + \underset{(0.25)}{0.36x_1x_3} + \underset{(0.25)}{0.42x_2x_3} \quad (9.29)$$

with $R^2 = 0.9487$. The experimental region is the three-component triangle and the reference mixture s is the centroid of the triangle with the coordinates $\mathbf{s} = (\frac{1}{3}, \frac{1}{3}, \frac{1}{3})$. The plot was generated using the DESIGN-EXPERT software package of STAT-EASE (1994).

In Figure 9.12, the letters A, B, and C stand for the fish types A = mullet (x_1), B = sheepshead (x_2), and C = croaker (x_3), respectively. Of the three fish types, mullet is the firmest (has the highest texture value) and croaker is the softest. For the patty cooking conditions that were used in generating these data however, sheepshead had the softest texture (1.42) and this is illustrated in Figure 9.12 by the low point of the curve B. Very little curvature or nonlinear blending is present in the texture values, as evidenced by the lack of curvature in the plots, with the exception perhaps between mullet and sheepshead (curves

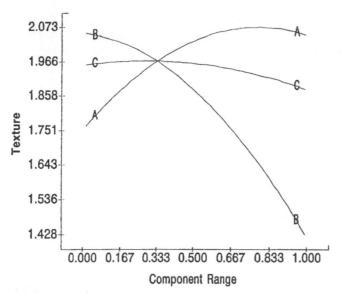

Figure 9.12   The response traces for A = mullet, B = sheepshead, and C = croaker over the ranges $0 \leq x_i \leq 1$.

A and *B*). This lack of nonlinear blending, particularly between mullet and croaker and between sheepshead and croaker, is reflected in the fitted equation Eq. 9.29 where the coefficient estimates 0.36 and 0.42 of the nonlinear blending terms between croaker and each of the other two fish types are not significantly different from zero. Mullet and sheepshead blend nonlinearly since the coefficient estimate 0.88 produces a t value of $t = 3.526$, which is significantly different from zero at the 0.0124 level.

In summary, the response trace plot is an effective tool for observing how estimated values of the response behave when changes are made in the proportion of each component separately along the component rays. We have shown a response trace plot for a three-component example, but the real utility of the response trace plot is that it can be drawn for many ($q > 3$) components; see, for example, Piepel (1982), who plotted the effects of 11 chemical components on leachability of glass used for waste vitrification. Furthermore, the response trace plot is a must when trying to determine the roles played by (the effects of) each of the components on the response and the experimental region is a highly constrained region inside the simplex.

In addition to mixture proportions, there are other variables called process variables, which represent external conditions whose settings can affect the blending properties of the mixture components. Some example of mixture experiments with process variables are:

Mixture Component Experiment	Process Variables
1. Three saltwater fish species (mullet, sheepshead, and croaker) blended to form sandwich patties. The measured responses are patty texture and shrinkage when fried.	Cooking temperatures (375°F and 425°F). Cooking time (5 min and 8 min)
2. Three gasoline components (Cat. Cracked, C_5-isomer and C_6-isomer) were blended to increase motor octane rating, thereby reducing motor knocking. The measured response was motor octane rating.	Camshaft speed of four cylinder engine (2500 and 5000 rpm)
3. Feeding trials in which pigs are fed proportions of three rations (A, B, and C). The measured response is the gain in weight over the trial period of six weeks.	Size or weight (30–40 lbs versus 70–80 lbs), age (6 months and 12 months old), lactation stage, etc.

Including process variables in mixture experiments involves the construction of designs utilizing the different settings of the process variables as well as the combined modeling of the mixture components and the process variables. We discuss both areas now.

9.4 INCLUDING PROCESS VARIABLES IN MIXTURE EXPERIMENTS

In mixture experiments with process variables, probably the most standard type of design is a lattice arrangement in the mixture components (a $\{q, m\}$ simplex lattice or simplex-centroid) set up at each point of a factorial arrangement in the levels of the process variables. For example, with three components having the proportions x_1, x_2, and x_3, suppose there are also two process variables, denoted by z_1 and z_2, and each process variable is to be studied at two levels. If a six-term quadratic model in x_1, x_2, and x_3 is to be fitted to data collected at the points of a $\{3, 2\}$ simplex lattice, the combined $\{3, 2\}$ lattice $\times 2^2$ factorial arrangement, displayed in Figure 9.13 can be used.

The terms in the most complete combined model in the mixture component proportions and the process variables are obtained by taking the product of the terms in the mixture component model with those in the process variable model. The six-term quadratic model in three components,

$$Y = \beta_1 x_1 + \beta_2 x_2 + \beta_3 x_3 + \beta_{12} x_1 x_2 + \beta_{13} x_1 x_3 + \beta_{23} x_2 x_3 + \varepsilon \qquad (9.30)$$

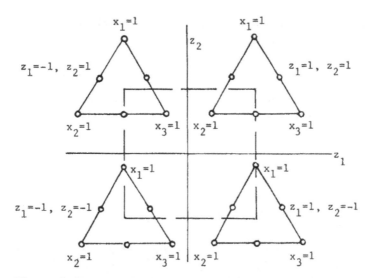

Figure 9.13   The 24 points of the combined {3, 2} simplex lattice by 2^2 factorial.

when combined with the four-term main and interaction effects model in z_1 and z_2,

$$Y = \alpha_0 + \alpha_1 z_1 + \alpha_2 z_2 + \alpha_{12} z_1 z_2 + \varepsilon \tag{9.31}$$

produces the 24-term combined model

$$
\begin{aligned}
Y &= \{(9.30)(9.31)\} \\
&= \sum_{i=1}^{3} \left[\gamma_i^0 + \sum_{l=1}^{2} \gamma_i^l z_l + \gamma_i^{12} z_1 z_2 \right] x_i \\
&\quad + \sum_{i<j}^{3} \left[\gamma_{ij}^0 + \sum_{l=1}^{2} \gamma_{ij}^l z_l + \gamma_{ij}^{12} z_1 z_2 \right] x_i x_j + \varepsilon
\end{aligned}
\tag{9.32}
$$

where $\gamma_i^l = \beta_i \alpha_l$, $\gamma_{ij}^l = \beta_{ij} \alpha_l$, $l = 0, 1, 2$, $\gamma_i^{12} = \beta_i \alpha_{12}$, and $\gamma_{ij}^{12} = \beta_{ij} \alpha_{12}$. The coefficients γ_i^0, $i = 1, 2, 3$, and γ_{ij}^0, $i < j$, are measures of the linear blending of component i and nonlinear blending of components i and j, respectively, when averaged over all the combinations of the levels of both process variables. The coefficients γ_i^l and γ_{ij}^l, $l = 1$ and 2, measure the main effect of process variable z_l on the linear blending of component i and the nonlinear blending of components i and j, respectively. Similarly, the coefficients γ_i^{12} and γ_{ij}^{12} represent the interaction effect of z_1 and z_2 on the linear blending of component i and nonlinear blending of components i and j, respectively. These meanings will be made clearer in Section 9.4.1.

When fitting the Scheffé-type mixture models, Eqs. 9.4 to 9.7, to data collected at the points of lattice arrangements which are set up at the 2^n factorial settings of n process variables, the number of design points can become excessive for large q ($q \geq 4$) and n ($n \geq 3$). A reduction in the total number of design points is achieved by performing only a subset (fraction) of the points in a complete lattice $\times 2^n$ design. The subset can be chosen from the complete factorial arrangement by selecting a 2^{n-p} ($p \geq 1$) fractional factorial design in the process variables, see for example, Cornell and Gorman (1984), or by selecting only a subset of the lattice points depending on the reduced form of the mixture model to be fitted. In this latter case, for example, two 10-point designs for fitting the following three-component quadratic model with one process variable (z), which is at two levels,

$$Y = \gamma_1^0 x_1 + \gamma_2^0 x_2 + \gamma_3^0 x_3 + \gamma_{12}^0 x_1 x_2 + \gamma_{13}^0 x_1 x_3 + \gamma_{23}^0 x_2 x_3$$
$$+ \gamma_1^1 x_1 z + \gamma_2^1 x_2 z + \gamma_3^1 x_3 z + \varepsilon$$

are displayed in Figure 9.14. The points of the two designs were chosen from the complete set of $12 = 6 \times 2$ candidate points using the DETMAX algorithm (Mitchell, 1974). These two designs are equivalent in terms of their $|\mathbf{X'X}|$ value. For additional reading on generated fractional lattice designs, see Piepel and Cornell (1987).

9.4.1 An Example of Modeling Three Mixture Components and One Process Variable

Three fish types were blended to form sandwich patties. The fish types were mullet (x_1), sheepshead (x_2), and croaker (x_3). Seven combinations of the fish were prepared (each patty weighing 100 g) and cooked at each of two oven temperatures (375° and 425°F) resulting in 14 treatment conditions. The response

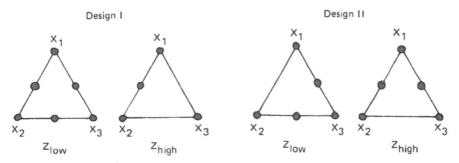

Figure 9.14 Two 10-point designs for fitting the quadratic by main effects model, $Y = \sum_{i=1}^{3} \gamma_i^0 x_i + \sum\sum_{i<j}^{3} \gamma_{ij}^0 x_i x_j + \sum_{i=1}^{3} \gamma_i^1 x_i z + \varepsilon$.

recorded on the patties was the average breaking force (texture), measured in grams of force, using the average of three Instron plunger readings on each patty. The texture of mullet is known to be firm (> 2000 g), that of the croaker soft (< 1000 g), and the texture of the sheepshead is somewhere in between firm and soft.

With each of the 14 treatment combinations, two replicate patties were prepared, tested, and their values averaged. The breaking force (grams $\times 10^{-3}$) values and the design settings are listed in Table 9.6. Also included in Table 9.6 are response designations to illustrate the formulas for calculating the estimates of the coefficients of the terms in a special cubic model.

Fitted to the simplex-centroid design data at the high cooking temperature (425°F) only, the special cubic model is

$$\hat{Y}(\mathbf{x}, 425) = 2.86x_1 + 1.60x_2 + 1.10x_3 - 2.80x_1x_2$$
$$\qquad\quad (0.05) \quad\ (0.05) \quad\ (0.05) \quad\ (0.25)$$
$$\qquad - 0.68x_1x_3 + 0.60x_2x_3 + 3.96x_1x_2x_3$$
$$\qquad\quad (0.25) \qquad (0.25) \qquad (1.76)$$
$$\tag{9.33}$$

where the estimates of the linear terms are the texture values at the pure fish blends, respectively. Since the texture values in Table 9.6 are averages of two patties, the variance of the estimates of the linear terms is $\sigma^2/2$ where σ^2 is the variance of the texture value of a single patty. The estimate of the coefficient

Table 9.6 Average Breaking Force Values of Fish Patties—Design Is a Simplex-Centroid $\times$ 2 Temperatures Arrangement

Mullet (x_1)	Sheepshead (x_2)	Croaker (x_3)	Cooking Temperature (°F)	Average Breaking Force (grams $\times 10^{-3}$)	Response Designations
1	0	0	375	1.84	
1	0	0	425	2.86	$\bar{Y}_1$
0	1	0	375	1.51	
0	1	0	425	1.60	$\bar{Y}_2$
0	0	1	375	0.67	
0	0	1	425	1.10	$\bar{Y}_3$
$\frac{1}{2}$	$\frac{1}{2}$	0	375	1.29	
$\frac{1}{2}$	$\frac{1}{2}$	0	425	1.53	$\bar{Y}_{12}$
$\frac{1}{2}$	0	$\frac{1}{2}$	375	1.42	
$\frac{1}{2}$	0	$\frac{1}{2}$	425	1.81	$\bar{Y}_{13}$
0	$\frac{1}{2}$	$\frac{1}{2}$	375	1.16	
0	$\frac{1}{2}$	$\frac{1}{2}$	425	1.50	$\bar{Y}_{23}$
$\frac{1}{3}$	$\frac{1}{3}$	$\frac{1}{3}$	375	1.59	
$\frac{1}{3}$	$\frac{1}{3}$	$\frac{1}{3}$	425	1.68	$\bar{Y}_{123}$

of the crossproduct term x_1x_2 is calculated using Eq. 9.8. For this data we get

$$b_{12} = 4\bar{Y}_{12} - 2(\bar{Y}_1 + \bar{Y}_2) = 4(1.53) - 2(2.86 + 1.60) = -2.80$$

The estimate -2.80 has a variance equal to $16(\sigma^2/2) + 4(\sigma^2/2 + \sigma^2/2) = 12\sigma^2$. Finally, the least squares formula for calculating the estimate b_{123}, of the coefficient of the special cubic term is

$$27\bar{Y}_{123} - 12(\bar{Y}_{12} + \bar{Y}_{13} + \bar{Y}_{23}) + 3(\bar{Y}_1 + \bar{Y}_2 + \bar{Y}_3)$$

which numerically gives

$$b_{123} = 27(1.68) - 12(1.53 + 1.81 + 1.50) + 3(2.86 + 1.60 + 1.10)$$
$$= 3.96$$

and this estimate has a variance equal to $1188\,(\sigma^2/2) = 594\sigma^2$. An estimate of σ^2, $s^2 = 0.0052$, was calculated from the two replicate patties at each of the seven treatments at the high cooking temperature. The quantities in parenthesis below the coefficient estimates in Eq. 9.33 are their estimated standard errors.

At the low cooking temperature (375°F), the special cubic model is

$$\hat{Y}\,(\mathbf{x}, 375) = \underset{(0.04)}{1.84x_1} + \underset{(0.04)}{1.51x_2} + \underset{(0.04)}{0.67x_3} - \underset{(0.21)}{1.54x_1x_2} \tag{9.34}$$
$$+ \underset{(0.21)}{0.66x_1x_3} + \underset{(0.21)}{0.28x_2x_3} + \underset{(1.50)}{8.55x_1x_2x_3}$$

where the between patties estimate of σ^2 is $s^2 = 0.0038$. If we combine the data from both cooking temperatures, the fitted equation for average breaking force is

$$\hat{Y}\,(\mathbf{x}, z) = \hat{\gamma}_1^0 x_1 + \hat{\gamma}_2^0 x_2 + \hat{\gamma}_3^0 x_3 + \hat{\gamma}_{12}^0 x_1x_2 + \hat{\gamma}_{13}^0 x_1x_3 + \hat{\gamma}_{23}^0 x_2x_3$$
$$+ \hat{\gamma}_{123}^0 x_1x_2x_3 + \hat{\gamma}_1^1 x_1 z + \hat{\gamma}_2^1 x_2 z + \hat{\gamma}_3^1 x_3 z + \hat{\gamma}_{12}^1 x_1x_2 z$$
$$+ \hat{\gamma}_{13}^1 x_1x_3 z + \hat{\gamma}_{23}^1 x_2x_3 z + \hat{\gamma}_{123}^1 x_1x_2x_3 z$$
$$= \underset{(0.03)}{2.35x_1} + 1.555x_2 + 0.885x_3 - \underset{(0.16)}{2.17x_1x_2} - 0.01x_1x_3$$
$$+ 0.44x_2x_3 + \underset{(1.16)}{6.255x_1x_2x_3} + \underset{(0.03)}{0.51x_1 z} + 0.045x_2 z$$
$$+ 0.215x_3 z - \underset{(0.16)}{0.63x_1x_2 z} - 0.67x_1x_3 z + 0.16x_2x_3 z$$
$$- \underset{(1.16)}{2.295x_1x_2x_3 z} \tag{9.35}$$

where z is the coded variable, $z = (T - 400°)/25°$, for cooking temperature.

The coefficient estimates in the combined model 9.35 are simple functions of the coefficient estimates in the models of Eqs. 9.33 and 9.34. For example, some of the coefficient estimates are

$$\hat{\gamma}_1^0 = \frac{2.86 + 1.84}{2} = 2.35 \qquad \hat{\gamma}_1^1 = \frac{2.86 - 1.84}{2} = 0.51$$

$$\hat{\gamma}_{12}^0 = \frac{-2.80 - 1.54}{2} = -2.17 \qquad \hat{\gamma}_{12}^1 = \frac{-2.80 + 1.54}{2} = -0.63$$

$$\hat{\gamma}_{123}^0 = \frac{3.96 + 8.55}{2} = 6.255 \qquad \hat{\gamma}_{123}^1 = \frac{3.96 - 8.55}{2} = -2.295$$

and the estimated variance of a breaking force value with the combined model is the average of the estimated variances with models 9.33 and 9.34, that is, $s^2 = (0.0052 + 0.0038)/2 = 0.0045$.

Having fitted the combined model, some approximate results from tests performed on the individual coefficient estimates and interpretations of these tests are as follows:

$\hat{\gamma}_{12}^0 = -2.17 < 0$

Since $t = -2.17/0.16 = -13.56$, we infer that synergistic blending (a softening of the texture) of mullet with sheepshead was present. Also, the significance of the test for $\hat{\gamma}_{12}^1 = -0.63$ (i.e., $t = -0.63/0.16 = -3.94$) implies that the synergistic effect of mullet with sheepshead was greater at the higher temperature.

$\hat{\gamma}_{13}^1 = -0.67 < 0$

Since $t = -0.67/0.16 = -4.19$, we see that raising the cooking temperature had a significant effect on the synergistic blending of mullet with croaker. Surprisingly, at low temperature the blending of the soft textured croaker with mullet did not produce as soft a patty as expected (since in (9.34) $b_{13} = 0.66 > 0$), but at the high cooking temperature, synergistic blending was observed (in Eq. 9.33, $b_{13} = -0.68 < 0$).

$\hat{\gamma}_{123}^0 = 6.255 > 0$

Since $t = 6.255/1.16 = 5.39$, the texture of the blend of the three fish (i.e., $x_1 = x_2 = x_3 = 1/3$) was higher than found with some of the two fish combinations. However, the nearly significant estimate $\hat{\gamma}_{123}^1 = -2.295$ ($t = -1.98$) implies that at the high cooking temperature, the difference between the firmness of the three-fish patties and the firmness of the two-fish patties was not as large as observed at the low cooking temperature. In other words, the difference between the texture of the three-fish patties and the two-fish patties is smaller when the patties are cooked at the high temperature than when cooked at the low temperature.

Multiresponse experiments were discussed in Chapter 7 where the data analysis methods (parameter estimation and response optimization) and the construction of designs focused on nonmixture settings, that is, the input variables were independent. In the case of investigating multiple responses in a mixture setting where the q component proportions are dependent, it is possible to transform or map from the q-component mixture system to a metric of $q - 1$ independent variables and then apply the techniques presented in Chapter 7 to the independent variables. Two types of linear transformations to $q - 1$ independent variables are discussed in Chapter 3 of Cornell (1990a).

In the next section we remain in the metric of the mixture components and present a graphical technique, called a *biplot*, for investigating multiple responses in mixture experiments. With the biplot we shall display the correlational structure among the multiple responses and, in addition, provide an approximate measure of the effects of the separate mixture components on each of the responses. A three-component plastics compounding experiment with five separate responses is presented to illustrate the usefulness of the biplot. A four-component plus one process variable photographic film coating experiment in which 14 color responses were biplotted is presented and discussed in detail by Smith and Cornell (1993).

9.5 BIPLOTS FOR MULTIPLE RESPONSES

Almost all mixture experiments have multiple ($r \geq 2$) response variables $Y_1, Y_2, \ldots, Y_r$ where some are perhaps more important than others. Some examples are,

1. Solubility Y_1, tint strength Y_2, porosity Y_3, and sheen Y_4 of a house paint consisting of pigment, vehicle, solvent, and additives.
2. Viscosity Y_1, chemical durability Y_2, and electrical conductivity Y_3 of borosilicate glass, which is a solid, stable medium used for the disposal of high-level radioactive nuclear wastes and which comprises up to nine major oxide components.
3. Tensile modulus Y_1, tensile strength Y_2, warp Y_3, flexural modulus Y_4, and flexural strength Y_5 of a plastic compound consisting of resin, glass fibers, and microspheres.

Usually the measured responses $Y_1, Y_2, \ldots, Y_r$ are correlated with one another (some positively and some negatively) so that finding a feasible mixture blend for which the values of all the response variables fall within certain specified limits simultaneously is not an easy task. Since one cannot optimize all the responses simultaneously, one must settle on some sort of compromise. Identifying the compromises so that tradeoffs can be made is possible through the use of a biplot.

In addition to displaying graphically the multiple responses, it is possible in the biplot to show approximately how each of the components affects each of the r responses. However, a supplementary plot, called a j-plot, which does not contain the response vectors in it, is better suited for illustrating the *composite effects* of the components where the composite effect of a component is a measure of how the particular component affects the entire group of responses. The steps taken in constructing a j-plot are described at the end of the example in the next section, while a figure of a j-plot for the example in the next section is presented by Smith and Cornell (1993).

The concept of a *component effect* in a multicomponent mixture system is quite simple. For a particular response Y, in which the linear blending model of Eq. 9.4 has been fitted and is considered to be adequate, the effect of component i is the difference in the estimated response values for blends in which component i is at its high and low proportion settings, respectively, that is, $\hat{Y}(\mathbf{x}_{i,\,high})$ − $\hat{Y}(\mathbf{x}_{i,\,low})$. In terms of a straight-line response trace (Section 9.3.3), the effect is the difference in the estimated response values at the ends of the trace where the difference may be negative or positive.

To calculate the effect of component i on some response Y_p, it is necessary first to determine the coordinates of the blends $\mathbf{x}_{i,\,high}$ and $\mathbf{x}_{i,\,low}$. If the experimental region is the whole simplex, $\mathbf{x}_{i,\,high}$ is $x_i = 1$, $x_j = 0$, $j \neq i$, and $\mathbf{x}_{i,\,low}$ is $x_i = 0$, $x_j = 1/(q − 1)$ for all $j \neq i$. When the experimental region is a constrained region inside the simplex, then the coordinates of $\mathbf{x}_{i,\,high}$ and $\mathbf{x}_{i,\,low}$ are found using Eqs. 9.27 and 9.28. In fact, if the x_i-ray intersects the boundary $x_i = U_i$ of the constrained region, then the coordinates of $\mathbf{x}_{i,\,high}$ are $x_i = U_i$, $x_j = s_j(1 − U_i)/(1 − s_i)$, for all $j \neq i$, and if the x_i-ray intersects the boundary $x_i = L_i$, the coordinates of $\mathbf{x}_{i,\,low}$ are $x_i = L_i$, $x_j = s_j(1 − L_i)/(1 − s_i)$, for all $j \neq i$. In this case, and having fitted the linear blending model $\hat{Y}(\mathbf{x}) = \sum_{i=1}^{q} b_i x_i$, the effect of component i on the response, denoted by E_i, is

$$E_i = b_i(U_i − L_i) − \sum_{j \neq i}^{q} b_j s_j(U_i − L_i)/(1 − s_i)$$

$$= R_i\left[b_i − \sum_{j \neq i}^{q} b_j s_j/(1 − s_i) \right] \tag{9.36}$$

where $R_i = U_i − L_i$ is the range for component i.

A $q \times 1$ vector of component effects on the response Y_p, for example, can be written using Eq. 9.36 as follows. Let the $q \times 1$ vector of coefficient estimates for the fitted first-order model on Y_p be denoted by $\mathbf{b}_p = (b_1, b_2, \ldots, b_q)'_p$. Then the $q \times 1$ vector of component effects is

$$\mathbf{E} = \mathbf{D}\mathbf{b}_p$$

$$= [d_{ij}]\mathbf{b}_p \tag{9.37}$$

where the elements of the $q \times q$ matrix $\mathbf{D}$ are $d_{ii} = R_i$ and $d_{ij} = -R_i s_j / (1 - s_i)$, $i \neq j$. For r responses $Y_1, Y_2, \ldots, Y_r$, we shall use the matrix $\mathbf{D}$ of coefficients in Eq. 9.37 to compute a $q \times r$ matrix of component effects.

Before presenting the methodology used in constructing a biplot, let us first remove the different units of scale of the r response variables $Y_1, Y_2, \ldots, Y_r$ by standardizing them. Let Y_{up} represent the uth observed value of the pth response variable, $u = 1, 2, \ldots, N$, $p = 1, 2, \ldots, r$ and let the number of mixture components be q. Define

$$z_{up} = \frac{Y_{up} - \bar{Y}_{\cdot p}}{\sqrt{\sum_{u=1}^{N} (Y_{up} - \bar{Y}_{\cdot p})^2}} \tag{9.38}$$

where $\bar{Y}_{\cdot p} = \sum_{u=1}^{N} Y_{up} / N$. Then z_{up} is the uth standardized value for the pth response. Letting $\mathbf{z}_p$ represent the $N \times 1$ vector of values of the pth standardized response, a $N \times r$ matrix of standardized responses is $\mathbf{Z} = [\mathbf{z}_1 : \mathbf{z}_2 : \cdots : \mathbf{z}_r]$.

Let us write the linear blending model for the pth standardized response in matrix notation as

$$\mathbf{z}_p = \mathbf{X}\boldsymbol{\beta}_p + \boldsymbol{\varepsilon}_p, \qquad p = 1, 2, \ldots, r \tag{9.39}$$

where $\mathbf{X}$ is an $N \times q$ matrix in which the uth row contains the mixture component proportions corresponding to the uth trial, $\boldsymbol{\beta}_p$ is a $q \times 1$ vector of coefficients for the linear blending terms, and $\boldsymbol{\varepsilon}_p$ is an $N \times 1$ vector of random errors. A $q \times r$ matrix of least squares estimates of the coefficients in the r separate models of Eq. 9.39 is (see Section 7.2.1)

$$\mathbf{B} = [\mathbf{b}_1 : \mathbf{b}_2 : \cdots : \mathbf{b}_r] \tag{9.40}$$
$$= (\mathbf{X}'\mathbf{X})^{-1}\mathbf{X}'\mathbf{Z}$$

Furthermore, by generalizing Eq. 9.40 to the multiple response setting, a $q \times r$ matrix of component effects is

$$\mathbf{E} = \mathbf{DB} \tag{9.41}$$

The component effects matrix $\mathbf{E}$ *in Eq. 9.41 is what we seek to biplot.*

The biplot is a method of displaying a lower rank (usually rank 2) approximation of a matrix whose rank is greater than 2. In our biplot, where the display will be two dimensional, the first step taken is to approximate the $q \times r$ effects matrix $\mathbf{E}$ in Eq. 9.41 with a $q \times r$ matrix $\mathbf{E}^*$ having rank 2. Once we have obtained the matrix $\mathbf{E}^*$, the biplot is then used to plot the rows and the columns of $\mathbf{E}^*$.

To define the matrix $\mathbf{E}^*$, first we write the matrix $\mathbf{E}$ in Eq. 9.41 as a product of three matrices. Suppose the rank of the matrix $\mathbf{E}$ is τ, where $\tau \leq \min(q - 1, r)$ since the rank of $\mathbf{D}$ is $q - 1$ and the rank of $\mathbf{B}$ is r. It is possible to write

E, using the singular-value decomposition, as

$$\mathbf{E} = \mathbf{P\Lambda Q}' \qquad (9.42)$$

where **P** is $q \times \tau$, $\mathbf{\Lambda}$ is $\tau \times \tau$, and **Q**′ is $\tau \times r$. The columns of **P** are orthonormal eigenvectors of **EE**′ and the columns of **Q** are orthonormal eigenvectors of **E**′**E**; that is, $\mathbf{P'P} = \mathbf{Q'Q} = \mathbf{I}$. The matrix $\mathbf{\Lambda}$ is a diagonal matrix of ordered singular values $l_1 \geq l_2 \geq \cdots \geq l_\tau > 0$. These singular values are the positive square roots of the nonzero eigenvalues of the square matrix **E**′**E** as well as of **EE**′.

The rank-two matrix **E***, which approximates the matrix **E** in Eq. 9.41, is defined by letting $\tau = 2$ in Eq. 9.42; that is,

$$\mathbf{E}^* = l_1 \mathbf{p}_1 \mathbf{q}_1' + l_2 \mathbf{p}_2 \mathbf{q}_2' \qquad (9.43)$$

$$= [\mathbf{p}_1 \colon \mathbf{p}_2] \begin{bmatrix} l_1 & 0 \\ 0 & l_2 \end{bmatrix} \begin{bmatrix} \mathbf{q}_1' \\ \mathbf{q}_2' \end{bmatrix}$$

In Eq. 9.43, $\mathbf{p}_1$ and $\mathbf{p}_2$ are the first two (leftmost two) columns of **P** in Eq. 9.41 and $\mathbf{q}_1'$ and $\mathbf{q}_2'$ are the top two rows of **Q**′ in Eq. 9.42. Equation 9.43 is the best rank-two approximation of the matrix **E** in the sense of minimizing the Euclidean norm of the matrix of residuals

$$\|\mathbf{E} - \mathbf{E}^*\|.$$

We are now ready to biplot the elements of the matrix **E***. To do this, Gabriel (1981) and Jackson (1991) discussed two types of factorizations that are possible with **E*** based on Eq. 9.43. Either factorization can be used, but we shall use the one referred to as the **GH**′ factorization; that is, $\mathbf{E}^* = \mathbf{GH}'$, where

$$\mathbf{G} = [\mathbf{p}_1 \colon \mathbf{p}_2] \qquad (9.44)$$

$$\mathbf{H} = [l_1 \mathbf{q}_1 \colon l_2 \mathbf{q}_2]$$

In Eq. 9.44, the row markers $r_1, r_2, \ldots, r_q$ for the biplot are the q rows of **G** (these markers will define the mixture component symbols in the plot) and the column markers $c_1, c_2, \ldots, c_r$ for the biplot, which are the r rows of **H**, define the response vectors.

In Eq. 9.44, the columns of **G** are orthonormal and therefore $\mathbf{G'G} = \mathbf{I}$. As a consequence,

$$\mathbf{E}^{*\prime}\mathbf{E}^* = \mathbf{HG'GH'} \qquad (9.45)$$

$$= \mathbf{HH}'$$

When the columns of **E** are mean corrected, then **HH**′ in Eq. 9.45 is $n - 1$ times the variance-covariance matrix of the elements in **E*** and the cosines between the column vectors of **H**′ are equal to the correlations between the columns of **E***. If the rank-two approximation of **E** by **E*** is good [see Smith and Cornell (1993) for a measure of goodness], then the correlations between the columns of $\mathbf{E}^*$ will closely approximate the correlations between the columns of $\mathbf{E}$, where these latter columns represent the r responses.

9.5.1 A Plastics-Compounding Example

The following example is taken from the paper by Smith and Cornell (1993). Virtually all plastics are complex formulations of polymer and additives, the latter being added to improve key properties of the plastic compound. In formulating the plastic compound, a virgin resin was studied alone and with each of two additives (glass fibers and microspheres). A lower bound of 80% of the mixture was set on the resin, allowing the two additives, individually and combined, to make up at most 20% of any mixture. These bounds, when expressed in terms of the component proportions x_1 (resin), x_2 (glass fibers), and x_3 (microspheres), are

$$0.80 \le x_1, \quad 0 \le x_2 \le 0.20, \quad 0 \le x_3 \le 0.20 \qquad (9.46)$$

The triangular factor space defined by the constraints in Eq. 9.46 is shown in Figure 9.15. Also shown in Figure 9.15 are seven distinct mixture blends that were run in the experiment. The centroid blend was replicated twice, making $N = 8$.

Five properties of the plastic compound were measured. They were tensile strength Y_1, tensile modulus Y_2, flexural strength Y_3, flexural modulus Y_4, and warp Y_5. Table 9.7 lists the values of the properties given as a percent of that of virgin resin alone along with the proportions of resin, glass fibers, and microspheres in each of the seven blends.

The first step in the analysis is to define the matrix of centered and standardized responses $\mathbf{Z}$. From Eq. 9.38 we have

$$
\mathbf{Z}_{8\times5} =
\begin{array}{ccccc}
\text{TS} & \text{TM} & \text{FS} & \text{FM} & \text{W} \\
\end{array}
\left[
\begin{array}{ccccc}
-0.2847 & -0.4726 & -0.3461 & -0.4086 & 0.1879 \\
0.3350 & -0.0018 & 0.2943 & 0.0199 & -0.1708 \\
-0.3974 & -0.3549 & -0.3461 & -0.3199 & 0.2528 \\
0.0292 & -0.0018 & 0.0598 & -0.0429 & 0.1650 \\
0.0131 & -0.0477 & 0.0237 & -0.1279 & 0.1460 \\
0.6569 & 0.7360 & 0.6551 & 0.7770 & -0.8500 \\
0.1016 & 0.2910 & 0.1319 & 0.2784 & -0.0448 \\
-0.4537 & -0.1482 & -0.4724 & -0.1759 & 0.3139 \\
\end{array}
\right]
\qquad (9.47)
$$

The positive and negative signs in the W column of $\mathbf{Z}$ have been reversed so that a high positive value (or low warp value) is preferred.

Calculating the component effects on each of the five responses is the next order of business. Listed in Table 9.8 are the estimated coefficients for resin (x_1), glass fibers (x_2), and microspheres (x_3) for each of the five standardized

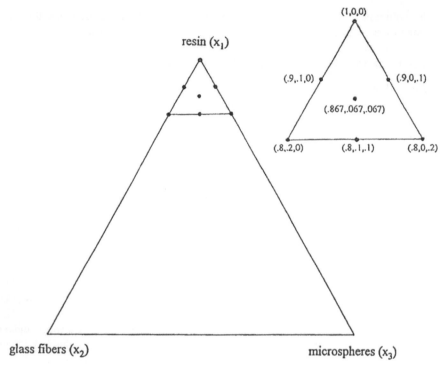

Figure 9.15 The experimental region and coordinate settings of the seven blends in the plastics-compounding experiment.

responses based on the linear blending model

$$z_p = \beta_1 x_1 + \beta_2 x_2 + \beta_3 x_3 + \varepsilon_p, \qquad p = 1, 2, 3, 4, 5$$

along with the corresponding R^2 value for each of the fitted models. A check on the adequacy of fit of each of the five fitted models did not suggest the need to include any nonlinear blending terms, so the linear blending model was considered adequate for all five responses. The matrix $\mathbf{D}$ in Eq. 9.37, for computing the component effects across the five standardized responses, is of the form

$$\mathbf{D} = \begin{bmatrix} 0.2000 & -0.1000 & -0.1000 \\ -0.1286 & 0.1385 & -0.0099 \\ -0.1286 & -0.0099 & 0.1385 \end{bmatrix} \qquad (9.48)$$

Table 9.7 Plastics Compounding Data

| Mixture Components | | | Properties (Y_p) | | | | |
Resin x_1	Glass Fibers x_2	Microspheres x_3	Tensile Strength (Y_1)	Tensile Modulus (Y_2)	Flexural Strength (Y_3)	Flexural Modulus (Y_4)	Warp (Y_5)
1	0	0	100	100	100	100	100
0.9	0.1	0	177	264	171	216	194
0.9	0	0.1	86	141	100	124	83
0.867	0.067	0.067	139	264	145	199	106
0.867	0.067	0.067	137	248	141	176	111
0.8	0.2	0	217	521	211	421	372
0.8	0.1	0.1	148	366	153	286	161
0.8	0	0.2	79	213	86	163	67

Table 9.8 Coefficient Estimates in the Linear Blending Models for Each of the Standardized Responses

	Coefficient Estimates			
Response	Resin (x_1)	Glass Fibers (x_2)	Microspheres (x_3)	R^2
Tensile strength, Y_1	−0.24	4.52	−1.38	0.98
Tensile modulus, Y_2	−0.53	5.59	1.35	0.98
Flexural strength, Y_3	−0.28	4.69	−1.10	0.98
Flexural modulus, Y_4	−0.49	5.44	0.95	0.96
Warp, Y_5	0.27	−4.48	1.03	0.89

Borrowing the coefficient estimates from Table 9.8, the effects matrix **E**, from Eq. 9.37, is of the form

$$\mathbf{E} = \begin{bmatrix} -0.3623 & 0.8008 & -0.4134 & -0.7381 & 0.3982 \\ 0.6713 & 0.8298 & 0.6955 & 0.8078 & -0.6643 \\ -0.2054 & 0.2001 & -0.1636 & 0.1414 & 0.1523 \end{bmatrix} \tag{9.49}$$

The rank of the matrix **E** is 2 since the rank of the **D** matrix in Eq. 9.48 is 2. To find the elements of the matrix **E***, the matrices **P**, Λ, and **Q**′ in Eq. 9.42 are

$$\mathbf{P} = \begin{bmatrix} -0.6115 & -0.4164 \\ 0.7912 & -0.3167 \\ -0.0048 & 0.8522 \end{bmatrix}, \quad \Lambda = \begin{bmatrix} 2.0751 & 0 \\ 0 & 0.4577 \end{bmatrix}$$

$$\mathbf{Q}' = \begin{bmatrix} 0.3632 & 0.5519 & 0.3874 & 0.5252 & -0.3710 \\ -0.5173 & 0.5269 & -0.4103 & 0.3757 & 0.3810 \end{bmatrix} \tag{9.50}$$

The rank-two approximation matrix **E*** is identical in form to **E** in Eq. 9.49, as it must be since the rank of **E** is 2, and therefore the biplot will be of the elements of **E** in Eq. 9.49.

For the **GH**′ factorization, the form of the matrices **G** and **H**′ are **G** = **P** in Eq. 9.50, and

$$\mathbf{H}' = \Lambda\mathbf{Q}' = \begin{bmatrix} 0.7537 & 1.1453 & 0.8038 & 1.0899 & -0.7698 \\ -0.2368 & 0.2412 & -0.1878 & 0.1720 & 0.1744 \end{bmatrix}$$

The biplot, shown in Figure 9.16, was constructed by plotting the rows of **G** and using an asterisk to signify the location of each of the three components in the plot, along with plotting the columns of **H**′ in the form of vectors.

The effect of a specific component on a particular response is estimated visually, by looking at the biplot of Figure 9.16 and proceeding as follows. From the symbol representing the component, draw a perpendicular projection onto

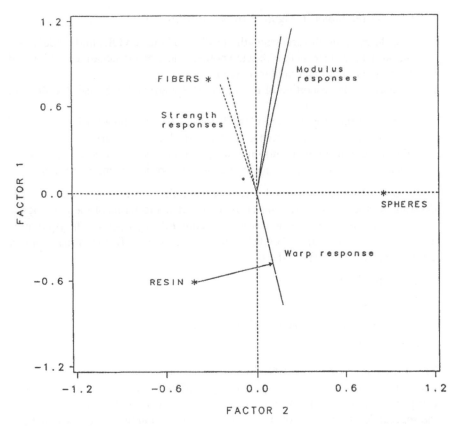

Figure 9.16 The biplot of the plastics-compounding experiment. The symbols for the components resin, fibers, and spheres are the rows of **G** and the response vectors are the columns of **H'**. The projection arrow from the resin component onto the warp vector helps to illustrate the effect of resin on the warp response.

the response vector of interest, extended if necessary. The effect is then estimated mentally by taking the product of the distance from the origin to the point of intersection of the projection (called the *d-length* of the projection) made onto the response vector and the length of the response vector projected upon. As an example, the row marker for RESIN has been projected onto the vector for the response Warp in Figure 9.16. The projection is made onto the positive side of the vector, indicating that RESIN affects Warp in a positive manner (the higher the proportion of RESIN, the greater the plastic warps, which is not desirable). The *d* length of the projection is the distance from the origin to the tip of the arrow.

Based on the biplot in Figure 9.16, we can infer that

1. Tensile strength, flexural strength, tensile modulus, and flexural modulus are all positively correlated with one another, but warp is negatively correlated with the strength and modulus responses.
2. Increasing the proportion of resin increases warp and decreases the strength and modulus responses.
3. Increasing the proportion of glass fibers is very favorable for increasing the strength and modulus responses and it also decreases warp.
4. Tensile and flexural moduli and warp increase while tensile and flexural strength decrease with an increase in the proportion of microspheres.

A more quantitative method for measuring the component effects and for determining the relative ordering of the components in terms of their composite effects on the five responses is possible by constructing a j-plot. The j-plot uses as vectors the row markers of the J matrix in another factorization (Gabriel 1971, 1981) of $E^* = JK'$, where

$$J = [l_1 p_1 : l_2 p_2] \tag{9.51}$$

$$K = [q_1 : q_2]$$

The columns of K are orthonormal, $K'K = I$, so that

$$E^* E^{*\prime} = JK'KJ' = JJ' \tag{9.52}$$

The q diagonal elements of $E^* E^{*\prime}$ approximate the q diagonal elements of EE', where the latter are equal to the squared lengths of the effects vectors for the mixture components. If the length of an effect vector is taken to be a relative measure of the magnitude of the effect of that component on the multiple responses, then the lengths of the row markers in J approximate the absolute magnitudes of the composite effects. For our example and from Eq. 9.51, the J matrix is

$$J = \begin{bmatrix} -1.2690 & -0.1906 \\ 1.6419 & -0.1450 \\ -0.0099 & 0.3900 \end{bmatrix} \begin{matrix} \text{resin} \\ \text{glass fibers} \\ \text{microspheres} \end{matrix} \tag{9.53}$$

The lengths of the row vectors or relative measures of resin, glass fibers, and microspheres are the square roots of the diagonal elements of JJ' ($=EE'$) and are 1.28, 1.65, and 0.39, respectively. As for the signs of the composite effects, it is not as easy to see from the elements of the J matrix but would be from the j-plot and it can be seen from Figure 9.16 that the composite effect of resin is opposite in sign to that of glass fibers while the composite effect of microspheres is somewhere in between those of resin and glass fibers.

In summary, the biplot display of an effects matrix of a q-component system on a set of r multiple responses is an extremely useful tool from the standpoint of studying the correlational structure of the responses when either or both of the following conditions hold: (a) the number of response variables is large (say $r > 4$) so that it is difficult even for the experienced analyst to understand the structure, and (b) the response variables are correlated with one another. This second condition is necessary with multiple responses to be able to approximate the response pattern in a plane (in two dimensions). Fortunately, most mixture experiments satisfy both of these conditions of having many response variables that are correlated, and thus the biplot is a valuable and essential tool when only linear blending is present. The development of the biplot when nonlinear blending is present among the components is a problem that remains to be solved.

9.6 SUMMARY

Many materials are mixtures of several components (ingredients). Characteristics, such as the quality of the product being manufactured, depend on the relative proportions of the components in the mixture. An example is the viscosity of a liquid detergent used in cleaning oil stains where the viscosity depends on the proportions of the components water, ethyl alcohol, urea, and sodium xylene sulphonate in the mixture.

Mixture experiments differ from standard factorial experiments, where in the latter the response varies depending on the actual amounts of each of the input variables, while in the former the mixture components represent proportionate amounts of the mixture. These proportions are by volume, weight, or mole fraction. The component proportions are nonnegative and, if expressed as fractions of the mixture, sum to unity. As a result of the constraint, $x_1 + x_2 + \cdots + x_q = 1$, the factor space takes the form of a $(q - 1)$-dimensional simplex where each x_i varies between 0 and 1.

For the purpose of studying the response over the entire simplex factor space, two classes of designs were presented: $\{q, m\}$ simplex lattices and the simplex-centroid designs. Two advantages of these designs are (1) simplicity of choosing the combinations of the proportions of the components to run, and (2) ease in estimating the mixture model coefficients when data are collected at the lattice points. Furthermore, with these designs equal coverage is given to each of the components, so that all pairs are studied with equal precision, all triplets are studied with equal precision, and so on.

In many mixture experiments, a subset or possibly all of the components must be present in each blend to produce an end result. Forcing component i to be present in all blends is equivalent to placing a lower bound L_i, on the compoent proportion, that is, $x_i \geq L_i$. When lower bounds are placed on one or more x_i, the factor space is a simplex located inside the original simplex. For the purpose of setting up designs and fitting models over the inner simplex region, pseudocomponents are suggested since both design construction and the fitting of polynomials are simplified when done in the pseudocomponent system.

When both lower and upper bounds ($L_i \le x_i \le U_i$) are forced on the component proportions, the factor space becomes more complicated in shape than the simplex. The factor space or constrained region is a convex polyhedron, and locating the coordinates of the extreme vertices of the region is the first order of business. Depending on the degree of the mixture model to be fitted, the vertices (or a subset of the total number of vertices) and convex combinations of the vertices are used as design points. When the number of components is large, say, for example, $q \ge 5$, locating the vertices of the constrained region requires a computer algorithm.

Process variables are external factors or conditions that can affect the blending properties of the mixture components. An example of a process variable is the cooking temperature when frying patties made from blends of pure ground beef with three different types of peanut soy meal. Including process variables (at two or more levels) in mixture experiments increases the size of the overall experimental program, but more often than not, the additional experimentation pays dividends in the end. For additional coverage of mixture designs other than the simplex lattices and the extreme vertices as well as of mixture models other than the Scheffé-type polynomials, the reader is directed to Cornell (1990a). Also discussed are types of transformations that take one from the dependent mixture component system to a system of $q - 1$ independent variables in which standard response surface designs, such as factorial and central composite designs, and standard polynomial models can be used.

EXERCISES

9.1   The Scheffé second-order polynomial in $q = 2$ components, whose proportions are denoted by x_1 and x_2, is

$$\eta = \beta_1 x_1 + \beta_2 x_2 + \beta_{12} x_1 x_2$$

Derive the least squares estimates b_1, b_2, and b_{12} of the parameters in the model, in terms of the response values, Y_1, Y_2, Y_3, and Y_4, where the response values are measured at the following blends:

(a) Y_1 at $x_1 = 1$, $x_2 = 0$
       Y_2, Y_3 at $x_1 = x_2 = \frac{1}{2}$
       Y_4 at $x_1 = 0$, $x_2 = 1$

(b) Y_1 at $x_1 = 1$, $x_2 = 0$
       Y_2 at $x_1 = \frac{2}{3}$, $x_2 = \frac{1}{3}$
       Y_3 at $x_1 = \frac{1}{3}$, $x_2 = \frac{2}{3}$
       Y_4 at $x_1 = 0$, $x_2 = 1$

In terms of providing minimum variance estimates of the parameters in the second-order model, which design, the one in (a), or the one in (b), is better?

9.2 The following data represent average numbers of mites ($\times 10^{-2}$) counted on plants layed out in triplicate where two types (A and B) of chemical pesticides were sprayed on the plants.

Pesticide Blend		Average Number of Mites ($\times 10^{-2}$)			
$A(x_1)$	$B(x_2)$	Y_1	Y_2	Y_3	$\bar{Y}$
100% (1)	0% (0)	3.8	3.0	3.7	3.5
50% ($\frac{1}{2}$)	50% ($\frac{1}{2}$)	1.8	2.2	2.0	2.0
0% (0)	100% (1)	4.0	4.7	4.8	4.5

(a) Plot the response average mite number versus each pesticide blend.
(b) Calculate the coefficient estimates for the fitted model

$$\hat{Y}(x) = b_1 x_1 + b_2 x_2 + b_{12} x_1 x_2$$

(c) The within blend estimate of σ^2 is $s^2 = 0.14$. Is there evidence of synergistic blending from using the two pesticides? Hint: Test $H_o : \beta_{12} = 0$ vs. $H_a : \beta_{12} \neq 0$.

9.3 The first six sets of numbers in the following data set were collected at the points (points 1–6) of a $\{3, 2\}$ simplex lattice. Points 7–10 were chosen as locations to collect observations for testing the fit of the second-order model that was fitted to the data at points 1–6.

Design Point	Component Proportions (x_1, x_2, x_3)	Observed Response Y_u	Check Point	(x_1, x_2, x_3)	Observed Response Y_u
1	(1, 0, 0)	11.0, 12.2	7	$(\frac{1}{3}, \frac{1}{3}, \frac{1}{3})$	18.2
2	(0, 1, 0)	8.6, 8.1	8	$(\frac{2}{3}, \frac{1}{6}, \frac{1}{6})$	15.4
3	(0, 0, 1)	16.4, 17.3	9	$(\frac{1}{6}, \frac{2}{3}, \frac{1}{6})$	9.6
4	$(\frac{1}{2}, \frac{1}{2}, 0)$	7.9, 7.4, 8.5	10	$(\frac{1}{6}, \frac{1}{6}, \frac{2}{3})$	16.2
5	$(\frac{1}{2}, 0, \frac{1}{2})$	13.4, 14.4, 15.0			
6	$(0, \frac{1}{2}, \frac{1}{2})$	15.9, 17.4, 16.6			

(a) Fit a second-order model to the data observed at points 1–6. Set up the ANOVA table ($N = 15$) and calculate an estimate of σ^2.
(b) With the fitted model in (a), predict the response value at the centroid (point 7) of the triangle. Compare the observed value $Y_{16} = 18.2$ to the value predicted with the fitted model. Based on the difference, $\hat{Y}_{16} - 18.2$, do you feel the special cubic model would be better than the quadratic model? If so, fit the special-cubic model to the data at points 1–7.

   (c) Fit a second-order model or a special-cubic model to all 19 observations. Set up the ANOVA table and test for lack of fit of the fitted
equation.

9.4 The following constraints are imposed on the three component proportions

$$0.20 \leq x_1 \leq 0.50 \quad 0.05 \leq x_2 \leq 0.65 \quad 0.15 \leq x_3 \leq 0.75$$

   (a) A first-order model is to be fitted to data collected from the constrained region. How many vertices does the region possess? List the
coordinates of the vertices in (x_1, x_2, x_3). If you could afford to perform 10 experiments, list the blends that you would recommend to
an experimenter keeping in mind that you would like to test for lack
of fit of the first-order model.

   (b) Suppose only the lower bounds are imposed, i.e., $0.20 \leq x_1$, $0.05 \leq$
x_2, $0.15 \leq x_3$. Define the pseudocomponents x_1', x_2', and x_3' using the
transformation in Eq. 9.15. Set up a $\{3, 2\}$ simplex lattice in the pseudocomponents and list the corresponding settings in x_1, x_2, and x_3.

9.5 A $\{2, 2\}$ simplex lattice in the two components, x_1 and x_2, was set up at
each of two levels (low and high) of a single process variable (z). The
data and the fitted model are:

(x_1, x_2):	$(1, 0)$	$(\frac{1}{2}, \frac{1}{2})$	$(0, 1)$
at $z = -1$	4.6, 4.8	5.0, 4.0	5.0, 5.6
at $z = +1$	2.6, 2.4	6.1, 7.1	4.0, 3.4

$$\hat{Y}(\mathbf{x}, z) = 3.6x_1 + 4.5x_2 + 6.0x_1x_2 - 1.1x_1z - 0.8x_2z + 8.0x_1x_2z$$

Answer the following questions posed by the experimenter and explain the
reasons for your answers:

   (a) Do the mixture components have an effect on the response behavior? If
so, explain.

   (b) Is the shape of the mixture surface different at the different levels of the
process variable z? If so, how does changing the level of z from low to
high affect the blending properties of the mixture components (i.e., how
does z affect the response to $x_1 = 1$, affect the response to $x_2 = 1$, affect
the nonlinear blending of x_1 and x_2)?

   (c) If a high value of the response is desirable, what settings in x_1, x_2, and
z do you recommend?

REFERENCES AND BIBLIOGRAPHY

Becker, N. G. (1978). Models and Designs for Experiments with Mixtures, *Austral. J.
Statist.*, **20**, 195–208.

Cornell, J. A. (1973). Experiments with Mixtures: A Review, *Technometrics*, **15**, 437–455.

Cornell, J. A. (1986). A Comparison Between Two Ten-Point Designs for Studying Three Component Mixture Systems, *J. Quality Tech.*, **18**, 1–15.

Cornell, J. A. (1990a). *Experiments with Mixtures: Designs, Models, and the Analysis of Mixture Data*, 2nd ed., New York: John Wiley.

Cornell, J. A. (1990b). *How to Run Mixture Experiments for Product Quality*, Vol. 5, rev. ed., Milwaukee, WI: Amer. Soc. Quality Control.

Cornell, J. A. and J. W. Gorman (1984). Fractional Design Plans for Process Variables in Mixture Experiments, *J. Quality Tech.*, **16**, 20–38.

Cox, D. R. (1971). A Note on Polynomial Response Functions for Mixtures, *Biometrika*, **58**, 155–159.

Crosier, R. B. (1984). Mixture Experiments: Geometry and Pseudocomponents, *Technometrics*, **26**, 209–216.

Gabriel, K. R. (1971). The Biplot-Graphic Display of Matrices with Applications to Principal Component Analysis, *Biometrika*, **58**, 453–467.

Gabriel, K. R. (1981). Biplot Display of Multivariate Matrices for Inspection of Data and Diagnosis. In *Interpreting Multivariate Data* (V. Barnett, ed.), New York: John Wiley, pp. 147–173.

Goel, B. S. (1980). Designs for Restricted Exploration in Mixture Experiments, *Biometrical J.*, **22**, 351–358.

Jackson, J. E. (1991). *Biplots: A User's Guide to Principal Components*, New York: John Wiley.

Kurotori, I. S. (1966). Experiments with Mixtures of Components Having Lower Bounds, *Industrial Quality Control*, **22**, 592–596.

McLean, R. A. and V. L. Anderson (1966). Extreme Vertices Design of Mixture Experiments, *Technometrics*, **8**, 447–454.

Mitchell, T. J. (1974). An Algorithm for the Construction of *D*-optimal Experimental Designs, *Technometrics*, **16**, 203–210.

Nigam, A. K., S. C. Gupta, and S. Gupta (1983). A New Algorithm for Extreme Vertices Designs for Linear Mixture Models, *Technometrics*, **25**, 367–371.

Piepel, G. F. (1982). Measuring Component Effects in Constrained Mixture Experiments, *Technometrics*, **24**, 29–39.

Piepel, G. F. (1990). *MIXSOFT Software for the Design and Analysis of Mixture Experiments, User's Guide, Version 2.0*. Richland, WA: MIXSOFT—Mixture Experiment Software.

Piepel, G. F. and J. A. Cornell (1987). Designs for Mixture-Amount Experiments, *J. Quality Tech.*, **19**, 11–28.

Scheffé, H. (1958). Experiments with Mixtures, *J. Roy. Statist. Soc.*, **B20**, 344–360.

Scheffé, H. (1963). The Simplex-Centroid Design for Experiments with Mixtures, *J. Roy. Statist. Soc.*, **B25**, 235–263.

Smith, W. F., Jr., and J. A. Cornell (1993). Biplot Displays for Looking at Multiple Response Data in Mixture Experiments, *Technometrics*, **35**, 337–350.

Snee, R. D. (1971). Design and Analysis of Mixture Experiments, *J. Quality Tech.*, **3**, 159–169.

Snee, R. D. and D. W. Marquardt (1974). Extreme Vertices Designs for Linear Mixture Models, *Technometrics*, **16**, 399–408.

STAT-EASE, Inc. (1994). *DESIGN-EXPERT, Version 4.0.6*, Minneapolis, MN: Author.

Thompson, W. O. and R. H. Myers (1968). Response Surface Designs for Experiments with Mixtures, *Technometrics*, **10**, 739–756.

Welch, W. J. (1985). ACED: Algorithms for the Construction of Experimental Designs, *The Amer. Statistician*, **39**, 146.

10

Nonlinear Response Surface Models

10.1 INTRODUCTION

The models that we have mainly considered in the previous chapters were linear in the parameters, hence the term *linear models*. Many mathematical models used in scientific research, however, contain parameters that are not expressed linearly. By definition, a model is said to be nonlinear if at least one of its parameters appears nonlinearly. For example, the models

$$\eta(x) = \theta_1 e^{-\theta_2 x} \tag{10.1}$$

$$\eta(x) = \theta_1 + \theta_2 e^{-\theta_3 x} \tag{10.2}$$

$$\eta(x) = \frac{1}{\theta_1 + \theta_2 x} \tag{10.3}$$

$$\eta(x) = \left(\frac{\theta_1}{\theta_1 - \theta_2}\right) (e^{-\theta_2 x} - e^{-\theta_1 x}) \tag{10.4}$$

where $\eta(x)$ denotes the mean response at x, are all nonlinear. Throughout this chapter, x denotes a conveniently coded value of an input variable. In model 10.1, θ_1 is a linear parameter, but θ_2 is not. In model 10.2 θ_1 and θ_2 are linear, and θ_3 is nonlinear. In models 10.3 and 10.4 both parameters, θ_1 and θ_2, appear nonlinearly. Sometimes the term *partially nonlinear* is used to describe a model in which some of the parameters are linear and some are nonlinear, such as models 10.1 and 10.2.

385

A nonlinear model is said to be intrinsically linear if it can be reduced to a linear model by a suitable reparameterization of the model. For example, the nonlinear model

$$\eta(x) = \theta_1 + e^{\theta_2}x$$

can be reduced to a linear model, $\eta(x) = \theta_1 + \gamma_1 x$, by transforming θ_2 to γ_1 using the transformation $\gamma_1 = e^{\theta_2}$.

Sometimes the phrase *intrinsically linear* is used with reference to a nonlinear model which can be reduced to a linear form by applying a transformation to the model itself. For example, if we consider the model given in Eq. 10.1, then a natural logarithmic transformation can reduce $\eta(x)$ to the linear form

$$\ln[\eta(x)] = \ln(\theta_1) - \theta_2 x$$

provided $\theta_1 > 0$. Such a transformation can change the structure and distribution of the error term associated with the model. To explain this, let Y and ε be the observed response and random error, respectively, for model 10.1. Then

$$\ln(Y) = \ln[\eta(x) + \varepsilon] \tag{10.5}$$

$$= \ln[\eta(x)] + \ln\left[1 + \frac{\varepsilon}{\eta(x)}\right]$$

The error term for the transformed model is now $\ln[1 + \varepsilon/\eta(x)]$, which in general has a distribution different from that of ε. For example, if ε satisfies the usual assumptions of normality, independence, and homogeneity of variance, the error term for model 10.5 will have a nonnormal distribution which depends on x through $\eta(x)$. Thus, the variance of this error term cannot be assumed to be constant as in the original model. Consequently, even if the mean $\eta(x)$ in a nonlinear model can be reduced to a linear form by a proper transformation, such a transformation should be used only if it can be demonstrated that the aforementioned assumptions with respect to the transformed model are not severely violated.

Nonlinear models have been used in many fields, particularly in biological and chemical sciences where the growth of a particular organism, or the yield that results from a chemical reaction, can be depicted by a nonlinear model. Draper and Smith (1981: Ch. 10) list several examples of growth models (see also Ratkowsky, 1983: Ch. 4). In chemistry, most chemical relationships are nonlinear. For example, the relationship between physical properties and biological potency of drug analogs is expressed nonlinearly (see Martin and Hackbarth, 1977). Also the relationship between the rate of reaction and the concentration of the reactants is nonlinear. Sometimes a nonlinear model can be derived as a solution to a differential or a difference equation. This equation is usually developed on the basis of certain assumptions made about the type of growth or the actual mechanism in a chemical reaction. Such models are called mechanistic. When a complete description of such a mechanism is not obtainable, or is too complex, a quasi-mechanistic or even an empirical nonlinear model can be used. This is particularly true in some industrial chemical engineering applications.

10.2 LEAST SQUARES ESTIMATES OF THE PARAMETERS IN A NONLINEAR MODEL

Consider the nonlinear model

$$Y = f(x_1, x_2, \ldots, x_k; \theta_1, \theta_2, \ldots, \theta_p) + \varepsilon \qquad (10.6)$$

where $x_1, x_2, \ldots, x_k$ are coded input variables; $\theta_1, \theta_2, \ldots, \theta_p$ are unknown parameters, and ε is a random error with mean zero and variance σ^2. We assume that f is continuous and has continuous first-order and second-order partial derivatives with respect to the parameters. The least squares estimates of the parameters are obtained as usual by minimizing the function

$$S(\theta_1, \theta_2, \ldots, \theta_p) = \sum_{u=1}^{N} [Y_u - f(x_{u1}, x_{u2}, \ldots, x_{uk}; \theta_1, \theta_2, \ldots, \theta_p)]^2 \quad (10.7)$$

where Y_u is the response value observed at the point $(x_{u1}, x_{u2}, \ldots, x_{uk})$; $u = 1$, $2, \ldots, N$. The associated error terms $\varepsilon_1, \varepsilon_2, \ldots, \varepsilon_N$ are assumed to be uncorrelated with variances equal to σ^2. The minimization of $S(\theta_1, \theta_2, \ldots, \theta_p)$ is carried out iteratively and can be very complex. There are several methods available for computing the least squares estimates. The most widely used methods are the Gauss-Newton method and its modified version by Hartley (1961), the steepest descent method, and the method developed by Marquardt (1963). All of these methods require that the partial derivatives of the mean response f in Eq. 10.6 with respect to the parameters be provided explicitly. More recently, Ralston and Jennrich (1978) developed an algorithm called DUD for "doesn't use derivatives," a derivative-free Gauss-Newton algorithm, which approximates f at each iteration by a function which agrees with f at $p + 1$ previous values of the parameter vector, $\boldsymbol{\theta} = (\theta_1, \theta_2, \ldots, \theta_p)'$.

All of the aforementioned methods require that initial values be specified for the nonlinear model's parameters. The convergence of any of these methods to the least squares estimates and the rate of convergence heavily depend on the choice of the initial values. Ratkowsky (1983: Ch. 8) described procedures for obtaining good initial values of the parameters. Lawton and Sylvestre (1971) introduced a method whereby the specification of initial values is required only for those parameters which appear nonlinearly in the model. The actual implementation of these methods can be easily carried out by using PROC NLIN in SAS (1989).

In this chapter we shall not be concerned with methods of obtaining least squares estimates of a nonlinear model's parameters. The description of these methods is widely available in the literature (see, for example, Bard, 1974; Chambers, 1973; Draper and Smith, 1981; and Kennedy and Gentle, 1980). Rather, the focus of attention in this chapter is on the design aspect of nonlinear models.

10.2.1 Properties of the Least Squares Estimators

Let $\theta = (\theta_1, \theta_2, \ldots, \theta_p)'$ be the vector of unknown parameters and $x_u = (x_{u1}, x_{u2}, \ldots, x_{uk})'$ be the vector of coded input variables evaluated at the uth experiment ($u = 1, 2, \ldots, N$). At this point, model 10.6 can be written as

$$Y_u = f(x_u, \theta) + \varepsilon_u \qquad u = 1, 2, \ldots, N \tag{10.8}$$

Let $\hat{\theta}_N$ be the least squares estimator of θ obtained from the data set $\{x_u', Y_u\}_{u=1}^{N}$. The statistical properties of $\hat{\theta}_N$ are different from those of the least squares estimator of the parameter vector in a linear model. Unlike the latter estimator, $\hat{\theta}_N$ is not unbiased or normally distributed, even under the assumption of normality on the random errors. Under normality, however, $\hat{\theta}_N$ and its linear least squares counterpart share the property of being maximum likelihood estimators of their respective parameter vectors. The amount of bias in $\hat{\theta}_N$ can be computed on the basis of a formula derived by Box (1971).

As $N \to \infty$, $\sqrt{N}(\hat{\theta}_N - \theta)$ has an asymptotic normal distribution with a mean equal to zero and a variance-covariance matrix $\sigma^2 V$. A consistent estimator of V is given by the $p \times p$ matrix

$$\hat{V}_N = \left[\frac{F'(\hat{\theta}_N)F(\hat{\theta}_N)}{N} \right]^{-1} \tag{10.9}$$

where $F(\hat{\theta}_N)$ is an $N \times p$ matrix whose (u, i)th element is

$$\left. \frac{\partial f(x_u, \theta)}{\partial \theta_i} \right|_{\theta = \hat{\theta}_N} \qquad u = 1, 2, \ldots, N \qquad i = 1, 2, \ldots, p$$

A consistent estimator of σ^2 is given by

$$\hat{\sigma}_N^2 = \frac{S(\hat{\theta}_N)}{N - p} \tag{10.10}$$

where $S(\hat{\theta}_N)$ is the residual sum of squares in Eq. 10.7 evaluated at $\hat{\theta}_N$. Furthermore, $(N - p)\hat{\sigma}_N^2/\sigma^2$ has an asymptotic chi-square distribution with $(N - p)$ degrees of freedom independently of the asymptotic distribution of $\hat{\theta}_N$. Thus, in large samples, $\hat{\theta}_N$ is approximately normally distributed with a mean θ and a variance-covariance matrix $[\mathbf{F}'(\boldsymbol{\theta})\mathbf{F}(\boldsymbol{\theta})]^{-1}\sigma^2$. Formal mathematical discussions of the asymptotic properties of $\hat{\boldsymbol{\theta}}_N$ can be found in Jennrich (1969) and Malinvaud (1980: Ch. 9).

10.3 TESTS OF HYPOTHESES AND CONFIDENCE REGIONS

Hypothesis testing and confidence regions concerning the parameters of a nonlinear model are based on either the asymptotic properties of the least squares estimators of the parameters, or on the likelihood ratio principle. We discuss both procedures here.

Let $\mathbf{A}\boldsymbol{\theta}$ be a linear vector function of the p parameters, where $\mathbf{A}$ is a matrix or order $q \times p$ and rank $q(\leq p)$, and let $\hat{\boldsymbol{\theta}}_N$ be the least squares estimator of $\boldsymbol{\theta}$. From the discussion in Section 10.2.1 it is easy to see that a $(1 - \alpha)100\%$ confidence region on $\mathbf{A}\boldsymbol{\theta}$ is approximately given by

$$\frac{[\mathbf{A}(\hat{\boldsymbol{\theta}}_N - \boldsymbol{\theta})]'(\mathbf{A}\hat{\mathbf{G}}_N\mathbf{A}')^{-1}[\mathbf{A}(\hat{\boldsymbol{\theta}}_N - \boldsymbol{\theta})]}{q\hat{\sigma}_N^2} \leq F_{\alpha,q,N-p} \tag{10.11}$$

where

$$\hat{\mathbf{G}}_N = [\mathbf{F}'(\hat{\boldsymbol{\theta}}_N)\mathbf{F}(\hat{\boldsymbol{\theta}}_N)]^{-1} \tag{10.12}$$

and $F_{\alpha,q,N-p}$ is the upper $\alpha 100\%$ point of the F-distribution with q and $N - p$ degrees of freedom. Thus, at the approximate α-level, the hypothesis

$$H_o : \mathbf{A}\boldsymbol{\theta} = \mathbf{m} \qquad \text{vs.} \qquad H_a : \mathbf{A}\boldsymbol{\theta} \neq \mathbf{m} \tag{10.13}$$

is rejected if, when $\mathbf{A}\boldsymbol{\theta}$ is replaced by $\mathbf{m}$, the left-hand side of Eq. 10.11 exceeds $F_{\alpha,q,N-p}$. In particular, if $\mathbf{A} = \mathbf{I}_p$, the hypothesis

$$H_o : \boldsymbol{\theta} = \boldsymbol{\theta}_o \qquad \text{vs.} \qquad H_a : \boldsymbol{\theta} \neq \boldsymbol{\theta}_o \tag{10.14}$$

is rejected at the approximate α-level if

$$\frac{(\hat{\boldsymbol{\theta}}_N - \boldsymbol{\theta}_o)'\hat{\mathbf{G}}_N^{-1}(\hat{\boldsymbol{\theta}}_N - \boldsymbol{\theta}_o)}{p\hat{\sigma}_N^2} > F_{\alpha,p,N-p} \tag{10.15}$$

Confidence regions and tests of hypotheses concerning subsets of the parameter vector $\boldsymbol{\theta}$ can also be obtained by a proper choice of the $\mathbf{A}$ matrix. For example, let $\boldsymbol{\theta}_1$ consist of the first q elements of $\boldsymbol{\theta}$. Then an approximate $(1 - \alpha)100\%$ confidence region on $\boldsymbol{\theta}_1$ is obtained from Eq. 10.11 by choosing $\mathbf{A} = [\mathbf{I}_q : \mathbf{0}]$, where $\mathbf{0}$ is a zero matrix of order $q \times (p - q)$. In particular, if we consider the ith element of $\boldsymbol{\theta}$, namely θ_i $(i = 1, 2, \ldots, p)$, the hypothesis

$$H_o : \theta_i = \theta_{io} \qquad \text{vs.} \qquad H_a : \theta_i \neq \theta_{io} \tag{10.16}$$

can be rejected at the approximate α-level if

$$\frac{(\hat{\theta}_{iN} - \theta_{io})^2}{\hat{g}_{Nii}\hat{\sigma}_N^2} > F_{\alpha,1,N-p} \tag{10.17}$$

or equivalently, if

$$\frac{|\hat{\theta}_{iN} - \theta_{io}|}{\sqrt{\hat{g}_{Nii}\hat{\sigma}_N^2}} > t_{\alpha/2,N-p} \tag{10.18}$$

where $\hat{\theta}_{iN}$ is the ith element of $\hat{\theta}_N$ and $\hat{g}_{Nii}$ is the ith diagonal element ($i = 1, 2, \ldots, p$) of $\hat{G}_N$. From Eq. 10.18 an approximate $(1 - \alpha)100\%$ confidence interval on θ_i is

$$\hat{\theta}_{iN} \pm t_{\alpha/2,N-p} \sqrt{\hat{\sigma}_N^2 \hat{g}_{Nii}} \qquad i = 1, 2, \ldots, p \tag{10.19}$$

The likelihood ratio principle provides an alternative procedure for deriving confidence regions and tests of hypotheses. Consider model 10.8 and let Y_1, $Y_2, \ldots, Y_N$ be a sample of N observations. Under normality assumptions, the likelihood function of this sample is

$$L_N(\mathbf{Y}; \boldsymbol{\theta}, \sigma^2) = (2\pi\sigma^2)^{-\frac{N}{2}} \exp\left\{-\frac{1}{2\sigma^2} \sum_{u=1}^{N} [Y_u - f(\mathbf{x}_u, \boldsymbol{\theta})]^2\right\}$$

By differentiating L_N with respect to both $\boldsymbol{\theta}$ and σ^2 and equating these derivatives to zero, we find that the maximum likelihood estimators of $\boldsymbol{\theta}$ and σ^2 are, respectively, $\hat{\theta}_N$, the least squares estimator of $\boldsymbol{\theta}$, and $\tilde{\sigma}_N^2$ which is expressed as

$$\tilde{\sigma}_N^2 = \frac{S(\hat{\theta}_N)}{N} \tag{10.20}$$

where $S(\boldsymbol{\theta})$ is defined in Eq. 10.7 and $S(\hat{\theta}_N)$ is the residual sum of squares for model 10.8. The maximum value of L_N is

$$\max L_N(\mathbf{Y}; \boldsymbol{\theta}, \sigma^2) = (2\pi\tilde{\sigma}_N^2)^{-\frac{N}{2}} \exp\left(\frac{-N}{2}\right) \tag{10.21}$$

Let $\hat{\theta}_N^0$ denote the least squares estimator of $\boldsymbol{\theta}$ obtained under the null hypothesis H_o given in Eq. 10.13. The corresponding maximum likelihood estimator of σ^2 and the maximum value of L_N are, respectively,

$$\tilde{\sigma}_N^{0^2} = \frac{S(\hat{\theta}_N^0)}{N} \tag{10.22}$$

where $S(\hat{\theta}_N^0)$ denotes the residual sum of squares under the null hypothesis, and

$$\max_{H_o} L_N(\mathbf{Y}; \boldsymbol{\theta}, \sigma^2) = (2\pi\tilde{\sigma}_N^{0^2})^{-\frac{N}{2}} \exp\left(\frac{-N}{2}\right) \tag{10.23}$$

The likelihood ratio is

$$\lambda(\tilde{\sigma}_N^2, \tilde{\sigma}_N^{0^2}) = \frac{\max_{H_o} L_N(\mathbf{Y}; \boldsymbol{\theta}, \sigma^2)}{\max L_N(\mathbf{Y}; \boldsymbol{\theta}, \sigma^2)}$$

$$= \left(\frac{\tilde{\sigma}_N^2}{\tilde{\sigma}_N^{0^2}}\right)^{\frac{N}{2}}$$

Under the null hypothesis H_o in Eq. 10.13, $-2\ln(\lambda)$ is asymptotically distributed as a chi-square random variable with q degrees of freedom. But,

$$-2\ln[\lambda(\tilde{\sigma}_N^2, \tilde{\sigma}_N^{0^2})] = N \ln \left(\frac{\tilde{\sigma}_N^{0^2}}{\tilde{\sigma}_N^2}\right)$$

$$= N \ln \left[1 + \frac{S(\hat{\boldsymbol{\theta}}_N^0) - S(\hat{\boldsymbol{\theta}}_N)}{S(\hat{\boldsymbol{\theta}}_N)}\right]$$

A first-order approximation of $-2\ln(\lambda)$ is given by

$$-2\ln[\lambda(\tilde{\sigma}_N^2, \tilde{\sigma}_N^{0^2})] \simeq N \left[\frac{S(\hat{\boldsymbol{\theta}}_N^0) - S(\hat{\boldsymbol{\theta}}_N)}{S(\hat{\boldsymbol{\theta}}_N)}\right]$$

which can be written as

$$-2\ln[\lambda(\tilde{\sigma}_N^2, \tilde{\sigma}_N^{0^2})] \simeq \frac{S(\hat{\boldsymbol{\theta}}_N^0) - S(\hat{\boldsymbol{\theta}}_N)}{\tilde{\sigma}_N^2}$$

Since $\tilde{\sigma}_N^2$ converges in probability to σ^2, Milliken and DeBruin (1978) proposed the following test statistic for testing $H_o: \mathbf{A}\boldsymbol{\theta} = \mathbf{m}$ vs. $H_a: \mathbf{A}\boldsymbol{\theta} \neq \mathbf{m}$:

$$T = \frac{N - p}{Nq} \left[\frac{S(\hat{\boldsymbol{\theta}}_N^0) - S(\hat{\boldsymbol{\theta}}_N)}{\tilde{\sigma}_N^2}\right]$$

$$= \frac{[S(\hat{\boldsymbol{\theta}}_N^0) - S(\hat{\boldsymbol{\theta}}_N)]/q}{S(\hat{\boldsymbol{\theta}}_N)/(N - p)} \tag{10.24}$$

They showed that, under H_o, T is approximately distributed as an F random variable with q and $N - p$ degrees of freedom. We note that this statistic is analogous to the F-statistic used to test H_o when the model is linear. In the latter case, T has, under H_o, the exact F-distribution with q and $N - p$ degrees of freedom. In particular, if H_o is given as $H_o: \boldsymbol{\theta} = \boldsymbol{\theta}_o$, then H_o can be rejected at the approximate α-level if

$$\frac{[S(\boldsymbol{\theta}_o) - S(\hat{\boldsymbol{\theta}}_N)]/p}{S(\hat{\boldsymbol{\theta}}_N)/(N - p)} \geq F_{\alpha, p, N - p} \tag{10.25}$$

From Eq. 10.25 we can obtain an approximate $(1 - \alpha)100\%$ confidence region on $\boldsymbol{\theta}$ that consists of all points $\boldsymbol{\theta}$ which satisfy

$$S(\boldsymbol{\theta}) \leq S(\hat{\boldsymbol{\theta}}_N) \left[1 + \left(\frac{p}{N - p} \right) F_{\alpha, p, N-p} \right] \tag{10.26}$$

There are no known general results which can be used to determine which testing procedure to employ for testing $H_o : \mathbf{A}\boldsymbol{\theta} = \mathbf{m}$; the one based on the asymptotic properties of the least squares estimators, or the one based on the likelihood ratio principle. Unlike linear models, with nonlinear models these two procedures are not equivalent. There are, however, some simulation results (Gallant, 1975) which indicate that, when H_o concerns only a parameter which enters the model nonlinearly, the likelihood ratio procedure has the higher power of the two procedures. However, if the parameter enters the model linearly, then the two procedures have the same power.

10.4 A NUMERICAL EXAMPLE

Vohnout and Jimenez (1975) conducted a study whose purpose was to develop methods for optimal utilization of tropical resources in livestock feeding. They investigated the relationship between Y = weight gain (in kilograms) per day per animal (heifer) and the input variables x_1 = stocking rate (total weight of animals in kilograms per hectare of available grass) and x_2 = intake of supplemental feed (in kilograms) per day per animal. The data are shown in Table 10.1. It was assumed that a nonlinear model of the form

$$Y = \theta_1 + \theta_2 \exp \left(\frac{-\theta_3}{x_1} - \theta_4 x_2 \right) + \varepsilon$$

would adequately represent the relationship between Y and x_1, x_2. The parameter θ_1 represents the maximum weight gain. Thus, to obtain the least squares estimates of θ_1, θ_2, θ_3, and θ_4 we may choose as an initial value of θ_1 the value of $\theta_{10} = 0.72$ since the largest observed response value in Table 10.1 is 0.715. Initial values of θ_2, θ_3, and θ_4 can be chosen as the least squares estimates of θ_2, θ_3, and θ_4 in the linearized model

$$\ln(\theta_{10} - Y) = \ln(-\theta_2) - \frac{\theta_3}{x_1} - \theta_4 x_2 + \varepsilon'$$

which results from the nonlinear model by replacing θ_1 with θ_{10} and then applying the natural logarithmic transformation. Using the data set in Table 10.1 we find $\theta_{20} = -1.08$, $\theta_{30} = 1263.42$, and $\theta_{40} = 0.038$. On the basis of these values and the use of the DUD method by Ralston and Jennrich (1978), which

Table 10.1 Weight Gain of Heifers

Stocking Rate, x_1 (kg/ha)	Supplemental Feed Intake, x_2 (kg/day per animal)	Weight Gain, Y (kg/day per animal)
242	5	0.715
665	1.5	0.591
687	8.5	0.592
1432	0	0.248
1530	10	0.428
1542	5	0.321
2108	1.5	0.124
2469	8.5	0.215
2570	5	0.190

Source: K. Vohnout and C. Jimenez (1975). Reproduced with permission of the American Society of Agronomy, Crop Science Society of America, and Soil Science Society of America.

is available in PROC NLIN in SAS (1989), it was found that the least squares estimates of the parameters in the nonlinear model are $\hat{\theta}_1 = 0.718$, $\hat{\theta}_2 = -1.205$, $\hat{\theta}_3 = 1374.418$, $\hat{\theta}_4 = 0.045$. The PROC NLIN computer code used to obtain these estimates is shown in Table 10.2. A portion of the corresponding SAS printout is displayed in Table 10.3.

In this example, the partial derivatives of the mean response with respect to θ_1, θ_2, θ_3, and θ_4 are

$$\frac{\partial f(\mathbf{x}_u, \boldsymbol{\theta})}{\partial \theta_1} = 1 \qquad u = 1, 2, \ldots, 9$$

$$\frac{\partial f(\mathbf{x}_u, \boldsymbol{\theta})}{\partial \theta_2} = \exp\left(\frac{-\theta_3}{x_{u1}} - \theta_4 x_{u2}\right) \qquad u = 1, 2, \ldots, 9$$

$$\frac{\partial f(\mathbf{x}_u, \boldsymbol{\theta})}{\partial \theta_3} = -\left(\frac{\theta_2}{x_{u1}}\right) \exp\left(\frac{-\theta_3}{x_{u1}} - \theta_4 x_{u2}\right) \qquad u = 1, 2, \ldots, 9$$

$$\frac{\partial f(\mathbf{x}_u, \boldsymbol{\theta})}{\partial \theta_4} = -\theta_2 x_{u2} \exp\left(\frac{-\theta_3}{x_{u1}} - \theta_4 x_{u2}\right) \qquad u = 1, 2, \ldots, 9$$

Hence, the matrix $\mathbf{F}(\hat{\boldsymbol{\theta}}_N)$ is the 9×4 matrix

$$\mathbf{F}(\hat{\boldsymbol{\theta}}_N) = \left[\frac{\partial f}{\partial \theta_1} \vdots \frac{\partial f}{\partial \theta_2} \vdots \frac{\partial f}{\partial \theta_3} \vdots \frac{\partial f}{\partial \theta_4}\right]$$

Table 10.2   The SAS/NLIN Computer Code for the Weight
Gain Model

```
Data;
Input x1 x2 Y;
Cards;
242      5.0   0.715
665      1.5   0.591
687      8.5   0.592
1432     0.0   0.248
1530    10.0   0.428
1542     5.0   0.321
2108     1.5   0.124
2469     8.5   0.215
2570     5.0   0.190
PROC NLIN Method = DUD;
Parameters Theta1 = 0.72 Theta2 = -1.08
            Theta3 = 1263.42 Theta4 = 0.038;
Model Y = Theta1 + Theta2*Exp(-Theta3/x1 - Theta4*x2);
Run;
```

---

with the estimates $\hat{\theta}_1$, $\hat{\theta}_2$, $\hat{\theta}_3$, and $\hat{\theta}_4$ substituted into each $\frac{\partial f}{\partial \theta_i}$, so that the matrix $\mathbf{F}'(\hat{\boldsymbol{\theta}}_N)\mathbf{F}(\hat{\boldsymbol{\theta}}_N)$ needed in setting up confidence regions is of the form

$$
\mathbf{F}'(\hat{\boldsymbol{\theta}}_N)\mathbf{F}(\hat{\boldsymbol{\theta}}_N) = \begin{bmatrix} 9 & 2.550 & 0.002 & 14.082 \\ 2.550 & 0.966 & 0.0006 & 4.939 \\ 0.002 & 0.0006 & 0 & 0.003 \\ 14.082 & 4.939 & 0.003 & 39.823 \end{bmatrix}
$$

The residual sum of squares is $S(\hat{\boldsymbol{\theta}}_N) = 0.0034$ with 5 degrees of freedom, hence the residual mean square is $\hat{\sigma}_N^2 = 0.0007$. The regression sum of squares has the value 1.653 with 4 degrees of freedom. Approximate 95% confidence intervals, based on Eq. 10.19, for θ_1, θ_2, θ_3, and θ_4 are (0.651, 0.786), (−1.467, −0.943), (908.061, 1840.774), and (0.025, 0.066), respectively. These intervals are referred to as asymptotic 95% confidence intervals in Table 10.3. We note that at the 5% level all four parameters are significantly different from zero.

Let us now test the following hypothesis:

$$
H_o: \begin{cases} \theta_1 = 0.9 \\ \theta_2 = -2.0 \\ \theta_3 = 1500.0 \\ \theta_4 = 0.5 \end{cases}
$$

Table 10.3 A Portion of the SAS/NLIN Printout for the Weight Gain Model

Nonlinear Least Squares Summary Statistics			Dependent Variable Y
Source	DF	Sum of Squares	Mean Square
Regression	4	1.6530074566	0.4132518642
Residual	5	0.0033925434	0.0006785087
Uncorrected Total	9	1.6564000000	
(Corrected Total)	8	0.3537582222	

Parameter	Estimate	Asymptotic Std. Error	Asymptotic 95% Confidence Interval	
			Lower	Upper
THETA1	0.718419	0.02616347	0.65116489	0.7856736
THETA2	−1.205192	0.10187890	−1.46707622	−0.9433077
THETA3	1374.417753	181.42324786	908.06120719	1840.7742993
THETA4	0.045455	0.00780346	0.02539574	0.0655140

Asymptotic Correlation Matrix

Corr	THETA1	THETA2	THETA3	THETA4
THETA1	1	0.3439789972	−0.747373545	−0.43767675
THETA2	0.3439789972	1	−0.810329983	−0.614741671
THETA3	−0.747373545	−0.810329983	1	0.4561967177
THETA4	−0.43767675	−0.614741671	0.4561967177	1

Note that the "Parameters" statement in Table 10.2 specifies initial values for the parameters in the model. Several initial values for each parameter can also be specified (see SAS, 1989, p. 1145).

By using the test described in inequality 10.15, we find that the value on the left side of this inequality is equal to 2751.3, a highly significant result. Another hypothesis that might be of interest is

$$H_o: \begin{cases} \theta_1 + \theta_2 = 0 \\ \theta_4 = 0 \end{cases}$$

Here $\theta_1 + \theta_2$ represents the weight loss when $x_2 = 0$ and as $x_1 \to +\infty$. This hypothesis is of the type considered in Eq. 10.13 with $\mathbf{A}$ being the 2×4 matrix

$$\mathbf{A} = \begin{bmatrix} 1 & 1 & 0 & 0 \\ 0 & 0 & 0 & 1 \end{bmatrix}$$

Under this hypothesis, the nonlinear model becomes

$$Y = \theta_1 \left[1 - \exp\left(\frac{-\theta_3}{x_1}\right)\right] + \varepsilon$$

The residual sum of squares for this reduced model is $S(\hat{\theta}_N^0) = 0.0413$. Hence, the value of the approximate F-statistic given in Eq. 10.24 is

$$T = \frac{(0.0413 - 0.0034)/2}{0.0034/5} = 27.87$$

This value is significant at a level not exceeding $0.005(F_{0.005,2,5} = 18.31)$.

10.5 DESIGNS FOR NONLINEAR MODELS

10.5.1 The Box-Lucas Criterion

The earliest and most commonly used criterion for the choice of design for a nonlinear model such as Eq. 10.8 is that of Box and Lucas (1959). This criterion depends on the assumption that in some neighborhood of θ_o, where θ_o is an initial value of the vector of parameters that is available before any data are obtained, the function $f(\mathbf{x}_u, \theta)$ in Eq. 10.8 is approximately linear in θ for $u = 1, 2, \ldots, N$. In this case, a first-order Taylor's expansion of $f(\mathbf{x}_u, \theta)$ in a neighborhood of θ_o yields the following approximation of $f(\mathbf{x}_u, \theta)$:

$$f(\mathbf{x}_u, \theta) \simeq f(\mathbf{x}_u, \theta_o) + \sum_{i=1}^{p} (\theta_i - \theta_{io}) \left.\frac{\partial f(\mathbf{x}_u, \theta)}{\partial \theta_i}\right|_{\theta=\theta_0} \qquad u = 1, 2, \ldots, N$$

$$(10.27)$$

In other words, when θ is close to θ_o, we have approximately

$$\mathbf{z} = \mathbf{F}(\theta_o)\boldsymbol{\psi} + \boldsymbol{\varepsilon} \qquad\qquad (10.28)$$

where $\mathbf{z} = (z_1, z_2, \ldots, z_N)'$ with $z_u = Y_u - f(\mathbf{x}_u, \theta_o)$, $(u = 1, 2, \ldots, N)$, $\mathbf{F}(\theta_o)$ is the $N \times p$ matrix whose (u, i)th element is

$$\left.\frac{\partial f(\mathbf{x}_u, \theta)}{\partial \theta_i}\right|_{\theta=\theta_o} \qquad u = 1, 2, \ldots, N; \qquad i = 1, 2, \ldots, p$$

$\boldsymbol{\psi} = \theta - \theta_o$, and $\boldsymbol{\varepsilon} = (\varepsilon_1, \varepsilon_2, \ldots, \varepsilon_N)'$. For the nonlinear model in (10.8), Box and Lucas (1959) proposed as a design those points that minimize the generalized variance of the least squares estimator, $\hat{\boldsymbol{\psi}}$, of $\boldsymbol{\psi}$ in the linear model (10.28). This is achieved by minimizing the determinant of the variance-covariance matrix $\mathbf{M}^{-1}(\mathbf{D}, \theta_o)\sigma^2$, where

$$\mathbf{M}(\mathbf{D}, \theta_o) = \mathbf{F}'(\theta_o)\mathbf{F}(\theta_o) \qquad\qquad (10.29)$$

The matrix $\mathbf{D}$ in Eq. 10.29 denotes the $N \times k$ design matrix (x_{uj}), $u = 1, 2, \ldots, N$; $j = 1, 2, \ldots, k$. In other words, the Box-Lucas design criterion chooses the design which maximizes the determinant of $\mathbf{M}(\mathbf{D}, \boldsymbol{\theta}_o)$. Such a design is D-optimal for the linear model 10.28 (see Section 12.2 in Chapter 12). It can therefore be described as being locally D-optimal for the nonlinear model 10.8 in the vicinity of $\boldsymbol{\theta}_o$.

The Box-Lucas criterion can be extended to obtain an optimal design for the precise estimation of only a subset of the parameters in a nonlinear model. To show this, let $\boldsymbol{\theta}$ be partitioned as $\boldsymbol{\theta} = (\boldsymbol{\theta}_1' : \boldsymbol{\theta}_2')'$, where $\boldsymbol{\theta}_1$ and $\boldsymbol{\theta}_2$ are of dimensions p_1 and p_2, respectively, and let $p = p_1 + p_2$. Correspondingly, the matrices $\mathbf{F}(\boldsymbol{\theta}_o)$ and $\mathbf{M}(\mathbf{D}, \boldsymbol{\theta}_o)$ can be partitioned as

$$\mathbf{F}(\boldsymbol{\theta}_o) = [\mathbf{F}_1(\boldsymbol{\theta}_o) : \mathbf{F}_2(\boldsymbol{\theta}_o)]$$

$$\mathbf{M}(\mathbf{D}, \boldsymbol{\theta}_o) = \begin{bmatrix} \mathbf{F}_1'(\boldsymbol{\theta}_o)\mathbf{F}_1(\boldsymbol{\theta}_o) & \mathbf{F}_1'(\boldsymbol{\theta}_o)\mathbf{F}_2(\boldsymbol{\theta}_o) \\ \mathbf{F}_2'(\boldsymbol{\theta}_o)\mathbf{F}_1(\boldsymbol{\theta}_o) & \mathbf{F}_2'(\boldsymbol{\theta}_o)\mathbf{F}_2(\boldsymbol{\theta}_o) \end{bmatrix}$$

where the (u, i)th elements of $\mathbf{F}_1(\boldsymbol{\theta}_o)$ and $F_2(\boldsymbol{\theta}_o)$ are of the form

$$\left. \frac{\partial f(\mathbf{x}_u, \boldsymbol{\theta})}{\partial \theta_i} \right|_{\boldsymbol{\theta}=\boldsymbol{\theta}_o} \qquad u = 1, 2, \ldots, N$$

with $i = 1, 2, \ldots, p_1$ for $\mathbf{F}_1(\boldsymbol{\theta}_o)$ and $i = p_1 + 1, p_1 + 2, \ldots, p_1 + p_2$ for $\mathbf{F}_2(\boldsymbol{\theta}_o)$. Suppose that the interest is in obtaining a precise estimate for $\boldsymbol{\theta}_1$ only. If the initial parameter vector $\boldsymbol{\theta}_o = (\boldsymbol{\theta}_{1o}' : \boldsymbol{\theta}_{2o}')'$ is partitioned the same way as $\boldsymbol{\theta}$, then the variance-covariance matrix of $\hat{\boldsymbol{\psi}}_1$, the least squares estimator of $\boldsymbol{\psi}_1 = \boldsymbol{\theta}_1 - \boldsymbol{\theta}_{1o}$, is approximately given by $M_{11}^{-1}(\mathbf{D}, \boldsymbol{\theta}_o)\sigma^2$, where

$$M_{11}(\mathbf{D}, \boldsymbol{\theta}_o) = \mathbf{F}_1'(\boldsymbol{\theta}_o)\mathbf{F}_1(\boldsymbol{\theta}_o) - \mathbf{F}_1'(\boldsymbol{\theta}_o)\mathbf{F}_2(\boldsymbol{\theta}_o)[\mathbf{F}_2'(\boldsymbol{\theta}_o)\mathbf{F}_2(\boldsymbol{\theta}_o)]^{-1}\mathbf{F}_2'(\boldsymbol{\theta}_o)\mathbf{F}_1(\boldsymbol{\theta}_o)$$
$$(10.30)$$

Hence, a locally D-optimal design for the estimation of $\boldsymbol{\theta}_1$ (or $\boldsymbol{\psi}_1$) is the one that minimizes the generalized variance of $\hat{\boldsymbol{\psi}}_1$, that is, minimizes $|M_{11}^{-1}(\mathbf{D}, \boldsymbol{\theta}_o)|$. This is equivalent to maximizing the determinant $|M_{11}(\mathbf{D}, \boldsymbol{\theta}_o)|$ with respect to the model's input variables. We note that this determinant can be written as

$$|M_{11}(\mathbf{D}, \boldsymbol{\theta}_o)| = \frac{|\mathbf{M}(\mathbf{D}, \boldsymbol{\theta}_o)|}{|M_{22}(\mathbf{D}, \boldsymbol{\theta}_o)|} \qquad (10.31)$$

where $M_{22}(\mathbf{D}, \boldsymbol{\theta}_o) = \mathbf{F}_2'(\boldsymbol{\theta}_o)\mathbf{F}_2(\boldsymbol{\theta}_o)$.

Example 10.1. Consider the example used by Box and Lucas (1959) of a consecutive first-order chemical reaction in which a raw material A reacts to form a product B, which, in turn, decomposes to form substance C. After time x has elapsed, the mean yield of the intermediate product B is given by

$$f(x, \boldsymbol{\theta}) = \frac{\theta_1}{\theta_1 - \theta_2} [e^{-\theta_2 x} - e^{-\theta_1 x}] \qquad (10.32)$$

where θ_1 and θ_2 are the rate constants for the reactions A $\to$ B and B $\to$ C, respectively. The partial derivatives needed to evaluate the matrix $\mathbf{M}(\mathbf{D}, \boldsymbol{\theta})$ in Eq. 10.29 are

$$\frac{\partial f(x, \boldsymbol{\theta})}{\partial \theta_1} = \left[\frac{\theta_1 x}{\theta_1 - \theta_2} + \frac{\theta_2}{(\theta_1 - \theta_2)^2}\right] e^{-\theta_1 x} - \frac{\theta_2}{(\theta_1 - \theta_2)^2} e^{-\theta_2 x} \qquad (10.33)$$

$$\frac{\partial f(x, \boldsymbol{\theta})}{\partial \theta_2} = \frac{-\theta_1}{(\theta_1 - \theta_2)^2} e^{-\theta_1 x} + \left[\frac{\theta_1}{(\theta_1 - \theta_2)^2} - \frac{\theta_1 x}{\theta_1 - \theta_2}\right] e^{-\theta_2 x} \qquad (10.34)$$

Box and Lucas sought a two-point design $\mathbf{D} = (x_1, x_2)'$. The determinant to be maximized is $|\mathbf{M}(\mathbf{D}, \boldsymbol{\theta})| = |\mathbf{F}'(\boldsymbol{\theta})\mathbf{F}(\boldsymbol{\theta})|$, where

$$\mathbf{F}(\boldsymbol{\theta}) = \begin{bmatrix} \dfrac{\partial f(x_1, \boldsymbol{\theta})}{\partial \theta_1} & \dfrac{\partial f(x_1, \boldsymbol{\theta})}{\partial \theta_2} \\[2ex] \dfrac{\partial f(x_2, \boldsymbol{\theta})}{\partial \theta_1} & \dfrac{\partial f(x_2, \boldsymbol{\theta})}{\partial \theta_2} \end{bmatrix}$$

In this example, $\mathbf{F}(\boldsymbol{\theta})$ is a square matrix. Hence,

$$|\mathbf{M}(\mathbf{D}, \boldsymbol{\theta}_o)| = |\mathbf{F}(\boldsymbol{\theta}_o)|^2 = \left[\frac{\partial f(x_1, \boldsymbol{\theta}_o)}{\partial \theta_1}\frac{\partial f(x_2, \boldsymbol{\theta}_o)}{\partial \theta_2} - \frac{\partial f(x_1, \boldsymbol{\theta}_o)}{\partial \theta_2}\frac{\partial f(x_2, \boldsymbol{\theta}_o)}{\partial \theta_1}\right]^2$$
$$(10.35)$$

The initial value $\boldsymbol{\theta}_o$ used by Box and Lucas was $\boldsymbol{\theta}_o = (0.7, 0.2)'$. By substituting the partial derivatives given by Eqs. 10.33 and 10.34 in 10.35 and then maximizing the ensuing function with respect to x_1 and x_2, it was found that the desired two-point locally D-optimal design consists of the reaction times $x_1 = 1.23$ and $x_2 = 6.86$ units. The corresponding maximum value of $|\mathbf{M}(\mathbf{D}, \boldsymbol{\theta}_o)|$ is 0.6568.

Now, let us suppose that we are interested in the precise estimation of θ_1. In this case we need to maximize the determinant $|\mathbf{M}_{11}(\mathbf{D}, \boldsymbol{\theta}_o)|$ in Eq. 10.31, which in this example has the form

$$|\mathbf{M}_{11}(\mathbf{D}, \boldsymbol{\theta}_o))| = \frac{|\mathbf{F}(\boldsymbol{\theta}_o)|^2}{\left[\dfrac{\partial f(x_1, \boldsymbol{\theta}_o)}{\partial \theta_2}\right]^2 + \left[\dfrac{\partial f(x_2, \boldsymbol{\theta}_o)}{\partial \theta_2}\right]^2}$$

where $\mathbf{F}(\boldsymbol{\theta}_o)$ is given in Eq. 10.35 with $\boldsymbol{\theta}_o = (0.7, 0.2)'$. The maximum value of this determinant is achieved when $x_1 = 1.17$ and $x_2 = 7.44$ (see Hill and Hunter, 1974: 428). Hence, $\mathbf{D} = (1.17, 7.44)'$ is a locally D-optimal design for the precise estimation of θ_1.

Example 10.2. This example is given in Box and Hunter (1965) and is also reported in Box (1970). It concerns the true rate, η, of the chemical reaction

$$P \to Q + R$$

where the reactant P is one of certain tertiary or long chain primary alcohols, the product Q is an olefin, and the product R is water. The rate η is described by the nonlinear model

$$\eta = \frac{\theta_1 \theta_3 x_1}{1 + \theta_1 x_1 + \theta_2 x_2} \tag{10.36}$$

where x_1 and x_2 are, respectively, the partial pressures of the reactant P and the product Q, θ_1, θ_2, and θ_3 are constants associated with the reactant P, the product Q, and the reaction rate, respectively. The initial values vector was taken to be $\theta_o = (2.9, 12.2, 0.69)'$. The experimental region over which $|\mathbf{M}(\mathbf{D}, \theta_o)|$ is maximized is described by the following constraints: $0 \leq x_1 \leq 3$, $0 \leq x_2 \leq 3$. In this example

$$\frac{\partial f(\mathbf{x}, \theta)}{\partial \theta_1} = \frac{\theta_3 x_1 (1 + \theta_2 x_2)}{(1 + \theta_1 x_1 + \theta_2 x_2)^2} \tag{10.37}$$

$$\frac{\partial f(\mathbf{x}, \theta)}{\partial \theta_2} = \frac{-\theta_1 \theta_3 x_1 x_2}{(1 + \theta_1 x_1 + \theta_2 x_2)^2} \tag{10.38}$$

$$\frac{\partial f(\mathbf{x}, \theta)}{\partial \theta_3} = \frac{\theta_1 x_1}{1 + \theta_1 x_1 + \theta_2 x_2} \tag{10.39}$$

For $N = 3$, the optimal three-point design for $\theta = (\theta_1, \theta_2, \theta_3)'$ is the one that maximizes $|\mathbf{M}(\mathbf{D}, \theta_o)|$, or equivalently, $|\mathbf{F}(\theta_o)|$, where $\mathbf{F}(\theta_o)$ is the 3×3 matrix

$$\mathbf{F}(\theta_o) = \begin{bmatrix} \dfrac{\partial f(\mathbf{x}_1, \theta)}{\partial \theta_1} & \dfrac{\partial f(\mathbf{x}_1, \theta)}{\partial \theta_2} & \dfrac{\partial f(\mathbf{x}_1, \theta)}{\partial \theta_3} \\[2ex] \dfrac{\partial f(\mathbf{x}_2, \theta)}{\partial \theta_1} & \dfrac{\partial f(\mathbf{x}_2, \theta)}{\partial \theta_2} & \dfrac{\partial f(\mathbf{x}_2, \theta)}{\partial \theta_3} \\[2ex] \dfrac{\partial f(\mathbf{x}_3, \theta)}{\partial \theta_1} & \dfrac{\partial f(\mathbf{x}_3, \theta)}{\partial \theta_2} & \dfrac{\partial f(\mathbf{x}_3, \theta)}{\partial \theta_3} \end{bmatrix}_{\theta = \theta_o}$$

where $\mathbf{x}_u = (x_{u1}, x_{u2})'$, $u = 1, 2, 3$. This optimal design was shown to consist of the points $\mathbf{x}_1 = (0.28, 0)'$, $\mathbf{x}_2 = (3.0, 0)'$, $\mathbf{x}_3 = (3.0, 0.795)'$.

10.5.2 The Problem of Design Dependency on the Parameters

One of the main difficulties associated with the Box-Lucas criterion is the dependency of the optimal design on the nonlinear model's parameters. In other words, one cannot determine the optimal design settings for estimating θ without having to specify an initial estimate, θ_o, of θ. This dependency on θ is an unappealing characteristic of nonlinear model designs and was most appropriately depicted by Cochran (1973): "You tell me the value of θ and I promise to design the best experiment for estimating θ." A more efficient approach would be to treat θ_o as an initial guess in a sequential procedure which starts by selecting a p-point optimal design on the basis of this guess, where p is the number of parameters in the model. A series of p experiments are then performed at the optimal design settings and an estimate of θ is obtained. Thereafter, the experiments can be designed sequentially by adding design points one at a time as in Box and Hunter (1965). At each stage a new estimate of θ is obtained and used to generate the next design point. This sequential procedure continues until some stability appears in the values of the parameter estimates and those of the design points. The closer the initial guess, θ_o, is to the true value of θ, the faster this stability is brought about. This explains the need for θ_o to be a "good guess" for θ.

We note that this sequential procedure can help reduce the dimensionality of the optimization problem associated with the determination of the locally D-optimal design. Only k variables need to be determined each time a new design point is added, where k is the number of input variables in the model. In a related work, Atkinson and Hunter (1968) showed that under certain conditions, when the number of experiments, N, is a multiple of the number of parameters, p, the optimal N-point design consists solely of replications of the p-point locally D-optimal design. In this case, the reduction in the number of dimensions from Nk to pk may represent considerable savings.

The dependency of the locally D-optimal design on θ is influenced by those parameters which appear nonlinearly in the model. For example, in the model

$$f(x, \theta) = \theta_1 + \theta_2 e^{-\theta_3 x}$$

the parameters θ_1 and θ_2 appear linearly and θ_3 nonlinearly. In this case, the design depends only on θ_3, and not on θ_1 or θ_2. This issue will be discussed in the next section.

10.5.3 Partially Nonlinear Models

In some nonlinear models, the locally D-optimal design for the estimation of the parameter vector θ may only depend on a proper subset of θ. This occurs when the model is partially nonlinear, that is, when some elements of θ appear linearly while others appear nonlinearly. For example, the model given in Eq. 10.36 is partially nonlinear since θ_3 is a linear parameter. Hill (1980) gave the following characterization of partial nonlinearity: If $\nabla f(\mathbf{x}, \theta)$ is a vector of dimension p whose ith element is the partial derivative of the mean response with respect to θ_i ($i = 1, 2, \ldots, p$), then model 10.8 is partially nonlinear if $\nabla f(\mathbf{x}, \theta)$

can be written as

$$\nabla f(\mathbf{x}, \boldsymbol{\theta}) = \mathbf{B}(\boldsymbol{\theta})\phi(\mathbf{x}, \tilde{\boldsymbol{\theta}}) \tag{10.40}$$

where $B(\boldsymbol{\theta})$ is a nonsingular matrix not involving $\mathbf{x}$ and $\phi(\mathbf{x}, \tilde{\boldsymbol{\theta}})$ is a vector of functions dependent on $\mathbf{x}$ and on a proper subset $\tilde{\boldsymbol{\theta}}$ of $\boldsymbol{\theta}$. In this case, Hill (1980) showed that the locally D-optimal design for the estimation of $\boldsymbol{\theta}$ depends only on the elements of $\tilde{\boldsymbol{\theta}}$.

For the partially nonlinear model in Eq. 10.36, the optimal design depends only on θ_1 and θ_2 but not on θ_3. It can be seen from Eqs. 10.37, 10.38, and 10.39 that for this model,

$$B(\boldsymbol{\theta}) = \begin{bmatrix} \theta_3 & 0 & 0 \\ 0 & -\theta_3 & 0 \\ 0 & 0 & 1 \end{bmatrix}$$

$$\tilde{\boldsymbol{\theta}} = (\theta_1, \theta_2)', \quad \mathbf{x} = (x_1, x_2)'$$

$$\phi(\mathbf{x}, \tilde{\boldsymbol{\theta}}) = \frac{[x_1(1 + \theta_2 x_2), \theta_1 x_1 x_2, \theta_1 x_1(1 + \theta_1 x_1 + \theta_2 x_2)]'}{(1 + \theta_1 x_1 + \theta_2 x_2)^2}$$

Hence, only the initial values of θ_1 and θ_2 need be specified for the construction of the optimal design for $\boldsymbol{\theta}$.

Now suppose we are interested in the precise estimation of only a subset, $\boldsymbol{\theta}_1$, of s of the p ($s < p$) elements of $\boldsymbol{\theta}$ ($\boldsymbol{\theta}_1$ can be considered as consisting of the first s elements of $\boldsymbol{\theta}$). Then it is not always true that the corresponding locally D-optimal design for the estimation of $\boldsymbol{\theta}_1$ will depend only on the elements of $\tilde{\boldsymbol{\theta}}$ when the model is partially nonlinear and satisfies Eq. 10.40. For this to be true we need in addition to Eq. 10.40 the following condition (Khuri, 1984): Let $\mathbf{B}(\boldsymbol{\theta})$ in Eq. 10.40 be partitioned as $[\mathbf{B}_1'(\boldsymbol{\theta}):\mathbf{B}_2'(\boldsymbol{\theta})]'$, where $\mathbf{B}_1(\boldsymbol{\theta})$ and $\mathbf{B}_2(\boldsymbol{\theta})$ are matrices of orders $s \times p$ and $(p - s) \times p$, respectively. If $\mathbf{B}_2(\boldsymbol{\theta})$ can be expressed as

$$\mathbf{B}_2(\boldsymbol{\theta}) = \mathbf{C}(\boldsymbol{\theta})[\mathbf{I}_{p-s}:\mathbf{K}] \tag{10.41}$$

where $\mathbf{C}(\boldsymbol{\theta})$ is a nonsingular matrix of order $(p - s) \times (p - s)$, $\mathbf{I}_{p-s}$ is the identity matrix of order $(p - s) \times (p - s)$, and $\mathbf{K}$ is a matrix of order $(p - s) \times s$ not involving $\boldsymbol{\theta}$, then the locally D-optimal design for the estimation of $\boldsymbol{\theta}_1$ depends only on the elements of $\tilde{\boldsymbol{\theta}}$.

For example, let us consider the following nonlinear model reported in Hill (1980) and used in chemical engineering to estimate the surface areas of certain solid catalysts:

$$f(x, \boldsymbol{\theta}) = \frac{\theta_1 x}{(1 - x)[\theta_2 + (1 - \theta_2)x]} \tag{10.42}$$

Let $\boldsymbol{\theta}_1 = \theta_1$, that is, the interest here is in obtaining a precise estimate of θ_1. It can be verified that $\nabla f(x, \boldsymbol{\theta})$ is of the form described in Eq. 10.40 with

$$\mathbf{B}(\boldsymbol{\theta}) = \begin{bmatrix} 0 & 1 \\ \theta_1 & 0 \end{bmatrix}$$

$$\bar{\boldsymbol{\theta}} = \theta_2$$

$$\phi(x, \bar{\boldsymbol{\theta}}) = \left(\frac{-x}{[\theta_2 + (1 - \theta_2)x]^2}, \frac{x}{(1 - x)[\theta_2 + (1 - \theta_2)x]} \right)'$$

Hence, $\mathbf{B}_2(\boldsymbol{\theta}) = (\theta_1, 0)$, which can be written as in Eq. 10.41 with $\mathbf{C}(\boldsymbol{\theta}) = \theta_1$, $\mathbf{I}_{p-s} = 1$, and $\mathbf{K} = 0$. Consequently, the locally D-optimal design for the estimation of θ_1 depends only on θ_2.

10.5.4 Parameter-Free Designs for Nonlinear Models

In this section we shall introduce a new criterion for the choice of a nonlinear model design that does not depend on the initial values of the parameters. This criterion is based on approximating the mean response $f(\mathbf{x}_u, \boldsymbol{\theta})$ in Eq. 10.8 with a Lagrange interpolating polynomial of a certain degree. We shall restrict our consideration of model 10.8 to the case of a single input variable x.

The Lagrangian Interpolation of the Mean Response. Let R denote the experimental region for the model

$$Y = f(x, \boldsymbol{\theta}) + \varepsilon \tag{10.43}$$

We suppose that the region R is of the form, $R = \{x : c \le x \le d\}$. Using the change of variable,

$$t = \frac{2x - c - d}{d - c} \tag{10.44}$$

the region R is reduced to the interval $-1 \le t \le 1$. If we substitute t for x in model 10.43, we obtain the model

$$Y = h(t, \boldsymbol{\theta}) + \varepsilon \tag{10.45}$$

We assume that

1. $h(t, \boldsymbol{\theta})$ has continuous derivatives up to order $r + 1$ with respect to t over the interval $-1 \le t \le 1$ for all $\boldsymbol{\theta}$, where r is such that $r + 1$ is a positive integer greater than or equal to p, the number of parameters in model 10.43. We also assume that r is large enough so that

$$\frac{\max\limits_{-1 \le t \le 1} |h_{r+1}(t, \boldsymbol{\theta})|}{2^r(r + 1)!} < \delta \tag{10.46}$$

for all θ in some parameter space, Ω, where

$$h_{r+1}(t, \theta) = \frac{\partial^{r+1} h(t, \theta)}{\partial t^{r+1}}$$

and δ is a small positive constant selected so that the Lagrangian interpolation of $h(t, \theta)$ meets a certain degree of accuracy.

2. For any set of distinct points, $\bar{z}_o, \bar{z}_1, \ldots, \bar{z}_r$ such that $-1 \le \bar{z}_o < \bar{z}_1 < \ldots < \bar{z}_r \le 1$, where r is the integer defined in assumption 1., the $p \times (r + 1)$ matrix,

$$\mathbf{U}(\theta) = [\nabla h(\bar{z}_0, \theta) : \nabla h(\bar{z}_1, \theta) : \ldots : \nabla h(\bar{z}_r, \theta)] \tag{10.47}$$

has rank p, where $\nabla h(\bar{z}_i, \theta)$ is the vector of partial derivatives of $h(\bar{z}_i, \theta)$ with respect to the elements of θ ($i = 0, 1, \ldots, r$).

Consider now the points $\bar{z}_0, \bar{z}_1, \ldots, \bar{z}_r$ described in assumption 2. There exists a unique polynomial $p_r(t, \theta)$, of degree r or less for which $p_r(\bar{z}_i, \theta) = h(\bar{z}_i, \theta)$ for $i = 0, 1, \ldots, r$, that is, $p_r(t, \theta)$ agrees with $h(t, \theta)$ at the points $\bar{z}_0, \bar{z}_1, \ldots, \bar{z}_r$. This polynomial can be written in the form

$$p_r(t, \theta) = \sum_{i=0}^{r} h(\bar{z}_i, \theta) l_i(t) \tag{10.48}$$

where

$$l_i(t) = \prod_{\substack{j=0 \\ j \ne i}}^{r} \frac{t - \bar{z}_j}{\bar{z}_i - \bar{z}_j} \qquad i = 0, 1, \ldots, r \tag{10.49}$$

is the ith Lagrange polynomial of degree r (see, e.g., Young and Gregory, 1972: Ch. 6; and DeBoor, 1978: Chs. 1 and 2). The polynomial $p_r(t, \theta)$ is called a Lagrange interpolating polynomial and the points $\bar{z}_0, \bar{z}_1, \ldots, \bar{z}_r$ are called interpolation points.

The polynomial $p_r(t, \theta)$ provides an approximation to the function $h(t, \theta)$ over the interval $-1 \le t \le 1$. This approximation is very sensitive to the choice of the interpolation points. It is possible for some functions that a poor choice of points leads to a polynomial approximation with a large error of approximation even for large values of r. Uniformly spaced interpolation points, in particular, should be avoided since the interpolation error can actually increase with r, as $p_r(t, \theta)$ may fail entirely to approximate $h(t, \theta)$ (see DeBoor, 1978: 22–26). The error resulting from the use of Lagrangian interpolation is, by definition, the difference between $h(t, \theta)$ and $p_r(t, \theta)$ and is given by

$$e_r(t, \theta) = \frac{\prod_{i=0}^{r} (t - \bar{z}_i) h_{r+1}(\zeta, \theta)}{(r + 1)!} \tag{10.50}$$

where $-1 < \zeta < 1$ (see DeBoor, 1978: 8; and Hartley, 1964: 351). An upper bound on $e_r(t, \theta)$ is given by

$$\max_{-1 \leq t \leq 1} |e_r(t, \theta)| \leq \frac{w(\bar{z}_0, \bar{z}_1, \ldots, \bar{z}_r) \max_{-1 \leq t \leq 1} |h_{r+1}(t, \theta)|}{(r + 1)!} \qquad (10.51)$$

where $w(\bar{z}_0, \bar{z}_1, \ldots, \bar{z}_r) = \max_{-1 \leq t \leq 1} |\Pi_{i=0}^r(t - \bar{z}_i)|$.

The interpolation points influence the accuracy of the approximation through the function $w(\bar{z}_0, \bar{z}_1, \ldots, \bar{z}_r)$. Using a remarkably useful result given in DeBoor (1978: 26–31, in particular, formulas 6, 7, 8, and 10), we can state that, in general, if Lagrangian interpolation is desired over the interval $a \leq t \leq b$, then the function $w(\bar{z}_0, \bar{z}_1, \ldots, \bar{z}_r)$ in Eq. 10.51 can attain a minimum value of $2(b - a)^{r+1}/4^{r+1}$ if the interpolation points are chosen as the zeros of a Chebyshev polynomial of the first kind and of degree $r + 1$ for the interval $a \leq t \leq b$. These zeros are given by the formula

$$\bar{z}_i = \frac{a + b}{2} - \left(\frac{a - b}{2}\right) \cos\left[\left(i + \frac{1}{2}\right)\frac{\pi}{r + 1}\right] \qquad i = 0, 1, \ldots, r$$

In our case $a = -1$ and $b = 1$, hence the minimum value of $w(\bar{z}_0, \bar{z}_1, \ldots, \bar{z}_r)$ is $1/2^r$ and is attained at the Chebyshev points

$$\bar{z}_i = \cos\left[\left(i + \frac{1}{2}\right)\frac{\pi}{r + 1}\right] \qquad i = 0, 1, \ldots, r \qquad (10.52)$$

Chebyshev points can be obtained geometrically be subdividing the semicircle over the interval $-1 \leq t \leq 1$ into $r + 1$ equal arcs and then projecting the midpoint of each arc onto the interval as can be seen from Figure 10.1 (see DeBoor, 1978: 26). Chebyshev points are thus optimal in the sense of minimizing the upper bound in Eq. 10.51 with respect to $\bar{z}_0, \bar{z}_1, \ldots, \bar{z}_r$. In other words, among all sets of interpolation points of size $r + 1$, Chebyshev points produce a polynomial approximation with a minimum upper bound on the maximum error of approximation. From Eq. 10.51 and the fact that the minimum value of $w(\bar{z}_0, \bar{z}_1, \ldots, \bar{z}_r)$ is $1/2^r$, we obtain an upper bound for the error of approximation,

$$\max_{-1 \leq t \leq 1} |e_r(t, \theta)| \leq \frac{\max_{-1 \leq t \leq 1} |h_{r+1}(t, \theta)|}{2^r(r + 1)!} \qquad (10.53)$$

The upper bound in Eq. 10.53 will be used to measure the degree of approximation accuracy and for the determination of the rate of convergence of $p_r(t, \theta)$ to $h(t, \theta)$, as $r \to \infty$. This convergence is guaranteed by assumption 1. since inequality 10.46 is equivalent to requiring that the upper bound in Eq. 10.53 go to zero for all θ as $r \to \infty$.

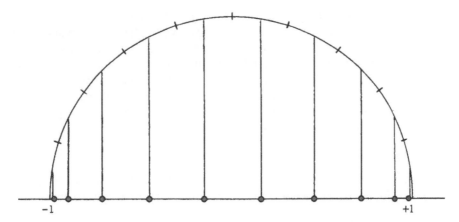

Figure 10.1 The Chebyshev points for the interval $[-1, 1]$, where $r = 9$.

Powell (1967) showed that the use of the Chebyshev points in the construction of the Lagrange interpolating polynomial $p_r(t, \theta)$ produces an approximation which for all practical purposes differs very little from the best possible approximation of $h(t, \theta)$ by polynomials of the same degree. More explicitly, let $p_r^*(t, \theta)$ be the best approximating polynomial of $h(t, \theta)$ of degree r over the interval $-1 \le t \le 1$, that is,

$$\max_{-1 \le t \le 1} |h(t, \theta) - p_r^*(t, \theta)| \le \max_{-1 \le t \le 1} |h(t, \theta) - p_r(t, \theta)| \tag{10.54}$$

for any other polynomial, $p_r(t, \theta)$, of the same degree and for all θ. Let $e_r^*(t, \theta)$ and $e_r(t, \theta)$ be the approximation errors associated with $p_r^*(t, \theta)$ and $p_r(t, \theta)$, respectively. Then $\max_{-1 \le t \le 1} |e_r^*(t, \theta)| \le \max_{-1 \le t \le 1} |e_r(t, \theta)|$. Furthermore, Powell (1967: Table 1, p. 406) showed that for $r \le 20$,

$$\max_{-1 \le t \le 1} |e_r(t, \theta)| \le 3.901 \max_{-1 \le t \le 1} |e_r^*(t, \theta)| \tag{10.55}$$

Inequality 10.55 indicates that the maximum error of approximation resulting from Lagrangian interpolation with Chebyshev interpolation points does not exceed the minimax error, namely, $\max_{-1 \le t \le 1} |e_r^*(t, \theta)|$ by more than a factor of four for $r \le 20$. Thus a Lagrange interpolating polynomial can be almost as efficient as the best approximating polynomial as long as the degree of the polynomial is not very high. This is a very useful result since the derivation of the best approximating polynomial can be tedious and complicated, whereas a polynomial approximation obtained by Lagrangian interpolation and the use of Chebyshev interpolation points is simple and easy to obtain.

Designs Based on Lagrangian Interpolation. Consider the Lagrange interpolating polynomial $p_r(t, \theta)$ given in Eq. 10.48 with interpolation points being of the Chebyshev type described in Eq. 10.52. The positive integer r is determined by requiring the upper bound in Eq. 10.53 to be less than some specified value $\delta > 0$ for all θ in some parameter space Ω. For such a value of r, $p_r(t, \theta)$ is approximately equal to $h(t, \theta)$ with an error of approximation not exceeding δ over the interval $-1 \leq t \leq 1$ and for all θ in Ω. Thus, an approximate representation of model 10.45 is given by

$$Y = p_r(t, \theta) + \varepsilon \tag{10.56}$$

The precise estimation of the parameters in model 10.56 will be used as a criterion for the development of the design. From Eqs. 10.48 and 10.56 we have

$$\nabla p_r(t, \theta) = \sum_{i=0}^{r} \frac{\partial h(\tilde{z}_i, \theta)}{\partial \theta} l_i(t) \tag{10.57}$$

where $\nabla p_r(t, \theta)$ is the vector of partial derivatives of $p_r(t, \theta)$ with respect to the elements of θ and $l_i(t)$ is defined in Eq. 10.49. In matrix form, Eq. 10.57 can be written as

$$\nabla p_r(t, \theta) = \mathbf{U}(\theta)\boldsymbol{v}(t) \tag{10.58}$$

where $\mathbf{U}(\theta)$ is the $p \times (r + 1)$ matrix given in Eq. 10.47, which by assumption 2. is of rank p, and $\boldsymbol{v}(t)$ is the vector of Lagrange polynomials, $\boldsymbol{v}(t) = [l_0(t), l_1(t), \ldots, l_r(t)]'$.

By comparing Eq. 10.58 with 10.40, $\nabla f(\mathbf{x}, \theta) = \mathbf{B}(\theta)\phi(\mathbf{x}, \tilde{\theta})$, we note that the Lagrangian interpolation scheme makes it possible to approximate model 10.45 with the pseudo partially nonlinear model given in Eq. 10.56. The use of *pseudo partially nonlinear* here refers to the fact that Eq. 10.58 is similar to Eq. 10.40, which characterizes partially nonlinear models. The difference is that unlike $\mathbf{B}(\theta)$ in Eq. 10.40, the matrix $\mathbf{U}(\theta)$ is, in general, a rectangular matrix. This similarity with partially nonlinear models will be utilized to develop a design which is free of θ. Hartley (1964) made similar use of Lagrange interpolation for the construction of exact confidence regions on the parameters of model 10.45.

Let $t_1, t_2, \ldots, t_N$ be the points on the interval $-1 \leq t \leq 1$ which correspond in a one-to-one manner through Eq. 10.44 to the points of the design $\mathbf{D} = (x_1, x_2, \ldots, x_N)'$ with x_u ($u = 1, 2, \ldots, N$) belonging to the experimental region $R = \{x : c \leq x \leq d\}$. Let $\mathbf{M}_l(\mathbf{D}, \theta)$ denote the matrix given in Eq. 10.29 with respect to model Eq. 10.56. From Eq. 10.58 we then have

$$\mathbf{M}_l(\mathbf{D}, \theta) = \mathbf{U}(\theta) \left[\sum_{u=1}^{N} \boldsymbol{v}(t_u)\boldsymbol{v}'(t_u) \right] \mathbf{U}'(\theta) \tag{10.59}$$

If $N \geq r + 1$ and at least $r + 1$ of the design points are distinct, then the $(r + 1) \times (r + 1)$ matrix,

$$\Lambda(t_1, t_2, \ldots, t_N) = \sum_{u=1}^{N} \boldsymbol{v}(t_u)\boldsymbol{v}'(t_u) \tag{10.60}$$

should be of rank $r + 1$ and thus nonsingular. To show this, suppose that $\Lambda(t_1, t_2, \ldots, t_N)$ is of rank less than $r + 1$. Then the $(r + 1) \times N$ matrix $\mathbf{G} = [\boldsymbol{v}(t_1):\boldsymbol{v}(t_2):\ldots:\boldsymbol{v}(t_N)]$ is of the same rank as $\Lambda(t_1, t_2, \ldots, t_N)$. In this case, there must exist a set of α_i $(i = 0, 1, \ldots, r)$ not all equal to zero such that

$$\sum_{i=0}^{r} \alpha_i l_i(t_u) = 0, \qquad u = 1, 2, \ldots, N \tag{10.61}$$

From Eq. 10.61 we conclude that the rth degree polynomial $\Sigma_{i=0}^{r} \alpha_i l_i(t)$ has N roots, namely, $t_1, t_2, \ldots, t_N$, which is impossible since $N \geq r + 1$ and at least $r + 1$ of $t_1, t_2, \ldots, t_N$ are distinct (a polynomial of degree r has at most r distinct roots).

Now, from Eqs. 10.59 and 10.60 we have

$$|\mathbf{M}_l(\mathbf{D}, \boldsymbol{\theta})| = |\mathbf{U}(\boldsymbol{\theta})\Lambda(t_1, t_2, \ldots, t_N)\mathbf{U}'(\boldsymbol{\theta})| \tag{10.62}$$

Let $\lambda_0(t_1, t_2, \ldots, t_N)$ and $\lambda_r(t_1, t_2, \ldots, t_N)$ denote the smallest and largest eigenvalues of $\Lambda(t_1, t_2, \ldots, t_N)$, respectively. These eigenvalues are positive since $\Lambda(t_1, t_2, \ldots, t_N)$ is nonsingular as was shown earlier and is thus positive definite. From Eq. 10.62 we conclude that

$$\lambda_0^p(t_1, t_2, \ldots, t_N)|\mathbf{U}(\boldsymbol{\theta})\mathbf{U}'(\boldsymbol{\theta})| \leq |\mathbf{M}_l(\mathbf{D}, \boldsymbol{\theta})| \leq \lambda_r^p(t_1, t_2, \ldots, t_N)|\mathbf{U}(\boldsymbol{\theta})\mathbf{U}'(\boldsymbol{\theta})| \tag{10.63}$$

We note that the determinant of the $p \times p$ matrix $\mathbf{U}(\boldsymbol{\theta})\mathbf{U}'(\boldsymbol{\theta})$ in Eq. 10.63 is not zero since by assumption 2. $\mathbf{U}(\boldsymbol{\theta})$ is of rank p.

To maximize the determinant of the matrix $\mathbf{M}_l(\mathbf{D}, \boldsymbol{\theta})$ independently of $\boldsymbol{\theta}$, we may choose our design, that is, $t_1, t_2, \ldots, t_N$, so that $\lambda_0^p(t_1, t_2, \ldots, t_N)$ in Eq. 10.63 is a maximum. This design, however, may be inefficient because the maximum value of $\lambda_0^p(t_1, t_2, \ldots, t_N)$ can be very small. A more efficient design can be found as follows: From Eq. 10.63 we have

$$|\mathbf{M}_l(\mathbf{D}, \boldsymbol{\theta})| = [\gamma\lambda_0^p(t_1, t_2, \ldots, t_N) + (1 - \gamma)\lambda_r^p(t_1, t_2, \ldots, t_N)]|\mathbf{U}(\boldsymbol{\theta})\mathbf{U}'(\boldsymbol{\theta})| \tag{10.64}$$

where $0 \leq \gamma \leq 1$. If we assume a noninformative uniform prior distribution on γ with the density

$$v(\gamma) = \begin{cases} 1 & 0 \leq \gamma \leq 1 \\ 0 & \text{elsewhere} \end{cases}$$

then from Eq. 10.64, the expected value of $|\mathbf{M}_l(\mathbf{D}, \boldsymbol{\theta})|$ with respect to γ is equal to

$$E_\gamma[|\mathbf{M}_l(\mathbf{D}, \boldsymbol{\theta})|] = \kappa(t_1, t_2, \ldots, t_N)|\mathbf{U}(\boldsymbol{\theta})\mathbf{U}'(\boldsymbol{\theta})| \tag{10.65}$$

where

$$\kappa(t_1, t_2, \ldots, t_N) = \frac{[\lambda_0^p(t_1, t_2, \ldots, t_N) + \lambda_r^p(t_1, t_2, \ldots, t_N)]}{2} \tag{10.66}$$

Consequently, our recommended parameter-free design is the one which maximizes $E_\gamma[|\mathbf{M}_l(\mathbf{D}, \boldsymbol{\theta})|]$ with respect to $t_1, t_2, \ldots, t_N$. This can be conveniently carried out by using the computer program SEARCH, which was written by Conlon (1985) and is based on Price's (1977) controlled random search procedure, to maximize $\kappa(t_1, t_2, \ldots, t_N)$ in Eq. 10.66.

In summary, the method of Lagrange interpolation essentially reduces the nonlinear model to the pseudo partially nonlinear model given in Eq. 10.56. The extended version of the local D-optimality criterion, which consists of maximizing the expected value of the determinant of $\mathbf{M}_l(\mathbf{D}, \boldsymbol{\theta})$ given in Eq. 10.65, is then applied to the latter model.

Example 10.3. Let us again consider the model in Eq. 10.32. Suppose that the experimental region is $R = \{x : 0 \le x \le d\}$, where d is some positive value specified by the experimenter (in this example we choose $d = 10$), and let the parameter space Ω be such that $0 \le \theta_1 \le 1$ and $0 \le \theta_2 \le 1$. Using the transformation given in Eq. 10.44, $t = (2x - c - d)/(d - c)$, with $c = 0$, model 10.32 becomes

$$h(t, \boldsymbol{\theta}) = \frac{\theta_1}{\theta_1 - \theta_2}[e^{-d\theta_2(t+1)/2} - e^{-d\theta_1(t+1)/2}] \tag{10.67}$$

with $-1 \le t \le 1$. It can be shown that for all $\boldsymbol{\theta}$ in Ω

$$\max_{-1 \le t \le 1}|h_{r+1}(t, \boldsymbol{\theta})| \le \left(\frac{d}{2}\right)^{r+1} \max(r + 1, d - r - 1) \tag{10.68}$$

where $h_{r+1}(t, \boldsymbol{\theta})$ is the $(r + 1)$th-order partial derivative of $h(t, \boldsymbol{\theta})$ with respect to t, and $\max(r + 1, d - r - 1)$ is the larger value of $r + 1$ and $d - r - 1$. Hence, from Eq. 10.53 the Lagrangian interpolation accuracy is determined by

$$\max_{-1 \le t \le 1}|e_r(t, \boldsymbol{\theta})| \le \frac{[(\frac{d}{2})^{r+1} \max(r + 1, d - r - 1)]}{2^r(r + 1)!} \tag{10.69}$$

for all $\boldsymbol{\theta}$ in Ω.

Suppose now we require that the upper bound in Eq. 10.69 (with $d = 10$) be less than or equal to $\delta = 0.05$. By computing this upper bound for several assignments of r, we obtain the values given in Table 10.4. From this table we

Table 10.4 Maximum
Error of Lagrange
Interpolation

r	Maximum Error (Formula 10.69)
1	50
2	36.46
3	19.53
4	8.14
5	4.07
6	1.70
7	0.61
8	0.19
9	0.05

note that the first value of r to satisfy the $\delta = 0.05$ requirement is $r = 9$. The Chebyshev points corresponding to this value of r are obtained from Eq. 10.52. These points are given in Table 10.5. By choosing N, the number of design points, to be equal to $r + 1 = 10$ (here all 10 design points must be distinct), the vector $\mathbf{v}(t)$ of Lagrange functions of degree 9 as well as the matrix $\Lambda(t_1, t_2, \ldots, t_{10})$ can be determined from Eqs. 10.49 and 10.60. The maximum value of $\kappa(t_1, t_2, \ldots, t_{10})$ in Eq. 10.66 with $p = 2$ is $\kappa = 17.457$ and the corresponding optimal values of t_u ($u = 1, 2, \ldots, 10$) are also given in Table 10.5.

Table 10.5 Chebyshev Points
and Optimal Design

Chebyshev Points	Optimal Design (in terms of t)
0.98769	0.99785
0.89101	0.99672
0.70711	0.99657
0.45399	0.99312
0.15643	0.90840
−0.15643	0.40876
−0.45399	0.21563
−0.70711	−0.19248
−0.89101	−0.72377
−0.98769	−0.86158

EXERCISES

10.1 Consider the nonlinear model given in Eq. 10.36, that is,

$$\eta = \frac{\theta_1 \theta_3 x_1}{1 + \theta_1 x_1 + \theta_2 x_2}$$

A total of five experiments were run, and the following data were collected:

x_1	x_2	Y
0.28	0.0	0.33
2.9	0.10	0.57
3.2	0.82	0.33
4.7	1.1	0.31
5.5	2.0	0.23

(a) Obtain the least squares estimates for θ_1, θ_2, and θ_3.
(b) Obtain approximate 95% confidence intervals on θ_1, θ_2, θ_3.
(c) Construct an approximate 95% confidence region for $\theta = (\theta_1, \theta_2, \theta_3)'$.
(d) Test the hypothesis

$$H_o: \begin{cases} \theta_1 + \theta_2 = 18.1 \\ \quad\quad \theta_3 = 2.5 \end{cases}$$

at the $\alpha = 0.5$ level of significance, using
(i) Equation 10.11.
(ii) Equation 10.24.
(e) Construct an approximate 99% confidence interval on the mean response η when $x_1 = 1.5$, $x_2 = 0.75$.

10.2 Consider the nonlinear model

$$\eta = \theta_1[x + \exp(-\theta_3 x)] - \theta_2 x \exp(-\theta_3 x)$$

(a) Apply Eq. 10.40 to verify that the locally D-optimal design for the estimation of $\theta = (\theta_1, \theta_2, \theta_3)'$ depends only on θ_3.
(b) Suppose now that we are interested in the precise estimation of only θ_1 and θ_2.
Show that the corresponding locally D-optimal design does not depend on θ_3 only.

10.3 Prove the inequality given in Eq. 10.68.

10.4 Suppose that it is of interest to construct several two-point locally *D*-optimal designs for the model

$$\eta = \frac{\theta_1}{\theta_1 - \theta_2} \left[\exp(-\theta_2 x) - \exp(-\theta_1 x) \right]$$

(a) Obtain these designs which correspond to the following values of the parameters:

$\theta_1 = 0.10$	$\theta_2 = 0.20$
$\theta_1 = 0.20$	$\theta_2 = 0.30$
$\theta_1 = 0.30$	$\theta_2 = 0.50$
$\theta_1 = 0.50$	$\theta_2 = 1.0$
$\theta_1 = 0.70$	$\theta_2 = 0.20$
$\theta_1 = 0.90$	$\theta_2 = 0.80$
$\theta_1 = 1.0$	$\theta_2 = 0.90$

(b) Make use of the tabulation of designs obtained in (a) to demonstrate the dependency of the locally *D*-optimal design on θ_1 and θ_2.

10.5 Consider again the model described in problem 10.4, which can also be written as in Eq. 10.67. Use the Chebyshev points in Table 10.5 and Eq. 10.48 to obtain a Lagrange interpolating polynomial of degree 9 that approximates the true mean response function, η, with an error not exceeding $\delta = 0.05$.

REFERENCES AND BIBLIOGRAPHY

Atkinson, A. C. and W. G. Hunter (1968). The Design of Experiments for Parameter Estimation, *Technometrics*, **10**, 271–289.

Bard, Y. (1974). *Nonlinear Parameter Estimation*, New York: Academic Press.

Box, G. E. P. and W. G. Hunter (1965). Sequential Design of Experiments for Nonlinear Models, *Proceedings of I.B.M. Scientific Computing Symposium in Statistics*, 113–137.

Box, G. E. P. and H. L. Lucas (1959). Design of Experiments in Nonlinear Situations, *Biometrika*, **46**, 77–90.

Box, M. J. (1970). Some Experiences with a Nonlinear Experimental Design Criterion, *Technometrics*, **12**, 569–589.

Box, M. J. (1971). Bias in Nonlinear Estimation, *J. Roy. Statist. Soc.*, **B33**, 171–190.

Chambers, J. M. (1973). Fitting Nonlinear Models: Numerical Techniques, *Biometrika*, **60**, 1–13.

Cochran, W. G. (1973). Experiments for Nonlinear Functions (R. A. Fisher Memorial Lecture), *J. Amer. Statist. Assoc.*, **68**, 771–781.

Conlon, M. (1985). Controlled Random Search Procedure for Function Minimization, Technical Report No. 236, Dept. of Statistics, Univ. of Florida, Gainesville, FL.

DeBoor, C. (1978). *A Practical Guide to Splines*, New York: Springer-Verlag.

Draper, N. R. and H. Smith (1981). *Applied Regression Analysis*, 2nd ed., New York: John Wiley.

Gallant, A. R. (1975). Testing a Subset of the Parameters of a Nonlinear Regression Model, *J. Amer. Statist. Assoc.*, **70**, 927–932.

Hartley, H. O. (1961). The Modified Gauss-Newton Method for the Fitting of Nonlinear Regression Functions by Least Squares, *Technometrics*, **3**, 269–280.

Hartley, H. O. (1964). Exact Confidence Regions for the Parameters in Nonlinear Regression Laws, *Biometrika*, **51**, 347–353.

Hill, P. D. H. (1980). *D*-optimal Designs for Partially Nonlinear Regression Models, *Technometrics*, **22**, 275–276.

Hill, W. J. and W. G. Hunter (1974). Design of Experiments for Subsets of Parameters, *Technometrics*, **16**, 425–434.

Jennrich, R. I. (1969). Asymptotic Properties of Nonlinear Least Squares Estimators, *Ann. Math. Statist.*, **40**, 633–643.

Kennedy, W. J. and J. E. Gentle (1980). *Statistical Computing*, New York: Marcel Dekker.

Khuri, A. I. (1984). A Note on *D*-optimal Designs for Partially Nonlinear Regression Models, *Technometrics*, **26**, 59–61.

Lawton, W. H. and E. A. Sylvestre (1971). Elimination of Linear Parameters in Nonlinear Regression, *Technometrics*, **13**, 461–467.

Malinvaud, E. (1980). *Statistical Methods of Econometrics*, 3rd ed., Amsterdam: North-Holland.

Marquardt, D. W. (1963). An Algorithm for Least Squares Estimation of Nonlinear Parameters, *J. Soc. Industrial and Applied Math.*, **11**, 431–441.

Martin, Y. C. and J. J. Hackbarth (1977). Examples of the Application of Nonlinear Regression Analysis to Chemical Data. *Chemometrics: Theory and Application*, A Symposium Sponsored by the Division of Computers in Chemistry (B. R. Kowalski, ed.), pp. 153–164.

Milliken, G. A. and R. L. DeBruin (1978). A Procedure to Test Hypotheses for Nonlinear Models, *Commun. Statist. Theor. Meth.*, **A7**, 65–79.

Powell, M. J. D. (1967). On the Maximum Errors of Polynomial Approximations Defined by Interpolation and by Least Squares Criteria, *Computer Journal*, **9**, 404–407.

Price, W. L. (1977). A Controlled Random Search Procedure for Global Optimization, *Computer Journal*, **20**, 367–370.

Ralston, M. L. and R. I. Jennrich (1978). DUD, A Derivative-Free Algorithm for Nonlinear Least Squares, *Technometrics*, **20**, 7–14.

Ratkowsky, D. A. (1983). *Nonlinear Regression Modeling*, New York: Marcel Dekker.

SAS Institute, Inc. (1989). SAS/STAT User's Guide, Vol. 2, Version 6, 4th ed., Cary, NC: Author.

Vohnout, K. and C. Jimenez (1975). Supplemental By-Product Feeds in Pasture-Livestock Feeding Systems in the Tropics. In: Tropical Forages in Livestock Production Systems, ASA Special Publication No. 24, (E. C. Doll and G. O. Mott, eds.), Madison, WI: American Society of Agronomy, pp. 71–82.

Young, D. M. and R. T. Gregory (1972). *A Survey of Numerical Mathematics*, Vol. 1, Reading, MA: Addison-Wesley.

11

Taguchi's Robust Parameter Design

11.1 INTRODUCTION

In the introduction to Chapter 1, it was stated that "response surface methodology (RSM) is a set of techniques that encompasses"; we now restate this more briefly:

1. Setting up a series of experiments.
2. Determining a mathematical model.
3. Determining the optimal settings of the experimental factors that produce the maximum (or minimum) value of the response.

In the preceding chapters the main emphasis was on designs and model forms where the focus of attention centered on "optimizing the mean or expected value of the response(s)." However, one may ask, "What about reducing the variation in the process or in the response values while trying to achieve a mean response value close to some target value? Isn't that a way to optimize the process also?" The obvious answer to these two questions is yes, which brings us to the topic of Taguchi's robust parameter design.

11.2 THE ROBUST PARAMETER DESIGN

In the early 1980s, Genichi Taguchi spearheaded a move which, through a series of designed experiments, investigates factors that affect the process mean and

413

the variation in the process. By introducing the parameter design approach, Taguchi emphasized making processes insensitive to variation caused by environmental and/or uncontrollable factors. Fundamental to this approach is the concept of *control* and *noise* variables.

11.2.1   Control and Noise Factors

Control variables or *control factors* are those the experimenter can easily control in any conducted experiment. *Noise variables* are factors the experimenter either overlooks or chooses not to directly control. They may be controlled at the research or development level, but not at the production level or in the field. Some examples of noise variables are environmental conditions such as ambient temperature and humidity, or time-related product characteristics such as deterioration, wear and shrinkage, or possibly functions of such uncontrollable conditions as customer usage or variation in raw materials. As a result, noise variables are thought to be primarily responsible for transmitting variation to the process that shows up as uncontrolled variation in the values of the response.

The parameter design approach, sometimes called *robust parameter design*, represents an attempt to fine-tune the process, by focusing on the quality of the product it produces, without controlling or removing the causes of variation in the product characteristic measured. In a parameter design, settings of both the control and noise variables are varied according to some plan (see Figure 11.1). The objective of varying the control and noise variable settings is to determine the levels of the controllable factors (Taguchi calls these parameters) for which the following process conditions are satisfied:

1. The process mean or mean response is directed to a specified target value.
2. The process is insensitive or robust to changes in the settings of the noise factors.

Failure to achieve either or both conditions, according to Taguchi, results in a loss generated by the product to society. Mathematically, the loss, in dollars, say, to a customer is measured by an amount that some measured product characteristic, Y, deviates from its target value τ and is expressed as

$$l(Y) = k(Y - \tau)^2 \tag{11.1}$$

where k is some constant. It is straightforward to show that this quadratic loss $l(Y)$ is expressible as $l(Y) = k(s_Y^2 + (\bar{Y} - \tau)^2)$, which is a function of the variation in response value s_Y^2 plus a measure of the error resulting from the deviation of the average response from the target value. Kackar (1985) presents a plot of the quadratic loss function in Eq. 11.1 and points out the breakdown of $l(Y)$ in terms of customer's tolerance and the manufacturer's tolerance.

To understand the ideas behind performing a parameter design experiment, let us first define a set of controllable variables as $X_1, X_2, \ldots, X_p$, and a set of

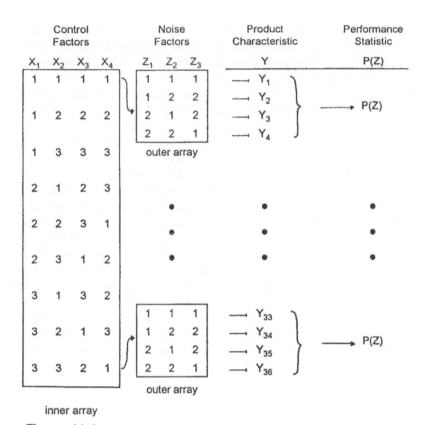

Figure 11.1 A schematic of a parameter design with four control factors and three noise factors. At each row of the inner array are the four rows of the outer array.

uncontrollable or noise variables as $Z_1, Z_2, \ldots, Z_q$. For the purpose of conducting the experiment, the settings of the Z_j's, $j = 1, 2, \ldots, q$, can be controlled. Next, let us consider a design matrix $\mathbf{X}$ of N_X rows, where the uth row of $\mathbf{X}$ specifies the settings of the coded values of the X_i's, $i = 1, 2, \ldots, p$, to be used in the uth trial. Similarly, we consider a matrix $\mathbf{Z}$ of N_Z rows containing the settings or levels of coded values of the Z_j's, $j = 1, 2, \ldots, q$. Then, the parameter design (PD) associated with these control and noise variables consists of crossing each row in $\mathbf{X}$ with $\mathbf{Z}$, that is PD $= \mathbf{X} \otimes \mathbf{Z}$ to produce $N = N_X \times N_Z$ experimental combinations. In parameter design terminology, the matrix $\mathbf{X}$ is called the *control or inner array* and the matrix $\mathbf{Z}$ is the *noise or outer array*. Shown in Figure 11.1 is an example of a parameter design with four ($p = 4$) control factors and three ($q = 3$) noise factors. In the figure, the inner array is an L_9 orthogonal array or a 1/9-th fraction of a 3^4 factorial consisting of the three coded levels (1, 2, and 3) for each of the X_i's, $i = 1, 2, 3, 4$. The outer array is a 2^{3-1} fractional factorial, where the coded levels are denoted by 1 and 2, in the Z_j's, $j = 1, 2, 3$. The total number of experimental trials to be performed with the crossed arrays in Figure 11.1 is $9 \times 4 = 36$.

Taguchi emphasized the use of orthogonal arrays in the PD [see, for example, Taguchi and Wu (1980) and Phadke et al. (1983)]. A great variety of these designs exist, the most commonly used being the 2^k factorials and 2^{k-p} fractional factorials that were discussed in Chapter 3, and the 3^k factorials (Chapter 4) and 3^{k-p} fractions. A drawback to the use of a 3^{k-p} fraction, particularly in the inner array, is the inability of the fraction to enable interactions between the control factors to be estimated. This is clearly explained by Hunter (1985). Taguchi's suggestion on the use of the L_9 orthogonal array shown in Figure 11.1 has drawn criticism from Box (1988), Hunter (1985), Shoemaker et al. (1991), and others.

Two methods have been suggested for analyzing data collected from a parameter design experiment. One method is to calculate a value of some type of performance statistic involving the noise variables at each setting of the control factors, and then, using these values, to estimate the effects of the control factors. The other method uses the individual data values to fit a model containing terms in the control and noise factors. In the latter case, for example, let us suppose that a response value Y is recorded at each setting of the control and noise factors, producing a total of $N = N_X \times N_Z$ observations. Then the observations are fitted using a model of the form

$$Y_u = \beta_0 + \sum_{i=1}^{p} \beta_i x_i + \sum_{j=1}^{q} \gamma_j z_j + \sum_{i=1}^{p} \sum_{j=1}^{q} \delta_{ij} x_i z_j + \varepsilon_u \qquad (11.2)$$

where the x_i and z_j are coded variables for the control and noise variables, respectively, and the cross-product terms $\delta_{ij} x_i z_j$ represent interaction terms between the control and noise factors only. Models containing interaction terms between the control factors as well as between the noise factors will be addressed later when discussing the use of combined arrays as an alternative design strategy to using crossed arrays.

11.2.2 Performance Statistics Defined as Signal-to-Noise Ratios

In place of fitting a model of the form in Eq. 11.2 in the response values, Taguchi advocated the calculation of a performance statistic $P(Z)$ at each setting of the control factors. The rationale behind the use of a performance statistic is to provide a measure of the behavior in the response values, when taken across the settings of the noise factors, that represents the quadratic loss associated with a particular setting of the control factors. To be more specific, let us define Y_{lu} to be the value of the response associated with the uth trial of the outer array at the lth setting of the control factors or inner array. The performance statistics $P(Z)$ suggested by Taguchi are called *signal-to-noise (S/N) ratios*. Three of the most popular $P(Z)$'s are

$$(S/N)_S = -10 \log \left(\frac{1}{N_Z} \sum_{u=1}^{N_Z} Y_{lu}^2 \right) \quad \text{"the smaller, the better,"}$$

$$l = 1, 2, \ldots, N_X \tag{11.3}$$

$$(S/N)_L = -10 \log \left(\frac{1}{N_Z} \sum_{u=1}^{N_Z} \frac{1}{Y_{lu}^2} \right) \quad \text{"the larger, the better,"}$$

$$l = 1, 2, \ldots, N_X \tag{11.4}$$

$$(S/N)_T = 10 \log \left(\frac{\bar{Y}_{l\cdot}^2}{s_l^2} \right) \quad \text{"target is best,"}$$

$$l = 1, 2, \ldots, N_X \tag{11.5}$$

where in Eq. 11.5,

$$\bar{Y}_{l\cdot} = \frac{1}{N_Z} \sum_{u=1}^{N_Z} Y_{lu} \quad \text{and} \quad s_l^2 = \sum_{u=1}^{N_Z} \frac{(Y_{lu} - \bar{Y}_{l\cdot})^2}{N_Z - 1}$$

Once a particular performance statistic or S/N ratio is chosen, and generally only one is chosen, there will be a single value for it at each setting of the control factors. The statistical analysis that is performed for investigating the effects of the control factors simply treats the value of the performance statistic as the response value at the particular setting of the control factors.

To determine the settings of the control factors for which the process is thought to be most insensitive to variation in the noise factors, one uses the signal-to-noise ratios in Eqs. 11.3 to 11.5 in the following way. In "the smaller, the better" case, the experimenter seeks to minimize the value of the process response, of which Y_{lu} is an observation, as if the target value is zero. Thus to minimize the quadratic loss function of Eq. 11.1, one tries to maximize the value of $(S/N)_S$ in Eq. 11.3. Similarly, in "the larger, the better" case, the experimenter tries to maximize the process output Y_{lu} or minimize $E(\sum_{u=1}^{N_Z}(1/Y_{lu}^2)$, which amounts to finding a maximum for $(S/N)_L$. In "the target is best" case, there are two situations involving the relationship between the process mean and process variance that are to be considered, although the same action is taken with respect to the control factors in the two situations. Before mentioning the two situations, however, suppose we classify the control factors as follows:

(i)   Those that affect both the mean and the variance of the process
(ii) Those that affect the variance only
(iii) Those that affect the mean only
(iv) Those that do not affect the mean or the variance

The first case to consider is when the process mean response and variance can be independently adjusted. The analysis then is a two-step procedure: First, one seeks the settings among those factors in (iii) for which the mean response (estimated by $\bar{Y}_{l\cdot}$) is close to the target value and second, one uses the factors in (ii) to maximize

$$(S/N)_{T\cdot} = -10 \log s_l^2$$

where s_l^2 is the same variance of the Y_{lu} values in the outer array at the lth setting of the control factors. Alternatively, one could first try to minimize the variability by maximizing the $(S/N)_{T\cdot}$ ratio using the factors in (i) and (ii), and then one would look for settings of the factors in (iii) for which the mean response is close to the target.

The second case to consider is when the process mean and process standard deviation cannot be independently adjusted because they are related. In this case, the maximum of $(S/N)_T$ in Eq. 11.5 is sought using the factors in (i) and (ii), after which the mean response is adjusted using the factors in (iii).

11.2.3   A Numerical Example

To illustrate the two-step procedure just discussed, Table 11.1 lists a set of 36 artificial response values for the parameter design shown in Figure 11.1. Included also in Table 11.1 are the values of the average response, sample variance, and the signal-to-noise ratios $(S/N)_S$, $(S/N)_L$, and $(S/N)_T$ at each of the nine combinations of the four control factors that make up the inner array.

A main-effects-only analysis is initiated by first plotting the average value of the signal-to-noise ratio against the levels of each control factor. Table 11.2 gives

Table 11.1 Artificial Data Values for the 36 Combinations of the Crossed Arrays and Summary Statistics Values for Each of the Nine Inner Array (IA) Combinations

IA	Control Factors				Noise Factors				Data Values								
l	X_1	X_2	X_3	X_4	Z_1	Z_2	Z_3		Y_{l1}	Y_{l2}	Y_{l3}	Y_{l4}	$\bar{Y}_{l\cdot}$	s_l^2	$(S/N)_S$	$(S/N)_L$	$(S/N)_T$
1	1	1	1	1	1	1	1	$\rightarrow Y_{l1}$	2	4	7	6	4.75	4.92	−14.191	10.449	6.617
2	1	2	2	2	1	2	2	$\rightarrow Y_{l2}$	5	3	8	10	6.50	9.67	−16.946	13.547	6.405
3	1	3	3	3	2	1	2	$\rightarrow Y_{l3}$	9	4	3	4	5.00	7.33	−14.843	12.068	5.327
4	2	1	2	3	2	2	1	$\rightarrow Y_{l4}$	3	3	1	6	3.25	4.25	−11.383	5.051	3.954
5	2	2	3	1					2	3	2	5	3.00	2.00	−10.212	7.884	6.532
6	2	3	1	2					4	9	7	7	6.75	4.25	−16.880	15.389	10.302
7	3	1	3	2					6	5	3	5	4.75	1.58	−13.757	12.618	11.539
8	3	2	1	3					8	6	3	6	5.75	4.25	−15.593	13.413	8.909
9	3	3	2	1					7	9	6	2	6.00	8.67	−16.284	11.100	6.184

Table 11.2 Average Values of $\bar{Y}_l$, s_l^2, and $(S/N)_T$ at the Levels of the Control Factors

| Control Factor | Statistic | Level | | |
		Low (1)	Middle (2)	High (3)
X_1	$\bar{Y}$	5.42	4.33	5.50
X_2		4.25	5.08	5.92
X_3		5.75	5.25	4.25
X_4		4.58	6.00	4.67
X_1	s^2	7.31	3.50	4.83
X_2		3.58	5.31	6.75
X_3		4.47	7.53	3.64
X_4		5.20	5.17	5.28
X_1	$(S/N)_T$	6.116	6.929	8.877
X_2		7.370	7.282	7.271
X_3		8.609	5.514	7.799
X_4		6.444	9.415	6.063

the average values of $\bar{Y}$, s^2 and $(S/N)_T$ at each level of X_i where the average here is taken across the levels of the remaining X_j, $j \neq i$. Figure 11.2 plots the average values of $(S/N)_T$ and $\bar{Y}$ at the levels of X_1, X_2, X_3, and X_4.

To find the "best" settings of the control factors let us suppose that a target value of 5.0 for the process mean output is sought along with a high value of $(S/N)_T$. From Figure 11.2a, we see that high average values of $(S/N)_T$ occur at $X_1 = 3$, $X_3 = 1$ or 3, and $X_4 = 2$. Across the levels of X_2, very little change occurs in the average value of $(S/N)_T$. This suggests that the $(S/N)_T$ ratio is insensitive to changes in X_2 or that $\bar{Y}_l$ and s_l are highly correlated when taken across the levels of X_2. Thus, for the "best or most robust" process conditions, from Figure 11.2a, we would set X_1 at its high level, X_3 at its low or high level, X_4 at its middle level. To adjust the process mean to the target value of 5.0, from Figure 11.2b, set X_2 at the middle level. Note from Table 11.1 that the row of the inner array that nearly matches the recommended combination is the $l = 7$ combination of $X_1 = 3$, $X_2 = 1$, $X_3 = 3$, and $X_4 = 2$. Having decided on the settings of the control variables thought to be best, the next order of business would be to perform some confirmatory trials at the two combinations $X_1 = 3$, $X_2 = 2$, $X_3 = 1$, $X_4 = 2$ and $X_1 = 3$, $X_2 = 2$, $X_3 = 3$, $X_4 = 2$, and then select the winner.

The numerical example that we have just discussed illustrates the simplicity of the analysis when using the signal-to-noise ratio as the response of interest. But when studying the behavior of the average value of $(S/N)_T$ and $\bar{Y}$ across the

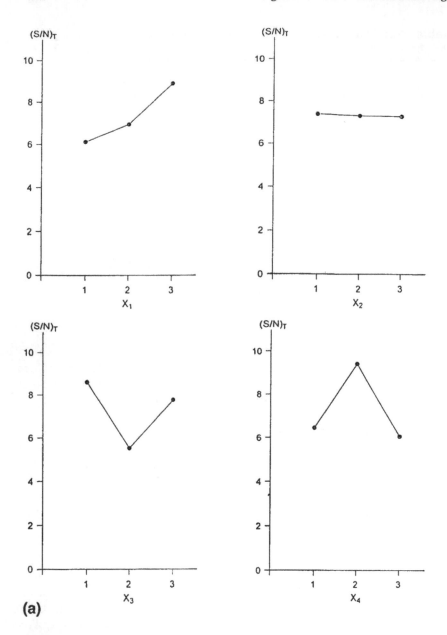

(a)

Figure 11.2 Plots of the average values of the (a) signal to noise ratio $(S/N)_T$ and (b) $\bar{Y}_i$. at the levels of the four control factors.

(b)

levels of the control factors, however, the inferences drawn are appropriate only if one is willing to assume that there are no interactions between the control factors. This assumption is highly questionable, particularly in some areas of application such as in the chemical industries, where it is well known that interactions among the control factors almost always exist. Also, Taguchi's assumption that certain control factor settings exist that result in the process variance being minimized suggests the presence of some interactions involving control and noise variables. Furthermore, the use of the signal-to-noise ratio does not guarantee that one will be successful in isolating those control factors that affect the process mean output and those that affect the process variance. For example, the signal-to-noise ratio $(S/N)_T$ in Eq. 11.5 can also be written as

$$(S/N)_T = 10(\log \overline{Y}_i^2 - \log s_i^2)$$

Thus when maximizing $(S/N)_T$, it is not clear whether one is selecting those control factors that maximize the quantity $\log \overline{Y}_i^2$ or one is selecting the control factors that minimize the quantity $\log s_i^2$. As a result, Taguchi's use of the S/N ratio as a process performance criterion has drawn criticism in the literature from Box (1985, 1988), Welch et al. (1990), Myers et al. (1992), and others, each of whom recommends modeling the response Y_{ul} rather than the S/N ratio. We shall not repeat the criticisms of others here except to mention two that were directed at Taguchi's PD and the use of the S/N ratio, which we feel inspired research on alternative response surface techniques for the parameter design experiment.

11.3 RESPONSE SURFACE ALTERNATIVES

Two criticisms of Taguchi's parameter design and the use of the S/N ratio are as follows:

1. The number of experimental trials $N = N_X \times N_Z$ in the crossed array is much larger than is needed, particularly if interactions among the control factors are ignored. On the other hand, if a small design is used in the inner array such as a saturated or near saturated Plackett-Burman design, the crossed array plan does not permit one to estimate the interactions between the control factors.
2. The use of the signal-to-noise ratio is a very inefficient approach for trying to achieve a thorough understanding of the process itself, particularly when it comes to trying to "identify" the control factors that affect only the process mean as distinguished from the control factors that affect only the process variance and those that affect both the process mean and variance.

These criticisms, and perhaps others, have led to the suggestions of response surface alternatives to Taguchi's parameter design approach. The suggested alternatives are the use of response surface designs of the types presented in Chapters 3 and 4 in place of the crossed arrays, and the fitting of a model of the form given in Eq. 11.2 to the individual data values. We address the modeling exercise next, after which we shall briefly mention some alternative designs. For additional details we recommend the articles by Myers et al. (1992) and Shoemaker et al. (1991).

Let us recall the model of Eq. 11.2 that expresses the process output Y as a function of the control and noise variables. For the present, let us assume that only linear effects of the control and noise variables are of interest along with interaction effects between the control and noise variables. In vector notation, the model of Eq. 11.2 is written as

$$Y_u = \beta_0 + \mathbf{x}_u'\boldsymbol{\beta} + \mathbf{z}_u'\boldsymbol{\gamma} + \mathbf{x}_u'\boldsymbol{\Delta}\mathbf{z}_u + \varepsilon_u, \qquad u = 1, 2, \ldots, N \qquad (11.6)$$

where

$\mathbf{x}_u' = (x_{u1}, x_{u2}, \ldots, x_{up})'$
$\mathbf{z}_u = (z_{u1}, z_{u2}, \ldots, z_{uq})'$
$\boldsymbol{\beta}$ is a $p \times 1$ vector of unknown coefficients representing the linear effects of the control variables
$\boldsymbol{\gamma}$ is a $q \times 1$ vector of unknown coefficients associated with the coded noise variables
$\boldsymbol{\Delta}$ is a $p \times q$ matrix of unknown coefficients associated with the interaction effects between the control and noise variables
ε_u is a random error term.

The x_{ui} and z_{uj} are coded variables representing the control and noise variables, respectively. In reality, the levels of the noise variables are random in the field stage even though the levels of the Z_j's are controlled for the purpose of conducting the experimental trials. The assumptions made regarding the random errors and the z_{uj}'s are

$$\varepsilon_u \sim \text{i.i.d. } N(0, \sigma^2), \qquad u = 1, 2, \ldots, N$$

and

$$E(\mathbf{z}_u) = \mathbf{0} \qquad \text{and} \qquad \text{Var}(\mathbf{z}_u) = \mathbf{V}, \qquad u = 1, 2, \ldots, N \qquad (11.7)$$

plus the $\mathbf{z}_u$ are independent of the ε_u. Then, from the properties of the ε_u's and z_{uj}'s, the expectation of the *process mean response surface* is

$$E(Y_u) = \beta_0 + \mathbf{x}_u'\boldsymbol{\beta} \qquad (11.8)$$

and the *process variance response surface* is

$$\begin{aligned}
\text{Var}(Y_u) &= \text{Var}\{[\boldsymbol{\gamma}' + \mathbf{x}_u'\boldsymbol{\Delta}]\mathbf{z}_u\} + \sigma^2 \\
&= [\boldsymbol{\gamma}' + \mathbf{x}_u'\boldsymbol{\Delta}]\mathbf{V}[\boldsymbol{\gamma}' + \mathbf{x}_u'\boldsymbol{\Delta}]' + \sigma^2 \qquad (11.9)
\end{aligned}$$

The process variance of Eq. 11.9 is a function of the levels of the control variables in $\mathbf{x}_u$. This is because of the part of the model, $\mathbf{x}_u'\boldsymbol{\Delta}\mathbf{z}_u$, in Eq. 11.6 representing the interaction effects between the control and noise variables. If we express the model without the interaction terms (i.e., let $\Delta = 0$ in Eq. 11.6), then the *process variance* is

$$\text{Var}(Y_u) = \gamma'V\gamma + \sigma^2$$

which does not depend on the values of the control variables.

The expressions in Eqs. 11.8 and 11.9 for the process mean and process variance response surfaces are easily generalized to include second-degree terms (e.g., $\beta_{11}x_i^2$ and $\beta_{il}x_ix_l$) in the control variables and to extend the interaction terms between the control and noise variables to include the second-degree control by linear noise interactions. All that is required in this case is to redefine x_u' and β in Eq. 11.6 as

$$x_u' = (x_{u1}, x_{u2}, \ldots, x_{up}, x_{u1}^2, \ldots, x_{up}^2, x_{u1}x_{u2}, \ldots, x_{up-1}x_{up})$$

$$\beta' = (\beta_1, \beta_2, \ldots, \beta_p, \beta_{11}, \ldots, \beta_{pp}, \beta_{12}, \ldots, \beta_{p-1,p})$$

so that $E(Y_u)$ in Eq. 11.8 is of degree 2. Equation 11.9 for $\text{Var}(Y_u)$ will be of degree 4 in the coded control variables.

The estimated process mean and variance response surfaces are obtained by first assuming the x_i's and z_j's are fixed factors and fitting the model of Eq. 11.6. From the fitted model containing the estimates b_0 for β_0, b for β, g for γ, and D for Δ, the estimates are substituted into Eqs. 11.8 and 11.9 to produce the equations

$$\hat{Y}_u = b_0 + x_u'b$$

and

$$\tilde{\text{Var}}(\hat{Y}) = [g' + x_u'D]V[g' + x_u'D]' + \hat{\sigma}^2$$

where $\hat{\sigma}^2$ is the error mean square taken from the ANOVA table. The tilde ($\tilde{}$) above $\tilde{\text{Var}}(\hat{Y})$ serves to remind us that the elements of the matrix V have to be known or at least estimated in order for $\tilde{\text{Var}}(\hat{Y})$ to be an estimate of $\text{Var}(\hat{Y})$. According to Myers et al. (1992), "In many cases the experimenter will have prior estimates of the variances and covariances that the noise variables will experience in the process or in the field. This is particularly true in cases where the process is not new and one is quite familiar with the dispersion properties of the noise variables."

A *design alternative* to the crossed-array structure of the PD is that of using a single array containing the settings of both the control and noise factors. The single or *combined array* approach has been proposed by many authors, including Welch et al. (1990), Lucas (1991), Shoemaker et al. (1991), Box and Jones (1992), and Myers et al. (1992). The combined array (CA) is used to model the response directly as a function of the control and noise variables as in Eq. 11.2. Using the combined array rather than the crossed arrays often results in a reduction in the number of experiments performed, and, on top of this, the CA offers considerably more flexibility in the estimation of the interactions of interest.

The following example illustrates the superiority of the CA over the crossed arrays, as is pointed out in Shoemaker et al. (1991). Suppose there are three control and three noise variables whose coded variables are denoted by x_1, x_2, x_3 and z_1, z_2, z_3, respectively. For Taguchi's PD, an obvious choice for the inner and outer arrays is a 2^{3-1} fractional factorial in the x_i's (with the defining relation $I = A_1A_2A_3$; see Section 3.3.2) crossed with the same 2^{3-1} fraction in the z_j's where in each array the fraction is of resolution III. The crossed arrays or PD, consisting of 16 combinations, is equivalent to a 2^{6-2} fraction having the defining relations $I = A_1A_2A_3 = A_4A_5A_6 = A_1A_2A_3A_4A_5A_6$. With this design however, the three two-factor interactions between the control factors are aliased with the main effects of the control factors. In fact, the six main effects of the x_i and z_j, which are estimated by the coefficient estimates b_i and g_j in the fitted model having the same form as the model of Eq. 11.2, are aliased with two-factor interactions.

An alternative to the crossed arrays is the combined array, again a 2^{6-2}, with the defining relations $I = A_1A_2A_3A_4 = A_4A_5A_6 = A_1A_2A_3A_5A_6$. This design has mixed resolution, however. It is of resolution III in the noise factors but is of resolution IV in the control factors. As a result, the three two-factor interactions between the control factors are estimated free of bias from the three interactions x_1z_1, x_2z_1, and x_3z_1 if the latter three interactions are zero. If one cannot assume the latter three interactions are zero but can assume x_1z_3, x_2z_3, and x_3z_3 are zero, one simply replaces A_4 in the defining relation $I = A_1A_2A_3A_4$ with A_6.

Other examples of combined arrays that are superior to crossed arrays, from the standpoint of allowing the estimation of control by control, control by noise, and noise by noise two-factor interactions are presented in Shoemaker et al. (1991). Myers et al. (1992) present some alternative second-order response surface designs for fitting second-order effects in the control factors that are preferred to using the Latin square type of orthogonal arrays suggested by Taguchi and pointed out as such by Hunter (1985). One example, in particular, for three control factors x_1, x_2 and x_3 in combination with two noise factors z_1 and z_2 is the following design matrix consisting of a CCD in x_1, x_2, and x_3 augmented with two columns for z_1 and z_2,

x_1	x_2	x_3	z_1	z_2
±1	±1	±1	±1	±1
−1	0	0	0	0
1	0	0	0	0
0	−1	0	0	0
0	1	0	0	0
0	0	−1	0	0
0	0	1	0	0
0	0	0	0	0

where the bottom row represents a set of $n_0 \geq 1$ center point replications. If the top row is represented by a 2^{5-1} fraction, the CA consists of $22 + n_0$ experimental runs, which when $n_0 < 14$, is lower in number than the 36 experimental trials that are performed when crossing the L_9 orthogonal array in x_1, x_2, x_3 with a 2^2 factorial in z_1 and z_2.

Before concluding this chapter with a numerical example illustrating the fit of a dual-response model, we think it only fair to mention that, like many others, we feel Taguchi has had a profound impact on the way process performance (a response surface problem) is measured today and will continue to be in the future. Taguchi's philosophy of recognizing that improved product quality begins with product and process design; that failure to recognize product variability during the manufacturing stage and in the hands of the customer can result in a loss to society, has most definitely awakened the scientific community. As a result, the dual-response approach of investigating both the process mean and variance has gained momentum as an important RSM tool for the improvement of product quality.

11.3.1   A Numerical Example Illustrating the Fit of a Dual Response Model

In Example 3.3 of Chapter 3 (see Section 3.4), the discussion centered on the results of a factorial experiment that was set up and run to determine the effects of X_1 = reaction temperature, X_2 = benzene/air ratio, X_3 = flow rate of reactants, and X_4 = height of the catalytic bed of particles at rest on the percent conversion (Y) of benzene to maleic anhydride. The total number of experimental trials was 20; the complete $2^4 = 16$ plus 4 center point replications. For the purpose of illustrating the fitting of a dual-response model, let us define the factors X_1 and X_2 as control factors but redefine factors X_3 $(=Z_1)$ and X_4 $(=Z_2)$ as noise variables in the experiment. In other words, even though the levels of X_3 and X_4 were controlled in the actual experimental trials performed, we will view the process mean percent conversion of benzene as a function of the reaction temperature and benzene/air ratio only while treating flow rate of reactants and bed height as noise variables. The combined array and percent conversion values are listed in Table 3.7.

The dual-response surface model, from Eq. 11.2, is

$$Y = \beta_0 + \beta_1 x_1 + \beta_2 x_2 + \beta_{12} x_1 x_2 + \gamma_1 z_1 + \gamma_2 z_2 + \delta_{11} x_1 z_1$$
$$+ \delta_{21} x_2 z_1 + \delta_{12} x_1 z_2 + \delta_{22} x_2 z_2 + \varepsilon \qquad (11.10)$$

We could also include in Eq. 11.10 the term $\gamma_{12} z_1 z_2$ representing the interaction between the noise variables. However, the analysis of the data and the subsequent discussion of the example in Chapter 3 did not single out this interaction as being important and therefore we have not included it in the model.

The final form of the fitted model, given by Eq. 3.30, but expressed now in x_1, x_2, z_1 and z_2, is

$$\hat{Y} = b_0 + b_1x_1 + b_2x_2 + b_{12}x_1x_2 + g_1z_1 + g_2z_2 + d_{12}x_1z_2$$

$$= 63.2325 + 9.2863x_1 - 4.6213x_2 + 0.8213x_1x_2$$

$$-5.4225z_1 + 5.2288z_2 - 2.2938x_1z_2 \tag{11.11}$$

Table 3.9 lists the results of the tests performed on the coefficient estimates in Eq. 11.11. If we view the four factors as being fixed factors, we would infer that all four main effects and both two-factor interactions are significant at a level lower than 0.02. More specifically, an interaction is present between the effects of the control factors, reaction temperature (x_1) and benzene/air ratio (x_2), and between reaction temperature and catalytic bed height (a noise factor).

In the context of estimating the dual-response surfaces for the process mean and process variance, we assume z_1 and z_2 are random variables with mean zero and variances $\sigma_{z_1}^2$ and $\sigma_{z_2}^2$, respectively. Further, assume z_1 and z_2 are uncorrelated so that $\mathbf{V} = \text{diag}\,(\sigma_{z_1}^2, \sigma_{z_2}^2)$. The model for the process mean is

$$E(Y_u) = \beta_0 + \beta_1x_{u1} + \beta_2x_{u2} + \beta_{12}x_{u1}x_{u2}$$

so that, by taking the coefficient estimates from Eq. 11.11, the prediction equation for the *process mean response surface* is

$$E(\hat{Y}) = 63.2325 + 9.2863x_1 - 4.6213x_2 + 0.8213x_1x_2 \tag{11.12}$$

The estimated surface depicted by Eq. 11.12 is not a plane. This is because the estimated rate of change of the surface, when measured in the direction of the x_1 axis, is expressed as $\partial\hat{Y}/\partial x_1 = 9.2863 + 0.8213x_2$, and the estimated rate depends on the value of x_2. Similarly, the estimated rate of change of the surface, when measured in the direction of the x_2 axis, is expressed as $\partial\hat{Y}/\partial x_2 = -4.6213 + 0.8213x_1$, and its value depends on the value of x_1. Note that since high values of percent conversion are desirable, the highest estimated response value is 76.3188 and this occurs at $x_1 = 1$, $x_2 = -1$, or by setting reaction temperature at 460°C and by setting the benzene/air ratio at 1.5.

The model for the process variance response surface is

$$\text{Var } \hat{Y} = \gamma_1^2\sigma_{z_1}^2 + \gamma_2^2\sigma_{z_2}^2 + \delta_{12}^2x_1^2\sigma_{z_2}^2 + 2\gamma_2\delta_{12}x_1\sigma_{z_2}^2 + \sigma^2$$

Again, borrowing the coefficient estimates g_1 for γ_1, g_2 for γ_2, and d_{12} for δ_{12} from Eq. 11.11, the equation for the estimated *process variance response surface* is

$$\widetilde{\text{Var}}(\hat{Y}) = (-5.4225)^2\hat{\sigma}_{z_1}^2 + (5.2288)^2\hat{\sigma}_{z_2}^2 + (-2.2938)^2x_1^2\hat{\sigma}_{z_2}^2$$

$$+ 2(5.2288)(-2.2938)x_1\hat{\sigma}_{z_2}^2 + \hat{\sigma}^2$$

$$= 29.4035\hat{\sigma}_{z_1}^2 + [5.2288 - 2.2938x_1]^2\hat{\sigma}_{z_2}^2 + \hat{\sigma}^2 \tag{11.13}$$

Note that the estimated process variance response surface is a function of the estimated variances of the noise variables and the error as well as a quadratic function of the setting of the control factor x_1. Within the boundaries of the experimental region, the value of the estimated process variance is minimized by setting $x_1 = 1$, or by setting temperature at the high level of 460°C. Furthermore, if we set $\hat{\sigma}_{z_1} = \hat{\sigma}_{z_2} = 1/2$, which amounts to equating the standard deviations of the noise variables to one-fourth the ranges for X_3 and X_4, respectively, and from Table 3.9, use residual mean square value 1.39 for $\hat{\sigma}^2$, the value of $\widetilde{\text{Var}}(\hat{Y})$ in Eq. 11.13 is 10.894.

To summarize this four-factor example where factors X_1 and X_2 were viewed as control factors and X_3 and X_4 were considered noise variables, the dual-response surface approach produced expressions for the process mean and process variance response surfaces in the form of Eqs. 11.12 and 11.13, respectively. The process mean was expressed as a function of the two control factors only while the process variance was expressed as a function of the variance components for the two noise variables and the error plus the setting for control factor X_1. From the expressions for the process mean and process variance, we were able to find the settings of the two control factors ($X_1 = 460°C$ and $X_2 = 1.5$) at which the estimated process mean attained a maximum value of 76.3188 while at the same time the process variance was minimum.

A final equation remains: "Would the use of Taguchi's signal-to-noise ratio $(S/N)_L$ give the same results?" We leave this as an exercise for the reader and provide the following table as a helpful aid.

Control Factors				
X_1	X_2	$\bar{Y}_i$.	s_i^2	$(S/N)_L$
420	1.5	59.52	107.5258	35.1733
460	1.5	76.45	41.5530	37.5971
420	2.5	48.64	120.6100	33.1936
460	2.5	68.85	65.4700	36.6195

REFERENCES AND BIBLIOGRAPHY

Box, G. E. P. (1985). Discussion of Off-Line Quality Control, Parameter Design, and the Taguchi Methods, *J. Quality Tech.*, **17**, 198–206.

Box, G. E. P. (1988). Signal to Noise Ratios, Performance Criteria, and Transformations (with discussion), *Technometrics*, **30**, 1–40.

Box, G. E. P., S. Bisgaard, and C. Fung (1988). An Explanation and Critique of Taguchi's Contributions to Quality Engineering, *Quality Reliability Eng.*, **4**, 123–131.

Box, G. E. P., and S. Jones (1992). Designing Products that Are Robust to the Environment, *Total Quality Management*, **3**, 265–282.

Hunter, J. S. (1985). Statistical Design Applied to Product Design, *J. Quality Tech.*, **17**, 210–221.

Kackar, R. N. (1985). Off-Line Quality Control, Parameter Design, and the Taguchi Method (with discussion), *J. Quality Tech.*, **17**, 176–188.

Kackar, R. N. (1986). Taguchi's Quality Philosophy: Analysis and Commentary, *Quality Progr.*, 21–29.

Leon, R. V., A. C. Shoemaker, and R. N. Kackar (1987). Performance Measure Independent of Adjustment: An Explanation and Extension of Taguchi's Signal to Noise Ratio, *Technometrics*, **29**, 253–285.

Lucas, J. M. (1991). Achieving a Robust Process Using Response Surface Methodology, Technical Report, E. I. DuPont, Wilmington, DE.

Myers, R. H., A. I. Khuri, and G. Vining (1992). Response Surface Alternatives to the Taguchi Robust Parameter Design Approach, *Amer. Statist.*, **46**, 131–139.

Phadke, M. S., R. N. Kackar, D. V. Speeney, and M. J. Greico (1983). Off-Line Quality Control for Integrated Circuit Fabrication Using Experimental Design, *Bell System Tech. J.*, **62**, 1273–1309.

Shoemaker, A. C., K. Tsui, and C. F. J. Wu (1991). Economical Experimentation Methods for Robust Design, *Technometrics*, **33**, 415–427.

Taguchi, G. (1978). Off-Line and On-Line Quality Control Systems, *Proceedings of the International Conference on Quality*, Tokyo, Japan.

Taguchi, G. (1986). *Introduction to Quality Engineering: Designing Quality into Products and Processes*, Kraus International Publications, White Plains, N.Y.

Taguchi, G. and Y. I. Wu (1980). Introduction to Off-Line Quality Control, Central Japan Quality Control Association, Nogoya, Japan (available from the American Supplier Institute, Dearborn, MI).

Vining, G. G. and R. H. Myers (1990). Combining Taguchi and Response Surface Philosophies: A Dual Response Approach, *J. Quality Tech.*, **22**, 38–45.

Welch, W. J., T. K. Yu, S. Kang, and J. Sacks (1990). Computer Experiments for Quality Control by Parameter Design, *J. Quality Tech.*, **22**, 15–22.

Additional Topics and Some Directions for Future Research

12.1 INTRODUCTION

In Chapters 3 and 4, the construction of fixed-point or discrete designs for estimating the parameters in first- and second-order polynomial models, respectively, was presented. In Chapter 6, additional suggestions were made regarding the spread of the design points for minimizing the separate properties, variance of $\hat{Y}(\mathbf{x})$ and the bias in $\hat{Y}(\mathbf{x})$, across the experimental region of interest. And while we have concentrated primarily on the integrated mean squared error of $\hat{Y}(\mathbf{x})$ as the major criterion for choosing a design, we would be remiss in our role as authors if we did not at least acknowledge other design criteria, in particular, criteria based on the theory of design optimality and design robustness. These criteria are described in Sections 12.2–12.4. In addition, several measures of rotatability are discussed in Section 12.5. Finally, some comments concerning future research directions are given in Section 12.6.

Optimal design theory was developed mainly after World War II. Kiefer (1958, 1959, 1960, 1961, 1962a, 1962b) is attributed to having provided the basic mathematical groundwork for optimal design theory. Presently there are two schools of thought regarding the application of the precepts of optimal design theory to the derivation of response surface designs: the "Kiefer school" and the "Box school." In the latter school, bias suspected of being present in the fitted model plays a significant role as was seen in Chapter 6. In the Kiefer school, however, bias is regarded as insignificant or that it does not exist. The main preoccupation in this school is, therefore, with designs that minimize the variance associated with $\hat{Y}(\mathbf{x})$.

12.2 VARIANCE MINIMIZING DESIGN CRITERIA

To introduce the theory of optimal design, let us recall the general linear model in k variables

$$\mathbf{Y} = \mathbf{X}\boldsymbol{\beta} + \boldsymbol{\varepsilon} \tag{12.1}$$

where $\mathbf{X}$ is an $N \times p$ $(k < p)$ matrix of design settings and functions of these settings. In estimating the elements of the $p \times 1$ parameter vector $\boldsymbol{\beta}$, it has been noted previously that the elements of $(\mathbf{X}'\mathbf{X})^{-1}$ are proportional to the variances and covariances of the elements in $\mathbf{b}$, the vector of least squares estimates of $\boldsymbol{\beta}$. It was also mentioned in Chapter 3 and later in Chapter 6 that the ideas associated with choosing a design are almost always centered on the selection of the particular settings of $\mathbf{X}$ that make the $p \times p$ matrix $(\mathbf{X}'\mathbf{X})^{-1}$ "small" in some sense.

To look at these ideas more closely, let us consider the p-dimensional confidence region for $\boldsymbol{\beta}$,

$$\{\boldsymbol{\beta} : (\boldsymbol{\beta} - \mathbf{b})'\mathbf{X}'\mathbf{X}(\boldsymbol{\beta} - \mathbf{b}) \leq c\} \tag{12.2}$$

where c is a fixed constant value. The region in Eq. 12.2 is ellipsoidal in shape. Directly proportional to the squared lengths of the p principal axes of the ellipsoid are the quantities $\psi_1, \psi_2, \ldots, \psi_p$, the eigenvalues of the matrix $(\mathbf{X}'\mathbf{X})^{-1}$. Therefore, to minimize the size of the confidence region 12.2, one needs only to concentrate on minimizing the magnitudes of the values of ψ_i, $i = 1, 2, \ldots, p$.

Directing our attention for the moment to the eigenvalues $\psi_1, \psi_2, \ldots, \psi_p$, let us denote the largest eigenvalue, the arithmetic mean of the eigenvalues, and the geometric mean of the eigenvalues, respectively, by

$$\max(\psi_1, \psi_2, \ldots, \psi_p) = \psi_{\max}$$

$$\frac{\sum_{i=1}^{p} \psi_i}{p} = \bar{\psi}_A \quad \text{(A for arithmetic)}$$

$$\left(\prod_{i=1}^{p} \psi_i\right)^{1/p} = \bar{\psi}_G \quad \text{(G for geometric)} \tag{12.3}$$

Using the single-valued functions in Eq. 12.3 of the p eigenvalues, some of the most popular criteria for selecting the design settings in $\mathbf{X}$ are based on:

$$\text{tr}(\mathbf{X}'\mathbf{X})^{-1} = p\bar{\psi}_A = \sum_{i=1}^{p} \psi_i \tag{12.4}$$

$$|\mathbf{X}'\mathbf{X}| = \overline{\psi_G}^{-p} = \prod_{i=1}^{p} \psi_i^{-1} \tag{12.5}$$

$$\psi_{max} \tag{12.6}$$

$$G = \max_{\mathbf{x} \in R}[\mathbf{f}'(\mathbf{x})(\mathbf{X}'\mathbf{X})^{-1}\mathbf{f}(\mathbf{x})] \tag{12.7}$$

where $\mathbf{f}'(\mathbf{x})$ is a $1 \times p$ vector whose elements correspond to the elements in a row of the matrix $\mathbf{X}$, but are valued at a point $\mathbf{x}$ in the experimental region R. When multiplied by σ^2, G in Eq. 12.7 is the maximum prediction variance over the experimental region.

With regard to the criteria in Eqs. 12.4–12.7, the desirability of a design increases as the values of each of $\overline{\psi_A}$, $\overline{\psi_G}$, ψ_{max}, and G decreases. In most practical situations, these quantities will decrease in value as the ranges of the k variables are spread out. Placing the design points on the boundaries of the experimental region R, in accordance with other optimal orientation of the points (such as symmetry) with respect to the design variable axes, is akin to producing optimal designs based on criteria 12.4–12.7. Designs based on these criteria are called A-, D-, E-, and G-optimal designs, respectively.

Criteria 12.4–12.7 were developed initially as continuous design criteria. The term *continuous design* is used mainly in Kiefer's school where a design is regarded as a probability measure, ζ, defined on the experimental region R. This measure satisfies the conditions

$$\zeta(\mathbf{x}) \geq 0 \qquad \text{for all } \mathbf{x} \in R$$

$$\int_R d\zeta(\mathbf{x}) = 1 \tag{12.8}$$

In particular, a collection of N points in R that are not necessarily distinct form a design measure, where

$$\zeta(\mathbf{x}_\ell) = \begin{cases} n_\ell/N & \text{if } \mathbf{x}_\ell \text{ is a design point } (\ell = 1, 2, \ldots, m) \\ 0 & \text{otherwise} \end{cases}$$

where n_ℓ denotes the number of observations being collected at the ℓth design point $(\ell = 1, 2, \ldots, m)$ and $\sum_{\ell=1}^{m} n_\ell = N$. Such a design measure is called a *discrete design* and is usually denoted by D_N. Discrete designs are the traditional designs considered in Box's school, as was seen in earlier chapters. Design measures other than discrete are called *continuous design measures*. For example, a design measure of the form

$$\zeta(\mathbf{x}_\ell) = \begin{cases} \kappa_\ell & \text{if } \mathbf{x}_\ell \text{ is a design point } (\ell = 1, 2, \ldots, m) \\ 0 & \text{otherwise} \end{cases}$$

where $\kappa_\ell \geq 0$ and $\sum_{\ell=1}^{m} \kappa_\ell = 1$ with at least one κ_ℓ being an irrational number (i.e., not expressible as a fraction), is a continuous design measure. We note that continuous design measures are not realizable in practice, but can be approximated fairly closely by a discrete design. The latter design is, therefore, referred to as an exact design measure.

In the continuous design framework, there are expressions that are analogous to the matrix $\mathbf{X}'\mathbf{X}$ and to the prediction variance $Var[\hat{Y}(\mathbf{x})]$ in the discrete design case. To show this, let us recall the form (Eq. 6.1) of the general linear model

$$Y_u = \mathbf{f}'(\mathbf{x}_u)\boldsymbol{\beta} + \varepsilon_u \qquad u = 1, 2, \ldots N \tag{12.9}$$

which, when expressed in matrix notation over N observations, is $\mathbf{Y} = \mathbf{X}\boldsymbol{\beta} + \boldsymbol{\varepsilon}$. With respect to model 12.9 and for a continuous design measure, ζ, a matrix which is analogous to $\mathbf{X}'\mathbf{X}/N$, called the *moment matrix* of the design ζ, is

$$\mathbf{M}(\zeta) = [m_{ij}(\zeta)] = \left[\int_R f_i(\mathbf{x}) f_j(\mathbf{x}) d\zeta(\mathbf{x}) \right] \tag{12.10}$$

where $f_i(\mathbf{x})$ is the ith element of $\mathbf{f}(\mathbf{x})$ evaluated at a point $\mathbf{x} \in R$ ($i = 1, 2, \ldots, p$). For an exact N-point design, $\mathbf{M}(\zeta) = \mathbf{X}'\mathbf{X}/N$. Also, in the continuous design case, a generalization of the expression for $\{(N/\sigma^2)Var[\hat{Y}(\mathbf{x})]\}$, the standardized prediction variance, is

$$d(\mathbf{x}, \zeta) = \mathbf{f}'(\mathbf{x})\mathbf{M}^{-1}(\zeta)\mathbf{f}(\mathbf{x}) \tag{12.11}$$

Let H denote the class of all design measures on R, and let $\lambda_i(\zeta)$ denote the ith eigenvalue of $\mathbf{M}(\zeta)$ for $\zeta \in H$. The continuous analogs of the A-, D-, E-, and G-optimality criteria are

1. A design measure is A-optimal if it maximizes $\sum_{i=1}^{p} \lambda_i(\zeta)$ over H.
2. A design measure is D-optimal if it maximizes $|\mathbf{M}(\zeta)|$ over H.
3. A design measure is E-optimal if it maximizes $\lambda_{(1)}(\zeta)$ over H, where $\lambda_{(1)}(\zeta)$ is the smallest eigenvalue of $\mathbf{M}(\zeta)$.
4. A design measure, ζ^*, is G-optimal if it minimizes over H the maximum standardized prediction variance over the experimental region R, that is,

$$\sup_{\mathbf{x} \in R} d(\mathbf{x}, \zeta^*) = \inf_{\zeta \in H} \left\{ \sup_{\mathbf{x} \in R} d(\mathbf{x}, \zeta) \right\}$$

where sup and inf are abbreviations for supremum and infimum, respectively. Kiefer and Wolfowitz (1960) proved that D-optimal and G-optimal designs as defined above are identical. Furthermore, they showed that a design measure, ζ^*, is G-optimal (or D-optimal) if and only if

$$\sup_{\mathbf{x} \in R} d(\mathbf{x}, \zeta^*) = p$$

where p is the number of parameters in the model.

One use of the continuous design measure in Eq. 12.8 involves the comparison of different types of response surface design configurations. Defining the experimental region as a k-dimensional hypersphere of radius 1, Lucas (1976) compared the central composite designs (Section 4.5.3 in Chapter 4) to the Box-Behnken designs (Section 4.5.2) and uniform shell designs (Section 4.5.8) for fitting a second-order model. For the case where the experimental region is a k-dimensional hypercube, Lucas compared the composite designs to the Hoke designs (Section 4.5.8), designs by Pesotchinsky (1975), and saturated designs due to Box and Draper (Section 4.5.8). For each design measure, ζ, the D- and G-efficiencies are defined as

$$D\text{-efficiency} = \left(\frac{|\mathbf{M}(\zeta)|}{\sup_{\zeta \in H}|\mathbf{M}(\zeta)|} \right)^{1/p}$$

$$G\text{-efficiency} = \frac{p}{\sup_{\mathbf{x} \in R} d(\mathbf{x}, \zeta)}$$

(12.12)

where p is the number of terms in the fitted model and H is the class of all design measures. The expressions in Eq. 12.12 were given for the CCD in Chapter 4 as Eqs. 4.27 and 4.28, respectively, along with particular values of D-efficiency and G-efficiency in Tables 4.4 and 4.5 for values of $k = 2, 3, 4,$ ..., 8. Each of the quantities in Eq. 12.12 is usually multiplied by 100 to enable efficiency to be expressed as percent.

Summarizing Lucas' findings, most of the designs had fairly high D-and G-efficiencies (say, $D > 80.0$ and $G > 50.0$) on the k-dimensional hypersphere. For $k = 3$, the uniform shell designs are as efficient as the CCD and the Box-Behnken designs, but the relative efficiency of the uniform shell designs to the others drops for $k \geq 4$. The superiority of the Box-Behnken and the CCD designs depends on the number of variables, as well as the number of points in the design, and there appears to be little difference between these designs (in terms of their D- and G-efficiencies) for $3 \leq k \leq 8$ on the k-dimensional hypersphere.

When the experimental region is a k-dimensional hypercube, the Hoke designs compared favorably with the "Best Composite Designs" for $k = 5, 6,$ and 7 factors, but the relative efficiency of the Hoke designs to the composite designs decreased for $k = 8, 9, 10,$ and 11. However, the Hoke designs contained fewer points than the best composite design offering a possible trade-off between design efficiency and size of the design. The Pesotchinsky (1975) designs compared favorably, as to efficiency, with the best composite design for $k = 4, 5, 6,$ and 7 factors, but did so while requiring more design points. The nearly saturated designs of Hoke (1974) and Box and Draper (1974) had the lowest D- and G-efficiencies, but this is to be expected in light of the fewer number of points and hence less coverage of the experimental region with the saturated designs.

12.3 DESIGN ROBUSTNESS AGAINST NONNORMALITY

Consider the linear model

$$\mathbf{Y} = \mathbf{X}\boldsymbol{\beta} + \boldsymbol{\varepsilon} \tag{12.13}$$

where $\mathbf{X}$ is $n \times p$ of rank p, $\boldsymbol{\beta}$ is a vector of unknown constant parameters, and $\boldsymbol{\varepsilon}$ is a random error vector. The assumptions usually made about $\boldsymbol{\varepsilon}$ require that it has a zero mean, a variance-covariance matrix given by $\sigma^2 \mathbf{I}_N$, and for the purpose of inference making with regard to $\boldsymbol{\beta}$, has the normal distribution. On the basis of these assumptions the F-ratios used to test the significance of the elements of $\boldsymbol{\beta}$ will have the F-distribution. Any departure from these assumptions will lead to approximate F-tests. The degree of approximation depends to a certain extent on the design used to fit the model. In other words, the choice of design can have an influence in reducing the sensitivity of the F-ratios to possible deviations from the usual assumptions. Such reduction in the sensitivity of the tests is referred to as *design-related robustness*, or just *design robustness*. Thus, by definition, a robust design in the above sense is a design which causes the F-ratios concerning the model's parameters to be insensitive to certain departures from the usual assumptions.

Robustness against nonnormality can be used as a criterion for the choice of design. Box and Watson (1962) introduced such a criterion for first-order models. The Box-Watson criterion is particularly useful in situations where it is desirable to have some protection against a possible violation of the normal theory assumption when the usual F-tests are used in testing the significance of the elements of $\boldsymbol{\beta}$. This situation is best described by Pearson (1931): "In practice the worker may either know that the variation in his populations is not exactly normal, or it may be that he has no means of being certain of the precise form of this variation at all. He, therefore, needs to feel some confidence that the test which he applies will not be invalid provided that the deviation from normality is not extreme."

Several other authors studied the impact of design on the robustness to nonnormality of the F-test in model 12.13. The work of Jensen et al. (1975) indicates that the class of orthogonal designs is robust in the case of a first-order model. Vuchkov and Solakov (1980) showed that designs with a quadratically balanced matrix, $\mathbf{X}$, and uniformly repeated runs at the design points, such as the continuous D-optimal designs with equal probability measures for all distinct design points, are very robust. The matrix $\mathbf{X}$ in Eq. 12.13 is quadratically balanced (Atiqullah, 1962) if the diagonal elements of the matrix $\mathbf{X}(\mathbf{X}'\mathbf{X})^{-1}\mathbf{X}'$ are all equal.

12.3.1 The Box-Watson Robustness Criterion

This criterion applies to first-order models of the form

$$Y_u = \beta_0 + \sum_{i=1}^{k} \beta_i x_{ui} + \varepsilon_u \qquad u = 1, 2, \ldots, N \tag{12.14}$$

where x_{ui} is the uth level of the ith input variable ($u = 1, 2, \ldots, N; i = 1, 2, \ldots, k$). In matrix form Eq. 12.14 can be written as

$$Y = \beta_0 1_N + D\tilde{\beta} + \varepsilon \tag{12.15}$$

where **D** is the $N \times k$ design matrix assumed to be of rank k and $\tilde{\beta} = (\beta_1, \beta_2, \ldots, \beta_k)'$. The elements of ε are assumed to be independently and identically distributed random variables with a zero mean and a variance σ^2. We also assume that the input variables have been corrected for their means so that $D'1_N = 0$. The number of center and noncenter points will be denoted by n_o and n ($>k$), respectively, thus, $N = n + n_o$.

Consider testing the hypothesis

$$H_o : \tilde{\beta} = 0 \qquad \text{versus} \qquad H_a : \tilde{\beta} \neq 0 \tag{12.16}$$

A test statistic for testing H_o is the F-ratio

$$T(D) = \frac{\dfrac{Y'D(D'D)^{-1}D'Y}{k}}{\dfrac{Y'(I_N - X(X'X)^{-1}X')Y}{N - k - 1}} \tag{12.17}$$

where $X = [1_N : D]$. Box and Watson (1962) showed that in nonnormal situations and under the null hypothesis H_o, $T(D)$ is distributed approximately as an F variate with degrees of freedom given by $\nu_1 = \theta k$ and $\nu_2 = \theta(N - k - 1)$, where θ is a correction factor which is equal to unity when $C_X = 0$, regardless of the distribution of the random error. The quantity C_X is a multivariate measure of kurtosis for the input variables and is given by the formula

$$C_X = \left[\frac{N(N-1)(N+1)}{k(N-k-1)(N-3)} \right] \left[d_0 - \frac{k(k+2)(N-1)}{N(N+1)} \right] \tag{12.18}$$

where $d_0 = \sum_{u=1}^{N} d_{uu}^2$ and d_{uu} is the uth diagonal element of $\Delta = D(D'D)^{-1}D'$ ($u = 1, 2, \ldots, N$). If **D** is chosen so that $C_X = 0$, or equivalently, $g(D) = 0$, where

$$g(D) = d_0 - \frac{k(k+2)(N-1)}{N(N+1)} \tag{12.19}$$

then the approximate distribution of $T(D)$ in Eq. 12.17 will have the same degrees of freedom as those that would have been obtained under the normality assumption. In other words, a design chosen as described above will lead to a robust test of H_o since it involves no degrees-of-freedom adjustment.

12.3.2 The Construction of a Design That Satisfies the Box-Watson Criterion for a Given Number of Experimental Runs

Box and Watson (1962) noted that d_0 in Eq. 12.18 is invariant to any nonsingular transformation of the columns of $\mathbf{D}$. We can, therefore, consider the columns of $\mathbf{D}$ to be orthogonal. The input variables are scaled so that $\mathbf{D}_1'\mathbf{D}_1 = n\mathbf{I}_k$, where $\mathbf{D}_1$ is the portion of $\mathbf{D}$ corresponding to the n noncenter points. Let $\mathbf{\Delta}_1 = \mathbf{D}_1(\mathbf{D}_1'\mathbf{D}_1)^{-1}\mathbf{D}_1'$, then $\mathbf{\Delta}_1$ is idempotent of rank k with eigenvalues consisting of k ones and $n - k$ zeros. Since $\mathbf{D}'\mathbf{1}_N = \mathbf{0}$, then $\mathbf{D}_1'\mathbf{1}_n = \mathbf{0}$ and $\mathbf{\Delta}_1\mathbf{1}_n = \mathbf{0}$. Thus, $\mathbf{1}_n$ is an eigenvector of $\mathbf{\Delta}_1$ for the eigenvalue zero. Also, because $\mathbf{\Delta}_1\mathbf{D}_1 = \mathbf{D}_1$, the k columns of $\mathbf{D}_1$ form an orthogonal set of eigenvectors of $\mathbf{\Delta}_1$ for the eigenvalue one. Thus, if $\mathbf{C} = [c_{ui}]$ is an orthogonal matrix of order $n \times n$ whose first column is $\mathbf{1}_n/\sqrt{n}$, then we can write

$$\mathbf{\Delta}_1 = \mathbf{C} \, \text{diag}(0, \mathbf{I}_k, \mathbf{0}_{n-k-1})\mathbf{C}' = \sum_{i=2}^{k+1} \mathbf{c}_i\mathbf{c}_i' \tag{12.20}$$

where $\mathbf{0}_{n-k-1}$ is a zero matrix of order $(n - k - 1) \times (n - k - 1)$, and $\mathbf{c}_i$ is the ith column of $\mathbf{C}$ ($i = 1, 2, \ldots, n$). From Eqs. 12.19 and 12.20 we obtain

$$g(\mathbf{D}) = \sum_{u=1}^{n} \left(\sum_{i=2}^{k+1} c_{ui}^2\right)^2 - \frac{k(k + 2)(N - 1)}{N(N + 1)} \tag{12.21}$$

If the number of center points, n_o, is zero, then Eq. 12.21 becomes

$$g(\mathbf{D}) = \sum_{u=1}^{n} \left(\sum_{i=2}^{k+1} c_{ui}^2\right)^2 - \frac{k(k + 2)(n - 1)}{n(n + 1)} \tag{12.22}$$

From Eq. 12.21 we conclude that a design $\mathbf{D}$ which satisfies the Box-Watson criterion, namely, $g(\mathbf{D}) = 0$, can be obtained if a solution to the equation

$$g(\mathbf{D}) = \sum_{u=1}^{n} \left(\sum_{i=2}^{k+1} c_{ui}^2\right)^2 - \frac{k(k + 2)(N - 1)}{N(N + 1)} = 0 \tag{12.23}$$

exists subject to the constraints

$$\sum_{u=1}^{n} c_{ui}^2 = 1 \qquad i = 2, 3, \ldots, k + 1$$

$$\sum_{u=1}^{n} c_{ui} = 0 \qquad i = 2, 3, \ldots, k + 1$$

$$\sum_{u=1}^{n} c_{ui}c_{ul} = 0 \qquad 2 \leq i < l \leq k + 1 \tag{12.24}$$

In practice, for a given number of experimental runs, the system of equations 12.23 and 12.24, may not have a solution. In this case we seek values of c_{ui} which minimize $g^2(\mathbf{D})$ subject to the equality constraints given in Eq. 12.24. We, therefore, use the method of Lagrange multipliers. This amounts to introducing the function

$$L = g^2(\mathbf{D}) + \sum_{i=2}^{k+1} \xi_i \left(\sum_{u=1}^{n} c_{ui}^2 - 1 \right) + \sum_{i=2}^{k+1} \mu_i \sum_{u=1}^{n} c_{ui}$$

$$+ \sum_{i<l=3}^{k+1} \eta_{il} \sum_{u=1}^{n} c_{ui} c_{ul} \tag{12.25}$$

where $\xi_2, \xi_3, \ldots, \xi_{k+1}$; $\mu_2, \mu_3, \ldots, \mu_{k+1}$; and the values of η_{il} ($2 \le i < l \le k + 1$) are Lagrange multipliers. The next step is to solve the equations

$$\frac{\partial L}{\partial c_{ui}} = 0 \qquad \begin{array}{l} i = 2, 3, \ldots, k + 1 \\ u = 1, 2, \ldots, n \end{array}$$

along with the equations in Eq. 12.24 for the values of c_{ui} and the Lagrange multipliers. These equations can be solved by applying the iterative method of Brown (1969) which is a variation of Newton's method. A computer documentation of Brown's method can be found in the International Mathematical and Statistical Libraries (IMSL) computer package (ZSYSTM subroutine). If the minimum value of $g^2(\mathbf{D})$ is equal to zero (or close enough to zero), then the Box-Watson criterion is considered fully satisfied, otherwise, the design will only approximately satisfy that criterion (this depends on how close the minimum value of $g^2(\mathbf{D})$ is to zero).

The above procedure can be used whenever the number of experimental runs, N, is fixed in advance. If, however, N is not fixed, then it is possible to have a design which fully satisfies the Box-Watson criterion, that is, satisfies Eq. 12.23. Khuri and Myers (1981) showed that in order for a solution of this equation to exist, we must necessarily have

$$\frac{k^2}{n} - \frac{k(k + 2)(N - 1)}{N(N + 1)} \le 0 \le \frac{k(n - 1)}{n} - \frac{k(k + 2)(N - 1)}{N(N + 1)} \tag{12.26}$$

In particular, if $n_o = 0$, then inequalities 12.26 become

$$\frac{-2k(n - k - 1)}{n(n + 1)} \le 0 \le \frac{k(n - 1)(n - k - 1)}{n(n + 1)}$$

The above inequalities are satisfied if $n \ge k + 1$. It follows that when $n_o = 0$, a necessary condition for the existence of a solution to Eq. 12.23 is that $n \ge k + 1$.

Khuri and Myers (1981) introduced two types of designs each of which can satisfy the Box-Watson criterion. One type consists of a pair of 2^{-m} fractions of a 2^k factorial design with m being such that $2^{k-m} \geq k + 1$. The levels of the ith factor in the first fraction are $\pm a$ and those in the second fraction are $\pm b$ ($i = 1, 2, \ldots, k$), where a^2 and b^2 are given by

$$a^2 = 1 - \left(\frac{2^{k-m+2} - 2k - 2}{k2^{k-m+1} + k}\right)^{1/2}$$

$$b^2 = 1 + \left(\frac{2^{k-m+2} - 2k - 2}{k2^{k-m+1} + k}\right)^{1/2} \tag{12.27}$$

Example 12.1   Consider fitting a first-order model in $k = 3$ input variables using a design which consists of $n = 5$ noncenter points. In this case, Eqs. 12.23 and 12.24 become

$$\sum_{u=1}^{5}\left(\sum_{i=2}^{4} c_{ui}^2\right)^2 - 2 = 0 \qquad \sum_{u=1}^{5} c_{ui}^2 = 1 \qquad i = 2, 3, 4$$

$$\sum_{u=1}^{5} c_{ui} = 0 \qquad i = 2, 3, 4 \qquad \sum_{u=1}^{5} c_{ui}c_{ul} = 0 \qquad 2 \leq i < l \leq 4 \tag{12.28}$$

Using the method of Lagrange multipliers as was described earlier, we find that columns 2, 3, and 4 of $\mathbf{C}$ (see Eq. 12.20) are columns 1, 2, and 3, respectively, of the matrix

$$\mathbf{C}_1 = \begin{bmatrix} 0.765 & 0.351 & 0.162 \\ -0.630 & 0.578 & 0.178 \\ -0.135 & -0.701 & 0.086 \\ 0 & -0.228 & 0.436 \\ 0 & 0 & -0.863 \end{bmatrix}$$

The corresponding design matrix $\mathbf{D}$ is then given by $\sqrt{5}\mathbf{C}_1$, that is,

$$\mathbf{D} = \begin{bmatrix} 1.711 & 0.785 & 0.362 \\ -1.408 & 1.292 & 0.399 \\ -0.303 & -1.567 & 0.193 \\ 0 & -0.510 & 0.975 \\ 0 & 0 & -1.929 \end{bmatrix}$$

If the number of experimental runs is not fixed, then we can use the designs introduced by Khuri and Myers (1981) that were mentioned earlier. For example, if we choose $n = 2^3$, then from Eq. 12.27 we have (with $m = 1$) that $a^2 = 0.456$ and $b^2 = 1.544$. Thus, a design consisting of two 1/2 fractions of a 2^3 factorial design with the levels in the first half being ±0.675 and those in the second half being ±1.243 satisfies Eq. 12.23.

12.4 DESIGN ROBUSTNESS TO MISSING OBSERVATIONS

Quite often in experimental work a situation may arise where some observations are lost or become unavailable. In such a situation one might consider dropping the missing response values along with the corresponding rows of X in model 12.13. Data anlysis can then proceed using the available data. Dropping rows of X, however, amounts to changing the design structure. For example, the design may be chosen so that it satisfies certain desirable properties, such as rotatability, *D*-optimality, etc. With rows of X missing there can be adverse effects on these properties. A design that guards against such effects is called robust to missing observations.

Herzberg and Andrews (1976) introduced a measure of robustness of a design, which they called a *probability of breakdown of a design*, given by

$$P[|\mathbf{X}'\mathbf{A}^2\mathbf{X}| = 0] \tag{12.29}$$

where $\mathbf{A}^2$ is a diagonal matrix with uth diagonal element

$$a_u^2 = \begin{cases} 0 & \text{with probability } \alpha(\mathbf{x}) \\ 1 & \text{with probability } 1 - \alpha(\mathbf{x}) \end{cases} \quad u = 1, 2, \ldots, N$$

and $\alpha(\mathbf{x})$ is the probability of losing an observation at $\mathbf{x}$, the losses at different points being independent. When a sufficient number of observations are lost, $|\mathbf{X}'\mathbf{A}^2\mathbf{X}| = 0$ and the model's parameters in Eq. 12.13 cannot all be estimated. The smaller the probability value in Eq. 12.29, the more robust the design. An alternative measure of robustness introduced by Herzberg and Andrews (1976) was the expected value of $|\mathbf{X}'\mathbf{A}^2\mathbf{X}|^{1/p}$, where p is the number of parameters in the model. This latter measure can be described as the average precision to be expected from a design.

McKee and Kshirsagar (1982) studied the effects of one or more missing observations on the parameter estimates and their variances for response surface designs arranged in blocks, in particular, the central composite design with orthogonal blocking. They concluded that, while the loss of data leads to an increase in the variance of the parameter estimates, such an effect is less severe when a center point is missing than when a factorial point or an axial point is missing.

Akhtar and Prescott (1986) examined the effects of missing observations on the *D*-optimality criterion. They considered a measure of loss of efficiency of a design defined as

$$\ell_{ij\ldots} = \frac{(|\mathbf{X}'\mathbf{X}| - |\mathbf{X}'\mathbf{X}|_{ij\ldots})}{|\mathbf{X}'\mathbf{X}|} \tag{12.30}$$

where $|\mathbf{X'X}|_{ij\ldots}$ is the same as $|\mathbf{X'X}|$, except that the rows of $\mathbf{X}$ indexed by i, j, ... are missing. A design is robust to these missing rows if $\ell_{ij\ldots}$ is as small as possible. Akhtar and Prescott (1986) showed that when a single observation is missing in a central composite design, say the ith observation, then

$$\ell_i = r_{ii}$$

where r_{ii} is the ith diagonal element of the matrix $[r_{ij}] = \mathbf{X(X'X)^{-1}X'}$. When the ith and jth observations are missing, then

$$\ell_{ij} = 1 - (1 - r_{ii})(1 - r_{jj}) + r_{ij}^2$$

A possible criterion for the choice of a design robust to missing observations, say m_0 observations, is to minimize the maximum value of $\ell_{ij\ldots}$ in Eq. 12.30 over all possible combinations of m_0 missing rows of $\mathbf{X}$. Akhtar and Prescott (1986) used this criterion to develop central composite designs robust to any single or any pair of missing observations.

An alternative approach to the problem of missing observations is to estimate the missing response values and then proceed with the analysis of data as originally planned, except for adjustments due to the loss of degrees of freedom. This approach was considered by Draper (1961). The advantage of this approach is that the design, which was originally chosen to satisfy certain optimal properties, remains intact.

Suppose that the vector of observations, $\mathbf{Y}$, in Eq. 12.13 is partitioned as $\mathbf{Y} = [\mathbf{Y_1'} : \mathbf{Y_2'}]'$, where $\mathbf{Y_1}$ is the vector of response values actually observed, and $\mathbf{Y_2}$ is a vector of m_0 missing observations. The matrix $\mathbf{X}$ in Eq. 12.13 can be accordingly partitioned as $\mathbf{X} = [\mathbf{X_1'} : \mathbf{X_2'}]'$. Assuming that $\mathbf{X_1}$ supports the estimation of all of the parameters in model 12.13, then the parameter estimates from the observed responses are the elements of

$$\mathbf{b} = (\mathbf{X_1'X_1})^{-1}\mathbf{X_1'Y_1} \tag{12.31}$$

The missing observations are estimated as

$$\mathbf{\hat{Y}_2} = \mathbf{X_2b}$$

Substituting $\mathbf{\hat{Y}_2}$ for $\mathbf{Y_2}$, the estimate of $\boldsymbol{\beta}$ in model 12.13 using the data vector, $\mathbf{Z} = [\mathbf{Y_1'} : \mathbf{\hat{Y}_2'}]'$, is $(\mathbf{X'X})^{-1}\mathbf{X'Z}$. Draper (1961) showed that

$$(\mathbf{X'X})^{-1}\mathbf{X'Z} = \mathbf{b}$$

Thus, the parameter estimates obtained when the missing observations are estimated are the same as those that would have been obtained from applying Eq.

12.31. The ANOVA table in this case is of the form

Source	d.f.	SS
Regression	$p - 1$	SSR
Residual	$N - m_0 - p$	SSE
Total	$N - m_0 - 1$	SST

The estimated variance-covariance matrix of $\mathbf{b}$ is given by $(\mathbf{X}_1'\mathbf{X}_1)^{-1}$MSE, where

$$\text{MSE} = \frac{\text{SSE}}{N - m_0 - p}$$

12.5 MEASURES OF ROTATABILITY

We may recall that in Section 2.8.3, rotatability was defined as a design property in which the prediction variance, $\text{Var}[\hat{Y}(\mathbf{x})]$, is a function of only the distance from the point $\mathbf{x}$ to the center of the design. This is a desirable invariance property that is easy to achieve in the case of first- and second-order designs, as was seen in Chapters 3 and 4, respectively. The constant prediction variance property is particularly appealing also when searching for the optimum response on concentric spheres, as in the method of ridge analysis (see Chapter 5).

There may be situations, however, in which rotatability is sacrificed for the sake of more important design properties, such as orthogonal blocking, uniform precision, or bias reduction. In this case, it would be desirable to preserve some degree of rotatability wherever possible. Previously, near rotatability was assessed by a visual examination of the contour plots of the prediction variance to determine whether they are close to being circular. Although this method is reasonable when the number of input variables is two, it certainly is infeasible in higher dimensions than two.

In the next section, a measure of rotatability is presented to enable us to quantify the degree of rotatability of any given design. This measure was initially introduced by Khuri (1988).

12.5.1 Derivation of a Measure of Rotatability

Consider fitting a response surface model of order d in k input variables, x_1, $x_2, \ldots, x_k$, over a spherical region of interest, R, using a design $\mathbf{D}$. The model can be written in vector form as in Eq. 12.13. From Appendix 2B we may recall that $\mathbf{D}$ is rotatable if and only if its design moments, that is, the elements of the moment matrix,

$$\mathbf{M} = \frac{1}{N}\mathbf{X}'\mathbf{X}$$

which are shown in Eq. 2.44, have the particular form given in Eq. 2B.1, where N is the number of experimental runs. Let us assume that the input variables have been coded so that

$$\sum_{u=1}^{N} x_{ui} = 0 \qquad i = 1, 2, \ldots, k \tag{12.32}$$

$$\frac{1}{N} \sum_{u=1}^{N} x_{ui}^2 = a_0 \qquad i = 1, 2, \ldots, k \tag{12.33}$$

where x_{ui} is the uth setting of the ith input variable, and a_0 is a positive constant.

Let $v(M)$ denote the vector consisting of the elements of M which are located above its diagonal as well as those along the diagonal. The elements of this vector are arranged so that the elements of the first row of the upper triangular half of M are listed first followed by those in the second row, and so forth. Note that the vector $v(M)$ is not scale-free since it depends on the coding used for the input variables. To remove this scale dependency, each design moment of order $\delta(= 0, 1, \ldots, 2d)$ is divided by τ^δ, where $\tau = (Na_0)^{1/2}$ and a_0 is the value in Eq. 12.33. This operation transforms $v(M)$ into a vector, denoted by $u(M)$, whose elements are scale-free.

Now, if the design D is rotatable, the elements of $v(M)$ corresponding to odd design moments are equal to zero and those corresponding to even design moments are of the form $\lambda_\delta h(\delta_1, \delta_2, \ldots, \delta_k)$, where by formula 2B.1

$$h(\delta_1, \delta_2, \ldots, \delta_k) = \frac{\Pi_{i=1}^{k}(\delta_i)!}{2^{\delta/2}\Pi_{i=1}^{k}(\delta_i/2)!} \tag{12.34}$$

and $\delta = \Sigma_{i=1}^{k} \delta_i$, $\delta = 0, 2, \ldots, 2d$. Let M_r denote the moment matrix when the design is rotatable. In this case, the vector $v(M_r)$ can be represented as

$$v(M_r) = \lambda_0\omega_0 + \lambda_2\omega_2 + \cdots + \lambda_{2d}\omega_{2d} \tag{12.35}$$

1where the elements of the vector ω_δ ($\delta = 0, 2, \ldots, 2d$) are in a one-to-one correspondence with the elements of $v(M_r)$: Those elements corresponding to design moments of order different from δ and odd design moments of order δ, are equal to zero; whereas those corresponding to even design moments of order δ have values given by formula 12.34. Thus, ω_δ depends only on the values of the δ_i's that designate the design moments of order $\delta(= 0, 2, \ldots, 2d)$ of a rotatable design.

From Eq. 12.34 we note that the elements of ω_0 are all zero except for the first element, which is equal to unity. Moreover, from Eq. 2B.1 it is easy to show that $\lambda_0 = 1$ and $\lambda_2 = a_0$ by the coding scheme of Eq. 12.33. Hence, $\lambda_2 = \tau^2/N$, since

$$\tau = (Na_0)^{1/2} = \left(\sum_{u=1}^{N} x_{ui}^2\right)^{1/2} \qquad i = 1, 2, \ldots, k$$

It follows that the vector $\mathbf{u}(\mathbf{M}_r)$ can be expressed as

$$\mathbf{u}(\mathbf{M}_r) = \boldsymbol{\omega}_0 + \frac{1}{N}\boldsymbol{\omega}_2 + \sum_{i=2}^{d} \theta_{2i}\boldsymbol{\omega}_{2i} \tag{12.36}$$

where $\theta_{2i} = \lambda_{2i}/\tau^{2i}$, $i = 2, 3, \ldots, d$.

Formula 12.36 provides a canonical representation for the moment matrix of a rotatable design. The values of $\lambda_4, \lambda_6, \ldots, \lambda_{2d}$ are considered design moments of orders 4, 6, $\ldots$, $2d$, respectively. Thus, $\theta_4, \theta_6, \ldots, \theta_{2d}$ represent scaled design moments. They can be chosen depending on whatever additional properties the rotatable design is required to have. Note that $\boldsymbol{\omega}_0, \boldsymbol{\omega}_2, \ldots, \boldsymbol{\omega}_{2d}$ are pairwise orthogonal; that is, $\boldsymbol{\omega}'_{2i}\boldsymbol{\omega}_{2j} = 0$, $i \neq j$. This follows from the fact that all the elements of $\boldsymbol{\omega}_{2i}$ are zero except for those corresponding to even design moments of order $2i$ ($i = 0, 1, \ldots, d$). For these latter elements, the corresponding elements of $\boldsymbol{\omega}_{2j}$, for $j \neq i$, must be zero by definition. Consequently, the vectors $\boldsymbol{\omega}_4, \boldsymbol{\omega}_6, \ldots, \boldsymbol{\omega}_{2d}$ form a basis for a $(d - 1)$-dimensional Euclidean space. Furthermore, since $\theta_{2i} \geq 0$ for $i = 2, 3, \ldots, d$, the vector $\sum_{i=2}^{d} \theta_{2i}\boldsymbol{\omega}_{2i}$ belongs to a convex cone K in this Euclidean space (a subset S is a convex cone if for any vectors, $\mathbf{v}_1$ and $\mathbf{v}_2$, in S and any nonnegative scalars, t_1 and t_2, the vector $t_1\mathbf{v}_1 + t_2\mathbf{v}_2$ belongs to S).

Now, suppose that for a given design $\mathbf{D}$, not necessarily rotatable, the input variables have been coded according to Eqs. 12.32 and 12.33. The corresponding vector $\mathbf{u}(\mathbf{M})$ can be written as

$$\mathbf{u}(\mathbf{M}) = \boldsymbol{\omega}_0 + \frac{1}{N}\boldsymbol{\omega}_2 + \mathbf{u}^*(\mathbf{M}) \tag{12.37}$$

where $\boldsymbol{\omega}_0$ and $\boldsymbol{\omega}_2$ are the same as in Eq. 12.36, which holds for a rotatable design with the same number N of experimental runs. The elements of $\mathbf{u}^*(\mathbf{M})$ are equal to the corresponding elements of $\mathbf{u}(\mathbf{M})$ except for those that correspond to design moments of order $\delta = 0$ and even design moments of order $\delta = 2$, which are equal to zero. If $\mathbf{u}^*(\mathbf{M})$ belongs to the convex cone K spanned by $\boldsymbol{\omega}_4, \boldsymbol{\omega}_6, \ldots, \boldsymbol{\omega}_{2d}$, then $\mathbf{D}$ must be rotatable; otherwise, it is not. Thus, to measure rotatability of $\mathbf{D}$, it is necessary to determine how well $\mathbf{u}^*(\mathbf{M})$ can be approximated with a vector that belongs to K. This is equivalent to finding nonnegative scalars, $\gamma_2, \gamma_3, \ldots, \gamma_d$, such that the squared Euclidean norm

$$Q_N(\mathbf{D}) = \left\|\mathbf{u}^*(\mathbf{M}) - \sum_{i=2}^{d} \gamma_i\boldsymbol{\omega}_{2i}\right\|^2$$

is minimum. It can be shown that the minimum value of $Q_N(\mathbf{D})$ is of the form

$$\min[Q_N(\mathbf{D})] = \|\mathbf{u}^*(\mathbf{M})\|^2 - \sum_{i=2}^{d} [\mathbf{u}^{*\prime}(\mathbf{M})\boldsymbol{\omega}_{2i}]^2/\|\boldsymbol{\omega}_{2i}\|^2 \tag{12.38}$$

and is attained by

$$\gamma_i = \frac{\mathbf{u}^{*\prime}(\mathbf{M})\boldsymbol{\omega}_{2i}}{\|\boldsymbol{\omega}_{2i}\|^2} \qquad i = 2, 3, \ldots, d \tag{12.39}$$

(see Appendix A in Khuri, 1988). Note that $\gamma_i \geq 0$ for $i = 2, 3, \ldots, d$ since $\mathbf{u}^{*\prime}(\mathbf{M})\boldsymbol{\omega}_{2i}$ is a positive linear combination of the even design moments of $\mathbf{D}$ of order $\delta = 2i$ $(i = 2, 3, \ldots, d)$. Hence the vector

$$\boldsymbol{v} = \sum_{i=2}^{d} \gamma_i \boldsymbol{\omega}_{2i}$$

where γ_i is given by Eq. 12.39, belongs to the convex cone K. Furthermore, $\boldsymbol{v}$ is the projection of $\mathbf{u}^*(\mathbf{M})$ on K, and

$$\|\boldsymbol{v}\|^2 = \sum_{i=2}^{d} \frac{[\mathbf{u}^{*\prime}(\mathbf{M})\boldsymbol{\omega}_{2i}]^2}{\|\boldsymbol{\omega}_{2i}\|^2}$$

This represents the portion of $\|\mathbf{u}^*(\mathbf{M})\|^2$ that can be attributed to rotatability. Hence, as a measure of rotatability we consider the value

$$\Phi_N(\mathbf{D}) = 100 \frac{\|\boldsymbol{v}\|^2}{\|\mathbf{u}^*(\mathbf{M})\|^2}$$

or equivalently,

$$\Phi_N(\mathbf{D}) = \frac{100}{\|\mathbf{u}^*(\mathbf{M})\|^2} \sum_{i=2}^{d} \frac{[\mathbf{u}^{*\prime}(\mathbf{M})\boldsymbol{\omega}_{2i}]^2}{\|\boldsymbol{\omega}_{2i}\|^2} \tag{12.40}$$

Note that $0 \leq \Phi_N(\mathbf{D}) \leq 100$, and $\Phi_N(\mathbf{D})$ is scale invariant since the elements of $\mathbf{u}(\mathbf{M})$, and hence $\mathbf{u}^*(\mathbf{M})$, are scale-free. Thus "contracting" or "expanding" the design by multiplying the input variables by a constant multiplier will not change its measure of rotatability. Furthermore, since an element of $\mathbf{u}^*(\mathbf{M})$ corresponding to a design moment of order δ $(= 3, 4, \ldots, 2d)$ is of the form

$$\frac{1}{\tau^\delta} \frac{1}{N} \sum_{u=1}^{N} x_{u_1}^{\delta_1} x_{u_2}^{\delta_2} \cdots x_{u_k}^{\delta_k} = \frac{1}{N} \left[\left(\sum_{u=1}^{N} x_{u_1}^{\delta_1} x_{u_2}^{\delta_2} \cdots x_{u_k}^{\delta_k} \right) \Big/ \left(\sum_{u=1}^{N} x_{ui}^{2} \right)^{\delta/2} \right]$$

$$i = 1, 2, \ldots, k$$

where $\sum_{i=1}^{k} \delta_i = \delta$, the value N drops out from the expression for $\Phi_N(\mathbf{D})$ in Eq. 12.40. Consequently, the addition of any number of center point replications does not change the value of $\Phi_N(\mathbf{D})$. This follows from the fact that $(\sum_{u=1}^{N} x_{u_1}^{\delta_1} x_{u_2}^{\delta_2} \cdots x_{u_k}^{\delta_k})/(\sum_{u=1}^{N} x_{ui}^{2})^{\delta/2}$ is invariant to such addition for $\delta = 3, 4, \ldots, 2d$.

The measure $\Phi_N(\mathbf{D})$ represents the percent rotatability that is inherent in $\mathbf{D}$. It is essentially similar to the coefficient of determination used in regression,

Measures of Rotatability **447**

which measures the percent of total variation in a data set that is attributed to the fitted model. If D is rotatable, then $u^*(M)$ must belong to the convex cone K mentioned earlier, hence $\Phi_N(D) = 100$. Thus a large value of $\Phi_N(D)$ indicates that D is near rotatable.

Khuri (1988) used the measure $\Phi_N(D)$ to increase the percent rotatability of a given nonrotatable design by augmenting it with additional runs. These runs can be obtained sequentially and one at a time. Each added point is selected by maximizing the percent rotatability of the augmented design (that is, the one whose last run is the added point). Increasing the percent rotatability is tantamount to "repairing" rotatability. This is particularly useful in situations where, because of technical limitations, some of the input variables may be set at levels slightly different from those required for rotatability. Also, it may be of interest to increase the percent rotatability of a nonrotatable design, which possesses certain desirable characteristics. This way, two or more design criteria can be nearly achieved. In another situation, a rotatable design may be modified by the introduction of new runs in order to acquire more information in certain regions of interest (see, for example, Littell and Mott, 1974). Such a modification may result in a sizable loss of rotatability.

12.5.2 Example

The purpose of this example is to demonstrate how to apply formula 12.40 to a particular design.

Consider fitting a second-order model in $k = 2$ input variables using a 3^2 factorial design. The levels of the ith factor are -1, 0, $1(i = 1, 2)$. We recall from Section 4.5.1 that this design is not rotatable. Let us therefore determine its percent rotatability.

The formulas in Eqs. 12.32 and 12.33 are satisfied with $N = 9$ and $a_0 = 6/9$. The moment matrix has the form

$$M = \tfrac{1}{9}X'X = \begin{pmatrix} 1 & x_1 & x_2 & x_1^2 & x_2^2 & x_1x_2 \\ 1 & 0 & 0 & \frac{6}{9} & \frac{6}{9} & 0 \\ 0 & \frac{6}{9} & 0 & 0 & 0 & 0 \\ 0 & 0 & \frac{6}{9} & 0 & 0 & 0 \\ \frac{6}{9} & 0 & 0 & \frac{6}{9} & \frac{4}{9} & 0 \\ \frac{6}{9} & 0 & 0 & \frac{4}{9} & \frac{6}{9} & 0 \\ 0 & 0 & 0 & 0 & 0 & \frac{4}{9} \end{pmatrix}$$

The vector $v(M)$ has 21 elements consisting of the upper triangular half of M, including the diagonal elements. The vector $u(M)$ is derived from $v(M)$ by dividing the elements corresponding to design moments of order δ (= 0, 1, 2, 3, 4) by τ^δ, where $\tau = (Na_0)^{1/2} = 6^{1/2}$. These two vectors along with ω_0, ω_2, ω_4, and $u^*(M)$ (see Eq. 12.37) are given in Table 12.1.

Table 12.1 The Elements of the Vectors $\mathbf{v}(M)$, $\mathbf{u}(M)$, $\boldsymbol{\omega}_0$, $\boldsymbol{\omega}_2$, $\boldsymbol{\omega}_4$, and $\mathbf{u}^*(M)$

Design moment	$\mathbf{v}(M)$	$\mathbf{u}(M)$	$\boldsymbol{\omega}_0$	$\boldsymbol{\omega}_2$	$\boldsymbol{\omega}_4$	$\mathbf{u}^*(M)$
[0]	1	1	1	0	0	0
[1]	0	0	0	0	0	0
[2]	0	0	0	0	0	0
[11]	$\frac{6}{9}$	$\frac{1}{9}$	0	1	0	0
[22]	$\frac{6}{9}$	$\frac{1}{9}$	0	1	0	0
[12]	0	0	0	0	0	0
[11]	$\frac{6}{9}$	$\frac{1}{9}$	0	1	0	0
[12]	0	0	0	0	0	0
[111]	0	0	0	0	0	0
[122]	0	0	0	0	0	0
[112]	0	0	0	0	0	0
[22]	$\frac{6}{9}$	$\frac{1}{9}$	0	1	0	0
[112]	0	0	0	0	0	0
[222]	0	0	0	0	0	0
[122]	0	0	0	0	0	0
[1111]	$\frac{6}{9}$	$\frac{6}{324}$	0	0	3	$\frac{6}{324}$
[1122]	$\frac{4}{9}$	$\frac{4}{324}$	0	0	1	$\frac{4}{324}$
[1112]	0	0	0	0	0	0
[2222]	$\frac{6}{9}$	$\frac{6}{324}$	0	0	3	$\frac{6}{324}$
[1222]	0	0	0	0	0	0
[1122]	$\frac{4}{9}$	$\frac{4}{324}$	0	0	1	$\frac{4}{324}$

Applying formula 12.40, we obtain

$$\Phi_9(\mathbf{D}) = \frac{100}{\|\mathbf{u}^*(M)\|^2} \left(\frac{\mathbf{u}^{*\prime}(M)\boldsymbol{\omega}_4}{\|\boldsymbol{\omega}_4\|} \right)^2$$

$$= 93.08.$$

We conclude that the 3^2 factorial design is 93.08% rotatable.

12.5.3 Other Measures of Rotatability

Draper and Guttman (1988) introduced a measure that provides information concerning the overall shape of specified prediction variance contours for only symmetric second-order designs. Their method is based on finding a value, denoted by φ, such that the contour of

$$\sum_{i=1}^{k} |x_i|^\varphi = 1$$

provides a good representation of a particular contour of the prediction variance. Here $x_1, x_2, \ldots, x_k$ represent the fitted second-order model's input variables. This particular contour is chosen as the one that goes through all axial points which are at a distance of one unit from the origin or center of the design. The design settings are scaled so that all the design points fall inside or onto a sphere (hypersphere, in general) of radius one. The value of φ thus found is then compared against a set of benchmark values. One advantage of this measure is that it can be used to determine how distorted the prediction variance contours are, in the sense of being nonspherical in appearance.

More recently, Draper and Pukelsheim (1990) introduced a measure denoted by Q^*, which is similar to Khuri's (1988) measure (see Eq. 12.40) in the sense that it is derived by comparing the form of the $X'X$ matrix for the design with the $X'X$ matrix for a rotatable design. Thus, it essentially measures the percent rotatability inherent in the design. Its primary advantage is its invariance when the design is rotated in the space of the input variables. It can also be used to repair rotatability. Khuri's (1988) as well as Draper and Guttman's (1988) measures are not invariant to design rotation. The measure Q^*, however, was developed for use with second-order designs. Draper and Pukelsheim remarked that their concept can be extended to higher-order designs.

A fourth measure of rotatability was proposed by Park et al. (1993) for only second-order designs. This measure is derived as follows: Suppose that the design is scaled so that all of its points lie inside or on the unit hypersphere. The average value of $(N/\sigma^2) \, \text{Var}[\hat{Y}(\mathbf{x})]$, over all the points on a hypersphere of radius ρ centered at the origin, is obtained, where σ^2 is the error variance and N is the number of experimental runs. Let this average value be denoted by $W_1(\rho)$. Next, the average value of the square of $\{(N/\sigma^2) \, \text{Var}[\hat{Y}(\mathbf{x})] - W_1(\rho)\}$ over the same hypersphere is taken, which results in the value $W_2(\rho)$. Let the quantity $R_k(\mathbf{D})$ be defined as

$$R_k(\mathbf{D}) = \frac{\int_0^1 \rho^{k-1} W_2(\rho) d\rho}{\int_0^1 \rho^{k-1} d\rho}$$

where k is the number of input variables. This quantity represents the average of the square of $\{(N/\sigma^2) \, \text{Var}[\hat{Y}(\mathbf{x})] - W_1(\rho)\}$ over the experimental region. Park et al.'s (1993) measure, denoted by $P_k(\mathbf{D})$, is given by

$$P_k(\mathbf{D}) = \frac{1}{1 + R_k(\mathbf{D})}$$

It takes values inside the interval $(0, 1]$, and is equal to one if and only if the design is rotatable. It is also invariant with respect to design rotation, just like the measure by Draper and Pukelsheim (1990).

12.6 SOME COMMENTS ON DIRECTIONS FOR FUTURE RESEARCH IN RSM

The vast majority of research in response surface methodology has concentrated on the design and analysis of single-response experiments. By contrast, the analysis of multiresponse experiments has received little attention in spite of the acute need for it. In many experimental situations, more often than not, data from several response variables are collected. A univariate analysis of such multiresponse data, that is, an analysis which considers the responses singly, forfeits information concerning any interrelationships among the responses. Instead, the response variables should be considered and studied as a whole using techniques that recognize the multivariate character of the data. In Chapter 7 we discussed optimization and the testing of lack of fit in a multiresponse situation. Further research in this area is sorely needed. For example, little is known about the choice of design for a multiresponse model. An efficient design for one response may not be efficient for some other response. Also, the problem of sequential experimentation can be quite challenging in a multiresponse environment. The decision to proceed from one stage to another depends on information from all the responses, and can be influenced by the extent of the interaction among the responses.

As was seen in Chapter 7, some multiresponse techniques were derived by proper extensions of the single-response case. In the design area, one possible extension of a single-response criterion can be obtained by using the *integrated mean squared error* (IMSE) criterion in Section 6.2. Kim and Draper (1994) presented an extension of this criterion to the case of two responses and a single input variable. They considered the minimization of the trace of $\mathbf{J}$, a 2×2 matrix representing a multiresponse version the IMSE of the predicted response (see Eq. 6.6), as a design criterion. This matrix depends on $\boldsymbol{\Sigma}$, the variance-covariance matrix for the responses, and Kim and Draper assumed $\boldsymbol{\Sigma}$ was known. A procedure that does not require such an assumption is therefore needed. There is also a need to extend Kim and Draper's work to models with more than two input variables and for cases with more than two responses.

The material in Chapter 8 concerning response surface models with block effects is another source of possible research into the multiresponse area. Recently, Khuri and Valeroso (1995) considered the effects of blocking on the estimation of the mean responses in a linear multiresponse model, and on the determination of optimum conditions concerning these responses. More work is needed to address the possibility of having random block effects in the model. In fact, the published literature on the use of models that contain fixed as well as random effects in response surface methodology is very limited at the present time even for the single-response case. Mixed-effects models, especially those that contain interactions involving fixed and random effects, can be very useful in identifying those input variables (control factors) that affect the prediction variance, as was seen in Chapter 11.

The general development of response surface methodology is based on certain assumptions concerning the form of the hypothesized model and the distribution of the error term in the model. Typically, the response data are assumed to be quantitative and have the normal distribution with a constant variance. In many experimental situations, however, such assumptions may not be valid. For example, in some biometric applications, the response may be categorical with either a nominal or an ordinal scale of measurement. In another situation, the response may represent count data such as the number of defects per lot. The use of traditional response surface methods in such situations is either infeasible or inappropriate. It would be more fitting to model these situations using *generalized linear models* (GLM), which can accommodate both quantitative and qualitative response data. Furthermore, fitting GLMs does not require making the same stringent assumptions that are made in the classical linear response models case. A detailed account of the theory of GLMs can be found in the book by McCullagh and Nelder (1989). Obviously, the use of GLMs would necessitate that existing response surface methods be modified and possibly supplemented. There have been only few attempts made in this direction (see Khuri, 1993; Myers et al. 1994). More work is definitely needed in this important area.

In Chapter 10 (Section 10.5.2), the dependency on the actual parameter values in a nonlinear model when constructing designs for fitting nonlinear models was addressed in detail. The construction of parameter-free designs using Lagrangian interpolation where the interpolation points are of the Chebyshev type was then discussed in Section 10.5.4. In both sections, the nonlinear model contained only a single independent variable x. Constructing designs for nonlinear models in two or more variables (or factors), for example x_1, x_2, and x_3, no doubt would lead to extreme difficulty in computing since the number of parameters in multifactor models is typically greater than the number of parameters in a single-factor model. Forcing all of the parameters in a multifactor model to converge in value simultaneously can be a painstaking chore; one that requires not only lengthy computing time but highly skilled programming as well.

Steinberg and Hunter (1984) and Myers et al. (1989) devote several pages to describing future directions for research in areas related to response surface methodology. We shall not echo their suggestions here but instead simply offer our support.

EXERCISES

12.1 Consider the material in Section 12.3.

(a) Show that d_0, the sum of squares of the diagonal elements of $\Delta = \mathbf{D}(\mathbf{D}'\mathbf{D})^{-1}\mathbf{D}'$ is equal to that of $\Delta_1 = \mathbf{D}_1(\mathbf{D}_1'\mathbf{D}_1)^{-1}\mathbf{D}_1'$.

(b)   Show that $d_0 \geq k^2/n$.

Hint: Make use of the inequality

$$\sum_{u=1}^{n} d_{1uu}^2 - \frac{1}{n}\left(\sum_{u=1}^{n} d_{1uu}\right)^2 \geq 0$$

where d_{1uu} is the uth diagonal element of $\mathbf{\Delta}_1$ ($u = 1, 2, \ldots, n$).

(c)   Show that $d_0 \leq k(n - 1)/n$.

Hint: From Eq. 12.20 we have

$$d_{1uu} = \sum_{i=2}^{k+1} c_{ui}^2 \qquad u = 1, 2, \ldots, n$$

Also, the sum of squares of the elements of the uth row of $\mathbf{C}$ is equal to one. Hence,

$$\frac{1}{n} + d_{1uu} \leq 1 \qquad u = 1, 2, \ldots, n$$

Now, use the inequality

$$d_{1uu}^2 \leq d_{1uu}\left(1 - \frac{1}{n}\right) \qquad u = 1, 2, \ldots, n$$

(d)   Make use of parts (b) and (c) to show that

$$\frac{k^2}{n} - \frac{k(k + 2)(N - 1)}{N(N + 1)} \leq g(\mathbf{D}) \leq \frac{k(n - 1)}{n}$$
$$- \frac{k(k + 2)(N - 1)}{N(N + 1)}$$

(e)   Deduce from part (d) that a necessary condition for a solution to $g(\mathbf{D}) = 0$ to exist is given by the double inequality of 12.26.

12.2   Suppose that a first-order design $\mathbf{D}$ consists of a pair of 2^{-m} fractions of a 2^k factorial design such that $2^{k-m} \geq k + 1$, where k is the number of input variables. The levels of the ith factor in the first fraction are $\pm a$ and those in the second fraction are $\pm b$ ($i = 1, 2, \ldots, k$). Show that if a^2 and b^2 are chosen as in Eq. 12.27, then $\mathbf{D}$ satisfies the Box-Watson criterion, that is, it provides a solution to Eq. 12.23.

12.3   Consider the hybrid design shown in Figure 4.10 (see Section 4.5.8). Measure the percent rotatability of this design by using formula 12.40.

12.4 Consider the following design for fitting a second-order model in two input variables:

x_1	x_2
-1.0	1.35
1.0	-1.25
-1.6	-0.85
1.0	1.0
-1.5	0.0
1.55	0.0
0.0	-1.0
0.0	1.55
0.55	0.30
0.0	0.0

(a) Compute its percent rotatability using formula 12.40.
(b) Augment this design with the point (x_{11}, x_{21}) to obtain the design $\tilde{D}$. Determine the values of x_{11} and x_{21} that maximize the percent rotatability of $\tilde{D}$, using formula 12.40, over the circular region

$$R = \{(x_1, x_2) | x_1^2 + x_2^2 \leq 4\}$$

What is the value of this maximum?

REFERENCES AND BIBLIOGRAPHY

Akhtar, M. and P. Prescott (1986). Response Surface Designs Robust to Missing Observations, *Commun. Statist. Simul.*, **15**, 345–363.

Atiqullah, M. (1962). The Estimation of Residual Variance in Quadratically Balanced Least-Squares Problems and the Robustness of the *F*-Test, *Biometrika*, **49**, 83–91.

Box, G. E. P. (1954). The Exploration and Exploitation of Response Surfaces: Some General Considerations and Examples, *Biometrics*, **10**, 16–60.

Box, G. E. P. (1982). Choice of Response Surface Design and Alphabetic Optimality, *Utilitas Mathematica*, **21B**, 11–55.

Box, G. E. P. and D. W. Behnken (1960). Some New Three-Level Designs for the Study of Quantitative Variables, *Technometrics*, **2**, 455–475.

Box, G. E. P. and G. S. Watson (1962). Robustness to Non-Normality of Regression Tests, *Biometrika*, **49**, 93–106.

Box, M. J. and N. R. Draper (1974). On Minimum-Point Second-Order Designs, *Technometrics*, **16**, 613–616.

Brown, K. M. (1969). A Quadratically Convergent Newton-Like Method Based Upon Gaussian Elimination, *Soc. Indust. Appl. Math. J. Numerical Anal.*, **6**, 560–569.

Doehlert, D. H. (1970). Uniform Shell Designs, *J. Roy. Statist. Soc.*, **C19**, 231–239.

Draper, N. R. (1961). Missing Values in Response Surface Designs, *Technometrics*, **3**, 389–398.

Draper, N. R. and I. Guttman (1988). An Index of Rotatability, *Technometrics*, **30**, 105–111.

Draper, N. R. and F. Pukelsheim (1990). Another Look at Rotatability, *Technometrics*, **32**, 195–202.

Herzberg, A. M. and D. F. Andrews (1976). Some Considerations in the Optimal Design of Experiments in Non-Optimal Situations, *J. Roy. Statist. Soc.*, **B38**, 284–289.

Hoke, A. T. (1974). Economical Second-Order Designs Based on Irregular Fractions of the 3^n Factorial, *Technometrics*, **16**, 375–384.

Jensen, D. R., L. S. Mayer, and R. H. Myers (1975). Optimal Designs and Large-Sample Tests for Linear Hypotheses, *Biometrika*, **62**, 71–78.

Khuri, A. I. (1988). A Measure of Rotatability for Response Surface Designs, *Technometrics*, **30**, 95–104.

Khuri, A. I. (1993). Response Surface Methodology Within the Framework of GLM, *J. Combinatorics Information System Sci.*, **18**, 193–202.

Khuri, A. I. and R. H. Myers (1981). Design Related Robustness of Tests in Regression Models, *Commun. Statist. Theor. Meth.*, **A10**, 223–235.

Khuri, A. I. and E. Valeroso (1995). Multiresponse Surface Models with Block Effects, Technical Report No. 486, Department of Statistics, University of Florida, Gainesville, FL.

Kiefer, J. (1958). On the Nonrandomized Optimality and the Randomized Nonoptimality of Symmetrical Designs, *Ann. Math. Statist.*, **29**, 675–699.

Kiefer, J. (1959). Optimum Experimental Designs (with discussion), *J. Roy. Statist. Soc.*, **B21**, 272–319.

Kiefer, J. (1960). Optimum Experimental Design V, with Applications to Systematic and Rotatable Designs, *Proceedings of the Fourth Berkeley Symposium on Mathematical Statistics and Probability*, **1**, 381–405.

Kiefer, J. (1961). Optimum Designs in Regression Problems II, *Ann. Math. Statist.*, **32**, 298–325.

Kiefer, J. (1962a). Two More Criteria Equivalent to D-Optimality of Designs, *Ann. Math. Statist.*, **33**, 792–796.

Kiefer, J. (1962b). An Extremum Result, *Canad. J. Math.*, **14**, 597–601.

Kiefer, J. and J. Wolfowitz (1960). The Equivalence of Two Extremum Problems, *Canad. J. Math.*, **12**, 363–366.

Kim, W. B. and N. R. Draper (1994). Choosing a Design for Straight Line Fits to Two Correlated Responses, *Statist. Sinica*, **4**, 275–280.

Littell, R. C. and G. O. Mott (1974). Computer Assisted Design and Analysis of Response Surface Experiments in Agronomy, *Soil and Crop Science Society of Florida Proceedings*, **34**, 94–97.

Lucas, J. M. (1976). Which Response Surface Design is Best, *Technometrics*, **18**, 411–417.

McCullagh, P. and J. A. Nelder (1989). *Generalized Linear Models*, 2nd ed., London: Chapman and Hall.

McKee, B. and A. M. Kshirsagar (1982). Effect of Missing Plots in Some Response Surface Designs, *Commun. Statist. Theor. Meth.*, **A11**, 1525–1549.

Myers, R. H., A. I. Khuri, and W. H. Carter, Jr. (1989). Response Surface Methodology: 1966–1988, *Technometrics*, **31**, 137–157.

Myers, W. R., R. H. Myers, and W. H. Carter, Jr. (1994). Some Alphabetic Optimal Designs for the Logistic Regression Model, *J. Statist. Plan. Inf.*, **42**, 57–77.

Park, S. H., J. H. Lim, and Y. Baba (1993). A Measure of Rotatability for Second Order Response Surface Designs, *Ann. Inst. Statist. Math.*, **45**, 655–664.

Pearson, E. S. (1931). Analysis of Variance in Cases of Non-Normal Variation, *Biometrika*, **23**, 114–133.

Pesotchinsky, L. L. (1975). D-Optimum and Quasi D-Optimum Second-Order Designs on a Cube, *Biometrika*, **62**, 335–340.

Steinberg, D. M. and W. G. Hunter (1984). Experimental Design: Review and Comment (with discussion and response), *Technometrics*, **26**, 71–130.

Vuchkov, I. N. and E. B. Solakov (1980). The Influence of Experimental Design on Robustness to Nonnormality of the F Test in Regression Analysis, *Biometrika*, **67**, 489–492.

Appendix

Table A.1   Percentage Points of the Student's *t*-Distribution

ν	$\alpha^a = 0.4$	0.25	0.1	0.05	0.025	0.01	0.005	0.0025	0.001	0.0005
1	0.325	1.000	3.078	6.314	12.706	31.821	63.657	127.32	318.31	636.62
2	0.289	0.816	1.886	2.920	4.303	6.965	9.925	14.089	22.327	31.598
3	0.277	0.765	1.638	2.353	3.182	4.541	5.841	7.453	10.214	12.924
4	0.271	0.741	1.533	2.132	2.776	3.747	4.604	5.598	7.173	8.610
5	0.267	0.727	1.476	2.015	2.571	3.365	4.032	4.773	5.893	6,869
6	0.265	0.718	1.440	1.943	2.447	3.143	3.707	4.317	5.208	5.959
7	0.263	0.711	1.415	1.895	2.365	2.998	3.499	4.029	4.785	5.408
8	0.262	0.706	1.397	1.860	2.306	2.896	3.355	3.833	4.501	5.041
9	0.261	0.703	1.383	1.833	2.262	2.821	3.250	3.690	4.297	4.781
10	0.260	0.700	1.372	1.812	2.228	2.764	3.169	3.581	4.144	4.587
11	0.260	0.697	1.363	1.796	2.201	2.718	3.106	3.497	4.025	4.437
12	0.259	0.695	1.356	1.782	2.179	2.681	3.055	3.428	3.930	4.318
13	0.259	0.694	1.350	1.771	2.160	2.650	3.012	3.372	3.852	4.221
14	0.258	0.692	1.345	1.761	2.145	2.624	2.977	3.326	3.787	4.140
15	0.258	0.691	1.341	1.753	2.131	2.602	2.947	3.286	3.733	4.073
16	0.258	0.690	1.337	1.746	2.120	2.583	2.921	3.252	3.686	4.015
17	0.257	0.689	1.333	1.740	2.110	2.567	2.898	3.222	3.646	3.965
18	0.257	0.688	1.330	1.734	2.101	2.552	2.878	3.197	3.610	3.922
19	0.257	0.688	1.328	1.729	2.093	2.539	2.861	3.174	3.579	3.883
20	0.257	0.687	1.325	1.725	2.086	2.528	2.845	3.153	3.552	3.850
21	0.257	0.686	1.323	1.721	2.080	2.518	2.831	3.135	3.527	3.819
22	0.256	0.686	1.321	1.717	2.074	2.508	2.819	3.119	3.505	3.792
23	0.256	0.685	1.319	1.714	2.069	2.500	2.807	3.104	3.485	3.767
24	0.256	0.685	1.318	1.711	2.064	2.492	2.797	3.091	3.467	3.745
25	0.256	0.684	1.316	1.708	2.060	2.485	2.787	3.078	3.450	3.725
26	0.256	0.684	1.315	1.706	2.056	2.479	2.779	3.067	3.435	3.707
27	0.256	0.684	1.314	1.703	2.052	2.473	2.771	3.057	3.421	3.690
28	0.256	0.683	1.313	1.701	2.048	2.467	2.763	3.047	3.408	3.674
29	0.256	0.683	1.311	1.699	2.045	2.462	2.756	3.038	3.396	3.659
30	0.256	0.683	1.310	1.697	2.042	2.457	2.750	3.030	3.385	3.646
40	0.255	0.681	1.303	1.684	2.021	2.423	2.704	2.971	3.307	3.551
60	0.254	0.679	1.296	1.671	2.000	2.390	2.660	2.915	3.232	3.460
120	0.254	0.677	1.289	1.658	1.980	2.358	2.617	2.860	3.160	3.373
∞	0.253	0.674	1.282	1.645	1.960	2.326	2.576	2.807	3.090	3.291

[a]The quantity α is the upper-tail area of the distribution for ν degrees of freedom.
Source: R. H. Myers (1990). *Classical and Modern Regression with Applications, 2nd ed.*, Boston: PWS-Kent. Reproduced with permission of the author and the publisher.

Table A.2 Percentage Points of the F Distribution with ν_1 and ν_2 Degrees of Freedom—Upper 10% Points

ν_2 \ ν_1	1	2	3	4	5	6	7	8	9	10	12	15	20	24	30	40	60	120	∞
1	39.86	49.50	53.59	55.83	57.24	58.20	58.91	59.44	59.86	60.19	60.71	61.22	61.74	62.00	62.26	62.53	62.79	63.06	63.33
2	8.53	9.00	9.16	9.24	9.29	9.33	9.35	9.37	9.38	9.39	9.41	9.42	9.44	9.45	9.46	9.47	9.47	9.48	9.49
3	5.54	5.46	5.39	5.34	5.31	5.28	5.27	5.25	5.24	5.23	5.22	5.20	5.18	5.18	5.17	5.16	5.15	5.14	5.13
4	4.54	4.32	4.19	4.11	4.05	4.01	3.98	3.95	3.94	3.92	3.90	3.87	3.84	3.83	3.82	3.80	3.79	3.78	3.76
5	4.06	3.78	3.62	3.52	3.45	3.40	3.37	3.34	3.32	3.30	3.27	3.24	3.21	3.19	3.17	3.16	3.14	3.12	3.10
6	3.78	3.46	3.29	3.18	3.11	3.05	3.01	2.98	2.96	2.94	2.90	2.87	2.84	2.82	2.80	2.78	2.76	2.74	2.72
7	3.59	3.26	3.07	2.96	2.88	2.83	2.78	2.75	2.72	2.70	2.67	2.63	2.59	2.58	2.56	2.54	2.51	2.49	2.47
8	3.46	3.11	2.92	2.81	2.73	2.67	2.62	2.59	2.56	2.54	2.50	2.46	2.42	2.40	2.38	2.36	2.34	2.32	2.29
9	3.36	3.01	2.81	2.69	2.61	2.55	2.51	2.47	2.44	2.42	2.38	2.34	2.30	2.28	2.25	2.23	2.21	2.18	2.16
10	3.29	2.92	2.73	2.61	2.52	2.46	2.41	2.38	2.35	2.32	2.28	2.24	2.20	2.18	2.16	2.13	2.11	2.08	2.06
11	3.23	2.86	2.66	2.54	2.45	2.39	2.34	2.30	2.27	2.25	2.21	2.17	2.12	2.10	2.08	2.05	2.03	2.00	1.97
12	3.18	2.81	2.61	2.48	2.39	2.33	2.28	2.24	2.21	2.19	2.15	2.10	2.06	2.04	2.01	1.99	1.96	1.93	1.90
13	3.14	2.76	2.56	2.43	2.35	2.28	2.23	2.20	2.16	2.14	2.10	2.05	2.01	1.98	1.96	1.93	1.90	1.88	1.85
14	3.10	2.73	2.52	2.39	2.31	2.24	2.19	2.15	2.12	2.10	2.05	2.01	1.96	1.94	1.91	1.89	1.86	1.83	1.80
15	3.07	2.70	2.49	2.36	2.27	2.21	2.16	2.12	2.09	2.06	2.02	1.97	1.92	1.90	1.87	1.85	1.82	1.79	1.76
16	3.05	2.67	2.46	2.33	2.24	2.18	2.13	2.09	2.06	2.03	1.99	1.94	1.89	1.87	1.84	1.81	1.78	1.75	1.72
17	3.03	2.64	2.44	2.31	2.22	2.15	2.10	2.06	2.03	2.00	1.96	1.91	1.86	1.84	1.81	1.78	1.75	1.72	1.69
18	3.01	2.62	2.42	2.29	2.20	2.13	2.08	2.04	2.00	1.98	1.93	1.89	1.84	1.81	1.78	1.75	1.72	1.69	1.66
19	2.99	2.61	2.40	2.27	2.18	2.11	2.06	2.02	1.98	1.96	1.91	1.86	1.81	1.79	1.76	1.73	1.70	1.67	1.63
20	2.97	2.59	2.38	2.25	2.16	2.09	2.04	2.00	1.96	1.94	1.89	1.84	1.79	1.77	1.74	1.71	1.68	1.64	1.61
21	2.96	2.57	2.36	2.23	2.14	2.08	2.02	1.98	1.95	1.92	1.87	1.83	1.78	1.75	1.72	1.69	1.66	1.62	1.59
22	2.95	2.56	2.35	2.22	2.13	2.06	2.01	1.97	1.93	1.90	1.86	1.81	1.76	1.73	1.70	1.67	1.64	1.60	1.57
23	2.94	2.55	2.34	2.21	2.11	2.05	1.99	1.95	1.92	1.89	1.84	1.80	1.74	1.72	1.69	1.66	1.62	1.59	1.55
24	2.93	2.54	2.33	2.19	2.10	2.04	1.98	1.94	1.91	1.88	1.83	1.78	1.73	1.70	1.67	1.64	1.61	1.57	1.53
25	2.92	2.53	2.32	2.18	2.09	2.02	1.97	1.93	1.89	1.87	1.82	1.77	1.72	1.69	1.66	1.63	1.59	1.56	1.52
26	2.91	2.52	2.31	2.17	2.08	2.01	1.96	1.92	1.88	1.86	1.81	1.76	1.71	1.68	1.65	1.61	1.58	1.54	1.50
27	2.90	2.51	2.30	2.17	2.07	2.00	1.95	1.91	1.87	1.85	1.80	1.75	1.70	1.67	1.64	1.60	1.57	1.53	1.49
28	2.89	2.50	2.29	2.16	2.06	2.00	1.94	1.90	1.87	1.84	1.79	1.74	1.69	1.66	1.63	1.59	1.56	1.52	1.48
29	2.89	2.50	2.28	2.15	2.06	1.99	1.93	1.89	1.86	1.83	1.78	1.73	1.68	1.65	1.62	1.58	1.55	1.51	1.47
30	2.88	2.49	2.28	2.14	2.05	1.98	1.93	1.88	1.85	1.82	1.77	1.72	1.67	1.64	1.61	1.57	1.54	1.50	1.46
40	2.84	2.44	2.23	2.09	2.00	1.93	1.87	1.83	1.79	1.76	1.71	1.66	1.61	1.57	1.54	1.51	1.47	1.42	1.38
60	2.79	2.39	2.18	2.04	1.95	1.87	1.82	1.77	1.74	1.71	1.66	1.60	1.54	1.51	1.48	1.44	1.40	1.35	1.29
120	2.75	2.35	2.13	1.99	1.90	1.82	1.77	1.72	1.68	1.65	1.60	1.55	1.48	1.45	1.41	1.37	1.32	1.26	1.19
∞	2.71	2.30	2.08	1.94	1.85	1.77	1.72	1.67	1.63	1.60	1.55	1.49	1.42	1.38	1.34	1.30	1.24	1.17	1.00

Source: Biometrika Tables for Statisticians, Vol. 1, Cambridge University Press, 1966, edited by E. S. Pearson and H. O. Hartley. Reproduced with permission of the *Biometrika* Trustees.

Table A.3 Percentage Points of the F Distribution with ν_1 and ν_2 Degrees of Freedom—Upper 5% Points

ν_2 \ ν_1	1	2	3	4	5	6	7	8	9	10	12	15	20	24	30	40	60	120	∞
1	161.4	199.5	215.7	224.6	230.2	234.0	236.8	238.9	240.5	241.9	243.9	245.9	248.0	249.1	250.1	251.1	252.2	253.3	254.3
2	18.51	19.00	19.16	19.25	19.30	19.33	19.35	19.37	19.38	19.40	19.41	19.43	19.45	19.45	19.46	19.47	19.48	19.49	19.50
3	10.13	9.55	9.28	9.12	9.01	8.94	8.89	8.85	8.81	8.79	8.74	8.70	8.66	8.64	8.62	8.59	8.57	8.55	8.53
4	7.71	6.94	6.59	6.39	6.26	6.16	6.09	6.04	6.00	5.96	5.91	5.86	5.80	5.77	5.75	5.72	5.69	5.66	5.63
5	6.61	5.79	5.41	5.19	5.05	4.95	4.88	4.82	4.77	4.74	4.68	4.62	4.56	4.53	4.50	4.46	4.43	4.40	4.36
6	5.99	5.14	4.76	4.53	4.39	4.28	4.21	4.15	4.10	4.06	4.00	3.94	3.87	3.84	3.81	3.77	3.74	3.70	3.67
7	5.59	4.74	4.35	4.12	3.97	3.87	3.79	3.73	3.68	3.64	3.57	3.51	3.44	3.41	3.38	3.34	3.30	3.27	3.23
8	5.32	4.46	4.07	3.84	3.69	3.58	3.50	3.44	3.39	3.35	3.28	3.22	3.15	3.12	3.08	3.04	3.01	2.97	2.93
9	5.12	4.26	3.86	3.63	3.48	3.37	3.29	3.23	3.18	3.14	3.07	3.01	2.94	2.90	2.86	2.83	2.79	2.75	2.71
10	4.96	4.10	3.71	3.48	3.33	3.22	3.14	3.07	3.02	2.98	2.91	2.85	2.77	2.74	2.70	2.66	2.62	2.58	2.54
11	4.84	3.98	3.59	3.36	3.20	3.09	3.01	2.95	2.90	2.85	2.79	2.72	2.65	2.61	2.57	2.53	2.49	2.45	2.40
12	4.75	3.89	3.49	3.26	3.11	3.00	2.91	2.85	2.80	2.75	2.69	2.62	2.54	2.51	2.47	2.43	2.38	2.34	2.30
13	4.67	3.81	3.41	3.18	3.03	2.92	2.83	2.77	2.71	2.67	2.60	2.53	2.46	2.42	2.38	2.34	2.30	2.25	2.21
14	4.60	3.74	3.34	3.11	2.96	2.85	2.76	2.70	2.65	2.60	2.53	2.46	2.39	2.35	2.31	2.27	2.22	2.18	2.13
15	4.54	3.68	3.29	3.06	2.90	2.79	2.71	2.64	2.59	2.54	2.48	2.40	2.33	2.29	2.25	2.20	2.16	2.11	2.07
16	4.49	3.63	3.24	3.01	2.85	2.74	2.66	2.59	2.54	2.49	2.42	2.35	2.28	2.24	2.19	2.15	2.11	2.06	2.01
17	4.45	3.59	3.20	2.96	2.81	2.70	2.61	2.55	2.49	2.45	2.38	2.31	2.23	2.19	2.15	2.10	2.06	2.01	1.96
18	4.41	3.55	3.16	2.93	2.77	2.66	2.58	2.51	2.46	2.41	2.34	2.27	2.19	2.15	2.11	2.06	2.02	1.97	1.92
19	4.38	3.52	3.13	2.90	2.74	2.63	2.54	2.48	2.42	2.38	2.31	2.23	2.16	2.11	2.07	2.03	1.98	1.93	1.88
20	4.35	3.49	3.10	2.87	2.71	2.60	2.51	2.45	2.39	2.35	2.28	2.20	2.12	2.08	2.04	1.99	1.95	1.90	1.84
21	4.32	3.47	3.07	2.84	2.68	2.57	2.49	2.42	2.37	2.32	2.25	2.18	2.10	2.05	2.01	1.96	1.92	1.87	1.81
22	4.30	3.44	3.05	2.82	2.66	2.55	2.46	2.40	2.34	2.30	2.23	2.15	2.07	2.03	1.98	1.94	1.89	1.84	1.78
23	4.28	3.42	3.03	2.80	2.64	2.53	2.44	2.37	2.32	2.27	2.20	2.13	2.05	2.01	1.96	1.91	1.86	1.81	1.76
24	4.26	3.40	3.01	2.78	2.62	2.51	2.42	2.36	2.30	2.25	2.18	2.11	2.03	1.98	1.94	1.89	1.84	1.79	1.73
25	4.24	3.39	2.99	2.76	2.60	2.49	2.40	2.34	2.28	2.24	2.16	2.09	2.01	1.96	1.92	1.87	1.82	1.77	1.71
26	4.23	3.37	2.98	2.74	2.59	2.47	2.39	2.32	2.27	2.22	2.15	2.07	1.99	1.95	1.90	1.85	1.80	1.75	1.69
27	4.21	3.35	2.96	2.73	2.57	2.46	2.37	2.31	2.25	2.20	2.13	2.06	1.97	1.93	1.88	1.84	1.79	1.73	1.67
28	4.20	3.34	2.95	2.71	2.56	2.45	2.36	2.29	2.24	2.19	2.12	2.04	1.96	1.91	1.87	1.82	1.77	1.71	1.65
29	4.18	3.33	2.93	2.70	2.55	2.43	2.35	2.28	2.22	2.18	2.10	2.03	1.94	1.90	1.85	1.81	1.75	1.70	1.64
30	4.17	3.32	2.92	2.69	2.53	2.42	2.33	2.27	2.21	2.16	2.09	2.01	1.93	1.89	1.84	1.79	1.74	1.68	1.62
40	4.08	3.23	2.84	2.61	2.45	2.34	2.25	2.18	2.12	2.08	2.00	1.92	1.84	1.79	1.74	1.69	1.64	1.58	1.51
60	4.00	3.15	2.76	2.53	2.37	2.25	2.17	2.10	2.04	1.99	1.92	1.84	1.75	1.70	1.65	1.59	1.53	1.47	1.39
120	3.92	3.07	2.68	2.45	2.29	2.17	2.09	2.02	1.96	1.91	1.83	1.75	1.66	1.61	1.55	1.50	1.43	1.35	1.25
∞	3.84	3.00	2.60	2.37	2.21	2.10	2.01	1.94	1.88	1.83	1.75	1.67	1.57	1.52	1.46	1.39	1.32	1.22	1.00

Source: Biometrika Tables for Statisticians, Vol. 1, Cambridge University Press, 1966, edited by E. S. Pearson and H. O. Hartley. Reproduced with permission of the *Biometrika* Trustees.

Table A.4 Percentage Points of the F Distribution with ν_1 and ν_2 Degrees of Freedom—Upper 2.5% Points

ν_2 \ ν_1	1	2	3	4	5	6	7	8	9	10	12	15	20	24	30	40	60	120	∞
1	647.8	799.5	864.2	899.6	921.8	937.1	948.2	956.7	963.3	968.6	976.7	984.9	993.1	997.2	1001	1006	1010	1014	1018
2	38.51	39.00	39.17	39.25	39.30	39.33	39.36	39.37	39.39	39.40	39.41	39.43	39.45	39.46	39.46	39.47	39.48	39.49	39.50
3	17.44	16.04	15.44	15.10	14.88	14.73	14.62	14.54	14.47	14.42	14.34	14.25	14.17	14.12	14.08	14.04	13.99	13.95	13.90
4	12.22	10.65	9.98	9.60	9.36	9.20	9.07	8.98	8.90	8.84	8.75	8.66	8.56	8.51	8.46	8.41	8.36	8.31	8.26
5	10.01	8.43	7.76	7.39	7.15	6.98	6.85	6.76	6.68	6.62	6.52	6.43	6.33	6.28	6.23	6.18	6.12	6.07	6.02
6	8.81	7.26	6.60	6.23	5.99	5.82	5.70	5.60	5.52	5.46	5.37	5.27	5.17	5.12	5.07	5.01	4.96	4.90	4.85
7	8.07	6.54	5.89	5.52	5.29	5.12	4.99	4.90	4.82	4.76	4.67	4.57	4.47	4.42	4.36	4.31	4.25	4.20	4.14
8	7.57	6.06	5.42	5.05	4.82	4.65	4.53	4.43	4.36	4.30	4.20	4.10	4.00	3.95	3.89	3.84	3.78	3.73	3.67
9	7.21	5.71	5.08	4.72	4.48	4.32	4.20	4.10	4.03	3.96	3.87	3.77	3.67	3.61	3.56	3.51	3.45	3.39	3.33
10	6.94	5.46	4.83	4.47	4.24	4.07	3.95	3.85	3.78	3.72	3.62	3.52	3.42	3.37	3.31	3.26	3.20	3.14	3.08
11	6.72	5.26	4.63	4.28	4.04	3.88	3.76	3.66	3.59	3.53	3.43	3.33	3.23	3.17	3.12	3.06	3.00	2.94	2.88
12	6.55	5.10	4.47	4.12	3.89	3.73	3.61	3.51	3.44	3.37	3.28	3.18	3.07	3.02	2.96	2.91	2.85	2.79	2.72
13	6.41	4.97	4.35	4.00	3.77	3.60	3.48	3.39	3.31	3.25	3.15	3.05	2.95	2.89	2.84	2.78	2.72	2.66	2.60
14	6.30	4.86	4.24	3.89	3.66	3.50	3.38	3.29	3.21	3.15	3.05	2.95	2.84	2.79	2.73	2.67	2.61	2.55	2.49
15	6.20	4.77	4.15	3.80	3.58	3.41	3.29	3.20	3.12	3.06	2.96	2.86	2.76	2.70	2.64	2.59	2.52	2.46	2.40
16	6.12	4.69	4.08	3.73	3.50	3.34	3.22	3.12	3.05	2.99	2.89	2.79	2.68	2.63	2.57	2.51	2.45	2.38	2.32
17	6.04	4.62	4.01	3.66	3.44	3.28	3.16	3.06	2.98	2.92	2.82	2.72	2.62	2.56	2.50	2.44	2.38	2.32	2.25
18	5.98	4.56	3.95	3.61	3.38	3.22	3.10	3.01	2.93	2.87	2.77	2.67	2.56	2.50	2.44	2.38	2.32	2.26	2.19
19	5.92	4.51	3.90	3.56	3.33	3.17	3.05	2.96	2.88	2.82	2.72	2.62	2.51	2.45	2.39	2.33	2.27	2.20	2.13
20	5.87	4.46	3.86	3.51	3.29	3.13	3.01	2.91	2.84	2.77	2.68	2.57	2.46	2.41	2.35	2.29	2.22	2.16	2.09
21	5.83	4.42	3.82	3.48	3.25	3.09	2.97	2.87	2.80	2.73	2.64	2.53	2.42	2.37	2.31	2.25	2.18	2.11	2.04
22	5.79	4.38	3.78	3.44	3.22	3.05	2.93	2.84	2.76	2.70	2.60	2.50	2.39	2.33	2.27	2.21	2.14	2.08	2.00
23	5.75	4.35	3.75	3.41	3.18	3.02	2.90	2.81	2.73	2.67	2.57	2.47	2.36	2.30	2.24	2.18	2.11	2.04	1.97
24	5.72	4.32	3.72	3.38	3.15	2.99	2.87	2.78	2.70	2.64	2.54	2.44	2.33	2.27	2.21	2.15	2.08	2.01	1.94
25	5.69	4.29	3.69	3.35	3.13	2.97	2.85	2.75	2.68	2.61	2.51	2.41	2.30	2.24	2.18	2.12	2.05	1.98	1.91
26	5.66	4.27	3.67	3.33	3.10	2.94	2.82	2.73	2.65	2.59	2.49	2.39	2.28	2.22	2.16	2.09	2.03	1.95	1.88
27	5.63	4.24	3.65	3.31	3.08	2.92	2.80	2.71	2.63	2.57	2.47	2.36	2.25	2.19	2.13	2.07	2.00	1.93	1.85
28	5.61	4.22	3.63	3.29	3.06	2.90	2.78	2.69	2.61	2.55	2.45	2.34	2.23	2.17	2.11	2.05	1.98	1.91	1.83
29	5.59	4.20	3.61	3.27	3.04	2.88	2.76	2.67	2.59	2.53	2.43	2.32	2.21	2.15	2.09	2.03	1.96	1.89	1.81
30	5.57	4.18	3.59	3.25	3.03	2.87	2.75	2.65	2.57	2.51	2.41	2.31	2.20	2.14	2.07	2.01	1.94	1.87	1.79
40	5.42	4.05	3.46	3.13	2.90	2.74	2.62	2.53	2.45	2.39	2.29	2.18	2.07	2.01	1.94	1.88	1.80	1.72	1.64
60	5.29	3.93	3.34	3.01	2.79	2.63	2.51	2.41	2.33	2.27	2.17	2.06	1.94	1.88	1.82	1.74	1.67	1.58	1.48
120	5.15	3.80	3.23	2.89	2.67	2.52	2.39	2.30	2.22	2.16	2.05	1.94	1.82	1.76	1.69	1.61	1.53	1.43	1.31
∞	5.02	3.69	3.12	2.79	2.57	2.41	2.29	2.19	2.11	2.05	1.94	1.83	1.71	1.64	1.57	1.48	1.39	1.27	1.00

Table A.5 Percentage Points of the F Distribution with ν_1 and ν_2 Degrees of Freedom—Upper 1% Points

ν_2 \ ν_1	1	2	3	4	5	6	7	8	9	10	12	15	20	24	30	40	60	120	∞
1	4052	4999.5	5403	5625	5764	5859	5928	5981	6022	6056	6106	6157	6209	6235	6261	6287	6313	6339	6366
2	98.50	99.00	99.17	99.25	99.30	99.33	99.36	99.37	99.39	99.40	99.42	99.43	99.45	99.46	99.47	99.47	99.48	99.49	99.50
3	34.12	30.82	29.46	28.71	28.24	27.91	27.67	27.49	27.35	27.23	27.05	26.87	26.69	26.60	26.50	26.41	26.32	26.22	26.13
4	21.20	18.00	16.69	15.98	15.52	15.21	14.98	14.80	14.66	14.55	14.37	14.20	14.02	13.93	13.84	13.75	13.65	13.56	13.46
5	16.26	13.27	12.06	11.39	10.97	10.67	10.46	10.29	10.16	10.05	9.89	9.72	9.55	9.47	9.38	9.29	9.20	9.11	9.02
6	13.75	10.92	9.78	9.15	8.75	8.47	8.26	8.10	7.98	7.87	7.72	7.56	7.40	7.31	7.23	7.14	7.06	6.97	6.88
7	12.25	9.55	8.45	7.85	7.46	7.19	6.99	6.84	6.72	6.62	6.47	6.31	6.16	6.07	5.99	5.91	5.82	5.74	5.65
8	11.26	8.65	7.59	7.01	6.63	6.37	6.18	6.03	5.91	5.81	5.67	5.52	5.36	5.28	5.20	5.12	5.03	4.95	4.86
9	10.56	8.02	6.99	6.42	6.06	5.80	5.61	5.47	5.35	5.26	5.11	4.96	4.81	4.73	4.65	4.57	4.48	4.40	4.31
10	10.04	7.56	6.55	5.99	5.64	5.39	5.20	5.06	4.94	4.85	4.71	4.56	4.41	4.33	4.25	4.17	4.08	4.00	3.91
11	9.65	7.21	6.22	5.67	5.32	5.07	4.89	4.74	4.63	4.54	4.40	4.25	4.10	4.02	3.94	3.86	3.78	3.69	3.60
12	9.33	6.93	5.95	5.41	5.06	4.82	4.64	4.50	4.39	4.30	4.16	4.01	3.86	3.78	3.70	3.62	3.54	3.45	3.36
13	9.07	6.70	5.74	5.21	4.86	4.62	4.44	4.30	4.19	4.10	3.96	3.82	3.66	3.59	3.51	3.43	3.34	3.25	3.17
14	8.86	6.51	5.56	5.04	4.69	4.46	4.28	4.14	4.03	3.94	3.80	3.66	3.51	3.43	3.35	3.27	3.18	3.09	3.00
15	8.68	6.36	5.42	4.89	4.56	4.32	4.14	4.00	3.89	3.80	3.67	3.52	3.37	3.29	3.21	3.13	3.05	2.96	2.87
16	8.53	6.23	5.29	4.77	4.44	4.20	4.03	3.89	3.78	3.69	3.55	3.41	3.26	3.18	3.10	3.02	2.93	2.84	2.75
17	8.40	6.11	5.18	4.67	4.34	4.10	3.93	3.79	3.68	3.59	3.46	3.31	3.16	3.08	3.00	2.92	2.83	2.75	2.65
18	8.29	6.01	5.09	4.58	4.25	4.01	3.84	3.71	3.60	3.51	3.37	3.23	3.08	3.00	2.92	2.84	2.75	2.66	2.57
19	8.18	5.93	5.01	4.50	4.17	3.94	3.77	3.63	3.52	3.43	3.30	3.15	3.00	2.92	2.84	2.76	2.67	2.58	2.49
20	8.10	5.85	4.94	4.43	4.10	3.87	3.70	3.56	3.46	3.37	3.23	3.09	2.94	2.86	2.78	2.69	2.61	2.52	2.42
21	8.02	5.78	4.87	4.37	4.04	3.81	3.64	3.51	3.40	3.31	3.17	3.03	2.88	2.80	2.72	2.64	2.55	2.46	2.36
22	7.95	5.72	4.82	4.31	3.99	3.76	3.59	3.45	3.35	3.26	3.12	2.98	2.83	2.75	2.67	2.58	2.50	2.40	2.31
23	7.88	5.66	4.76	4.26	3.94	3.71	3.54	3.41	3.30	3.21	3.07	2.93	2.78	2.70	2.62	2.54	2.45	2.35	2.26
24	7.82	5.61	4.72	4.22	3.90	3.67	3.50	3.36	3.26	3.17	3.03	2.89	2.74	2.66	2.58	2.49	2.40	2.31	2.21
25	7.77	5.57	4.68	4.18	3.85	3.63	3.46	3.32	3.22	3.13	2.99	2.85	2.70	2.62	2.54	2.45	2.36	2.27	2.17
26	7.72	5.53	4.64	4.14	3.82	3.59	3.42	3.29	3.18	3.09	2.96	2.81	2.66	2.58	2.50	2.42	2.33	2.23	2.13
27	7.68	5.49	4.60	4.11	3.78	3.56	3.39	3.26	3.15	3.06	2.93	2.78	2.63	2.55	2.47	2.38	2.29	2.20	2.10
28	7.64	5.45	4.57	4.07	3.75	3.53	3.36	3.23	3.12	3.03	2.90	2.75	2.60	2.52	2.44	2.35	2.26	2.17	2.06
29	7.60	5.42	4.54	4.04	3.73	3.50	3.33	3.20	3.09	3.00	2.87	2.73	2.57	2.49	2.41	2.33	2.23	2.14	2.03
30	7.56	5.39	4.51	4.02	3.70	3.47	3.30	3.17	3.07	2.98	2.84	2.70	2.55	2.47	2.39	2.30	2.21	2.11	2.01
40	7.31	5.18	4.31	3.83	3.51	3.29	3.12	2.99	2.89	2.80	2.66	2.52	2.37	2.29	2.20	2.11	2.02	1.92	1.80
60	7.08	4.98	4.13	3.65	3.34	3.12	2.95	2.82	2.72	2.63	2.50	2.35	2.20	2.12	2.03	1.94	1.81	1.73	1.60
120	6.85	4.79	3.95	3.48	3.17	2.96	2.79	2.66	2.56	2.47	2.34	2.19	2.03	1.95	1.86	1.76	1.66	1.53	1.38
∞	6.63	4.61	3.78	3.32	3.02	2.80	2.64	2.51	2.41	2.32	2.18	2.04	1.88	1.79	1.70	1.59	1.47	1.32	1.00

Source: Biometrika Tables for Statisticians, Vol. 1, Cambridge University Press, 1966, edited by E. S. Pearson and H. O. Hartley. Reproduced with permission of the *Biometrika* Trustees.

Table A.6 Percentage Points of the F Distribution with ν_1 and ν_2 Degrees of Freedom—Upper 0.5% Points

ν_2 \ ν_1	1	2	3	4	5	6	7	8	9	10	12	15	20	24	30	40	60	120	∞
1	16211	20000	21615	22500	23056	23437	23715	23925	24091	24224	24426	24630	24836	24940	25044	25148	25253	25359	25465
2	198.5	199.0	199.2	199.2	199.3	199.3	199.4	199.4	199.4	199.4	199.4	199.4	199.4	199.5	199.5	199.5	199.5	199.5	199.5
3	55.55	49.80	47.47	46.19	45.39	44.84	44.43	44.13	43.88	43.69	43.39	43.08	42.78	42.62	42.47	42.31	42.15	41.99	41.83
4	31.33	26.28	24.26	23.15	22.46	21.97	21.62	21.35	21.14	20.97	20.70	20.44	20.17	20.03	19.89	19.75	19.61	19.47	19.32
5	22.78	18.31	16.53	15.56	14.94	14.51	14.20	13.96	13.77	13.62	13.38	13.15	12.90	12.78	12.66	12.53	12.40	12.27	12.14
6	18.63	14.54	12.92	12.03	11.46	11.07	10.79	10.57	10.39	10.25	10.03	9.81	9.59	9.47	9.36	9.24	9.12	9.00	8.88
7	16.24	12.40	10.88	10.05	9.52	9.16	8.89	8.68	8.51	8.38	8.18	7.97	7.75	7.65	7.53	7.42	7.31	7.19	7.08
8	14.69	11.04	9.60	8.81	8.30	7.95	7.69	7.50	7.34	7.21	7.01	6.81	6.61	6.50	6.40	6.29	6.18	6.06	5.95
9	13.61	10.11	8.72	7.96	7.47	7.13	6.88	6.69	6.54	6.42	6.23	6.03	5.83	5.73	5.62	5.52	5.41	5.30	5.19
10	12.83	9.43	8.08	7.34	6.87	6.54	6.30	6.12	5.97	5.85	5.66	5.47	5.27	5.17	5.07	4.97	4.86	4.75	4.64
11	12.23	8.91	7.60	6.88	6.42	6.10	5.86	5.68	5.54	5.42	5.24	5.05	4.86	4.76	4.65	4.55	4.44	4.34	4.23
12	11.75	8.51	7.23	6.52	6.07	5.76	5.52	5.35	5.20	5.09	4.91	4.72	4.53	4.43	4.33	4.23	4.12	4.01	3.90
13	11.37	8.19	6.93	6.23	5.79	5.48	5.25	5.08	4.94	4.82	4.64	4.46	4.27	4.17	4.07	3.97	3.87	3.76	3.65
14	11.06	7.92	6.68	6.00	5.56	5.26	5.03	4.86	4.72	4.60	4.43	4.25	4.06	3.96	3.86	3.76	3.66	3.55	3.44
15	10.80	7.70	6.48	5.80	5.37	5.07	4.85	4.67	4.54	4.42	4.25	4.07	3.88	3.79	3.69	3.58	3.48	3.37	3.26
16	10.58	7.51	6.30	5.64	5.21	4.91	4.69	4.52	4.38	4.27	4.10	3.92	3.73	3.64	3.54	3.44	3.33	3.22	3.11
17	10.38	7.35	6.16	5.50	5.07	4.78	4.56	4.39	4.25	4.14	3.97	3.79	3.61	3.51	3.41	3.31	3.21	3.10	2.98
18	10.22	7.21	6.03	5.37	4.96	4.66	4.44	4.28	4.14	4.03	3.86	3.68	3.50	3.40	3.30	3.20	3.10	2.99	2.87
19	10.07	7.09	5.92	5.27	4.85	4.56	4.34	4.18	4.04	3.93	3.76	3.59	3.40	3.31	3.21	3.11	3.00	2.89	2.78
20	9.94	6.99	5.82	5.17	4.76	4.47	4.26	4.09	3.96	3.85	3.68	3.50	3.32	3.22	3.12	3.02	2.92	2.81	2.69
21	9.83	6.89	5.73	5.09	4.68	4.39	4.18	4.01	3.88	3.77	3.60	3.43	3.24	3.15	3.05	2.95	2.84	2.73	2.61
22	9.73	6.81	5.65	5.02	4.61	4.32	4.11	3.94	3.81	3.70	3.54	3.36	3.18	3.08	2.98	2.88	2.77	2.66	2.55
23	9.63	6.73	5.58	4.95	4.54	4.26	4.05	3.88	3.75	3.64	3.47	3.30	3.12	3.02	2.92	2.82	2.71	2.60	2.48
24	9.55	6.66	5.52	4.89	4.49	4.20	3.99	3.83	3.69	3.59	3.42	3.25	3.06	2.97	2.87	2.77	2.66	2.55	2.43
25	9.48	6.60	5.46	4.84	4.43	4.15	3.94	3.78	3.64	3.54	3.37	3.20	3.01	2.92	2.82	2.72	2.61	2.50	2.38
26	9.41	6.54	5.41	4.79	4.38	4.10	3.89	3.73	3.60	3.49	3.33	3.15	2.97	2.87	2.77	2.67	2.56	2.45	2.33
27	9.34	6.49	5.36	4.74	4.34	4.06	3.85	3.69	3.56	3.45	3.28	3.11	2.93	2.83	2.73	2.63	2.52	2.41	2.29
28	9.28	6.44	5.32	4.70	4.30	4.02	3.81	3.65	3.52	3.41	3.25	3.07	2.89	2.79	2.69	2.59	2.48	2.37	2.25
29	9.23	6.40	5.28	4.66	4.26	3.98	3.77	3.61	3.48	3.38	3.21	3.04	2.86	2.76	2.66	2.56	2.45	2.33	2.21
30	9.18	6.35	5.24	4.62	4.23	3.95	3.74	3.58	3.45	3.34	3.18	3.01	2.82	2.73	2.63	2.52	2.42	2.30	2.18
40	8.83	6.07	4.98	4.37	3.99	3.71	3.51	3.35	3.22	3.12	2.95	2.78	2.60	2.50	2.40	2.30	2.18	2.06	1.93
60	8.49	5.79	4.73	4.14	3.76	3.49	3.29	3.13	3.01	2.90	2.74	2.57	2.39	2.29	2.19	2.08	1.96	1.83	1.69
120	8.18	5.54	4.50	3.92	3.55	3.28	3.09	2.93	2.81	2.71	2.54	2.37	2.19	2.09	1.98	1.87	1.75	1.61	1.43
∞	7.88	5.30	4.28	3.72	3.35	3.09	2.90	2.74	2.62	2.52	2.36	2.19	2.00	1.90	1.79	1.67	1.53	1.36	1.00

Source: *Biometrika Tables for Statisticians*, Vol. 1, Cambridge University Press, 1966, edited by E. S. Pearson and H. O. Hartley. Reproduced with permission of the *Biometrika* Trustees.

General Bibliography

Addelman, S. (1961). Irregular Fractions of the 2^n Factorial Experiments, *Technometrics*, **3**, 479–496.

Addelman, S. (1963). Techniques for Constructing Fractional Replicate Plans, *J. Amer. Statist. Assoc.*, **58**, 45–71.

Addelman, S. (1969). Sequences of Two-Level Fractional Factorial Plans, *Technometrics*, **11**, 477–509.

Addelman, S. (1972). Recent Developments in the Design of Factorial Experiments, *J. Amer. Statist. Assoc.*, **67**, 103–111.

Akhtar, M. and P. Prescott (1986). Response Surface Designs Robust to Missing Observations, *Commun. Statist. Simul.*, **15**, 345–363.

Allen, D. M. and F. B. Cady (1981). *Analyzing Experimental Data by Regression*, Belmont, CA: Lifetime Learning Publications.

Atiqullah, M. (1962). The Estimation of Residual Variance in Quadratically Balanced Least-Squares Problems and the Robustness of the F-Test, *Biometrika*, **49**, 83–91.

Atkinson, A. C. and W. G. Hunter (1968). The Design of Experiments for Parameter Estimation, *Technometrics*, **10**, 271–289.

Bard, Y. (1974). *Nonlinear Parameter Estimation*, New York: Academic Press.

Becker, N. G. (1978). Models and Designs for Experiments with Mixtures, *Austral. J. Statist.*, **20**, 195–208.

Belsley, D. A., E. Kuh, and R. E. Welsch (1980). *Regression Diagnostics: Identifying Influential Data and Sources of Collinearity*, New York: John Wiley.

Biles, W. E. (1975). A Response Surface Method for Experimental Optimization of Multi-Response Processes, *Ind. Eng. Chem.*, *Process Des. Dev.*, **14**, 152–158.

Bishop, T., B. Petersen, and D. Trayser (1982). Another Look at the Statistician's Role in Experimental Planning and Design, *Amer. Statist.*, **36**, 387–389.

Bose, R. C. and N. R. Draper (1959). Second Order Rotatable Designs in Three Dimensions, *Ann. Math. Statist.*, **30**, 1097–1112.

Box, G. E. P. (1952). Multi-Factor Designs of First Order, *Biometrika*, **39**, 49–57.

Box, G. E. P. (1954). The Exploration and Exploitation of Response Surfaces: Some General Consideration and Examples, *Biometrics*, **10**, 16–60.

Box, G. E. P. (1982). Choice of Response Surface Design and Alphabetic Optimality, *Utilitas Mathematica*, **21B**, 11–55.

Box, G. E. P. (1985). Discussion of Off-Line Quality Control, Parameter Design, and the Taguchi Methods, *J. Quality Tech.*, **17**, 198–206.

Box, G. E. P. (1988). Signal to Noise Ratios, Performance Criteria, and Transformations (with discussion), *Technometrics*, **30**, 1–40.

Box, G. E. P. (1992). Sequential Experimentation and Sequential Assembly of Designs, *Quality Eng.*, **5**, 321–330.

Box, G. E. P. and D. W. Behnken (1960). Some New Three-Level Designs for the Study of Quantitative Variables, *Technometrics*, **2**, 455–475.

Box, G. E. P. and D. R. Cox (1964). An Analysis of Transformations (with discussion), *J. Roy. Statist. Soc.*, **B26**, 211–252.

Box, G. E. P. and N. R. Draper (1959). A Basis for the Selection of a Response Surface Design, *J. Amer. Statist. Assoc.*, **54**, 622–654.

Box, G. E. P. and N. R. Draper (1963). The Choice of a Second Order Rotatable Design, *Biometrika*, **50**, 335–352.

Box, G. E. P. and N. R. Draper (1965). The Bayesian Estimation of Common Parameters from Several Responses, *Biometrika*, **52**, 355–365.

Box, G. E. P. and N. R. Draper (1975). Robust Designs, *Biometrika*, **62**, 347–352.

Box, G. E. P. and N. R. Draper (1987). *Empirical Model-Building and Response Surfaces*, New York: John Wiley.

Box, G. E. P. and J. S. Hunter (1954). A Confidence Region for the Solution of a Set of Simultaneous Equations with an Application to Experimental Design, *Biometrika*, **41**, 190–199.

Box, G. E. P. and J. S. Hunter (1957). Multifactor Experimental Designs for Exploring Response Surfaces, *Ann. Math. Statist.*, **28**, 195–241.

Box, G. E. P. and J. S. Hunter (1961a). The 2^{k-p} Fractional Factorial Designs, Part I, *Technometrics*, **3**, 311–351.

Box, G. E. P. and J. S. Hunter (1961b). The 2^{k-p} Fractional Factorial Designs, Part II, *Technometrics*, **3**, 449–458.

Box, G. E. P. and W. G. Hunter (1965). Sequential Design of Experimenta for Nonlinear Models, *Proceedings of IBM Scientific Computing Symposium in Statistics*, 113–137.

Box, G. E. P. and S. Jones (1992). Designing Products that Are Robust to the Environment, *Total Quality Manag.*, **3**, 265–282.

Box, G. E. P. and H. L. Lucas (1959). Design of Experiments in Nonlinear Situations, *Biometrika*, **46**, 77–90.

Box, G. E. P. and G. S. Watson (1962). Robustness to Non-Normality of Regression Tests, *Biometrika*, **49**, 93–106.

Box, G. E. P. and K. B. Wilson (1951). On the Experimental Attainment of Optimum Conditions (with discussion), *J. Roy. Statist. Soc.*, **B13**, 1–45.

Box, G. E. P. and P. V. Youle (1955). The Exploration and Exploitation of Response Surfaces: An Example of the Link Between the Fitted Surface and the Basic Mechanism of the System, *Biometrics*, **11**, 287–323.

Box, G. E. P., S. Bisgaard, and C. Fung (1988). An Explanation and Critique of Taguchi's Contributions to Quality Engineering, *Quality Reliability Eng.*, **4**, 123–131.

Box, G. E. P., W. G. Hunter, and J. S. Hunter (1978). *Statistics for Experimenters: An Introduction to Design, Data Analysis, and Model Building*, New York: John Wiley.

Box, G. E. P., W. G. Hunter, J. F. MacGregor, and J. Erjavec (1973). Some Problems Associated with the Analysis of Multiresponse Data, *Technometrics*, **15**, 33–51.

Box, M. J. (1970). Some Experiences with a Nonlinear Experimental Design Criterion, *Technometrics*, **12**, 569–589.

Box, M. J. (1971). Bias in Nonlinear Estimation, *J. Roy. Statist. Soc.*, **B33**, 171–190.

Box, M. J. and N. R. Draper (1971). Factorial Designs, the $|X'X|$ Criterion and Some Related Matters, *Technometrics*, **13**, 731–742.

Box, M. J. and N. R. Draper (1974). On Minimum-Point Second-Order Designs, *Technometrics*, **16**, 613–616.

Box, M. J., N. R. Draper, and W. G. Hunter (1970). Missing Values in Multiresponse Nonlinear Model Fitting, *Technometrics*, **12**, 613–620.

Bradley, R. A. (1958). Determination of Optimum Operating Conditions by Experimental Methods: Part I, Mathematics and Statistics Fundamental to the Fitting of Response Surfaces, *Industrial Quality Control*, **15**, 16–20.

Brown, K. M. (1969). A Quadratically Convergent Newton-Like Method Based Upon Gaussian Elimination, *Soc. Industrial Appl. Math. J. Numerical Analysis*, **6**, 560–569.

Carroll, C. W. (1961). The Created Response Surface Technique for Optimizing Nonlinear, Restrained Systems, *Operations Res.*, **9**, 169–185.

Chambers, J. M. (1973). Fitting Nonlinear Models: Numerical Techniques, *Biometrika*, **60**, 1–13.

Chatterjee, S. and B. Price (1977). *Regression Analysis by Example*, New York: John Wiley.

Cochran, W. G. (1973). Experiments for Nonlinear Functions (R. A. Fisher Memorial Lecture), *J. Amer. Statist. Assoc.*, **68**, 771–781.

Cochran, W. G. and G. M. Cox (1957). *Experimental Designs*, 2nd ed., New York: John Wiley.

Coleman, D. E. and D. C. Montgomery (1993). A Systematic Approach to Planning for a Designed Industrial Experiment (with discussion), *Technometrics*, **35**, 1–27.

Conlon, M. (1985). Controlled Random Search Procedure for Function Minimization, Technical Report No. 236, Dept. of Statistics, University of Florida, Gainesville.

Conlon, M. and A. I. Khuri (1992). Multiple Response Optimization, Technical Report, Dept. of Statistics, University of Florida, Gainesville.

Connor, W. S. and M. Zelen (1959). Fractional Factorial Experimental Designs for Factors at Three Levels. National Bureau of Standards, Washington, DC, Applied Mathematics Series, No. 54.

Cornell, J. A. (1973). Experiments with Mixtures: A Review, *Technometrics*, **15**, 437–455.

Cornell, J. A. (1986). A Comparison Between Two Ten-Point Designs for Studying Three-Component Mixture Systems, *J. Quality Tech.*, **18**, 1–15.

Cornell, J. A. (1990a). *Experiments with Mixtures, Designs, Models, and the Analysis of Mixture Data*, 2nd ed., New York: John Wiley.

Cornell, J. A. (1990b). How to Apply Response Surface Methodology, *The ASQC Basic References in Quality Control: Statistical Techniques*, **8**, rev. ed., Milwaukee, WI: ASQC.

Cornell, J. A. (1990c). *How to Run Mixture Experiments for Product Quality*, Vol. 5, rev. ed., Milwaukee, WI: ASQC.

Cornell, J. A. and J. W. Gorman (1984). Fractional Design Plans for Process Variables in Mixture Experiments, *J. Quality Tech.*, **16**, 20–38.

Cornell, J. A. and D. C. Montgomery (1996). Interaction Models as Alternatives to Low-Order Polynomials, *J. Quality Tech.*, **28**, 163–176.

Cox, D. R. (1971). A Note on Polynomial Response Functions for Mixtures, *Biometrika*, **58**, 155–159.

Crosier, R. B. (1984). Mixture Experiments: Geometry and Pseudocomponents, *Technometrics*, **26**, 209–216.

Daniel, C. (1959). Use of Half-Normal Plots in Interpreting Factorial Two-Level Experiments, *Technometrics*, **1**, 311–341.

Daniel, C. and F. S. Wood (1980). *Fitting Equations to Data*, 2nd ed., New York: John Wiley.

Davies, O. L. (1956). *The Design and Analysis of Industrial Experiments*, 2nd ed., London: Oliver and Boyd.

Davies, O. L. and W. A. Hay (1950). The Construction and Uses of Fractional Factorial Designs in Industrial Research, *Biometrics*, **6**, 233–249.

DeBaun, R. M. (1959). Response Surface Designs for Three Factors at Three Levels, *Technometrics*, **1**, 1–8.

DeBoor, C. (1978). *A Practical Guide to Splines*, New York: Springer-Verlag.

Derringer, G. and R. Suich (1980). Simultaneous Optimization of Several Response Variables, *J. Quality Tech.*, **12**, 214–219.

Doehlert, D. H. (1970). Uniform Shell Designs, *J. Roy. Statist. Soc.*, **C19**, 231–239.

Doehlert, D. H. and V. L. Klee (1972). Experimental Designs Through Level Reduction of the *d*-Dimensional Cuboctahedron, *Discrete Math.*, **2**, 309–334.

Draper, N. R. (1960). Second Order Rotatable Designs in Four or More Dimensions, *Ann. Math. Statist.*, **31**, 23–33.

Draper, N. R. (1961). Missing Values in Response Surface Designs, *Technometrics*, **3**, 389–398.

Draper, N. R. (1963). "Ridge Analysis" of Response Surfaces, *Technometrics*, **5**, 469–479.

Draper, N. R. (1982). Center Points in Second-Order Response Surface Designs, *Technometrics*, **24**, 127–133.

Draper, N. R. and I. Guttman (1988). An Index of Rotatability, *Technometrics*, **30**, 105–111.

Draper, N. R. and A. M. Herzberg (1971). On Lack of Fit, *Technometrics*, **13**, 231–241.

Draper, N. R. and W. G. Hunter (1966). Design of Experiments for Parameter Estimation in Multiresponse Situations, *Biometrika*, **53**, 525–533.

Draper, N. R. and W. E. Lawrence (1965). Designs Which Minimize Model Inadequacies: Cuboidal Regions of Interest, *Biometrika*, **52**, 111–118.

Draper, N. R. and F. Pukelsheim (1990). Another Look at Rotatability, *Technometrics*, **32**, 195–202.

Draper, N. R. and H. Smith (1981). *Applied Regression Analysis*, 2nd ed., New York: John Wiley.

Dubova, I. S. and V. V. Fedorov (1972). Tables of Optimum Designs II (Saturated *D*-Optimal Designs on a Cube). Preprint No. 40, in Russian. Issued by Interfaculty Laboratory of Statistical Methods, Moscow University.

Dykstra, O. (1960). Partial Duplication of Response Surface Designs, *Technometrics*, **2**, 185–195.

Fedorov, V. V. (1972). *Theory of Optimal Experiments*, New York: Academic Press.

Finney, D. J. (1945). The Fractional Replication of Factorial Arrangements, *Ann. Eugenics*, **12**, 291–301.

Fletcher, R. and M. J. D. Powell (1963). A Rapidly Convergent Descent Method for Minimization, *Computer J.*, **6**, 163–168.

Fonseca, R., E. Chornet, C. Roy, and M. Grandbois (1980). Statistical Study on the Batch Conversion of Estevan Lignite into Oils Using Carbon Monoxide and Hydrogen Mixtures, *Canad. J. Chem. Eng.*, **58**, 458–465.

Foster, F. G. (1957). Upper Percentage Points of the Generalized Beta Distribution II, *Biometrika*, **44**, 441–453.

Fuguitt, R. E. and J. E. Hawkins (1947). Rate of Thermal Isomerization of α-Pinene in the Liquid Phase, *J. Amer. Chem. Soc.*, **69**, 319–322.

Gabriel, K. R. (1971). The Biplot-Graphic Display of Matrices with Applications to Principal Component Analysis, *Biometrika*, **58**, 453–467.

Gabriel, K. R. (1981). Biplot Display of Multivariate Matrices for Inspection of Data and Diagnosis. In: *Interpreting Multivariate Data* (V. Barnett, ed.), New York: John Wiley, pp. 147–173.

Gallant, A. R. (1975). Testing a Subset of the Parameters of a Nonlinear Regression Model, *J. Amer. Statist. Assoc.*, **70**, 927–932.

Goel, B. S. (1980). Designs for Restricted Exploration in Mixture Experiments, *Biometrical J.*, **22**, 351–358.

Graybill, F. A. (1961). *An Introduction to Linear Statistical Models*, Vol. 1, New York: McGraw-Hill.

Gunst, R. F. and R. L. Mason (1980). *Regression Analysis and Its Application: A Data-Oriented Approach*, New York: Marcel Dekker.

Hahn, G. J. (1977). Some Things Engineers Should Know About Experimental Design, *J. Quality Tech.*, **9**, 13–20.

Hahn, G. J. (1984). Experimental Design in a Complex World, *Technometrics*, **26**, 19–31.

Harrington, E. C. (1965). The Desirability Function, *Industrial Quality Control*, **21**, 494–498.

Hartley, H. O. (1959). Smallest Composite Designs for Quadratic Response Surfaces, *Biometrics*, **15**, 611–624.

Hartley, H. O. (1961). The Modified Gauss-Newton Method for the Fitting of Nonlinear Regression Functions by Least Squares, *Technometrics*, **3**, 269–280.

Hartley, H. O. (1964). Exact Confidence Regions for the Parameters in Nonlinear Regression Laws, *Biometrika*, **51**, 347–353.

Herzberg, A. M. (1966). Cylindrically Rotatable Designs, *Ann. Math. Statist.*, **37**, 242–247.

Herzberg, A. M. and D. F. Andrews (1976). Some Considerations in the Optimal Design of Experiments in Non-Optimal Situations, *J. Roy. Statist. Soc.*, **B38**, 284–289.

Hicks, C. R. (1982). *Fundamental Concepts of Design of Experiments*, 3rd ed., New York: Holt, Rinehart and Winston.

Hill, P. D. H. (1980). *D*-Optimal Designs for Partially Nonlinear Regression Models, *Technometrics*, **22**, 275–276.

Hill, W. J. and W. G. Hunter (1966). A Review of Response Surface Methodology: A Literature Survey, *Technometrics*, **8**, 571–590.

Hill, W. J. and W. G. Hunter (1974). Design of Experiments for Subsets of Parameters, *Technometrics*, **16**, 425–434.

Hoadley, A. and J. Kettenring (1990). Communications Between Statisticians and Engineers/Physical Scientists, (with commentary), *Technometrics*, **32**, 243–274.

Hoke, A. T. (1974). Economical Second-Order Designs Based on Irregular Fractions of the 3^n Factorial, *Technometrics*, **16**, 375–384.

Hunter, J. S. (1958). Determination of Optimum Operating Conditions by Experimental Methods: Part II-1, Models and Methods, *Industrial Quality Control*, **15**, 16–24.

Hunter, J. S. (1959a). Part II-2, Models and Methods, *Industrial Quality Control*, **16**, 7–15.

Hunter, J. S. (1959b). Part II-3, Models and Methods, *Industrial Quality Control*, **16**, 6–14.

Hunter, J. S. (1985). Statistical Design Applied to Product Design, *J. Quality Tech.*, **17**, 210–221.

Jackson, J. E. (1991). *Biplots: A User's Guide to Principal Components*, New York: John Wiley.

Jennrich, R. I. (1969). Asymptotic Properties of Nonlinear Least Squares Estimators, *Ann. Math. Statist.*, **40**, 633–643.

Jensen, D. R., L. S. Mayer, and R. H. Myers (1975). Optimal Designs and Large-Sample Tests for Linear Hypotheses, *Biometrika*, **62**, 71–78.

John, P. W. M. (1961). The Three-Quarter Replicates of 2^4 and 2^5 Designs, *Biometrics*, **17**, 319–321.

John, P. W. M. (1962). Three Quarter Replicates of 2^n Designs, *Biometrics*, **18**, 172–184.

Kackar, R. N. (1985). Off-Line Quality Control, Parameter Design, and the Taguchi Method (with discussion), *J. Quality Tech.*, **17**, 176–188.

Kackar, R. N. (1986). Taguchi's Quality Philosophy: Analysis and Commentary, *Quality Progress*, 21–29.

Kackar, R. N. and D. A. Harville (1981). Unbiasedness of Two-Stage Estimation and Prediction Procedures for Mixed Linear Models, *Commun. Statist. Theor. Meth.*, **A10**, 1249–1261.

Kakwani, N. C. (1967). The Unbiasedness of Zellner's Seemingly Unrelated Regression Equations Estimators, *J. Amer. Statist. Assoc.*, **62**, 141–142.

Kaplan, W. (1952). *Advanced Calculus*, Reading, MA: Addison-Wesley.

Karson, M. J. (1970). Design Criterion for Minimum Bias Estimation of Response Surfaces, *J. Amer. Statist. Assoc.*, **65**, 1565–1572.

Karson, M. J., A. R. Manson, and R. J. Hader (1969). Minimum Bias Estimation and Experimental Design for Response Surfaces, *Technometrics*, **11**, 461–475.

Kempthorne, O. (1952). *The Design and Analysis of Experiments*, New York: John Wiley.

Kennedy, W. J. and J. E. Gentle (1980). *Statistical Computing*, New York: Marcel Dekker.

Khuri, A. I. (1984). A Note on *D*-Optimal Designs for Partially Nonlinear Regression Models, *Technometrics*, **26**, 59–61.

Khuri, A. I. (1985). A Test for Lack of Fit of a Linear Multiresponse Model, *Technometrics*, **27**, 213–218.

Khuri, A. I. (1986). Exact Tests for the Comparison of Correlated Response Models with an Unknown Dispersion Matrix, *Technometrics*, **28**, 347–357.

Khuri, A. I. (1988). A Measure of Rotatability for Response Surface Designs, *Technometrics*, **30**, 95–104.

Khuri, A. I. (1990a). The Effect of Response Scaling on the Detection of Linear Dependencies Among Multiresponse Data, *Metrika*, **37**, 217–231.

Khuri, A. I. (1990b). Multiresponse Rotatability, *J. Statist. Plan. Inf.*, **25**, 1–6.

Khuri, A. I. (1992). Response Surface Models with Random Block Effects, *Technometrics*, **34**, 26–37.

Khuri, A. I. (1993a). *Advanced Calculus with Applications in Statistics*, New York: John Wiley.

Khuri, A. I. (1993b). Response Surface Methodology Within the Framework of GLM, *J. Combinatorics, Information and System Sciences*, **18**, 193–202.

Khuri, A. I. (1994). Effect of Blocking on the Estimation of a Response Surface, *J. Appl. Statist.*, **21**, 305–316.

Khuri, A. I. (1996). Response Surface Models with Mixed Effects, *J. Quality Tech.*, **28**, 177–186.

Khuri, A. I. and M. Conlon (1981). Simultaneous Optimization of Multiple Responses Represented by Polynomial Regression Functions, *Technometrics*, **23**, 363–375.

Khuri, A. I. and J. A. Cornell (1981). Lack of Fit Revisited, Technical Report No. 167, Dept. of Statistics, University of Florida, Gainesville.

Khuri, A. I. and R. H. Myers (1979). Modified Ridge Analysis, *Technometrics*, **21**, 467–473.

Khuri, A. I. and R. H. Myers (1981). Design Related Robustness of Tests in Regression Models, *Commun. Statist. Theor. Meth.*, **A10**, 223–235.

Khuri, A. I. and H. Sahai (1985). Variance Components Analysis: A Selective Literature Survey, *Inter. Statist. Rev.*, **53**, 279–300.

Khuri, A. I. and E. Valeroso (1995). Multiresponse Surface Models with Block Effects, Technical Report No. 486, Dept. of Statistics, University of Florida, Gainesville.

Kiefer, J. (1958). On the Nonrandomized Optimality and the Randomized Nonoptimality of Symmetrical Designs, *Ann. Math. Statist.*, **29**, 675–699.

Kiefer, J. (1959). Optimum Experimental Designs (with discussion), *J. Roy. Statist. Soc.*, **B21**, 272–319.

Kiefer, J. (1960). Optimum Experimental Design V, with Applications to Systematic and Rotatable Designs, *Proc. Fourth Berkeley Symposium on Mathematical Statistics and Probability*, **1**, 381–405.

Kiefer, J. (1961). Optimum Designs in Regression Problems II, *Ann. Math. Statist.*, **32**, 298–325.

Kiefer, J. (1962a). Two More Criteria Equivalent to *D*-Optimality of Designs, *Ann. Math. Statist.*, **33**, 792–796.

Kiefer, J. (1962b). An Extremum Result, *Canad. J. Math.*, **14**, 597–601.

Kiefer, J. and J. Wolfowitz (1960). The Equivalence of Two Extremum Problems, *Canad. J. Math.*, **12**, 363–366.

Kim, W. B. and N. R. Draper (1994). Choosing a Design for Straight Line Fits to Two Correlated Responses, *Statist. Sinica*, **4**, 275–280.

Kizer, O., C. Chavarie, C. Laguerie, and D. Cassimatis (1978). Quadratic Model of the Behavior of a Fluidized Bed Reactor: Catalytic Oxidation of Benzene to Maleic Anhydride, *Canad. J. Chem. Eng.*, **56**, 716–724.

Kupper, L. L. and E. F. Meydrech (1973). A New Approach to Mean Squared Error Estimation of Response Surfaces, *Biometrika*, **60**, 573–579.

Kupper, L. L. and E. F. Meydrech (1974). Experimental Design Considerations Based on a New Approach to Mean Square Error Estimation of Response Surfaces, *J. Amer. Statist. Assoc.*, **69**, 461–463.

Kurotori, I. S. (1966). Experiments with Mixtures of Components Having Lower Bounds, *Industrial Quality Control*, **22**, 592–596.

Lawton, W. H. and E. A. Sylvestre (1971). Elimination of Linear Parameters in Nonlinear Regression, *Technometrics*, **13**, 461–467.

Leon, R. V., A. C. Shoemaker, and R. N. Kackar (1987). Performance Measure Independent of Adjustment: An Explanation and Extension of Taguchi's Signal to Noise Ratio, *Technometrics*, **29**, 253–285.

Levy, M. S. and J. W. Neill (1990). Testing for Lack of Fit in Linear Multiresponse Models Based on Exact or Near Replicates, *Commun. Statist. Theor. Meth.*, **A19**, 1987–2002.

Lind, E. E., J. Goldin, and J. B. Hickman (1960). Fitting Yield and Cost Response Surfaces, *Chem. Eng. Progress*, **56**, 62–68.

Littell, R. C. and G. O. Mott (1974). Computer Assisted Design and Analysis of Response Surface Experiments in Agronomy, *Soil and Crop Science Society of Florida Proceedings*, **34**, 94–97.

Lucas, J. M. (1974). Optimum Composite Designs, *Technometrics*, **16**, 561–567.

Lucas, J. M. (1976). Which Response Surface Design Is Best, *Technometrics*, **18**, 411–417.

Lucas, J. M. (1977). Design Efficiencies for Varying Numbers of Center Points, *Biometrika*, **64**, 145–147.

Lucas, J. M. (1991). Achieving a Robust Process Using Response Surface Methodology, Technical Report, E. I. DuPont, Wilmington, DE.

Malinvaud, E. (1980). *Statistical Methods of Econometrics*, 3rd ed., Amsterdam: North-Holland.

Mardia, K. V. (1971). The Effect of Nonnormality on Some Multivariate Tests and Robustness to Nonnormality in the Linear Model, *Biometrika*, **58**, 105–121.

Margolin, B. H. (1969). Results of Factorial Designs of Resolution IV for the 2^n and $2^n 3^m$ Series, *Technometrics*, **11**, 431–444.

Marquardt, D. W. (1963). An Algorithm for Least Squares Estimation of Nonlinear Parameters, *J. Soc. Industrial Appl. Math.*, **11**, 431–441.

Martin, Y. C. and J. J. Hackbarth (1977). Examples of the Application of Nonlinear Regression Analysis to Chemical Data. In: *Chemometrics: Theory and Application, A Symposium Sponsored by the Division of Computers in Chemistry* (B. R. Kowalski, ed.), pp. 153–164.

McCullagh, P. and J. A. Nelder (1989). *Generalized Linear Models*, 2nd ed., London: Chapman and Hall.

McKee, B. and A. M. Kshirsagar (1982). Effect of Missing Plots in Some Response Surface Designs, *Commun. Statist. Theor. Math.*, **A11**, 1525–1549.

McLean, D. D., D. J. Pritchard, D. W. Bacon, and J. Downie (1979). Singularities in Multiresponse Modelling, *Technometrics*, **21**, 291–298.

McLean, R. A. and V. L. Anderson (1966). Extreme Vertices Design of Mixture Experiments, *Technometrics*, **8**, 447–454.

McLean, R. A. and V. L. Anderson (1984). *Applied Factorial and Fractional Designs*, New York: Marcel Dekker.

Mead, R. and D. J. Pike (1975). A Review of Response Surface Methodology from a Biometric Point of View, *Biometrics*, **31**, 803–851.

Mehta, J. S. and M. N. Das (1968). Asymmetric Rotatable Designs and Orthogonal Transformations, *Technometrics*, **10**, 313–322.

Milliken, G. A. and R. L. DeBruin (1978). A Procedure to Test Hypotheses for Nonlinear Models, *Commun. Statist. Theor. Meth.*, **A7**, 65–79.

Mitchell, T. J. (1974). An Algorithm for the Construction of D-Optimal Experimental Designs, *Technometrics*, **16**, 203–210.

Montgomery, D. C. (1991). *Design and Analysis of Experiments*, 3rd ed., New York: John Wiley.

Montgomery, D. C. and E. A. Peck (1992). *Introduction to Linear Regression Analysis*, 2nd ed., New York: John Wiley.

Morrison, D. F. (1976). *Multivariate Statistical Methods*, 2nd ed., New York: McGraw-Hill.

Myers, R. H. (1976). *Response Surface Methodology*, Blacksburg, VA: Author.

Myers, R. H. (1990). *Classical and Modern Regression with Applications*, 2nd ed., Boston, MA: PWS-Kent.

Myers, R. H. and W. H. Carter, Jr. (1973). Response Surface Techniques for Dual Response Systems, *Technometrics*, **15**, 301–317.

Myers, R. H. and A. I. Khuri (1979). A New Procedure for Steepest Ascent, *Commun. Statist. Theor. Meth.*, **A8**, 1359–1376.

Myers, R. H. and D. C. Montgomery (1995). *Response Surface Methodology*, New York: John Wiley.

Myers, R. H., A. I. Khuri, and W. H. Carter, Jr. (1989). Response Surface Methodology: 1966–1988, *Technometrics*, **31**, 137–157.

Myers, R. H., A. I. Khuri, and G. Vining (1992). Response Surface Alternatives to the Taguchi Robust Parameter Design Approach, *Amer. Statist.*, **46**, 131–139.

Myers, W. R., R. H. Myers, and W. H. Carter, Jr. (1994). Some Alphabetic Optimal Designs for the Logistic Regression Model, *J. Statist. Plan. Inf.*, **42**, 57–77.

Nelder, J. A. and R. Mead (1965). A Simplex Method for Function Minimization, *Computer J.*, **7**, 308–313.

Nigam, A. K., S. C. Gupta, and S. Gupta (1983). A New Algorithm for Extreme Vertices Designs for Linear Mixture Models, *Technometrics*, **25**, 367–371.

Olsson, D. M. (1974). A Sequential Simplex Program for Solving Minimization Problems, *J. Quality Tech.*, **6**, 53–57.

Olsson, D. M. and L. S. Nelson (1975). The Nelder-Mead Simplex Procedure for Function Minimization, *Technometrics*, **17**, 45–51.

Ott, R. L. and J. A. Cornell (1974). A Comparison of Methods Which Utilize the Integrated M.S.E. Criterion for Constructing Response Surface Designs, *Commun. Statist. Theor. Meth.*, **A3**, 1053–1068.

Park, S. H., J. H. Lim, and Y. Baba (1993). A Measure of Rotatability for Second Order Response Surface Designs, *Ann. Inst. Statist. Math.*, **45**, 655–664.

Pearson, E. S. (1931). Analysis of Variance in Cases of Non-Normal Variation, *Biometrika*, **23**, 114–133.

Peng, K. C. (1967). *The Design and Analysis of Scientific Experiments*, Reading, MA: Addison-Wesley.

Pesotchinsky, L. L. (1975). *D*-Optimum and Quasi *D*-Optimum Second-Order Designs on a Cube, *Biometrika*, **62**, 335–340.

Phadke, M. S., R. N. Kackar, D. V. Speeney, and M. J. Greico (1983). Off-Line Quality Control for Integrated Circuit Fabrication Using Experimental Design, *Bell System Tech. J.*, **62**, 1273–1309.

Piepel, G. F. (1982). Measuring Component Effects in Constrained Mixture Experiments, *Technometrics*, **24**, 29–39.

Piepel, G. F. (1990). *MIXSOFT Software for the Design and Analysis of Mixture Experiments, User's Guide*, Version 2.0, MIXSOFT—Mixture Experiment Software, Richland, WA.

Piepel, G. F. and J. A. Cornell (1987). Designs for Mixture-Amount Experiments, *J. Quality Tech.*, **19**, 11–28.

Plackett, R. L. and J. P. Burman (1946). The Design of Optimum Multifactorial Experiments, *Biometrika*, **33**, 305–325.

Powell, M. J. D. (1962). An Iterative Method for Finding Stationary Value of a Function of Several Variables, *Computer J.*, **5**, 147–151.

Powell, M. J. D. (1967). On the Maximum Errors of Polynomial Approximations Defined by Interpolation and by Least Squares Criteria, *Computer J.*, **9**, 404–407.

Price, W. L. (1977). A Controlled Random Search Procedure for Global Optimization, *Computer J.*, **20**, 367–370.

Ralston, M. L. and R. I. Jennrich (1978). DUD, A Derivative-Free Algorithm for Nonlinear Least Squares, *Technometrics*, **20**, 7–14.

Ratkoe, B. L., A. Hedayat, and W. T. Federer (1981). *Factorial Designs*, New York: John Wiley.

Ratkowsky, D. A. (1983). *Nonlinear Regression Modeling*, New York: Marcel Dekker.

Richert, S. H., C. V. Morr, and C. M. Cooney (1974). Effect of Heat and Other Factors Upon Foaming Properties of Whey Protein Concentrates, *J. Food Sci.*, **39**, 42–48.

Rizvi, S. S. H. (1994). *Supercritical Fluid Processing of Food and Biomaterials*, Blackie Academic & Professional, New York: Chapman & Hall.

Roquemore, K. G. (1976). Hybrid Designs for Quadratic Response Surfaces, *Technometrics*, **18**, 419–423.

Rosenbrock, H. (1960). An Automatic Method for Finding the Greatest or Least Value of a Function, *Computer J.*, **3**, 175–184.

Roy, S. N., R. Gnanadesikan, and J. N. Srivastava (1971). *Analysis and Design of Certain Quantitative Multiresponse Experiments*, Oxford: Pergamon Press.

SAS Institute, Inc. (1989). *SAS/STAT User's Guide*, Vol. 2, Version 6, 4th ed., Cary, NC: Author.

SAS Institute, Inc. (1990). SAS/IML Software, Version 6, 1st ed., Cary, NC: Author.

SAS Institute, Inc. (1992). *SAS Technical Report P-229, SAS/STAT Software* (Release 6.07), Cary, NC: Author.

Scheffé, H. (1958). Experiments with Mixtures, *J. Roy. Statist. Soc.*, **B20**, 344–360.

Scheffé, H. (1963). The Simplex-Centroid Design for Experiments with Mixtures, *J. Roy. Statist. Soc.*, **B25**, 235–263.

Schmidt, R. H., B. L. Illingworth, J. C. Deng, and J. A. Cornell (1979). Multiple Regression and Response Surface Analysis of the Effects of Calcium Chloride and Cysteine on Heat-Induced Whey Protein Gelation, *J. Agric. Food Chem.*, **27**, 529–532.

Searle, S. R. (1971). *Linear Models*, New York: John Wiley.

Searle, S. R. (1982). *Matrix Algebra Useful for Statistics*, New York: John Wiley.

Searle, S. R., G. Casella, and C. E. McCulloch (1992). *Variance Components*, New York: John Wiley.

Seber, G. A. F. (1977). *Linear Regression Analysis*, New York: John Wiley.

Seber, G. A. F. (1984). *Multivariate Observations*, New York: John Wiley.

Shelton, J. T., A. I. Khuri, and J. A. Cornell (1983). Selecting Check Points for Testing Lack of Fit in Response Surface Models, *Technometrics*, **25**, 357–365.

Shoemaker, A. C., K. Tsui, and C. F. J. Wu (1991). Economical Experimentation Methods for Robust Design, *Technometrics*, **33**, 415–427.

Smith, W. F. Jr. and J. A. Cornell (1993). Biplot Displays for Looking at Multiple Response Data in Mixture Experiments, *Technometrics*, **35**, 337–350.

Snee, R. D. (1971). Design and Analysis of Mixture Experiments, *J. Quality Tech.*, **3**, 159–169.

Snee, R. D. (1973). Some Aspects of Nonorthogonal Data Analysis: Part I. Developing Prediction Equations, *J. Quality Tech.*, **5**, 67–79.

Snee, R. D. and D. W. Marquardt (1974). Extreme Vertices Designs for Linear Mixture Models, *Technometrics*, **16**, 399–408.

Spendley, W., G. R. Hext, and F. R. Himsworth (1962). Sequential Application of Simplex Designs in Optimization and Evolutionary Operation, *Technometrics*, **4**, 441–461.

Stablein, D. M., W. H. Carter, Jr., and G. L. Wampler (1983). Confidence Regions for Constrained Optima in Response-Surface Experiments, *Biometrics*, **39**, 759–763.

STAT-EASE, Inc. (1994). *DESIGN-EXPERT*, Version 4.0.6, Minneapolis, MN: Author.

Steinberg, D. M. and W. G. Hunter (1984). Experimental Design: Review and Comment (with discussion and response), *Technometrics*, **26**, 71–130.

Stewart, G. W. (1986). Collinearity, Scaling, and Rounding Error. In: *Proceedings of the Seventeenth Symposium on the Interface, Computer Science and Statistics* (D. M. Allen, ed.), Amsterdam: Elsevier Science Publishers, pp. 193–198.

Stewart, W. E. and J. P. Sorensen (1976). Sensitivity and Regression of Multicomponent Reactor Models. In: *Fourth International Symposium on Chemical Reaction Engineering*, Frankfurt: DECHEMA, I-12–I-20.

Stewart, W. E. and J. P. Sorensen (1981). Bayesian Estimation of Common Parameters from Multiresponse Data with Missing Observations, *Technometrics*, **23**, 131–141.

Taguchi, G. (1978). Off-Line and On-Line Quality Control Systems, *Proc. International Conference on Quality*, Tokyo, Japan.

Taguchi, G. (1986). *Introduction to Quality Engineering: Designing Quality into Products and Processes*, White Plains, NY: Kraus International Publications.

Taguchi, G. and Y. I. Wu (1980). Introduction to Off-Line Quality Control, Central Japan Quality Control Assoc., Nogoya, Japan (available from the American Supplier Institute, Dearborn, MI).

Thompson, W. O. (1973). Secondary Criterion in the Selection of Minimum Bias Designs in Two Variables, *Technometrics*, **15**, 319–328.

Thompson, W. O. and R. H. Myers (1968). Response Surface Designs for Experiments with Mixtures, *Technometrics*, **10**, 739–756.

Tseo, C. L., J. C. Deng, J. A. Cornell, A. I. Khuri, and R. H. Schmidt (1983). Effect of Washing Treatment on Quality of Minced Mullet Flesh, *J. Food Sci.*, **48**, 163–167.

Van der Sluis, A. (1969). Condition Numbers and Equilibrium of Matrices, *Numerische Mathematik*, **14**, 14–23.

Vining, G. G. and R. H. Myers (1990). Combining Taguchi and Response Surface Philosophies: A Dual Response Approach, *J. Quality Tech.*, **22**, 38–45.

Vohnout, K. and C. Jimenez (1975). Supplemental By-Product Feeds in Pasture-Livestock Feeding Systems in the Tropics. In: *Tropical Forages in Livestock Production Systems*, ASA Special Publication No. 24 (E. C. Doll and G. O. Mott, eds.), Madison, WI: American Society of Agronomy, pp. 71–82.

Vuchkov, I. N. and E. B. Solakov (1980). The Influence of Experimental Design on Robustness to Nonnormality of the F Test in Regression Analysis, *Biometrika*, **67**, 489–492.

Weisberg, S. (1980). *Applied Linear Regression*, New York: John Wiley.

Welch, W. J. (1985). ACED: Algorithms for the Construction of Experimental Designs, *Amer. Statist.*, **39**, 146.

Welch, W. J., T. K. Yu, S. Kang, and J. Sacks (1990). Computer Experiments for Quality Control by Parameter Design, *J. Quality Tech.*, **22**, 15–22.

Wijesinha, M. C. (1984). Design of Experiments for Multiresponse Models, Unpublished Ph.D. Thesis, Dept. of Statistics, University of Florida, Gainesville.

Wijesinha, M. C. and A. I. Khuri (1987a). The Sequential Generation of Multiresponse D-Optimal Designs When the Variance-Covariance Matrix Is Not Known, *Commun. Statist. Simul. Comput.*, **B16**, 239–259.

Wijesinha, M. C. and A. I. Khuri (1987b). Construction of Optimal Designs to Increase the Power of the Multiresponse Lack of Fit Test, *J. Statist. Plan. Inf.*, **16**, 179–192.

Wijesinha, M. C. and A. I. Khuri (1991). Robust Designs for First-Order Multiple Design Multivariate Models, *Commun. Statist. Theor. Meth.*, **A20**, 2987–2999.

Wynn, H. P. (1970). The Sequential Generation of *D*-Optimum Experimental Designs, *Ann. Math. Statist.*, **41**, 1655–1664.

Young, D. M. and R. T. Gregory (1972). *A Survey of Numerical Mathematics*, Vol. 1, Reading, MA: Addison-Wesley.

Zellner, A. (1962). An Efficient Method of Estimating Seemingly Unrelated Regressions and Tests for Aggregation Bias, *J. Amer. Statist. Assoc.*, **57**, 348–368.

Solutions to Selected Exercises

CHAPTER 2

2.1 (a) $E(\mathbf{b}) = (\mathbf{X'X})^{-1}\mathbf{X'}E[\mathbf{Y}]$
$= \boldsymbol{\beta} + (\mathbf{X'X})^{-1}\mathbf{X'X_2}\boldsymbol{\beta_2}$

$\mathrm{Var}(\mathbf{b}) = (\mathbf{X'X})^{-1}\sigma^2$

(b) $E(s^2) = \sigma^2 + \dfrac{\boldsymbol{\beta_2'}\mathbf{X_2'}[\mathbf{I} - \mathbf{X(X'X)^{-1}X'}]\mathbf{X_2}\boldsymbol{\beta_2}}{N - p}$

2.2 (a) $\hat{Y}(X) = 70.03 - 0.27X$

(b)

Source	d.f.	S.S.	M.S.	F
Regression	1	205.2014	205.2014	$41.9 > F_{0.05,1,10} = 4.96$
Residual	10	49.0011	4.9001	Reject $H_0: \beta_1 = 0$
Total	11	254.2025		

(c) $\hat{Y}(225) = 70.03 - 0.27(225) = 9.28$

$\widehat{\mathrm{Var}}[\hat{Y}(225)] = 4.9001\left[\dfrac{1}{12} + \dfrac{(5)^2}{2800}\right] = 0.452$

$\therefore$ A 95% confidence interval for $\mu_{Y|225}$ is
$9.28 \pm 2.228 \sqrt{0.452} \Rightarrow (7.78, 10.78)$

(d) Pure Error S.S. = 1.7067 with 7 d.f.

Lack of Fit S.S. = 49.0011 − 1.7067 = 47.2944 with 3 d.f.

$$F = \frac{47.2944/3}{1.7067/7} = 64.66 > F_{0.05,3,7} = 4.35 \qquad \text{Reject LOF is zero and}$$
fit a quadratic model.

Quadratic model: $\hat{Y}(X) = 624.41 - 5.33X + 0.0115X^2$

$$F = \frac{\text{Lack of Fit M.S.}}{\text{Pure Error M.S.}} = 2.81 < F_{0.10,2,7} = 3.26 \qquad \text{Do not reject.}$$

(e)   With the quadratic model, % impurities is minimized at $X = 231.7°F$, the minimum being approximately $\hat{Y} = 6.2\%$.

2.3   (a)   To test $H_0: \beta_1 = 0$, $F = 33.89 > F_{0.01,1,8} = 11.26$ Reject H_0
To test $H_0: \beta_2 = 0$, $F = 1.01 < F_{0.10,1,8} = 3.46$ Do not reject H_0
Thus V_1 is more important than V_2 by itself.

(b)   To test $H_0: \beta_2 = 0$ given $\beta_1 V_1$ is in the model,

$$F = \frac{\text{SSR}(\beta_2 | \beta_1, \beta_0)/1}{\text{Residual M.S.}} = \frac{0.5132}{1.465} = 0.35 < F_{0.10,1,7} = 3.59$$

Thus V_2 is not important given V_1 is in the model, or, $\mathbf{Y} = \mathbf{X}_3 \boldsymbol{\beta}_3 + \boldsymbol{\varepsilon}_3$ is not a better model than $\mathbf{Y} = \mathbf{X}_1 \boldsymbol{\beta}_1 + \boldsymbol{\varepsilon}_1$, but $\mathbf{Y} = \mathbf{X}_3 \boldsymbol{\beta}_3 + \boldsymbol{\varepsilon}_3$ is a better model than $\mathbf{Y} = \mathbf{X}_2 \boldsymbol{\beta}_2 + \boldsymbol{\varepsilon}_2$.

$$F = \frac{\text{SSR}(\beta_2, \beta_{12} | \beta_0, \beta_1)/2}{\text{Residual M.S.}} = \frac{7.8472/2}{2.92/6} = 15.07 > F_{0.01,2,6} = 10.92$$

Model $\mathbf{Y} = \mathbf{X}_4 \boldsymbol{\beta}_4 + \boldsymbol{\varepsilon}_4$ is a better model than $\mathbf{Y} = \mathbf{X}_1 \boldsymbol{\beta}_1 + \boldsymbol{\varepsilon}_1$

2.4   (a) $\mathbf{1}'\mathbf{r} = \mathbf{1}'(\mathbf{I} - \mathbf{R})\mathbf{Y}$ where $\mathbf{R} = \mathbf{X}(\mathbf{X}'\mathbf{X})^{-1}\mathbf{X}'$ and let $\mathbf{X} = [\mathbf{1}:\mathbf{X}_1]$.
To show $\mathbf{1}'\mathbf{r} = 0$, we need only to show that $\mathbf{1}'\mathbf{R} = \mathbf{1}'$. Now, since $\mathbf{X}'\mathbf{R} = \mathbf{X}'\mathbf{X}(\mathbf{X}'\mathbf{X})^{-1}\mathbf{X}' = \mathbf{X}'$ where $\mathbf{X}' = [\mathbf{1}:\mathbf{X}_1]'$, then

$$\begin{bmatrix} \mathbf{1}' \\ \mathbf{X}'_1 \end{bmatrix} \mathbf{R} = \begin{bmatrix} \mathbf{1}' \\ \mathbf{X}'_1 \end{bmatrix}, \text{ i.e., } \mathbf{1}'\mathbf{R} = \mathbf{1}'.$$

(b) $\mathbf{X}'\mathbf{r} = \mathbf{X}'(\mathbf{I} - \mathbf{R})\mathbf{Y} = (\mathbf{X}' - \mathbf{X}')\mathbf{Y} = \mathbf{0}$

2.5 $\mathbf{b} = (\mathbf{X}'\mathbf{X})^{-1}\mathbf{X}'\mathbf{Y} = \mathbf{R}\mathbf{Y}$ is a linear function of the Ys.

$$s^2 = \frac{1}{N - p} \mathbf{Y}'(\mathbf{I} - \mathbf{R})\mathbf{Y} \text{ is a quadratic function of the Ys.}$$

$\mathbf{b}$ and s^2 are independent since $\mathbf{R}(\mathbf{I} - \mathbf{R}) = (\mathbf{R} - \mathbf{R}\mathbf{R}) = (\mathbf{R} - \mathbf{R}) = \mathbf{0}$

2.6   (b)   yes, since $\hat{\beta}_1$ in $\mathbf{b}_1$ is equal to $\hat{\beta}_1$ in $\mathbf{b}_3$ and $\hat{\beta}_2$ in $\mathbf{b}_2$ is equal to $\hat{\beta}_2$ in $\mathbf{b}_3$.

(c)   Let $a_1 = \dfrac{A_1 - 1.5}{0.5}$ and $a_2 = \dfrac{A_2 - 1.5}{0.5}$. Then $\mathbf{X}'_3\mathbf{X}_3 = \text{diag}(8, 8, 4)$ and $\hat{Y}(\mathbf{a}) = 48.125 + 2.125a_1 + 1.5a_2$ where the estimates 2.125 and 1.5 are $\hat{\beta}_1/2$ and $\hat{\beta}_2/2$, respectively, in $\mathbf{b}_3 = (\hat{\beta}_0, \hat{\beta}_1, \hat{\beta}_2)'$, and $\mathbf{a} = (a_1, a_2)'$.

(d) $F = \dfrac{[\text{SSR}(\beta_1, \beta_2, \beta_{12}, \beta_{22}, \beta_{122}|\beta_0) - \text{SSR}(\beta_1, \beta_2|\beta_0)]/3}{[\mathbf{Y'Y} - \text{SSR}(\beta_1, \beta_2, \beta_{12}, \beta_{22}, \beta_{122}|\beta_0)]/2}$

$= \dfrac{[87.375 - 45.125]/3}{(91.375 - 87.375)/2} = 7.04 < F_{0.05,3,2} = 19.16$ Do not reject H_0

Would you not reject $H_0: \beta_{12} = \beta_{22} = \beta_{122} = 0$ when you found the fitted model $\hat{Y}_4$ possessed an $R^2 = 0.9562$ while the fitted model $\hat{Y}_2$ had an $R^2 = 0.4938$? Comment.

CHAPTER 3

3.1 (a) Consider the defining relation

$A_1A_2A_3A_4 = I$

This gives rise to the following one-half fraction and corresponding response values:

Treatment Combination	x_1	x_2	x_3	x_4	y
(1)	−1	−1	−1	−1	53.3
a_1a_2	1	1	−1	−1	78.9
a_1a_3	1	−1	1	−1	75.4
a_1a_4	1	−1	−1	1	67.5
a_2a_3	−1	1	1	−1	71.3
a_2a_4	−1	1	−1	1	72.8
a_3a_4	−1	−1	1	1	71.4
$a_1a_2a_3a_4$	1	1	1	1	81.7

$\hat{Y} = 71.538 + 4.338x_1 + 4.638x_2 + 3.413x_3 + 1.813x_4$

(b) $A_1 = A_2A_3A_4$
$A_2 = A_1A_3A_4$
$A_3 = A_1A_2A_4$
$A_4 = A_1A_2A_3$
$A_1A_2 = A_3A_4$
$A_1A_3 = A_2A_4$
$A_1A_4 = A_2A_3$

(c)

			Treatment Combination					
Effect	(1)	a_1a_2	a_1a_3	a_1a_4	a_2a_3	a_2a_4	a_3a_4	$a_1a_2a_3a_4$
A_1	−	+	+	+	−	−	−	+
A_2	−	+	−	−	+	+	−	+
A_3	−	−	+	−	+	−	+	+
A_4	−	−	−	+	−	+	+	+

$$\hat{A}_1 = 8.675 \qquad SS_{A_1} = 2\hat{A}_1^2 = 150.511$$

$$\hat{A}_2 = 9.275 \qquad SS_{A_2} = 2\hat{A}_2^2 = 172.051$$

$$\hat{A}_3 = 6.825 \qquad SS_{A_3} = 2\hat{A}_3^2 = 93.161$$

$$\hat{A}_4 = 3.625 \qquad SS_{A_4} = 2\hat{A}_4^2 = 26.281$$

(d)

Source	d.f.	SS
A_1	1	150.511
A_2	1	172.051
A_3	1	93.161
A_4	1	26.281
Residual	3	80.974

3.2 (a) $k = 4$, $N = 5$

$$a_i = \left[\frac{cN}{i(i + 1)} \right]^{1/2}, \qquad i = 1, 2, 3, 4$$

choosing $c = 1$, we get

$$a_1 = \left(\frac{5}{2} \right)^{1/2}, \; a_2 = \left(\frac{5}{6} \right)^{1/2}, \; a_3 = \left(\frac{5}{12} \right)^{1/2}, \; a_4 = \left(\frac{5}{20} \right)^{1/2}$$

x_1	x_2	x_3	x_4
$-a_1$	$-a_2$	$-a_3$	$-a_4$
a_1	$-a_2$	$-a_3$	$-a_4$
0	$2a_2$	$-a_3$	$-a_4$
0	0	$3a_3$	$-a_4$
0	0	0	$4a_4$

(b) $k = 5$, $N = 6$

$$a_i = \left[\frac{6}{i(i + 1)} \right]^{1/2}, \qquad i = 1, 2, \ldots, 5$$

x_1	x_2	x_3	x_4	x_5
$-a_1$	$-a_2$	$-a_3$	$-a_4$	$-a_5$
a_1	$-a_2$	$-a_3$	$-a_4$	$-a_5$
0	$2a_2$	$-a_3$	$-a_4$	$-a_5$
0	0	$3a_3$	$-a_4$	$-a_5$
0	0	0	$4a_4$	$-a_5$
0	0	0	0	$5a_5$

3.3 The first row of the design is

x_1	x_2	x_3	x_4	x_5	x_6	x_7	x_8	x_9	x_{10}	x_{11}
1	1	−1	1	1	1	−1	−1	−1	1	−1

The next ten rows are generated from this row by shifting it cyclically one place to the right ten times. A row of −1's is added as row 12.

3.4 (a) $MSE = \dfrac{1}{N - k - 1}\, \mathbf{Y'AY}$,

where $\mathbf{A} = \mathbf{I} - \mathbf{X(X'X)}^{-1}\mathbf{X'}$.

Since $E(\mathbf{Y}) = \boldsymbol{\eta} = \mathbf{X}\boldsymbol{\beta} + \mathbf{X_2}\boldsymbol{\beta_2}$, then

$$E(MSE) = \frac{1}{N - k - 1}\, [\boldsymbol{\eta'}\mathbf{A}\boldsymbol{\eta} + tr(\mathbf{A}\sigma^2)]$$

$$= \frac{1}{N - k - 1}\, [\boldsymbol{\beta_2'}\mathbf{X_2'}\mathbf{A}\mathbf{X_2}\boldsymbol{\beta_2} + (N - k - 1)\sigma^2]$$

$$= \sigma^2 + \frac{\boldsymbol{\beta_2'}\mathbf{X_2'}\mathbf{A}\mathbf{X_2}\boldsymbol{\beta_2}}{N - k - 1}$$

$$\text{Bias} = \frac{\boldsymbol{\beta_2'}\mathbf{X_2'}\mathbf{A}\mathbf{X_2}\boldsymbol{\beta_2}}{N - k - 1}$$

(b) (i) $E(\mathbf{b}) = \mathbf{(X'X)}^{-1}\mathbf{X'}\boldsymbol{\eta}$

$$= \boldsymbol{\beta} + \mathbf{(X'X)}^{-1}\mathbf{X'}\mathbf{X_2}\boldsymbol{\beta_2}$$

$$= \boldsymbol{\beta}$$

(ii) $\text{Bias} = \dfrac{\boldsymbol{\beta_2'}\mathbf{X_2'}\mathbf{X_2}\boldsymbol{\beta_2}}{N - k - 1} \geq \dfrac{\boldsymbol{\beta_2'}\mathbf{X_2'}\mathbf{A}\mathbf{X_2}\boldsymbol{\beta_2}}{N - k - 1}$, since $\mathbf{I} - \mathbf{A} = \mathbf{X(X'X)}^{-1}\mathbf{X'}$ is positive semidefinite.

Using Eq. 3.25,

$$E(SS_{LOF}) = E[(N - k - 1)MSE] - (n_0 - 1)\sigma^2$$

$$= (N - k - n_0)\sigma^2 + \boldsymbol{\beta_2'}\mathbf{X_2'}\mathbf{X_2}\boldsymbol{\beta_2}$$

$$\geq (N - k - n_0)\sigma^2 + \boldsymbol{\beta_2'}\mathbf{X_2'}\mathbf{A}\mathbf{X_2}\boldsymbol{\beta_2}$$

The right-hand side of the inequality is the expected value of SS_{LOF} in general, that is, when $\mathbf{X'X_2} \neq \mathbf{0}$. Hence, lack of fit is more easily detected when $\mathbf{X'X_2} = \mathbf{0}$.

3.5 The true mean response is

$$\boldsymbol{\eta} = \mathbf{X}\boldsymbol{\beta} + \mathbf{X_2}\boldsymbol{\beta_2}$$

where $\boldsymbol{\beta}_2$ consists of the parameters of the quadratic model excluding $\beta_0, \beta_1, \ldots, \beta_k$. Let $\mathbf{Y}$ be partitioned into $\mathbf{Y}_1$ and $\mathbf{Y}_0$, where $\mathbf{Y}_1$ is the vector of observations at the factorial points and $\mathbf{Y}_0$ is the vector of observations at the center. The matrices $\mathbf{X}$ and $\mathbf{X}_2$ are partitioned accordingly as $\mathbf{X} = [\mathbf{X}_1' : \mathbf{X}_0']'$, $\mathbf{X}_2 = [\mathbf{X}_{12}' : \mathbf{X}_{02}']'$. Note that $\mathbf{X}_{02} = \mathbf{0}$.

(a) $E(\bar{Y}_0) = \dfrac{1}{n_0} E(\mathbf{1}_{n_0}' \mathbf{Y}_0)$

$\qquad\quad = \dfrac{1}{n_0} \mathbf{1}_{n_0}' (\mathbf{X}_0 \boldsymbol{\beta} + \mathbf{X}_{02} \boldsymbol{\beta}_2)$

$\qquad\quad = \dfrac{1}{n_0} (n_0 \beta_0) = \beta_0$

(b) $E(\bar{Y}_1) = \dfrac{1}{n_1} E(\mathbf{1}_{n_1}' \mathbf{Y}_1)$, where $n_1 = N - n_0 = 2^k$. Hence,

$\qquad E(\bar{Y}_1) = \dfrac{1}{n_1} \mathbf{1}_{n_1}' (\mathbf{X}_1 \boldsymbol{\beta} + \mathbf{X}_{12} \boldsymbol{\beta}_2)$

$\qquad\qquad\quad = \dfrac{1}{n_1} \left(n_1 \beta_0 + n_1 \sum_{i=1}^{k} \beta_{ii} \right)$

$\qquad\qquad\quad = \beta_0 + \sum_{i=1}^{k} \beta_{ii}$

since the columns of $\mathbf{X}_1$, except for the first one, are orthogonal to $\mathbf{1}_{n_1}$; the first k columns of $\mathbf{X}_{12}$ are equal to $\mathbf{1}_{n_1}$, and the remaining columns are orthogonal to $\mathbf{1}_{n_1}$.

(c) $E(\bar{Y}_1 - \bar{Y}_0) = \sum_{i=1}^{k} \beta_{ii}$

$\qquad E(L_2) = \dfrac{n_0 n_1}{N} E(\bar{Y}_1 - \bar{Y}_0)^2$

$\qquad\qquad\quad = \dfrac{n_0 n_1}{N} \{ \mathrm{Var}(\bar{Y}_1 - \bar{Y}_0) + [E(\bar{Y}_1 - \bar{Y}_0)]^2 \}$

$\qquad\qquad\quad = \dfrac{n_0 n_1}{N} \left[\sigma^2 \left(\dfrac{1}{n_1} + \dfrac{1}{n_0} \right) + \left(\sum_{i=1}^{k} \beta_{ii} \right)^2 \right]$

$\qquad\qquad\quad = \sigma^2 + \dfrac{n_0 n_1}{N} \left(\sum_{i=1}^{k} \beta_{ii} \right)^2$

This agrees with formula (3.28) since in this case $c = n_1$.

3.6 (a) $SS_E = 496.285$ with 14 d.f.
$SS_{PE} = 7.087$ with 2 d.f.
$SS_{LOF} = 489.198$ with 12 d.f.

$$F = \frac{MS_{LOF}}{MS_{PE}} = 11.5046,\text{ which is significant at the 10\% level}$$

$(F_{0.10,12,2} = 9.41)$.

(b) $\bar{Y}_0 = 81.767$
$\bar{Y}_1 = 74.3125$

$$L_2 = \frac{3 \times 16}{19} (\bar{Y}_1 - \bar{Y}_0)^2 = 140.386\text{ with 1 d.f.}$$

$L_1 = SS_{LOF} - L_2 = 348.812$ with 11 d.f.

$$\frac{L_2}{MS_{PE}} = 39.618,\text{ which is significant at the 5\% level }(F_{0.05,1,2} = 18.51).$$

$$\frac{L_1/11}{MS_{PE}} = 8.949.\text{ This is not significant at the 10\% level }(F_{0.10,11,2} >$$

8.949).

3.7 (a) $\hat{Y}(x) = 49.0333 - 0.0333x_1 + 11.6625x_2 + 5.0875x_3 + 4.7375x_4$

$$+ 4.3125x_5 + 3.5625x_6 - 0.6875x_7$$

$$= b_0 + b_1x_1 + b_2x_2 + b_3x_3 + b_4x_4 + \cdots + b_7x_7$$

Eq. 3.24 $\hat{Y}(x) = 47.9833 + 1.4375x_1 + b_2x_2 + \cdots + b_7x_7$

$$= b_0' + b_1'x_1 + b_2x_2 + \cdots + b_7x_7$$

Only the estimates b_0 ($\neq b_0'$) and b_1 ($\neq b_1'$) (salted vs unsalted) differ in the two models.

(b) $E(MSE) = \sigma^2 + g(\beta_{12}, \ldots, \beta_{67}, \beta_{22} + \cdots + \beta_{77})$ so that $MSE = 43.64$ is an estimate of σ^2 plus model lack of fit.

(c) $\hat{Y}(x) = 56.075 - 0.850x_1 + 9.950x_2 + 3.250x_3 + 4.775x_4 - 1.750x_8$
$+ 1.825x_9$

Using the combined estimate, $\hat{\sigma}^2 = \dfrac{76.985 + 14.045}{3} = 30.34$, the estimated standard error of a coefficient estimate in either model Eq. 3.24 and (c) is $(30.34 \div 8)^{1/2} = 1.947$. Thus at the 0.10 level of significance, any coefficient estimate with an absolute value of $1.947t_{0.05,3} = 1.947(2.353) = 4.581$ or greater would be considered associated with an active factor. In Eq. 3.24, b_2, b_3, and b_4 would be declared nonzero and in (c) b_2 and b_4 would be declared nonzero.

(d)   Fitted to the $N = 20$ values, the first-order model in x_1 through x_4 is

$$\hat{Y}(\mathbf{x}) = 51.85 - 0.36x_1 + 10.81x_2 + 4.17x_3 + 4.76x_4$$
$$(-0.23) \quad (6.07) \quad\quad (2.34) \quad (2.64)$$

where the numbers in parentheses below the coefficient estimates are the calculated t values, based on $MSE = 50.69$ with 15 d.f. At the 0.05 level of significance, $t_{0.025,15} = 2.131$ so that b_2, b_3, and b_4 would be declared nonzero. Also, since one can show that $b_2 > b_3$ or b_4, factor X_2 or oil extraction pressure, is the most important factor among the three factors, oil extraction pressure, temperature, and time. Among the other five factors, separation pressure (X_5) and separation temperature (X_6) appear to be the most active based on the magnitudes of the estimates b_5 and b_6 relative to b_7, b_8, and b_9.

CHAPTER 4

4.1   (a)   Choose first a scale factor f such that $[ii] = 1$, $i = 1, 2$. Thus

$$\frac{f^2(6g^2 + 2h^2)}{13} = 1$$

$$f = \left(\frac{13}{9 + 2h^2}\right)^{1/2}, \text{ since } g = \sqrt{1.5}.$$

Hence,

$$[1122] = \frac{4f^4 g^4}{13}$$

$$[iiii] = \frac{f^4(6g^4 + 2h^4)}{13}, \qquad i = 1, 2$$

$$\kappa = \frac{6g^4 + 2h^4}{4g^4} = \frac{3}{2} + \frac{1}{2}\left(\frac{h}{g}\right)^4$$

$$= \frac{3}{2} + \frac{h^4}{2(1.5)^2}.$$

(c)   For the augmented design to be rotatable, $\kappa = 3$, that is,

$$\frac{3}{2} + \frac{h^4}{2(1.5)^2} = 3$$

$$h = \sqrt{1.5}\ 3^{1/4}$$
$$= 1.6118$$

4.2

	x_1	x_2	x_3	x_4	x_5
	c_1	c_2	0	0	0
	0	0	c_1	c_2	0
	0	c_1	0	0	c_2
	c_1	0	c_2	0	0
	0	0	0	c_1	c_2
	0^*	0^*	0^*	0^*	0^*
	...	...	...	...	...
	0	c_1	c_2	0	0
	c_1	0	0	c_2	0
	0	0	c_1	0	c_2
	c_1	0	0	0	c_2
	0	c_1	0	c_2	0
	0^*	0^*	0^*	0^*	0^*

where $c_1 = (-1, 1, -1, 1)'$, $c_2 = (-1, -1, 1, 1)'$, $0 = (0, 0, 0, 0)'$, $0^* = (0, 0, 0)'$.

This design is not rotatable since

$$[iijj] = \frac{4}{46}, \ i \neq j$$

$$[iiii] = \frac{16}{46}$$

Hence, $[iiii] \neq 3[iijj]$.

4.3   (a) Here, $F = 16$, hence,

$$\alpha = 16^{1/4} = 2$$

    (b) Using formula (4.23), we get

$$n_0 = 4\sqrt{F} + 4 - 2k$$

$$= 14$$

4.4   Scale the design settings so that $[ii] = 1, \ i = 1, 2$.

    (a) $\text{Bias}(b_1) = E(b_1) - \beta_1$

$$= \frac{N}{(4 + 2\alpha^2)^2} [(4 + 2\alpha^4)\beta_{111} + 4\beta_{122}]$$

       $\text{Bias}(b_2) = E(b_2) - \beta_2$

$$= \frac{N}{(4 + 2\alpha^2)^2} [(4 + 2\alpha^4)\beta_{222} + 4\beta_{112}]$$

    where $N = 8 + n_0$, b_i is the least-squares estimate of β_i $(i = 1, 2)$.

(b)   This follows directly from the bias expressions in part (a).

4.5   The corresponding ANOVA table is

Source	Degrees of Freedom	Sum of Squares
Linear terms	5	4323.7654
Quadratic terms	5	1238.9446
Crossproduct terms	10	883.3013
Lack of fit	6	205.7451
Pure error	6	33.8400

The test for lack of fit has the F-value:

$$F = \frac{205.7451/6}{33.84/6} = 6.08$$

The corresponding P-value is 0.0225. We conclude that there is a significant lack of fit associated with the fitted second-order model.

4.6   Augmenting the second-order model with the terms $\beta_{133}x_1x_3^2$ and $\beta_{233}x_2x_3^2$, the fitted model is

$$\hat{Y}(x) = 8.27 - 0.65x_1 - 5.11x_2 - 7.59x_3 - 0.29x_1^2 - 0.66x_2^2 + 3.21x_3^2$$
$$\qquad\ (0.42)\quad (0.42)\quad (0.29)\quad (0.43)\quad (0.43)\quad (0.43)$$
$$- 0.71x_1x_2 + 0.08x_1x_3 + 2.63x_2x_3 - 0.37x_1x_3^2 + 4.94x_2x_3^2$$
$$\ (0.42)\qquad (0.42)\qquad (0.42)\qquad (0.59)\qquad (0.59)$$

with $R^2 = 0.9657$, $R_A^2 = 0.9543$ and $MSE = 2.074$ with 33 d.f.

The above model fits the data better than Eq. 4.17 does since one or both of the coefficients β_{133} and β_{233} is/are different from zero, coupled with an increase in R^2 with the above model of 0.0736 and a decrease in the value of MSE from 6.157 to 2.074. If we remove from the above model those second- and third-degree terms whose coefficient estimates have a t-value less than 2.0 (where 2.0 is chosen arbitrarily), the resulting fitted model is

$$\hat{Y}(x) = 7.73 - 0.83x_1 - 5.11x_2 - 7.59x_3 + 3.28x_3^2 + 2.63x_2x_3 + 4.94x_2x_3^2$$
$$\qquad\ (0.30)\quad (0.42)\quad (0.30)\quad (0.44)\quad (0.42)\quad (0.60)$$

with $R^2 = 0.9595$, $R_A^2 = 0.9532$ and $MSE = 2.13$ with 38 d.f.

CHAPTER 5

5.1   (a)   A twisted plane

(b)   The slope of the twisted plane in (a) in the X_i direction is

$$\frac{\partial \eta}{\partial X_i} = \beta_i + \beta_{12}X_j, \qquad i, j = 1, 2, \ j \neq i, \text{ is a linear function of the value of } X_j.$$

5.2 The fitted model is

$$\hat{Y}(\mathbf{x}) = 61.6875 + 3.4375x_1 + 9.8125x_2 + 0.5125x_1x_2$$

with $MSE = 7.9588$ having 4 d.f. A test of $H_0:\beta_{12} = 0$ vs $H_a:\beta_{12} \neq 0$ at the 0.05 level of significance is

$$t = \frac{0.5125}{\sqrt{7.9588/8}} = 0.51 \text{ vs } t_{0.025,4} = 2.776$$

The square of the value of $t = 0.51$ is 0.2601 which is the value of the calculated F-statistic for testing lack of fit of the first-order model of Eq. 5.4.

5.3 Suppose n_1 observations are collected at the n points ($n_1 = rn$, where r is a positive integer) of a first-order orthogonal design and $n_0 > 1$ observations are collected at the center of the design. Define the averages of the n_1 and n_0 observations as $\bar{Y}_1$ and $\bar{Y}_0$, respectively. Next, consider the normal equations associated with fitting a second-order polynomial,

$$\eta = \beta_0 + \sum_{i=1}^{k} \beta_i x_i + \sum_{i<j}^{k}\sum \beta_{ij} x_i x_j + \sum_{i=1}^{k} \beta_{ii} x_i^2$$

to the data at the n design points and center of the design. If one selects the 1st and $(n + 1)^{st}$ normal equations and solves the two equations for β_0 and $\sum_{i=1}^{k} \beta_{ii}$, one obtains

$$b_0 = \bar{Y}_0 \quad \text{and} \quad \sum_{i=1}^{k} b_{ii} = \bar{Y}_1 - \bar{Y}_0$$

5.5 If $\mathbf{x}_0 = -\dfrac{\mathbf{B}^{-1}\mathbf{b}}{2}$, then $\hat{Y}_0 = b_0 + \mathbf{x}_0'\mathbf{b} + \mathbf{x}_0'\mathbf{B}\mathbf{x}_0$

$$= b_0 + \mathbf{x}_0'\mathbf{b} + \mathbf{x}_0'\mathbf{B}\left(-\frac{\mathbf{B}^{-1}\mathbf{b}}{2}\right)$$

$$= b_0 + \mathbf{x}_0'\mathbf{b}/2$$

5.6 (a) $\hat{Y}(\mathbf{x}) = 16.365 + 1.675x_1 + 2.765x_2 - 2.463x_1^2 - 1.931x_2^2 - 0.334x_1x_2$
with $R^2 = 0.9515$

Source	d.f.	S.S.	M.S.
Regression	5	204.705	40.9410
Residual	12	10.425	0.8687
Total	17	215.130	

(b) $\hat{Y}(x) = 16.369 + 2.369x_1 + 3.920x_2 - 4.938x_1^2 - 3.875x_2^2 - 0.670x_1x_2$
with $R^2 = 0.9522$

Source	d.f.	S.S.	M.S.
Regression	5	204.857	40.9714
Residual	12	10.273	0.8561
Total	17	215.130	

Note the difference in the Regression S.S. of the two ANOVA tables, owing to the values of x_1 and x_2.

(c) Pure Error S.S. = 4.7855 with 9 d.f.

In (a), $F_{LOF} = 3.54 < F_{0.05,3,9} = 3.86$

In (b), $F_{LOF} = 3.44 < F_{0.05,3,9} = 3.86$

Both models (a) and (b) do not possess significant lack of fit at the 0.05 level.

(d) For (a), $x_0 = \begin{bmatrix} 0.2932 \\ 0.6906 \end{bmatrix}$; while for (b), $x_0 = \begin{bmatrix} 0.2068 \\ 0.4879 \end{bmatrix}$. In each case, $\hat{Y}_0 = 17.57$.

(e) The surface has a maximum at the stationary point. The settings of fertilizers 1 and 2 that produce the maximum estimated yield are $F_1 = 95.26$ and $F_2 = 23.45$.

CHAPTER 6

6.2 The all-bias design is one for which the following holds: $A = \mu_{11}^{-1}\mu_{12}$ where $A = (X_1'X_1)^{-1}X_1'X_2$. This forces $\frac{1}{N}(X_1'X_1) = \mu_{11}$ and $\frac{1}{N}(X_1'X_2) = \mu_{12}$. When the region of interest is a sphere of radius ρ and $d_1 = 1$ and $d_2 = 2$, from Eqs. 6.27 and 6.28,

$$\mu_{11} = \begin{bmatrix} 1 & 0' \\ 0 & \left(\frac{\rho^2}{k+2}\right)I_k \end{bmatrix}, \mu_{12} = \frac{\rho^2}{k+2}\begin{bmatrix} 1_k' & 0' \\ 0 & 0 \end{bmatrix}$$

For $k = 3$ suppose we select a design matrix of the form shown in Eq. 6.37,

$$D = c\begin{bmatrix} F \\ -F \end{bmatrix}$$

so that the design points lie on a sphere of radius $c\sqrt{3}$. With $\mathbf{X}_1 = [\mathbf{1}:\mathbf{D}]$,

$$\mathbf{X}_1'\mathbf{X}_1 = \begin{bmatrix} N & \mathbf{0}' \\ \mathbf{0} & Nc^2\mathbf{I}_3 \end{bmatrix}$$

Equating $\frac{1}{N}(\mathbf{X}_1'\mathbf{X}_1) = \boldsymbol{\mu}_{11}$ forces

$$c^2 = \frac{\rho^2}{k+2} = \frac{1}{5} \qquad \text{since } \rho = 1$$

Thus with a design matrix of the form $\mathbf{D}$ above, the points lie on a sphere of radius $c\sqrt{3} = \sqrt{3/5}$.

6.3 With the double simplex design shown,

$$\mathbf{X}_1'\mathbf{X}_1 = \text{diag}(6, 3\rho^2, 3\rho^2), \quad \mathbf{X}_1'\mathbf{X}_2 = \begin{bmatrix} 3\rho^2 & 3\rho^2 & 0 \\ 0 & 0 & 0 \\ 0 & 0 & 0 \end{bmatrix}$$

Since $E(\mathbf{b}_1) = \boldsymbol{\beta}_1 + (\mathbf{X}_1'\mathbf{X}_1)^{-1}\mathbf{X}_1'\mathbf{X}_2\boldsymbol{\beta}_2$, then

$$E(b_0) = \beta_0 + \frac{\rho^2}{2}(\beta_{11} + \beta_{22})$$

$$E(b_1) = \beta_1$$

$$E(b_2) = \beta_2$$

The bias $\rho^2(\beta_{11} + \beta_{22})$ in b_0 and the unbiasedness of b_1 and b_2 are the same for this design as with the design in Eq. 6.46.

6.4 To ensure $\mathbf{A} = \boldsymbol{\mu}_{11}^{-1}\boldsymbol{\mu}_{12}$, we want a second-order design for which $\frac{1}{N}(\mathbf{X}_1'\mathbf{X}_1) = \boldsymbol{\mu}_{11}$ and $\frac{1}{N}(\mathbf{X}_1'\mathbf{X}_2) = \boldsymbol{\mu}_{12}$. Let us select the "cube" plus "octa-hedron" design in Eq. 6.50. With the cuboidal region of interest R, the region moment matrices are identical in form to those in Section 6.2.4 where in Eq. 6.55, $u\rho^2 = 1/3$ and $v\rho^2 = 1/9$, with $u = 1/(k+2)$ and $v = 1/(k+2)(k+4)$. Similarly, the design moment matrices with the "cube" plus "octahedron" design are as shown in Section 6.50 where in $\frac{1}{N}(\mathbf{X}_1'\mathbf{X}_1)$, $\lambda_2 = m_1$ and $\lambda_4 = m_3$. Thus setting $m_1 = 1/3$, $m_3 = 1/9$ and $m_2 - m_3 = (1/9)(4/5)$, then $m_2 = 1/5$ or $m_2 = (9/5)m_3$.

6.5 The true model is Eq. 6.74 and the fitted model is $\hat{Y}(x) = b_0 + b_1x = \mathbf{x}_1'\mathbf{b}_1$. The matrices $(\mathbf{X}'\mathbf{X})^{-1}$, $\boldsymbol{\mu}_{11}$, and $\tilde{\mathbf{A}}$ are

$$(\mathbf{X}'\mathbf{X})^{-1} = \frac{1}{2c}\begin{bmatrix} b & 0 & -a \\ 0 & \frac{c}{a} & 0 \\ -a & 0 & 2 \end{bmatrix}, \ \boldsymbol{\mu}_{11} = \begin{bmatrix} 1 & 0 \\ 0 & 1/3 \end{bmatrix}, \tilde{\mathbf{A}} = \begin{bmatrix} 1 & 0 & 1/3 \\ 0 & 1 & 0 \end{bmatrix}$$

where in $(\mathbf{X}'\mathbf{X})^{-1}$, $a = l_1^2 + l_2^2$, $b = l_1^4 + l_2^4$, and $c = 2b - a^2$. Thus

$$V = 4 \text{ trace } \{\tilde{\mathbf{A}}(\mathbf{X}'\mathbf{X})^{-1}\tilde{\mathbf{A}}'\mu_{11}\} = \frac{2}{9}\left(\frac{9b - 6a + 2}{2b - a^2}\right) + \frac{2}{3a}$$

In terms of a and b, V is minimized at $(a, b) = (4/3, 16/9)$ with min. $V = 7/4$. In terms of l_1 and l_2 ($l_2 \neq l_1$), V is minimized at $l_1 = 0$ and $l_2 = 2/\sqrt{3}$, attaining a value of min. $V = 7/4$. Note, of the four-point designs of the type considered, $a^2/2 \leq b \leq a^2$ with $b = a^2 2$ when $l_1 = l_2$ (a singular design), and $b = a^2$ when $l_1 = 0$, $l_2 \neq 0$. Thus our four-point design is really a three-point design with single observations at $x = \pm 2/\sqrt{3}$ and two observations at $x = 0$.

CHAPTER 7

7.1   The smallest eigenvalue of the $\mathbf{DD}'$ matrix, where $\mathbf{D}$ is defined in Eq. 7.15, is 0.00519. Since the data are rounded off to two decimal places, then $\delta = 0.005$. Hence, $\mu_\lambda \approx \dfrac{N-1}{3}\delta^2 = 0.0001$.

Using formula (7.17),

$$\sigma_\lambda \leq 0.0004$$

Thus, 0.00519 is at least 12 standard deviations away from μ_λ. We conclude that 0.00519 does not correspond to a zero eigenvalue of $\mathbf{DD}'$, hence, no linear relationships exist among the four responses.

7.2   Multiplying Eqs. 7.23 and 7.24 by 0.922 and 0.390, and by 0.222 and 2.18, respectively, and then adding in each case the resulting equations, we obtain

$$0.0298Y_1 - 0.0034Y_2 + 0.0212Y_3 + 1.0Y_4 + 0.0084Y_5 = 3.0919$$
$$1.0Y_1 + 1.0214Y_2 + 0.9126Y_3 + 1.0Y_4 + 0.9596Y_5 = 100.0$$

The latter two equations are approximately of the form given in Eqs. 7.25 and 7.26.

7.3   (a)   In this case, $N = 21$, $\nu_{PE} = 4$, $\nu_{LOF} = N - \rho - \nu_{PE} = 9$, where ρ is the rank of $\mathbf{W}$ and is equal to 8. The value of Roy's largest root test statistic is $e_{max}(\mathbf{G}_2^{-1}\mathbf{G}_1) = 76.456$. Using Foster's (1957) tables, the 10% critical value for this test is $\lambda_{0.10} = 127.205$. Therefore, no significant lack of fit can be detected at the 10% level.

(b) Since the lack of fit test is not significant, it is not necessary to do any further analysis. Nevertheless, we now give a listing of the values of $e_{max}(G_2^{-1}G_1)$ using all nonempty subsets of the three responses.

Subset	$e_{max}(G_2^{-1}G_1)$	10% Critical Value
Y_1, Y_2, Y_3	76.456	127.205
Y_1, Y_2	61.267	127.205
Y_1, Y_3	9.523	127.205
Y_2, Y_3	75.270	127.205
Y_1	5.914	127.205
Y_2	62.052	127.205
Y_3	3.873	127.205

We note that Y_2 is a major contributor to the lack of fit variation.

7.5 (a) The parameter estimates for the four response models are given in the following table:

Model Term	$\hat{Y}_1$	$\hat{Y}_2$	$\hat{Y}_3$	$\hat{Y}_4$
Constant	1.880	21.257	17.933	50.711
x_1	−0.097	5.661	0.744	2.436
x_2	−0.001	−0.172	−0.012	0.631
x_3	0.009	−1.227	−1.071	0.269
x_1^2	−0.103	8.168	3.452	−3.151
x_2^2	0.001	0.443	0.801	0.643
x_3^2	0.003	0.997	1.579	0.572
$x_1 x_2$	−0.026	2.659	1.800	0.089
$x_1 x_3$	−0.004	−0.659	2.100	0.476
$x_2 x_3$	−0.006	0.334	−0.575	−0.226

(b) (i)

		Location of Optimum		
Response	Individual Optimum	x_1	x_2	x_3
$\hat{Y}_1$	max. = 1.930	−0.499	0.200	1.653
$\hat{Y}_2$	min. = 19.083	−0.598	1.616	0.136
$\hat{Y}_3$	min. = 17.073	−0.728	1.183	1.033
$\hat{Y}_4$	max. = 54.171	0.330	1.705	0.078

(iii) The distance function $\rho_1[\hat{Y}(x), \hat{\phi}]$, where $\hat{\phi} = (1.930, 19.083, 17.073, 54.171)'$, attains the minimum value 2.654 at $x_1 = -0.446$, $x_2 = 1.480$, $x_3 = 0.776$. The corresponding simultaneous optima at this location are

$$\text{Max } \hat{Y}_1 = 1.925$$

$$\text{Min } \hat{Y}_2 = 19.578$$

$$\text{Min } \hat{Y}_3 = 17.568$$

$$\text{Max } \hat{Y}_4 = 51.408$$

7.6 (a) The value of Roy's largest root test statistic is $e_{max}(S_h S_e^{-1}) = 2.425$. The corresponding 1% critical value is 0.25. The hypothesis H_0 can therefore be rejected.

(b) This is equivalent to testing the hypothesis

$$H_0' : \beta_{11} = \beta_{21} = \beta_{31}$$

The value of Roy's largest root test statistic for H_0' is 0.104. The 10% critical value is 0.088. Thus, at the 10% level of significance we can conclude that the rate of change of log(weight) with respect to log(girth) is not the same at weeks 4, 6, and 7.

CHAPTER 8

8.1 Each block consists of a first-order orthogonal design. Furthermore, the fraction of the total sum of squares for variable x_i ($i = 1, 2, 3, 4$) in each block is equal to $\frac{4}{12} = \frac{1}{3}$. This is also equal to the fraction of the total number of experimental runs in each block.

8.2 Consider the factorial portion as one block and the axial portion as another block. Using Eq. 8.10, we get

$$\alpha = \left[\frac{F(2k + n_{0A})}{2(F + n_{0F})} \right]^{1/2}$$

$$2 = \left[\frac{16(10 + n_{0A})}{2(16 + n_{0F})} \right]^{1/2}$$

We must therefore have

$$n_{0F} - 2n_{0A} = 4 \tag{1}$$

Since $n_{0F} + n_{0A} = 6$, solving these two equations does not yield positive integer solutions for n_{0F} and n_{0A}. Hence, the design does not block orthogonally.

Now, let us consider increasing the number of center points so that the design blocks orthogonally. Denote the number of center points by n_0. Then,

$$n_{0F} + n_{0A} = n_0 \tag{2}$$

From Eqs. 1 and 2, we get

$$n_{0A} = \frac{n_0 - 4}{3}$$

Thus, $n_0 - 4$ must be a multiple of 3. For example, if $n_0 = 7$, then $n_{0A} = 1$ and $n_{0F} = 6$.

8.3   Consider a central composite design consisting of the following three blocks: Block 1 consists of a one-half fraction of 2^3 with three center point replications. Block 2 consists of the second one-half fraction of 2^3 with three center point replications. Block 3 consists of the six axial points plus one center point. In order to satisfy Eq. 8.9, the axial parameter α must satisfy the equation

$$\frac{2\alpha^2}{8 + 2\alpha^2} = \frac{1}{3}$$

Thus, $\alpha = \sqrt{2}$.

8.4   The original design is rotatable since $[iiii] = 1$, $i = 1, 2, 3$, and $[iijj] = \frac{8}{24} = \frac{1}{3}$, $i \neq j$, that is, $[iiii] = 3[iijj]$. Furthermore, the design blocks orthogonally. It is therefore rotatable as a blocked design.

8.6   Under $H_0: \sigma_\delta^2 = 0$, the variance-covariance matrix, Σ, of $\mathbf{Y}$ in Eq. 8.34 is equal to $\sigma_\varepsilon^2 \mathbf{I}_N$. Now, according to Searle (1971, Theorem 2, p. 57), $R(\delta|\zeta)/\sigma_\varepsilon^2$ has the chi-squared distribution with $b - 1$ degrees of freedom if the matrix $\mathbf{W}(\mathbf{W}'\mathbf{W})^{-1}\mathbf{W}' - \mathbf{U}(\mathbf{U}'\mathbf{U})^{-1}\mathbf{U}'$ is idempotent of rank $b - 1$. To verify that this is the case, consider

$$[\mathbf{W}(\mathbf{W}'\mathbf{W})^{-1}\mathbf{W}' - \mathbf{U}(\mathbf{U}'\mathbf{U})^{-1}\mathbf{U}']^2 = \mathbf{W}(\mathbf{W}'\mathbf{W})^{-1}\mathbf{W}'$$
$$- \mathbf{W}(\mathbf{W}'\mathbf{W})^{-1}\mathbf{W}'\mathbf{U}(\mathbf{U}'\mathbf{U})^{-1}\mathbf{U}'$$
$$- \mathbf{U}(\mathbf{U}'\mathbf{U})^{-1}\mathbf{U}'\mathbf{W}(\mathbf{W}'\mathbf{W})^{-1}\mathbf{W}'$$
$$+ \mathbf{U}(\mathbf{U}'\mathbf{U})^{-1}\mathbf{U}' \qquad (3)$$

Since the columns of $\mathbf{U}$ are spanned by the columns of $\mathbf{W}$, we can write $\mathbf{U} = \mathbf{W}\mathbf{C}$ for some matrix $\mathbf{C}$. Hence,

$$\mathbf{W}(\mathbf{W}'\mathbf{W})^{-1}\mathbf{W}'\mathbf{U} = \mathbf{W}\mathbf{C}$$
$$= \mathbf{U}$$

By making the proper substitution in Eq. 3, we get

$$[\mathbf{W}(\mathbf{W}'\mathbf{W})^{-1}\mathbf{W}' - \mathbf{U}(\mathbf{U}'\mathbf{U})^{-1}\mathbf{U}']^2 = \mathbf{W}(\mathbf{W}'\mathbf{W})^{-1}\mathbf{W}' - \mathbf{U}(\mathbf{U}'\mathbf{U})^{-1}\mathbf{U}'$$

This proves that the matrix under consideration is idempotent. Its rank is equal to its trace, namely,

$$tr[\mathbf{W}(\mathbf{W}'\mathbf{W})^{-1}\mathbf{W}'] - tr[\mathbf{U}(\mathbf{U}'\mathbf{U})^{-1}\mathbf{U}'] = p + b - (p + 1)$$
$$= b - 1$$

CHAPTER 9

9.1 (a) $b_1 = Y_1$, $b_2 = Y_4$, $b_{12} = 2(Y_2 + Y_3 - Y_1 - Y_4)$

(b) $b_1 = \frac{1}{20}(19Y_1 + 3Y_2 - 3Y_3 + Y_4)$, $b_2 = \frac{1}{20}(Y_1 - 3Y_2 + 3Y_3 + 19Y_4)$
$b_{12} = \frac{9}{4}(Y_2 + Y_3 - Y_1 - Y_4)$

(a) $\text{Var}(b_i) = \sigma^2 \quad i = 1, 2$	(b) $\text{Var}(b_i) = .95\sigma^2 \quad i = 1, 2$
$\text{Var}(b_{12}) = 16\sigma^2$	$\text{Var}(b_{12}) = 20.25\sigma^2$

In terms of estimating the nonlinear blending of components one and two with b_{12}, design (a) is preferred to (b) since $16\sigma^2 < 20.25\sigma^2$.

9.2 (b) $\hat{Y}(x) = 3.5x_1 + 4.5x_2 - 8.0x_1x_2$

(c) $\hat{\text{Var}}(b_{12}) = (24/3)MSE = 1.12$

To test $H_0:\beta_{12} = 0$ vs $H_a:\beta_{12} \neq 0$,

$$t = \frac{-8.0}{\sqrt{1.12}} = -7.56 < t_{0.025,6} = -2.447$$

Therefore, reject H_0 and conclude that the joint blending of the chemical pesticides A and B is beneficial. Stated another way, the average response (2.0) at the 50-50% blend of A and B is significantly ($P < 0.025$) lower in value than the average response (4.0) due to linear blending of A and B.

9.3 (a) $\hat{Y}(x) = 11.60x_1 + 8.35x_2 + 16.85x_3 - 8.17x_1x_2 + 0.17x_1x_3 + 16.13x_2x_3$ with the ANOVA table

Source	d.f.	S.S.	M.S.	F
Regression	5	196.639	39.328	82.5
Residual	9	4.290	$0.4767 = \hat{\sigma}^2$	
Total	14	200.929		

(b) $\hat{Y}(1/3, 1/3, 1/3) = 13.17$. This value is lower than the observed value $Y_{16} = 18.2$ by 5.03 units suggesting the special cubic model ought to be fitted to the data at points 1 to 7. When this is done, we get

$$\hat{Y}(x) = 11.60x_1 + 8.35x_2 + 16.85x_3 - 8.17x_1x_2 + 0.17x_1x_3$$
$$+ 16.13x_2x_3 + 135.80x_1x_2x_3$$

Now, a test of $H_0:\beta_{123} = 0$ vs $H_A:\beta_{123} \neq 0$ based on the fit of the special cubic model to the data at points 1 to 7 produces a t-value of 6.61 which exceeds $t_{0.025,9} = 2.262$ causing us to reject $H_0:\beta_{123} = 0$ and conclude the special cubic model, with the extra term, is better than the quadratic model. Note that we would get the same result had we fitted the quadratic model to the data at points 1 to 7 only to find the quadratic model suffered from lack of fit.

(c) From the ANOVA table in (a), Pure Error S.S. = 4.29. Using as an estimate of σ^2, $MSE = 0.476$, both the fitted quadratic and special cubic models suffer from lack of fit when fitted to the $N = 20$ data values. From these results and recognizing the configuration of design points supports fitting the special quartic model of Eq. 9.11, we proceed to fit the special quartic model. Upon fitting the special quartic model we observe that some second- and higher-degree terms have coefficient estimates with t values less than unity. Dropping these less important terms from the model and refitting, we obtain the reduced model

$$\hat{Y}(\mathbf{x}) = 11.52x_1 + 8.25x_2 + 16.81x_3 - 8.26x_1x_2 + 16.06x_1x_3$$
$$+ 445.76x_1^2x_2x_3 - 149.00x_1x_2^2x_3$$

with $R^2 = 0.9638$ and $R_A^2 = 0.9457$.

The terms $445.76x_1^2x_2x_3$ and $-149.00x_1x_2^2x_3$ suggest there is a mound and a depression in the surface shape, respectively, at the locations $(x_1, x_2, x_3) = (2/3, 1/6, 1/6)$ and $(1/6, 2/3, 1/6)$.

9.4 (a) The constrained region has four extreme vertices located at the blends $(x_1, x_2, x_3) = (0.50, 0.35, 0.15)$, $(0.50, 0.05, 0.45)$, $(0.20, 0.05, 0.75)$, and $(0.20, 0.65, 0.15)$. One suggestion for a design is the four vertices, the midpoints of the longest two sides and the overall centroid, giving a total of seven distinct points. Three points could be replicated a second time to provide an estimate of σ^2 with 3 d.f. for use in the denominator of the F test for lack of fit.

(b) Define the pseudocomponents

$$x_1' = \frac{x_1 - 0.20}{0.60}, \quad x_2' = \frac{x_2 - 0.05}{0.60}, \quad x_3' = \frac{x_3 - 0.15}{0.60}$$

Then corresponding to the settings of a {3, 2} simplex lattice in x_1', x_2', and x_3', the settings in x_1, x_2, and x_3 are

x_1'	x_2'	x_3'	x_1	x_2	x_3
1	0	0	.80	.05	.15
0	1	0	.20	.65	.15
0	0	1	.20	.05	.75
.5	.5	0	.50	.35	.15
.5	0	.5	.50	.05	.45
0	.5	.5	.20	.35	.45

9.5 (a) Based on a within-blend estimate of σ^2 at the six mixture components-process variable combination of $s^2 = 0.233$ with 6 d.f., there is evidence of nonlinear blending of components one and two from the results of the test of $H_0: \beta_{12} = 0$, i.e.,

$$t = \frac{6.0}{1.183} = 5.07 > t_{0.025,6} = 2.447.$$

(b) At $z = -1$, $\hat{Y}(x) = 4.7x_1 + 5.3x_2 - 2.0x_1x_2$. The estimate $b_{12} = -2.0$ is not different from zero.

At $z = +1$, $\hat{Y}(x) = 2.5x_1 + 3.7x_2 + 14.0x_1x_2$. The estimate $b_{12} = 14.0$ is significantly ($P < 0.01$) different from zero. Thus, it is at the high level ($z = +1$) of the process variable that nonlinear blending is present as detected in (a). Changing the level of z from low (-1) to high ($+1$) causes the shape of the mixture surface to change from that of linear blending (a straight line) to that of nonlinear blending (two lines or a curve).

(c) At $z = +1$, $\hat{Y}(x) = 2.5x_1 + 3.7x_2 + 14.0x_1x_2 = 3.7 + 12.8x_1 - 14.0x_1^2$. Taking $\dfrac{\partial \hat{Y}(x)}{\partial x_1} = 12.8 - 28.0x_1 = 0$ and solving for x_1 produces an estimated maximum response value of $\hat{Y} = 6.626$ at the blend $x_1 = 0.457$ and $x_2 = 0.543$.

CHAPTER 10

10.1 (a) $\hat{\theta}_1 = 2.849$, $\hat{\theta}_2 = 15.20$, $\hat{\theta}_3 = 0.745$.

(b)

Parameter	Asymptotic Std. Error	Asymptotic 95% Confidence Interval	
		Lower	Upper
θ_1	0.6279	0.1473	5.5510
θ_2	2.6148	3.9496	26.4510
θ_3	0.0589	0.4920	0.9984

(c) The partial derivatives of the mean response with respect to θ_1, θ_2, and θ_3 are

$$\frac{\partial \eta(x_u, \theta)}{\partial \theta_1} = \frac{\theta_3 x_{u1}(1 + \theta_2 x_{u2})}{(1 + \theta_1 x_{u1} + \theta_2 x_{u2})^2}$$

$$\frac{\partial \eta(x_u, \theta)}{\partial \theta_2} = \frac{-\theta_1 \theta_3 x_{u1} x_{u2}}{(1 + \theta_1 x_{u1} + \theta_2 x_{u2})^2}$$

$$\frac{\partial \eta(x_u, \theta)}{\partial \theta_3} = \frac{\theta_1 x_{u1}}{1 + \theta_1 x_{u1} + \theta_2 x_{u2}}, \qquad u = 1, 2, 3, 4, 5$$

The matrix $F(\hat{\theta})$ is the 5×3 matrix

$$F(\hat{\theta}) = \left[\frac{\partial \eta}{\partial \theta_1} : \frac{\partial \eta}{\partial \theta_2} : \frac{\partial \eta}{\partial \theta_3} \right]_{\theta = \hat{\theta}}$$

The approximate 95% confidence region for θ is given by

$$\frac{(\hat{\theta} - \theta)'F'(\hat{\theta})F(\hat{\theta})(\hat{\theta} - \theta)}{3\hat{\sigma}_5^2} \leq F_{0.05,3,2}$$

where $\hat{\sigma}_5^2 = 0.000646$ is the residual mean square for the fitted model with 2 degrees of freedom. Note that

$$F'(\hat{\theta})F(\hat{\theta}) = \begin{bmatrix} 0.0178 & -0.0023 & 0.1369 \\ -0.0023 & 0.0004 & -0.0169 \\ 0.1369 & -0.0169 & 1.2432 \end{bmatrix}$$

(d) (i) Applying formula (10.11) with

$$A = \begin{bmatrix} 1 & 1 & 0 \\ 0 & 0 & 1 \end{bmatrix}, \quad \hat{G} = [F'(\hat{\theta})F(\hat{\theta})]^{-1}$$

$$A\theta = m$$

$$= \begin{bmatrix} 18.1 \\ 2.5 \end{bmatrix}$$

the test statistic has the value

$$\frac{(A\hat{\theta} - m)'[A\hat{G}A']^{-1}(A\hat{\theta} - m)}{2\hat{\sigma}_5^2} = 486.174$$

which is significant at the 5% level.

(ii) Under H_0, the model becomes

$$\eta = \frac{2.5\theta_1 x_1}{1 + \theta_1 x_1 + (18.1 - \theta_1)x_2}$$

$$\eta = \frac{2.5\theta_1 x_1}{1 + 18.1 x_2 + \theta_1(x_1 - x_2)}$$

The residual sum of squares for this reduced model is $S(\hat{\theta}^0) = 0.070836$ with 4 degrees of freedom. From part (a), the residual sum of squares for the original model is $S(\hat{\theta}) = 0.001292$. The value of the approximate F-statistic in Eq. 10.24 is

$$T = \frac{(0.070836 - 0.001292)/2}{0.001292/2}$$

$$= 53.827$$

(e) Hint: Obtain an approximate 99% confidence region on θ as in part (c). Let $L_{0.01}$ and $U_{0.01}$ denote, respectively, the minimum and maximum values of

$$\eta = \frac{1.5\theta_1\theta_3}{1 + 1.5\theta_1 + 0.75\theta_2}$$

over this confidence region. Then, the interval $(L_{0.01}, U_{0.01})$ is a confidence interval on the mean response, when $x_1 = 1.5$ and $x_2 = 0.75$, with an approximate confidence coefficient greater than or equal to 0.99.

10.2 (a) The parameters θ_1 and θ_2 are linear, but θ_3 is a nonlinear parameter. The partial derivatives of the mean response η with respect to θ_1, θ_2 and θ_3 are

$$\frac{\partial \eta}{\partial \theta_1} = x + \exp(-\theta_3 x)$$

$$\frac{\partial \eta}{\partial \theta_2} = -x \exp(-\theta_3 x)$$

$$\frac{\partial \eta}{\partial \theta_3} = (-\theta_1 x + \theta_2 x^2) \exp(-\theta_3 x)$$

In this case, $\mathbf{B}(\theta)$ and $\phi(x, \tilde{\theta})$ in Eq. 10.40 are of the form

$$\mathbf{B}(\theta) = \begin{bmatrix} 1 & 0 & 0 \\ 0 & 1 & 0 \\ 0 & \theta_1 & \theta_2 \end{bmatrix}$$

$\phi(x, \tilde{\theta}) = [x + \exp(-\theta_3 x), -x \exp(-\theta_3 x), x^2 \exp(-\theta_3 x)]'$, where $\tilde{\theta} = \theta_3$. It follows that the locally D-optimal design for the estimation of θ depends only on θ_3.

(b) The condition given by Eq. 10.41 is not satisfied since $B_2(\theta) = (0, \theta_1, \theta_2)$ cannot be expressed as $\mathbf{C}(\theta)[1:K]$, where $\mathbf{K}$ is a matrix of order 1×2 not involving θ. This condition, however, is only sufficient for the dependency of the locally D-optimal design on θ_3, but is not necessary. In this case, we need to examine the determinant of the matrix $\mathbf{M}_{11}(\mathbf{D}, \theta)$ given in Eq. 10.31, and determine the nature of its dependency on the elements of θ.

Using part (a), we have

$$|\mathbf{M}(\mathbf{D}, \theta)| = |\mathbf{B}(\theta)\mathbf{Q}(\mathbf{D}, \theta_3)\mathbf{B}'(\theta)|$$

where $\mathbf{Q}(\mathbf{D}, \theta_3)$ is the 3×3 matrix

$$Q(\mathbf{D}, \theta_3) = \sum_{u=1}^{N} \Phi(x_u, \theta_3)\Phi'(x_u, \theta_3)$$

Hence,

$$|\mathbf{M}(\mathbf{D}, \boldsymbol{\theta})| = \theta_2^2|Q(\mathbf{D}, \theta_3)|$$

Now, the matrix $\mathbf{M}_{22}(\mathbf{D}, \boldsymbol{\theta})$ corresponding to θ_3 has the determinant

$$
\begin{aligned}
|\mathbf{M}_{22}(\mathbf{D}, \boldsymbol{\theta})| &= |\mathbf{B}_2(\boldsymbol{\theta})Q(\mathbf{D}, \theta_3)\mathbf{B}_2'(\boldsymbol{\theta})| \\
&= (0, \theta_1, \theta_2)Q(\mathbf{D}, \theta_3)(0, \theta_1, \theta_2)' \\
&= \sum_{u=1}^{N} (-\theta_1 x_u + \theta_2 x_u^2)^2 \exp(-2\theta_3 x_u) \\
&= \sum_{u=1}^{N} (\theta_1^2 x_u^2 - 2\theta_1\theta_2 x_u^3 + \theta_2^2 x_u^4) \exp(-2\theta_3 x_u)
\end{aligned}
$$

$$|\mathbf{M}_{11}(\mathbf{D}, \boldsymbol{\theta})| = \frac{|\mathbf{M}(\mathbf{D}, \boldsymbol{\theta})|}{|\mathbf{M}_{22}(\mathbf{D}, \boldsymbol{\theta})|}$$

Hence, the design $\mathbf{D}$ that maximizes $|\mathbf{M}_{11}(\mathbf{D}, \boldsymbol{\theta})|$ depends on θ_3 and on the ratio θ_1/θ_2.

10.3 The $(r + 1)$th-order partial derivative of $h(t, \boldsymbol{\theta})$ with respect to t is given by

$$
\begin{aligned}
h_{r+1}(t, \boldsymbol{\theta}) = (-1)^{r+1}\left(\frac{d}{2}\right)^{r+1} &\left(\frac{\theta_1}{\theta_1 - \theta_2}\right)\left\{\theta_2^{r+1} \exp\left[-\frac{d\theta_2(t + 1)}{2}\right]\right. \\
&\left. - \theta_1^{r+1} \exp\left[-\frac{d\theta_1(t + 1)}{2}\right]\right\}
\end{aligned}
$$

Let $\xi(t, \theta) = \theta^{r+1} \exp\left[-\dfrac{d\theta(t + 1)}{2}\right]$. By applying the Mean Value Theorem of differential calculus to $\xi(t, \theta)$ with respect to θ over the interval $[\theta_1, \theta_2]$, we get

$$
\begin{aligned}
\xi(t, \theta_2) - \xi(t, \theta_1) &= (\theta_2 - \theta_1)\frac{\partial \xi(t, \theta)}{\partial \theta}\bigg|_{\theta=\theta_c} \\
&= (\theta_2 - \theta_1)\theta_c^r \exp\left[-\frac{d\theta_c(t + 1)}{2}\right] \\
&\quad \times \left[r + 1 - \frac{d\theta_c(t + 1)}{2}\right]
\end{aligned}
$$

where $\theta_1 < \theta_c < \theta_2$ (if $\theta_1 < \theta_2$). Hence,

$$|h_{r+1}(t, \boldsymbol{\theta})| = \theta_1 \left(\frac{d}{2}\right)^{r+1} \theta_c^r \exp\left[-\frac{d\theta_c(t+1)}{2}\right]\left|r + 1 - \frac{d\theta_c(t+1)}{2}\right|$$

Since $0 \le \theta_c \le 1$ and $-1 \le t \le 1$, then $0 \le \theta_c(t + 1) \le 2$, and

$$|h_{r+1}(t, \boldsymbol{\theta})| \le \left(\frac{d}{2}\right)^{r+1}\left|r + 1 - \frac{d\theta_c(t+1)}{2}\right|$$

Furthermore,

$$\max_{-1\le t\le 1} \left|r + 1 - \frac{d\theta_c(t+1)}{2}\right| = \begin{cases} r + 1, & \text{if } d \le 2(r + 1) \\ d - r - 1, & \text{if } d > 2(r + 1) \end{cases}$$

Thus,

$$\max_{-1\le t\le 1} |h_{r+1}(t, \boldsymbol{\theta})| \le \left(\frac{d}{2}\right)^{r+1} \max(r + 1, d - r - 1)$$

10.4 The following table gives two-point locally *D*-optimal designs corresponding to the given values of θ_1 and θ_2:

θ_1	θ_2	x_1	x_2	$|M(D, \boldsymbol{\theta})|$
0.10	0.20	4.579	16.314	2.8375
0.20	0.30	2.651	9.491	0.5778
0.30	0.50	1.691	6.008	0.0748
0.50	1.00	0.916	3.256	0.0045
0.70	0.20	1.230	6.860	0.6568
0.90	0.80	0.739	2.806	0.0096
1.00	0.90	0.661	2.503	0.0060

CHAPTER 12

12.1 (a) This follow from the fact that

$$D = \begin{bmatrix} D_1 \\ 0_1 \end{bmatrix}$$

where D_1 corresponds to the noncenter points and 0_1 is a zero matrix of order $n_0 \times k$. Hence,

$$D(D'D)^{-1}D' = \text{diag}[D_1(D_1'D_1)^{-1}D_1', 0_2]$$

where 0_2 is a zero matrix of order $n_0 \times n_0$. Thus the sum of squares of the diagonal elements of $D(D'D)^{-1}D'$ is the same as that of $\Delta_1 = D_1(D_1'D_1)^{-1}D_1'$.

(b)   We have that $d_0 = \sum_{u=1}^{n} d_{1uu}^2$, and

$$\sum_{u=1}^{n} d_{1uu} = tr(\mathbf{\Delta}_1) = tr[\mathbf{D}_1'\mathbf{D}_1(\mathbf{D}_1'\mathbf{D}_1)^{-1}] = k$$

Using the hint, we conclude that $d_0 \geq \dfrac{k^2}{n}$.

(c)   Using the hint, we have

$$d_0 = \sum_{u=1}^{n} d_{1uu}^2 \leq \sum_{u=1}^{n} d_{1uu}\left(1 - \frac{1}{n}\right)$$

Hence,

$$d_0 \leq \frac{k(n-1)}{n}$$

(d)   From parts (b) and (c) we have the double inequality

$$\frac{k^2}{n} \leq d_0 \leq \frac{k(n-1)}{n}$$

Using Eq. 12.19 we obtain

$$\frac{k^2}{n} - \frac{k(k+2)(N-1)}{N(N+1)} \leq g(\mathbf{D}) \leq \frac{k(n-1)}{n} - \frac{k(k+2)(N-1)}{N(N+1)}$$

12.2   The columns c_2, c_3, ..., c_{k+1} of $\mathbf{C}$ in Eq. 12.20 are the same as the k columns of $\dfrac{1}{[2^{k-m}(a^2+b^2)]^{1/2}} \mathbf{D}_1$, that is, for $i = 2, 3, \ldots, k+1$,

$$c_{ui} = \frac{\pm a}{[2^{k-m}(a^2+b^2)]^{1/2}}, \ u = 1, 2, \ldots, 2^{k-m}$$

$$c_{ui} = \frac{\pm b}{[2^{k-m}(a^2+b^2)]^{1/2}}, \ u = 2^{k-m}+1, \ldots, 2^{k-m+1}$$

Hence, using the notation in Exercise 12.1,

$$d_{1uu} = \sum_{i=2}^{k+1} c_{ui}^2$$

$$= \frac{ka^2}{2^{k-m}(a^2+b^2)}, \ u = 1, 2, \ldots, 2^{k-m}$$

$$= \frac{kb^2}{2^{k-m}(a^2+b^2)}, \ u = 2^{k-m}+1, \ldots, 2^{k-m+1}$$

Note that $\sum_{u=1}^{n} d_{1uu} = k$, as it should be. Furthermore,

$$d_0 = \sum_{u=1}^{n} d_{1uu}^2 = \frac{k^2(a^4 + b^4)}{2^{k-m}(a^2 + b^2)^2}$$

To satisfy the Box-Watson criterion, a and b must satisfy Eq. 12.23, that is,

$$\frac{k^2(a^4 + b^4)}{2^{k-m}(a^2 + b^2)^2} = \frac{k(k + 2)(2^{k-m+1} - 1)}{2^{k-m+1}(2^{k-m+1} + 1)}$$

It can be verified that this equation is satisfied by the values of a^2 and b^2 given in Eq. 12.27.

12.3 The percent rotatability for this design is 99.40.

12.4 (a) The percent rotatability is 80.65.

   (b) $x_{11} = -0.1188$, $x_{21} = -1.8593$. The percent rotatability for the design $\tilde{D}$ is 89.99. This represents an increase of 9.34 in the percent rotatability beyond the value 80.65 for the initial design.

Index